Essentials
of
Physical
Geography
Today

Essentials of Physical Geography Today

THEODORE M. OBERLANDER
University of California, Berkeley

ROBERT A. MULLER
Louisiana State University

RANDOM HOUSE NEW YORK

Library of Congress Cataloging in Publication Data
Oberlander, Theodore.
Essentials of physical geography today.
Based on: Physical geography today / Robert A. Muller,
Theodore M. Oberlander. 2nd ed.
Bibliography: p.
Includes index.
1. Physical geography. I. Muller, Robert A. II. Muller,
Robert A. Physical geography today. III. Title.
GB55.023   910′.02   81–10557
ISBN 0-394-32543-5   AACR2

Manufactured in the United States of America

Text and cover design: Rodelinde Albrecht

Cover art: "The Vale of St. Thomas" painting by
Frederick E. Church. Wadsworth Atheneum, Hartford.

Preface

The world can be viewed in many dissimilar ways. Artists, physicists, economists, historians, and geologists all have their own unique perspectives and are likely to see the world differently. Geographers have an especially comprehensive view of the earth—one that focuses on the patterns of natural and man-made phenomena on the earth's surface, the causes of these phenomena, and the interrelationships between them. This book centers its attention upon physical geography, which is mainly concerned with the natural world as it is seen from the human perspective. Physical geography is the geography of the human environment, and is more than a mere composite of other physical sciences such as meteorology, climatology, biology, pedology, and geology. Physical geography not only describes the natural phenomena at the earth's surface but also, and more importantly, seeks explanations as to how and why physical processes act as they do to produce the natural phenomena that have significance to life on our planet. The scope of physical geography extends from the atmosphere through the hydrosphere and biosphere to the upper portion of the lithosphere, but concentrates upon the surface region where land, sea, and air meet and interact, and where life flourishes. Human impacts upon the natural systems operating at this interface are an important consideration in physical geography, and are touched upon in almost every chapter of our text.

The processes that influence surface phenomena involve both energy and materials. Energy continuously cascades through physical systems ranging in scale from the global biosphere to single microbes. Materials such as carbon, water, and oxygen flow through these systems in never-ending cycles that involve inputs, storages, outputs, and quasiequilibrium states. A basic emphasis of the study of physical geography concerns the ways in which the various physical systems interact and respond to one another, constantly exchanging energy and materials. While it is our desire to organize the complexity of our planet by utilizing the concepts of energy systems, it is not our intention to permit these concepts to diminish the fascination derived through consideration of the individual wonders of the real world. Despite our utilization of generalizing theory, our focus is on the world itself, in all its wondrous but comprehensible complexity.

Our aim has not been to introduce and define every term used by physical geographers in their professional work, nor to touch on every phenomenon that enters into the realm of physical geography. Rather than a light skimming of a very large surface, we have sought to develop an understanding of the most essential facts and relationships by extended treatment of those topics that seem to be of paramount importance. Some of these have not previously been explored in beginning textbooks in physical geography.

The text has been restructured and rewritten from the earlier *Physical Geography Today* (second edition) to be manageable within a single academic term. Together with a selection of additional readings it can also serve for a two-term sequence. The present volume omits the most dispensable verbiage included in *Physical Geography Today* and enlarges upon a number of topics that seemed inadequately treated in that volume. We have not truncated the larger book by simply deleting paragraphs, but have edited and rewritten the text line by line to maintain articulation of all of its parts, large and small. The present text is more compact than that of *Physical Geography Today*, but we believe it is a full treatment of the essential topics of today's physical geography. Several new illustrations appear here, and case studies have been utilized in every chapter to highlight human concerns with varying environmental phenomena. The appendixes provide discussions of some of the basic tools of the geographer, including maps and images resulting from remote sensing techniques. *Physical Geography Today*'s chapter opening art, which has proved popular with both students and instructors, has been retained.

ACKNOWLEDGMENTS

We wish once more to acknowledge our debt to the many reviewers listed in the second edition of *Physical Geography Today*, whose substantive comments remain part of the fabric of the present work. Additional reviewers of this latest effort were Tom Bishofberger, San Joaquin Delta College; Robert Christopherson, American River College; Charles E. Dick, Lakewood Community College; C. Gregory Knight, Pennsylvania State University; Harry Lane, Tennessee Technical University; Thomas R. Lewis, Manchester Community College; William D. Miller, Monroe Community College; Johnnie D. Shaw, United States Military Academy; and Thomas P. Templeton, Mesa Community College. We have greatly benefited by their assistance and hope that they will approve the use to which we have put their aid. A note of thanks is also due to Professor Robert Reed of the Department of Geography, University of California, Berkeley, for the facts behind the case study on the Manila galleons. Our particular gratitude is extended to Virginia Joyner, who collated reviewers' comments and thoughtfully as well as indefatigably edited the text of this work—enduring the thorny moods of authors protective of every nuance of their phrasing and organization. At Random House itself, we worked with several individuals, notably Barry Fetterolf, who coordinated the overall effort; Fred Burns, assisted by Jim Kwalwasser, who handled editorial matters; Suzanne Loeb, who handled the thousand aspects of the book production; and Suzanne Thibodeau, who assisted us in making major decisions concerning content and organization. It would be neither fair nor wise to overlook the many ways in which the assistance of our wives, Jeanne and Lucille, helped bring this undertaking to fruition. And, of course, we must acknowledge the contributions of our students, our teaching assistants, and the countless persons whose paths we have crossed to our benefit in the preparation of this book—as well as those whose needs have had to wait while project-related deadlines and emergencies were being met.

Overview

Contents

Essentials of Physical Geography Today

Photographic "Illumination" © by Glen Heller, 1973. (Courtesy, Images Gallery, New York)

The earth has been the scene of continuous change since its formation about 5 billion years ago. And it continues to change today because of the dynamic interplay of energy with air, water, and land. Physical geography is the study of the processes that have shaped the surface of the earth.

CHAPTER 1
Evolution of the Earth

The study of geography is more than merely description of the earth and the natural processes and human activities occurring on its surface. Geography is concerned with explaining *how* and *why* physical and human phenomena vary from place to place. The two broadest subdivisions within the field are *physical geography*, which focuses upon the natural processes that create diversity on the earth, and *human geography*, which is concerned with all aspects of human activity on our planet. However, human activities clearly affect natural processes and the physical environment, and physical processes and features exert many influences on human activities. Thus physical and human geography cannot be entirely separated.

The earth's surface is a constantly changing arena in which energy and materials continually interact. Here energy from the sun and from the earth's interior meet air, rock, soil, water, and a host of life forms. All of these are themselves linked by the physical processes that shape our environment. Because physical geography studies the natural processes that shape the human environment, its subject matter is extremely di-

Climatology Meteorology Hydrology Pedology Botany Ecology

Astronomy

Geodesy

Cartography

Geology

Geomorphology

Physical Oceanography

Physical Geography

Figure 1.1 (opposite)
Physical geography draws upon the specialized knowledge of many disciplines. The unique contribution of physical geography is its focus on the interactions of the varying phenomena that combine to give every place its particular character. (Tom Lewis)

verse, including most of the natural sciences (Figure 1.1). The focus is on interactions, such as the effect of solar energy on atmospheric motion, the role of water in the development of soil, or the influence of vegetation on erosion processes. Similarly, human geography studies the interplay of physical, cultural, historical, and economic influences on human activities throughout the world. Geography's emphasis on the interactions among various physical and cultural systems provides a unique point of view, one that no other single field of science can offer.

Physical geography, which is the subject of this book, is the original environmental science, traditionally concerned with the interaction between mankind and the physical environment. The spiraling rate at which we are using the earth's resources—extracting water, food, fuel, and raw materials—is increasingly affecting our environment and the natural processes that maintain it. In many cases human activities have triggered unintended changes in natural systems, affecting the earth's capacity to support life. Knowledge of the linkages in environmental systems can reduce this danger.

As this chapter makes clear, the present is part of a continuum of change on earth that goes back some 5 billion years. The changes that have occurred have left a variety of traces. These include the deposits made by geological processes, fossilized plants and animals, datable materials of various types, and even the record of the earth's changing magnetic field. Such clues permit a generally accepted reconstruction of the principal events in the history of the earth. This chapter will outline these events, which have created the major physical systems that are the subject of later chapters.

FORMATION OF THE EARTH

Birth of the Universe

Before the earth could be formed, the universe had to come into being. Scientific evidence suggests that the universe did not always exist, but was created at a definite point in time. Most scientists believe that before this time all the matter and energy in the universe was squeezed into a single nucleus, or "cosmic egg." Its internal energy eventually caused this nucleus to explode in what has been called the "big bang," which threw atomic particles outward in all directions. During this expansion the existing elements were formed by fusion of atomic protons, electrons, and neutrons.

As yet there were no stars. These formed millions of years later out of vast clouds of atoms that were condensed by the gravitational attraction of neighboring bits of matter. During this process of consolidation, the squeezed clouds of atoms became so heated that self-sustaining nuclear reactions caused them to glow. In this way they became stars—like our sun. These stars clustered by the billions in enormous rotating galaxies resembling pinwheels (Figure 1.2, see p. 6). The galaxies are so far apart their light takes years to pass the voids between them.

The big-bang hypothesis is supported scientifically by the discovery that the universe seems to contain an "echo" of the original explosion. This is in the form of a weak background radiation from a very low temperature source. Furthermore, the light received from distant galaxies indicates that they are rushing away from one another as the universe continues to expand like a balloon.

The present apparent rate of expansion of the universe suggests that the big bang occurred some 10 to 20 billion years ago. To realize how long ago this was in relation to later events, think of 10 billion years as a 24-hour day, with the universe originating at midnight, 24 hours ago. Each second of such a day would represent 100,000 actual years. Our sun's energy output

Figure 1.2
Our own galaxy, the Milky Way, seen from our position in one of its spiral arms. The Milky Way is a disk composed of billions of stars and is about 100,000 light years in diameter. One light year equals about 9.6×10^{12} kilometers, or 6×10^{12} miles (meaning 10 is multiplied by itself 12 times before being multiplied by the other number). The total number of galaxies observable in space is estimated to be in the billions. (Photo from Big Bear Solar Observatory, Sweden)

suggests that it became active some 6 billion years ago. This would be almost 10 hours after the beginning of the 24-hour day initiated by the big bang.

Formation of the Planets

Although we can observe billions of stars, the only planets we can detect are the nine circling our own sun. Since our sun seems to be an average star, it is assumed that many, and perhaps most, stars have similar planetary systems. These are invisible, even through the most powerful telescopes, because planets give off only reflected light, which is extremely faint compared to the light emitted by the furnace-like stars.

The most widely accepted explanation of the birth of our planetary system is that of Harold Urey, an American chemist and winner of the Nobel Prize. Urey proposed that our solar system began as a rapidly rotating, disk-shaped cloud of atoms. As the center of the cloud condensed to form the sun, the outer portions broke into separate eddies that themselves condensed into spinning swarms of solid matter called *planetesimals*. These were smaller than the present planets. Gravitational attraction then drew the separate swarms of planetesimals together to form the existing planets. The planets' moons are regarded as leftover planetesimal masses. Several types of evidence indicate that the earth formed in this manner about 4.6 billion years ago, or about 1 P.M. in our 24-hour day.

Urey's theory is favored by two simple facts. First, all the planets circle the sun in the same direction, presumably the direction of spin of the original cloud of atoms. Second, the orbits of all the planets lie in approximately the same plane, thought to be the plane of the disk-like cloud of atoms (Figure 1.3).

The rotation given to the earth at its creation

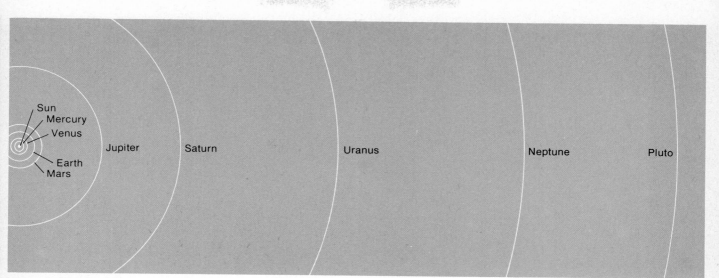

Figure 1.3
This diagram shows the distances of the planets from the sun in correct scale. The planets themselves are too small to be shown at the scale of their orbits. Between the orbits of Mars and Jupiter are swarms of asteroids, or minor planets, which may be fragments of earlier planets. The chunks of matter that form the asteroid belt range from dust particles to one lump that is as large as the British Isles.

The orbits of the planets lie nearly in the same plane, and every planet revolves about the sun in the same direction, which is consistent with the hypothesis that planetary bodies originally condensed from a whirling disk of gas. The inner planets, Mercury, Venus, Earth, and Mars, together with Pluto, are called *terrestrial* planets. They have solid bodies surrounded by relatively thin gaseous atmospheres. The outer planets, Jupiter, Saturn, Uranus, and Neptune, are largely gaseous, with thick hydrogen-rich atmospheres. (Doug Armstrong)

is what produces our cycle of daylight and darkness. The earth's rotation is also a primary control of atmospheric motion and thereby of weather and climate. The fact that the axis of rotation is oblique to the earth's orbit around the sun is the basis for our seasons, as will be seen in Chapter 3.

Structure of the Earth

According to Urey's model, the earth was formed by the collision of planetesimals at moderate temperatures. Our planet's hot interior is thought to have developed slowly after the formation of the earth. Heating resulted from the *radioactive decay* of elements in the earth's interior. This process continues today; however, the heat generated is transported upward and lost at the earth's surface. The temperature of the planet's outer layers seems to have been stable for at least 2 billion years. Radioactive decay affects the atoms of certain unstable heavy elements (such as uranium) that change by casting off electrons or particles from atomic nuclei. This releases energy that heats surrounding material. As the early earth's interior heated, solid material began to melt and move. Light material slowly rose toward the surface, while heavy material gradually sank toward the center. Eventually three concentric shells formed. These can be recognized by the way in which earthquake shock waves pass through the earth (Figure 1.4, see pp. 8–9).

Extending about half way to the surface from

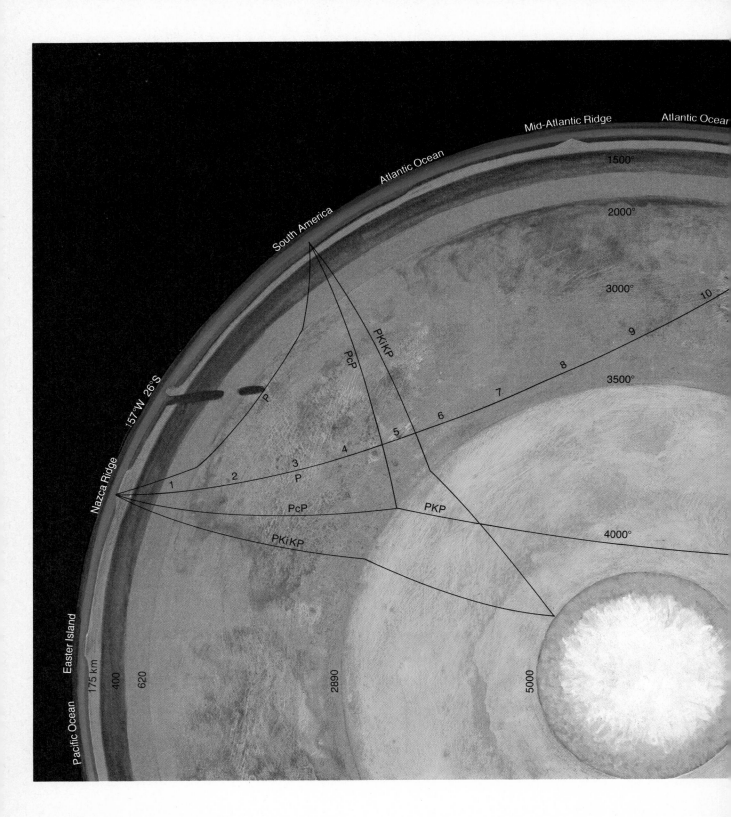

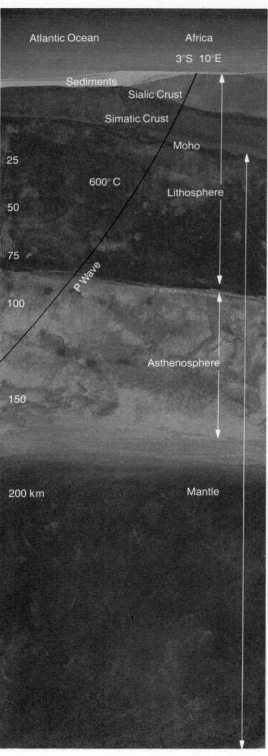

Atlantic Ocean   Africa

3°S 10°E

Sediments

Sialic Crust

Simatic Crust

Moho

25

600°C   Lithosphere

50

75

P Wave

100

Asthenosphere

150

200 km   Mantle

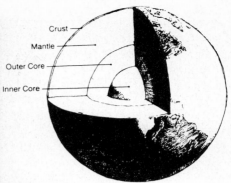

Crust

Mantle

Outer Core

Inner Core

Figure 1.4

The earth in cross section. The earth may be visualized as a series of concentric shells having different properties. The *crust* varies in thickness from about 10 km (6 miles) under the oceans to 40 km (25 miles) or more under the continents. The second shell, or *mantle*, extends to a depth of about 2,900 km (1,800 miles), at which point the *outer core* begins. At 5,000 km (3,000 miles) below the surface is the *inner core*, believed to be an alloy of iron and nickel (that possibly includes some silicon or sulfur).

The boundary between the crust and upper mantle is referred to as the Mohorovičić discontinuity (after its discoverer) or *Moho*. The outer 200 km (125 miles) of the earth may be separated into two classes of material: the zone of low strength and low seismic velocity is called the *asthenosphere*; the more rigid layers that overlie it, which include both the crust and the upper mantle, are called the *lithosphere*. *Lithospheric plates* may be visualized as rigid slabs that are free to move on the low-strength asthenosphere. It is important to note that lithospheric plates include both crust and mantle and that a single plate may include both oceanic and continental crust. These concepts are discussed further in Chapter 11.

Our knowledge of the inner constitution of the earth has come primarily from a study of the shock waves generated by earthquakes. These waves, called *seismic waves*, travel through the earth at speeds that vary according to the properties of the material through which they pass. The numbers on the path of the wave that travels to a point 3°S 10°E on the African coast represent the time elapsed in minutes. (Robert Kinyon/Millsap & Kinyon after R. Phinney)

the earth's center is the *core*, which has a radius of 3,400 kilometers (km), or 2,100 miles. The core has such a high density that it is thought to be composed of a mixture of iron and nickel. The physical state of this material is uncertain, since the combination of pressure and temperature at the center of the earth cannot be duplicated in laboratories.

Surrounding the core is the *mantle*, which is some 2,900 km (1,800 miles) thick. It is composed of high-density rock material dominated by silica, magnesium, and iron. Temperatures in the mantle are high enough to melt this mate-

rial, but pressures are so great at these depths that most of the mantle is rigid.

The earth's outermost shell is called the *crust*. This is the only portion of the earth to which humans have had access. The crust is from 10 to 70 km (6 to 40 miles) thick. It is composed of the solid rock that forms the conti-

Figure 1.5
The appearance of the earth's surface before the advent of life on the land may have been similar to this raw volcanic wasteland on the slopes of Haleakala Crater on the island of Maui, Hawaii. (Butch Higgins)

nents and ocean floors. The structure of the crust is complex. As we shall see in Chapter 11, the continents consist of scabs of low-density rock imbedded in a continuous layer of higher-density rock that forms the floors of the oceans and passes under the continents. The rock of the continental areas is diverse in origin (Chapter 11) and in many areas has been severely disturbed by vertical and horizontal crustal motions. The rock underlying the deep sea floors is mainly lava produced by submarine volcanic eruptions.

The Early Atmosphere

Since the earth and sun are thought to have formed from the same cloud of atoms, they should initially have had similar compositions. But the gaseous element neon, which is abundant in the sun and other stars, is now absent from the film of gases, known as the *atmosphere*, that surrounds the earth. This indicates that the earth lost its original gaseous atmosphere. This loss may have been caused by heating when the original planetesimals condensed, even before the planets were formed.

The loss of the earth's original atmosphere was only temporary. The first volcanic eruptions on the land began feeding new gases to the surface, producing a new atmosphere. Its composition would at first have been similar to the composition of volcanic gases, being dominated by carbon dioxide, with some water vapor, nitrogen, sulfurous gases, and only traces of free oxygen. The planet Venus has such an atmosphere at present—one quite incapable of supporting life.

The Early Landscape

The earth's surface 3 or 4 billion years ago would have resembled the volcanic landscape in Figure 1.5. Flows of molten rock and blankets of volcanic ash covered the land. The ground was completely barren, and no living organisms were present. Water (condensed from clouds of volcanic steam) fell as rain, helping to decay rock by chemical processes. Large and small rock fragments and mineral grains littered the ground. There was no soil, and rainwater flowed off, cutting deep gullies and carrying rock particles and dissolved minerals to the growing seas. Between rains, windstorms swept sand and dust across the naked landscape.

With no vegetative protection, erosional forms in rainy regions must have been spectacular. Hillslopes were steep, rocky, rutted with gullies, and bordered by piles of rock debris. The combination of high rainfall and lack of covering vegetation and soil would have given these early landscapes an appearance seen nowhere on earth today.

LIFE ON THE EARTH

About 3 billion years ago—with only a bit more than 7 hours left on our 24-hour clock—the history of the earth took a unique turn. Life appeared. How this occurred remains a puzzle. Experiments show that when mixtures of the simple volcanic gases present in the earth's primitive atmosphere are subjected to electric discharges, complex molecules are formed. These include amino acids, which are the building blocks of the proteins that are found in every living cell. In the early atmosphere, lightning and ultraviolet radiation from the sun could have provided the energy to break apart and recombine molecules. The new, more complex molecules would have been moved to the oceans during rainfalls. In time the oceans may have become an "organic broth," rich with complex molecules that were not "living" but had the potential to be organic building blocks.

Eventually an unusual combination of molecules appeared that was able to absorb energy from its surroundings and to reproduce itself. This combination of molecules was the first "living" organism.

The Fossil Record

We can only make guesses about the origin of life on earth, for there are no known traces of the first life forms. Our understanding of life in the remote past is based on the *fossil record*. *Fossils* are mineral replacements of the remains of plants and animals that have been buried by deposits of mud, silt, and sand (Figure 1.6). As these deposits accumulated to great depths, they became compacted and cemented to form layers of rock. By this time the remains of life forms within the deposits had been replaced with resistant minerals that preserve the original forms permanently. Obviously, in undisturbed sedimentary rock formations, the oldest life forms are found as fossils in the lowest layers of rock. Thus an erosional trench through a thick mass of sedimentary rock, as at the Grand Canyon of the Colorado River, gives us a catalogue of the life forms that have inhabited the region through tens to hundreds of millions of years.

The catalogue is not complete, however, for most fossils represent only hard bones, shells, and woody plant parts. The soft tissues of organisms usually disappear before burial and fossilization. Furthermore, if sediment is not accumulating at a location, there is no possibility of natural burial. Finally, much of the fossil record at a locality may have been removed by natural erosion of fossil-bearing rocks.

Organic Evolution

The earliest living organisms preserved in the fossil record had already evolved quite far from the earth's initial life forms. The earliest fossilized organisms resemble the blue-green algae seen on the surface of ponds today. These early plants produced energy by *photosynthesis*, a process that uses solar energy to produce carbohydrates, such as sugars and starches, from carbon dioxide and water, with oxygen given off as a by-product. This brings up an important question: how did complex plants, animals, and, eventually, human beings arise from the earliest simple forms of life?

An answer came from the British naturalists Charles Darwin and Alfred Russel Wallace, both of whom published theories of *organic evolution* in 1859. According to the various theories of evolution, organisms do not reproduce exact copies of themselves. In fact, the offspring of

Figure 1.6
This fossil is exceptional because the soft tissues of plants and animals are rarely preserved. This squidlike organism has been so well preserved in rock that even its ink and ink sac are identifiable to the expert. (Werner Wetzel)

any one pair of parents may vary considerably. In a given environment some variations have survival value. They may increase the organism's ability to gather food, escape predators that would use *it* for food, or withstand environmental stress. The favored organism has an above-average chance to survive and pass along the helpful variation. This affects the evolution of the species as a whole, which thus slowly changes by the process of *natural selection.*

If an organism finds an environmental niche to which it is extremely well adapted, further changes offer no new benefits and the species may not evolve further. An example is the opossum, which is virtually unchanged over the past 60 million years. On the other hand, natural selection may produce change in a species over the span of a few decades. For example, the English peppered moth, a light-colored moth with dark speckles, became predominantly dark in color between 1850 and 1900 as industrialization altered its physical environment. The darker coloration apparently helped the moth hide from predatory birds in the newly soot-covered towns and nearby woodlands.

In time the changes in species resulting from the process of natural selection may produce entirely new species—groups of mutually fertile individuals that differ from similar groups in some constant way. According to Darwin and Wallace, and most subsequent scientists, this is the process that has produced the amazing number of plant and animal species that populate the earth today. All of these can be traced back through time to the first primitive organisms formed in the earth's seas.

Interaction of Life with the Environment

The appearance of green plants 3 billion years ago—at 5 P.M. on the 24-hour clock—set in motion new processes that changed the face of the earth in several ways. For a long time plants were confined to the seas, where water shielded the organisms from the sun's cell-destroying ultraviolet radiation. But the simple marine plants had to remain in the sunlit upper portions of the sea to carry out the vital photosynthesis process. During photosynthesis water molecules are split into hydrogen and oxygen atoms. Plant cells use some of the oxygen and release the rest. Little of the oxygen released by the first plants remained in the atmosphere. Most of it combined with iron-bearing minerals in decomposed rock, turning the rock various shades of brown or yellow. Only when the exposed minerals had been oxidized could the oxygen released by photosynthesis begin to accumulate in the atmosphere. This appears to have required some 2 billion years—about 5 hours on our 24-hour clock. Today the atmosphere is more than 20 percent oxygen, most of it released by green plants through the ages.

At 10:30 P.M. on the 24-hour clock the fossil record reveals the sudden appearance of complex, shelled, multicellular marine animals. How they developed from preceding plant life is still unknown. What is clear is that these animals evolved the ability to make use of the oxygen being released by marine plants, as well as the ability to use plant material for food. Once these adaptations appeared, the natural selection process quickly expanded the range of possible variations, resulting in an explosion of oxygen-breathing animal species.

Soon marine animals and plants were working together to alter the earth's environment. Aquatic plants extracted carbon from carbon dioxide to produce carbohydrates as part of the photosynthesis process. Tiny aquatic animals that fed on these plants built hard shells by combining this carbon with calcium weathered from rocks and washed from the land. When these animals died, their remains drifted to the sea floor to form a growing deposit of calcium carbonate sediment, which was gradually transformed into limestone rock. Limestone, one of the world's most abundant rock types, is the chief storehouse of the carbon that once dominated the atmosphere. In many places the carbon derived from the remains of marine animals combined with hydrogen to form petroleum or natural gas.

As these processes removed more and more carbon from sea waters, it was replaced by atmospheric carbon dioxide. Thus, while biological processes were adding oxygen to the atmosphere, they were also removing carbon dioxide from it. This changed the atmosphere significantly—reversing the proportions of oxygen and carbon dioxide. Today carbon dioxide is present in the atmosphere in only minor amounts, about 320 molecules for every million molecules of air.

The growing presence of oxygen in the earth's atmosphere led to the formation of *ozone*, a molecule composed of three atoms of oxygen, as opposed to the oxygen molecule's two atoms. Ozone is created in the upper atmosphere by high energy solar radiation. The radiation splits apart oxygen molecules, and the atoms recombine into ozone molecules. Ozone is important because it absorbs much of the biologically destructive ultraviolet radiation streaming from the sun. The development of an ozone layer high in the atmosphere to screen out lethal radiation was required before the earth's land surfaces could become habitable to living organisms. Not until about 400 million years ago—only one hour before midnight on the 24-hour clock—did plants begin to colonize the exposed land surfaces. Small invertebrate animals, such as insects and scorpion-like arachnids, appeared soon afterward, with larger amphibious vertebrates moving onto the land some 50 million years after the first plants arrived.

When plants moved onto the land, the processes of rock decay, erosion, and sediment movement were altered. The prying action of plant roots and the chemical action of plant acids helped break down rocks. And organic material combined with the tiny rock particles to form the first true soils. The blanket of soil protected by vegetation softened the previously harsh angular outlines of landscapes. Shielded by a plant cover, the land was at last able to offer resistance to the erosive action of rain and flowing water. As a result, the rate of sediment production decreased markedly.

The development of land vegetation removed still more carbon from the atmosphere. Vigorous plant growth in swampy coastal areas led to the accumulation of thick masses of undecomposed plant remains. These eventually were buried by sediment and transformed into coal. An enormous amount of carbon taken from the atmosphere is now fixed in mineral form in the world's coal fields.

While plant and animal life has completely transformed our planet, the nonliving environment has at times taken its toll on the living world. Climatic changes due to astronomical or other causes have affected the evolutionary development of plants and animals. So has geological activity such as the uplift of mountain ranges and changes in the sizes and positions of the continents and oceans (see Chapter 11).

Some 130 million years ago giant reptiles known as dinosaurs (Greek: "terrible lizards") ruled the earth. These huge creatures, descendants of earlier small amphibians, were dominant until about 65 million years ago—just 10 minutes before midnight on the 24-hour clock. Then, suddenly, they disappeared, to be replaced by the ancestors of modern mammals (Figure 1.7). Not until about 4 million years ago—only half a minute before midnight—did the first human-like creatures appear, presumably our own ancestors.

Figure 1.7 (opposite)
The geologic time scale. To the left are shown selected life forms and the approximate times when they first appeared or were dominant. A general trend from less complex to more complex life forms reflects the processes of organic evolution. Note that the time scales on the figure are not evenly spaced but emphasize the most recent 600 million years. (John Dawson after *Adventures in Earth History*, edited by Preston Cloud, W. H. Freeman Company. Copyright © 1970)

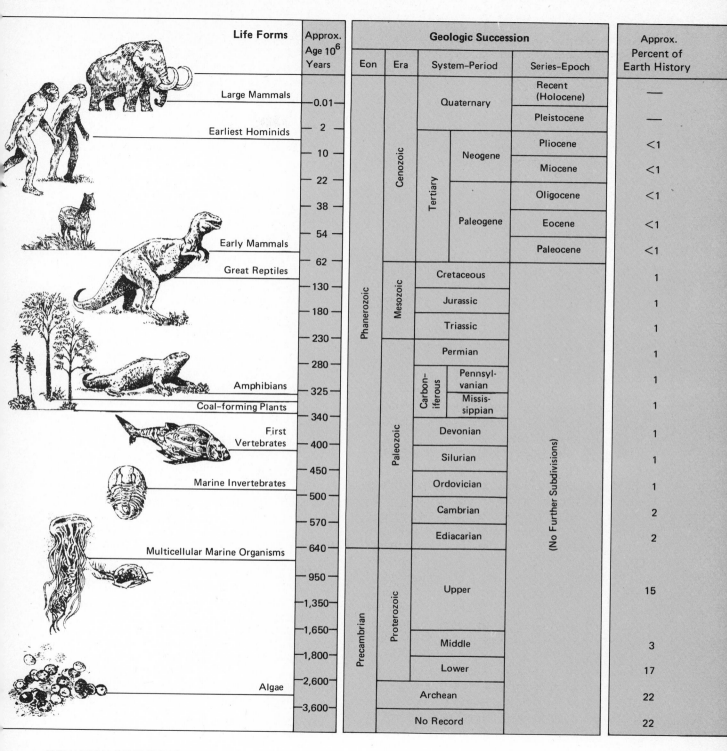

Life Forms	Approx. Age 10^6 Years	Geologic Succession				Approx. Percent of Earth History
		Eon	Era	System–Period	Series–Epoch	
Large Mammals	0.01	Phanerozoic	Cenozoic	Quaternary	Recent (Holocene)	—
					Pleistocene	—
Earliest Hominids	2			Tertiary — Neogene	Pliocene	<1
	10				Miocene	<1
	22					
	38			Tertiary — Paleogene	Oligocene	<1
	54				Eocene	<1
Early Mammals	62				Paleocene	<1
Great Reptiles	130		Mesozoic	Cretaceous	(No Further Subdivisions)	1
	180			Jurassic		1
	230			Triassic		1
	280		Paleozoic	Permian		1
Amphibians	325			Carboniferous — Pennsylvanian		1
Coal-forming Plants	340			Carboniferous — Mississippian		1
First Vertebrates	400			Devonian		1
	450			Silurian		1
Marine Invertebrates	500			Ordovician		1
	570			Cambrian		2
Multicellular Marine Organisms	640			Ediacarian		2
	950	Precambrian	Proterozoic	Upper		15
	1,350					
	1,650			Middle		3
	1,800			Lower		17
Algae	2,600			Archean		22
	3,600			No Record		22

EVOLUTION OF THE EARTH

Although we are newcomers on the earth, we are not intruders here. Like all our fellow organisms we have evolved with the earth and are adapted to it. Our respiration depends on the gases in the current atmosphere, and our eyes respond to those wavelengths of solar radiation that reach the earth's surface in the greatest abundance. Our interactions with the earth's environment will continue to affect our progress as a species.

SUMMARY

Physical geography is concerned with the interacting processes that have created the physical environment we inhabit. This environment is the result of continuous change since the formation of the earth almost 5 billion years ago. The universe itself appears to have originated explosively some 10 to 20 billion years ago. The planets, including earth, formed by the collision of matter in orbit around the sun. The earth's internal structure consists of a core, a surrounding mantle, and a thin outer crust that contains the continents and ocean basins.

The earth today is far different from what it was in the remote past. The earth's early atmosphere was dominated by gases vented during volcanic eruptions. Ancient land surfaces were barren and rutted by water erosion. The advent of life about 3 billion years ago initiated the changes resulting in our present environment. Life originated in the sea, where plant photosynthesis began to extract carbon dioxide from the atmosphere and to replace it with free oxygen. This made animal life possible. It also led to the formation of an ozone shield against ultraviolet radiation, which permitted life to exist on the land.

Fossils preserved in sedimentary rock record the evolution of organisms on the earth. The carbon removed from the atmosphere by biological activity has been stored in the form of limestone, coal, oil, and natural gas. In the course of biological evolution, many organisms have disappeared after a period of dominance. The mammals are relatively recent arrivals on the scene and are the dominant vertebrates today. Although the human species emerged but a moment ago in geologic time, it already has developed the capability to change the environments of all other organisms.

APPLICATIONS

1. Where within North America would you look for landscapes most similar to those of 500 million years ago? To those of 3 billion or more years ago?
2. What are the ages of the rocks present in the area of your campus? Around your home? Do these rocks contain fossils? If so, what life forms are represented? Where in your area is a good collection of local fossils on display?
3. For what purely physical reasons would life not be likely on either Venus or Mars?
4. What are some clear examples of "natural selection," by which specific life forms have become adapted to particular environments? Is the human species, *Homo sapiens*, adapted to a particular environment?

FURTHER READING

Abell, George O. *Exploration of the Universe.* 3rd ed. New York: Holt, Rinehart and Winston (1975), 738 pp. This introductory text often used in astronomy courses is especially useful for recent ideas about the formation of the universe and solar system.

Cloud, Preston. *Cosmos, Earth, and Man.* New Haven: Yale University Press (1978), 372 pp. A lucid account of the history of the universe and planet earth, emphasizing the major changes that have occurred due to the origin and evolution of life, by a leading earth scientist.

Dott, Robert H., Jr. and **Roger L. Batten**. *Evolution of the Earth*. New York: McGraw-Hill (1971), 649 pp. This introductory text traces the evolution of crustal features, flora, and fauna from a geological perspective.

Eicher, Donald L. *Geologic Time*. Englewood Cliffs, N.J.: Prentice-Hall (1968), 150 pp. This brief monograph, part of the Foundations of Earth Science series, surveys each of the methods used in dating crustal materials and features.

The Solar System, A Scientific American Book. San Francisco: W. H. Freeman (1975), 145 pp. One of a series of reprints of original articles in *Scientific American*. A good overview of the planets including the earth and our moon, with recent ideas on their origin and evolution.

CASE STUDY: **Dating the Earth**

Before 1800 people generally believed that the earth's landscape had been formed by a series of violent catastrophes that had lasted only a few thousand years. Toward the end of the eighteenth century, James Hutton, a Scottish naturalist, took a fresh look at the mountains and streams of his native land and interpreted what he saw as the products of erosion and mountain uplift. He theorized that these slow processes, given enough time, could have produced the landscapes of the British Isles, and concluded that the eras of geologic time must be much longer than previously thought. Not long afterward, Charles Darwin concluded that evolution required time spans on the order of hundreds of millions of years.

To check the ideas of Hutton and Darwin, scientists in the nineteenth century attempted to develop an accurate way to measure geologic time. The fossil record gave the succession of the geologic eras, but it did not indicate their actual ages. Therefore scientists measured the rate of sediment deposition to calculate how long it would have taken to build up the known thicknesses of sedimentary rock layers. The accuracy of this method was uncertain because the rate of sedimentation has varied over time and the record contains many gaps.

The accurate natural timekeeper that scientists sought was found early in the twentieth century when it was discovered that the passage of time could be measured using the decay rates of *radioactive isotopes*. An isotope is one of a number of atomically different forms of a single chemical element, such as uranium 238 and uranium 234 (the number representing the mass of each of its atoms). A radioactive isotope is one that decays spontaneously into another element

at a certain predictable rate. The time required for half the initial amount of a radioactive isotope to decay is called the isotope's *half-life*. The half-life is unaffected by changes in temperature, pressure, or other external factors.

Suppose a rock initially contained a certain amount of uranium 238 when it solidified from the molten state. Uranium 238 decays with a half-life of 4.5×10^9 years through a chain of steps to the stable end-product lead 206. Each uranium atom that decays eventually results in the formation of one atom of lead. Since the rate of decay is known, measurement of the relative numbers of uranium 238 and lead 206 atoms in the rock allows the elapsed time to be computed. For example, if the rock contains equal amounts of uranium 238 and lead 206, the elapsed time since the rock solidified must have been one half-life, or 4.5×10^9 years.

If the rock initially contained lead 206, or is younger than about 60 million years, the uranium 238/lead 206 method will not be accurate. It is therefore sometimes necessary to use more than one radioactive element to establish a date. The most useful radiometric clocks for geologic time are listed in the accompanying table. The decay of potassium 40 to argon 40 has been particularly useful as an age indicator, since potassium is found in several of the minerals composing common rock types formed by volcanic activity. The age of the volcanic rock sets a minimum age limit for the deposit found under it, and a maximum age limit for the deposit above it. The potassium/argon method has been used to map the ages of the sea floors (Chapter

Half-Lives of Radioactive Elements

RADIOACTIVE CLOCK	HALF-LIFE (YEARS)
Rubidium 87/Strontium 87	5.0×10^{10}
Uranium 238/Lead 206	4.5×10^9
Potassium 40/Argon 40	1.3×10^9
Uranium 235/Lead 207	0.7×10^9
Uranium 234/Thorium 230	248×10^3
Carbon 14	5730

Source: Gray, Dwight E. (ed.). 1972. *American Institute of Physics Handbook*. 3rd ed. New York: McGraw-Hill.

11) and to date the earliest ancestors of the human race discovered in the volcanic East African region.

Using radiometric dating methods, it is possible to make a fairly accurate determination of the age of the earth. Among the oldest rocks on earth are some found in Greenland. Their age, according to radiometric measurements, is about 3.8×10^9 years, which means that the earth is at least 3.8×10^9 years old. However, since erosion may have destroyed all traces of the earliest rocks, it is likely that the earth is considerably older. According to currently accepted models of the creation of the solar system, meteorites were formed at the same time as the earth. Radiometric dating of meteorites should therefore be an accurate method of determining the earth's true age. This method results in an age of 4.6 billion years. Direct measurements of the total uranium and lead content of the earth's crust also point to an age of 4.6 billion years.

Carbon 14 is another radioactive isotope that has been important for dating geologic events within the past 40,000 years or so. Carbon is a useful substance for age determination since it is a constituent of virtually all plant and animal matter. Carbon 14—a radioactive form of carbon that is continuously formed in the earth's atmosphere by the action of cosmic rays on nitrogen 14—makes up a constant proportion of all the carbon ingested by plants and animals. When an organism dies, it ceases to take in carbon. The amount of carbon 14 in its cells steadily declines due to radioactive decay, but the amount of ordinary carbon remains constant. Thus, measuring the ratio of radioactive carbon to ordinary carbon allows the age of the sample to be inferred. Carbon 14 dating has provided highly accurate data about relatively recent events in the history of the earth. For example, by dating wood from trees that were killed by the advance of glacial ice, it has been possible to fix the date of the ice's advance through a particular area. Carbon 14 dating can also be applied to charcoal, shells, bone, and the calcium carbonate in soils.

A number of other dating techniques are useful for measuring relatively short time spans of thousands of years. One method uses *varves*, which are annual sediment layers deposited in lakes and on the sea floor. Coarse sediment is deposited seasonally during periods of streamflow, with fine sediment settling out during the dry season or when water surfaces are frozen. In Scandinavia, varve analysis has been extended more than 10,000 years into the past, and correlation of varve patterns from different localities has allowed scientists to reconstruct the history of glacial recession in northern Europe.

Tree-ring analysis involves counting the annual growth rings of trees and measuring their relative widths. Logs used in the construction of ancient Indian pueblos have been dated by correlating their rings with ring sequences of known age. The ages of many kinds of topographic surfaces of recent origin can be determined by counting the rings of the largest trees growing on them. Other dating techniques are applicable in special circumstances, and the number of chronological indicators will surely increase with further research.

Ceaseless flows of energy cause continuous change on the earth. Energy streaming from the sun and from the earth's interior powers the earth's dynamic physical systems, which constantly interact to maintain the environments we inhabit.

The Starry Night by Vincent Van Gogh, 1889. (Collection, the Museum of Modern Art, New York; acquired through the Lillie P. Bliss bequest)

CHAPTER 2
Energy and the Earth's Systems

Energy is involved in every process and is the source of all change. But energy will not produce change unless there is something for it to act upon. The moon receives energy from the sun, but having no atmosphere, no water, and no life, the moon has changed little over billions of years. The earth, however, has a number of active physical systems—the atmosphere, the oceans, landforms, soils, and the plant and animal realms—all powered by energy in some form. The ways in which these systems respond to energy and interchange it are major themes of physical geography. To set the stage for later discussions, this chapter deals with energy—its sources and the forms it takes and its effect on the earth's physical systems.

SOURCES OF ENERGY

The most important source of energy on the earth is radiation from the sun. It provides the power to drive most of the physical processes important to life. These include the movements of the atmosphere that produce weather and drive the oceanic circulation. Solar energy also causes the growth of plants, which, in turn, support the animal world.

Two other sources of energy reside within the earth itself. One is gravitational force, which exerts a toward-the-center-of-the-earth pull on all objects on or near the earth's surface. This causes flows of air and water and falls of earth and rock. A second internal source of energy is the heat generated within the earth by the radioactive decay process. The sudden shock of an earthquake or the eruption of a volcano reminds us of unseen forces below the earth's surface. These forces work over long spans of time to lift mountain ranges and even change the position of continents.

FORMS OF ENERGY

The total amount of energy in the universe remains constant. Energy cannot be created, nor can it be destroyed. It can only be converted from one form to another. The forms of energy most significant to natural processes on the earth's surface are radiant energy, heat energy, gravitational energy, kinetic energy, and chemical energy, each of which is discussed in the following paragraphs (see also Figure 2.1, pp. 24-25).

Radiant energy from the sun heats the atmosphere and the earth's surface. It does this both directly and indirectly, as we shall see in Chapter 3. Solar radiation reaches the earth in about 8⅓ minutes, traveling at a speed of about 300,000 km (186,000 miles) per second. Most of this radiation is in the range of wavelengths of visible light, which our eyes are adapted to sense. The invisible longer-wavelength thermal radiation can be felt as heat, and the invisible shorter-wavelength ultraviolet radiation is what causes sunburn. Solar radiation varies in intensity from place to place on the earth and from time to time. It is the seasonal geographical variation of solar radiation that creates the earth's weather and climate systems.

Heat energy is the energy produced by the random motion of the atoms and molecules of substances. The hotter a substance, the more vigorous the motion of its atoms. This motion must be generated by an input of energy in some other form, such as radiant energy. There is a distinction between temperature, which is a measure of molecular motion, and heat energy, which refers to the total energy in a volume of a substance. A cup of hot coffee at 50°C (122°F) has a higher temperature than a bathtub of warm water at 35°C (95°F), but there is actually more heat energy in the bathtub water because of its larger volume. As we shall see later in this chapter, there is also an important difference between sensible heat and latent heat. *Sensible heat* is created by molecular motion and can be measured directly with a thermometer. *Latent heat* is stored energy that cannot be measured directly. It becomes sensible heat only when released by a change of state of a substance, as when water vapor condenses to liquid water or liquid water freezes to ice.

Gravitational energy is the potential energy an object has due to its elevation. A large boulder on a hillside or parcel of air aloft in the atmosphere has the potential to move downward if its support, or the force holding it aloft, is lost. Gravitational energy is proportionate to altitude and mass. A large boulder near the top of a mountain has more potential energy—and will make a bigger splash in a lake below it—than a smaller rock at the same elevation or a boulder the same size farther downhill.

Kinetic energy is energy of motion. The higher the speed of an object, the greater the energy it possesses. The size of the splash produced by a boulder that falls into a lake is actually a consequence of the kinetic energy it develops as its potential energy is lost. For a given velocity, kinetic energy is proportional to the mass of the object. If equal volumes of air and water are moving at the same speed, the volume of water will have much greater kinetic energy because of its much greater mass. This is most apparent along coasts, where the destructive force of wind-driven ocean waves is obvious.

Chemical energy is energy stored in the electrical bonds that hold together the molecules, atoms, and parts of atoms of substances. When substances react chemically, chemical energy is released, absorbed, or converted to other forms of energy. The energy produced by a nuclear reactor is a result of the recombination of subatomic particles. In plant photosynthesis, solar radiation generates chemical energy, which is stored in plant tissues.

Energy Transformations

Standing in the sunlight, you feel warm because solar radiant energy is exciting molecules of your skin, creating heat energy there. Running uses stored chemical energy to produce energy of motion. Coasting downhill on a bicycle, your speed increases as you descend because your initial gravitational (or potential) energy is being converted to kinetic energy.

As noted earlier, energy is never destroyed but is constantly being converted from one form to another. Some of these are of little use. When a snowball hits a wall, its kinetic energy is abruptly transformed. Its atoms, and some of those of the wall, are violently jostled, producing heat energy that diffuses into the air. In fact, some heat is generated in every process that involves the transfer of energy. This heat—or random motion of atoms—merely warms the surrounding environment and is not otherwise usable. Since energy transfers always generate a certain amount of heat that is quickly dissipated, usable energy can never be transferred with perfect efficiency.

Energy and Work

All processes in the physical world can be viewed in terms of energy and work (Figure 2.1, see pp. 24-25). Energy is the capacity for doing work or producing change. Heat energy within

the earth produces volcanoes, warps the crust, uplifts mountains, and moves the continents and ocean floors. Chemical energy stored in vegetation helps to form soils and support animal life. The energy in some vegetation becomes stored as fossil fuels, the main source of power for industrial society.

Radiant energy causes molecules of water to move from the earth to the atmosphere, where they have potential (or gravitational) energy. When this water falls as rain, it takes on kinetic energy, which on impact with the soil causes soil particles to break up or splash to a new location. Some of the water falling on the land runs into streams. As a stream flows toward lower altitudes, its gravitational and kinetic energy are used in the work of altering the landscape—eroding the land and transporting and depositing sediment. In cold climates and at high altitudes, precipitation may take the form of snow. Accumulated snow produces glaciers, which are thick sheets of moving ice that likewise alter the landscape by erosion and the deposition of sediment.

By producing temperature differences on the earth's surface, radiant energy also causes the motion of the atmosphere. The resulting wind, in turn, transfers energy to the ocean, producing ocean currents and waves. The kinetic energy of waves modifies shorelines by eroding rocks and transporting sand to and from beaches. More important still, water in both the seas and the atmosphere continually transports energy from warmer to cooler regions of the earth, helping to equalize the earth's energy distribution.

WATER: AN ENERGY CONVERTER

Water plays a critical role in converting energy from one form to another, in moving energy from place to place, and in energy storage. Let us look now at the unique properties of water that enable it to do this work.

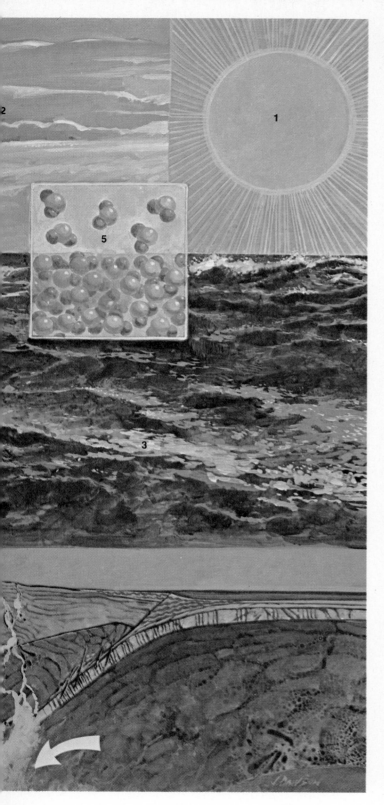

Figure 2.1
This painting interprets some of the forms of energy important in physical geography, and illustrates ways in which energy interacts with systems on the earth.

Most change on the earth is caused by two sources of energy: solar radiation (1) and the earth's internal heat (14). Solar energy drives the motion of the atmosphere (2) by causing temperature differences on the earth's surface. The wind in turn transfers energy to the ocean, producing ocean currents and waves (3). Ocean currents and the moving atmosphere help distribute energy from the hot equatorial regions of the earth to the cool polar regions. The energy of waves actively shapes coastlines (4) by eroding rocks and by transporting sand to and from beaches.

On the seas and on the land, energy from the sun frees water molecules to enter the atmosphere by evaporation (5). Water vapor is transported by the moving atmosphere and returns to the earth's surface as precipitation (6). Some of the water falling on the land runs into streams (7). As a stream flows toward lower altitudes, its gravitational, or potential, energy causes erosion of the land and transportation and deposition of sediment. In cold climates or at high altitudes, winter precipitation takes the form of snow (8). Accumulated snow produces glaciers (9), which are moving masses of ice that erode the land beneath.

Vegetation (10) absorbs solar radiant energy and transforms it to stored chemical energy by the process of photosynthesis. Some of the stored energy is passed on to animals (11), which ultimately depend on green plants as their source of food energy. Vegetation, moisture, and rock materials interact to form soils (12). The chemical energy in some vegetation and tiny marine animals becomes stored as fossil fuels, which constitute the main power source for modern industrial society (13).

The energy released by radioactive decay (15) creates heat within the earth. This geothermal energy can be tapped by man and may become an important source of power. Uneven heating below the earth's crust causes motion within the asthenosphere (Figure 1.4). At "hot spots" (17) currents rise and spread. This moves the continents and ocean floors, causing the crust to be wrinkled in some places and broken by faults (16) in others. Where large segments of the crust are forced together (18) the heavier segment, usually an oceanic plate, descends into the mantle and melts. The molten material rises back toward the surface to create volcanic eruptions (19). (John Dawson)

Ability to Change Phase

All substances can take three physical forms—solid, liquid, and gaseous. These are called *phases* of the substance. Water is the only common chemical compound that occurs in all three phases under the conditions prevailing at the earth's surface. Moreover, our planet is the only one in the solar system that has the proper range of temperatures to permit water to be seen in all three phases.

For a substance to change phase, heat energy must be either absorbed or released by the molecules of the substance. Changes from gaseous to liquid and from liquid to solid phases release energy, producing sensible heat that can be measured by a thermometer. Such phase changes can be seen in water vapor that condenses to liquid water, in liquid water that freezes to ice, and in water vapor that is transformed directly to ice. Changes from solid to liquid and from liquid to gaseous phases absorb energy, creating latent heat in the resulting substance. In this case the molecules that are changing phase do not themselves change in temperature. Examples of these phase changes are liquid water that evaporates, ice that melts, and ice that vaporizes without producing water. The energy released or absorbed in these phase changes is commonly measured in units called

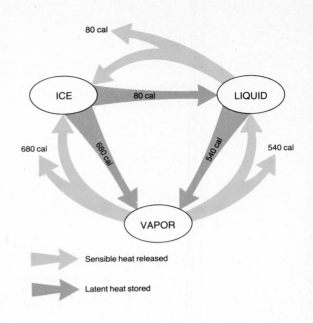

gram calories, or simply *calories*. One gram calorie (with a small "c") is the amount of energy required to raise the temperature of a gram of water from 14.5°C to 15.5°C. (The Calories, with a capital "C", used in relation to food energy in diets, are kilocalories, equal to 1,000 gram calories.) See Table 2.1 and Figure 2.2.

Because phase changes in water occur at temperatures that are commonly found on the

Table 2.1

Heat Transfers Associated with Phase Changes of Water

PHASE CHANGE	HEAT TRANSFER	TYPE OF HEAT
Liquid water to water vapor	540–590 calories absorbed	Latent heat of vaporization
Ice to liquid water	80 calories absorbed	Latent heat of fusion
Ice to water vapor	680 calories absorbed	Latent heat of sublimation
Water vapor to liquid water	540–590 calories released	Latent heat of condensation
Liquid water to ice	80 calories released	Latent heat of fusion
Water vapor to ice	680 calories released	Latent heat of condensation

Figure 2.2 (opposite)

Figure 2.2 (opposite)
Significant energy changes occur when water changes from one physical state to another. Energy must be added to change water from a state in which the water molecules are tightly bound to a state in which they are more loosely bound. To change 1 gram of water from solid ice directly into gaseous water vapor requires adding approximately 680 calories of energy, called the *latent heat of sublimation*. The change from solid to liquid requires 80 calories per gram, called the *latent heat of fusion*. The energy required to change 1 gram of liquid water to gaseous vapor, the *latent heat of vaporization*, varies slightly with the temperature of the water. For water at 15°C the latent heat of vaporization is 590 calories per gram, and for water at 100°C it is 540 calories per gram.

Energy is released when a substance changes from a state in which its molecules are loosely bound to a state in which they are more tightly bound. The amount of energy that must be released is numerically equal to the latent heats described above; for example, 590 calories must be removed from 1 gram of water vapor at 15°C to form 1 gram of liquid water at the same temperature.

earth, latent heat is an important factor in the earth's temperature balance. In moist climates and over warm seas, much of the incoming solar energy is utilized for evaporation rather than heating, resulting in significant local cooling. Atmospheric circulation transports this energy in its latent vaporized form to other regions, where it is released by the condensation that creates clouds. Where there is little water to be evaporated, most incoming solar radiation is converted to sensible heat, causing local temperatures to be higher than those in moist areas.

High Heat Capacity

In addition to its readiness to change phase, water also has an unusual capacity to act as a reservoir of heat energy. Water requires a large input of energy to become warmed, but once warmed it cools more slowly than most other substances. The amount of heat energy required to raise the temperature of one gram of a substance by 1°C at normal atmospheric pressure is known as a substance's *specific heat*. The specific heat of water (1 calorie per gram of water per degree Celsius) is about 5 times that of soil or air. This means that one gram of water must absorb 5 times more heat energy than one gram of soil or air to increase in temperature an equal amount.

The capacity of a substance to absorb heat in relation to its volume is its *heat capacity*. Because of the very high density of water relative to air, the heat capacity of water is many times that of air, and is 2 to 3 times that of dry soil. Its high heat capacity means that water can absorb and store large amounts of energy without changing temperature greatly. By contrast, small amounts of energy will produce large changes in the temperature of land surfaces or air. This helps to explain why temperatures vary much less in the oceans than on the land.

THE ATMOSPHERE: AN ENERGY TRANSPORTER

The earth's atmosphere is a major focus in physical geography because it is the chief means of transporting energy and moisture over the earth's surface. To understand how the mixture of gases forming the atmosphere behaves, it is necessary to know something about the behavior of gases in general.

Unlike the molecules in solids and liquids, the molecules of gases are free to move independently of one another. They frequently collide with solid surfaces and with each other. Each collision exerts a push on the molecule or surface that is struck. The total force that these molecular collisions exert on a given area of surface at any point in time is called the *pressure* of the gas.

The pressure exerted by the earth's atmosphere decreases rapidly with altitude. At an altitude of 5 to 6 km (3 to 3.6 miles) the atmospheric pressure is only half the pressure at sea

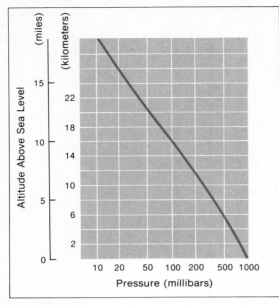

Figure 2.3
The pressure exerted by air in the atmosphere decreases rapidly with increased altitude. As the graph indicates, average pressure at sea level is approximately 1,000 millibars, but at an altitude of 5 km (3 miles) it falls to about 540 millibars. Note that the pressure is shown on a *logarithmic scale*, which is useful when graphing a quantity that varies over a wide range. The pressure at sea level can be thought of as the weight per unit area of a column of the earth's atmosphere. The graph shows the standard value of atmospheric pressure at each altitude; actual atmospheric pressure at a particular location on a particular day can be a few percent higher or lower than the standard value. If the air is denser than usual, the pressure is higher than normal, and if the air is less dense, the pressure is lower than normal. (Doug Armstrong adapted from *Handbook of Geophysics and Space Environment*, edited by Shea L. Valley, Air Force Cambridge Research Laboratories, U.S. Air Force, 1965)

Figure 2.4 (opposite)
(a) The pressure, volume, and temperature of a fixed amount of gas are interdependent; a change in one of the three quantities is always accompanied by a change in one or both of the others. If a gas is confined in a sealed box so that its volume is constant, changing the temperature of the gas changes the pressure the gas exerts on the walls of the box. The gas molecules move more rapidly at high temperatures and therefore collide more forcefully and more frequently with the walls. For gases under normal conditions, pressure increases with increased temperature if the volume is held constant.

(b) A parcel of air in the atmosphere is unconfined. If the parcel is heated, it must expand its volume in order to maintain a constant pressure. The volume of the parcel increases with increased temperature for a gas held at constant pressure. The greater the volume occupied by the gas molecules in the parcel, the lower the density of the gas. Density therefore decreases with increased temperature for a gas at constant pressure.

(c) The transfer of energy to and from parcels of air in the atmosphere can produce motion. A parcel of air that is hotter than the surrounding atmosphere is less dense than the atmosphere; in such a situation, the upward buoyant force on the parcel exceeds the downward weight force, and the parcel rises. Conversely, if a parcel of air is cooler than the surrounding atmosphere, it is denser. The downward weight force in this case exceeds the upward buoyant force, and the parcel descends. Upward motion causes a horizontal inflow of the surrounding air at the surface, and downward motion causes horizontal outflow at the surface. (Tom Lewis)

level (Figure 2.3). A barometer measures atmospheric pressure in terms of the height of the mercury column that exerts the same downward pressure as the atmosphere. Under average conditions at sea level, this is 760 millimeters (mm),

or 29.9 inches. Those who study the atmosphere express pressure in *millibars*—a unit of pressure equal to 1,000 dynes per square centimeter. (A *dyne* is the force required to cause a mass of one gram to accelerate by 1 centimeter per second in each second of time.) Standard atmospheric pressure at sea level is about 1,000 millibars.

The pressure, temperature, and volume of a given amount of gas are dependent on one another. A change in one always causes a change

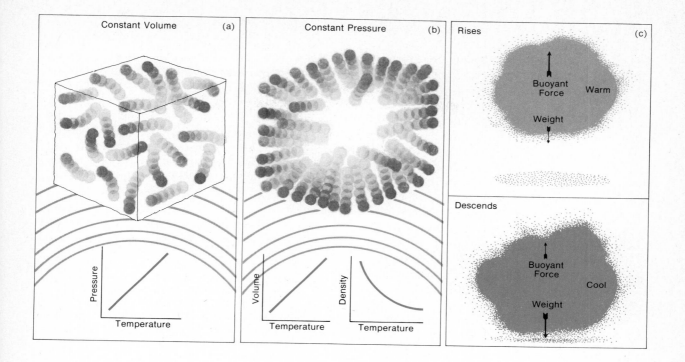

in either or both of the others (Figure 2.4). A rise in temperature will increase the volume of an unconfined gas. But if a gas is sealed in a container so it cannot expand, a rise in temperature will cause a rise in the pressure of the gas.

If a volume of air close to the earth's surface is heated, it will expand without changing its pressure. This expanded air is less dense than the surrounding cooler air because it has fewer molecules per unit of space. The heated volume of air has an upward buoyant force that exceeds the downward gravitational force on the air, and the warm air rises like a hot-air balloon. Differences in the density of air thus cause sensible heat to be transferred upward in the lower atmosphere. This vertical transfer of air and heat energy in turn sets air in motion horizontally, producing wind.

The earth's wind systems transfer warm air from areas that receive greater amounts of solar radiation to areas that receive lesser amounts. Winds also carry the latent heat present in wa-

ter vapor from regions where evaporation has occurred to far-off areas where the vapor condenses back to water or ice. The circulation of the atmosphere and its significance in energy transfer are the subjects of Chapter 4.

THE EARTH'S PHYSICAL SYSTEMS

A system is any collection of interacting objects. The earth as a whole can be thought of as a single physical system. However, the earth contains so many phenomena interacting in so many ways that to understand them it is necessary to subdivide the world into smaller systems. None of these is truly independent; all interact to some degree.

The environment at the earth's surface is divisible into four major systems. These are the *atmosphere*, the film of gases surrounding the solid surface of the earth; the *hydrosphere*, including all the waters of the earth; the *lithosphere*, composed of the solid material forming the outer shell of the earth; and the *biosphere*, which includes all life forms present on our planet.

All systems have some common features. All have energy sources and contain materials of some type. Most are *open systems*, with inputs of energy and materials from outside, transformations of energy and materials within the system, storage capacity, and outputs of energy and materials that become the inputs for other systems (Figure 2.5). In *closed systems* there are no external inputs and no outputs to other systems. Closed systems are entirely self-contained, like a battery-powered wristwatch. When its internal energy source exhausts itself, a closed system ceases to function.

When we look at the inputs and outputs of any system, we become aware of the interde-

pendence of different systems. Think for a moment of waves breaking on a beach. The inputs are wave energy and sand particles. The outputs are relocated sand and the form of the beach itself.

Daily changes in sand flow and beach form are the direct result of wave action against the shore, as discussed in Chapter 16. However, by tracing the inputs and outputs further we can see that the beach system is part of a much larger picture. The kinetic energy of the waves was originally produced by the force of wind against the water surface hundreds or thousands of kilometers out to sea. The wind resulted from geographical variations in atmo-

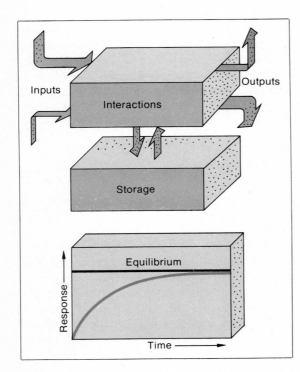

Figure 2.5
A system is any collection of interacting objects. Physical geography is concerned with a wide variety of systems, including landforms, soils, vegetation, and the atmosphere. Thinking about the world in terms of systems is useful because systems have many features in common, some of which are illustrated in this schematic diagram. The inputs to a system represent energy and material received from outside the system's boundaries. The inputs are transformed by system interactions to new forms that are either placed in storage for a time or transferred to other systems as outputs of material and energy. Consider a field of corn as an example of a system: the inputs include radiant energy from the sun, carbon dioxide from the atmosphere, and moisture and nutrients from the soil. Part of the input is stored as chemical energy in the process of photosynthesis. Outputs include the water vapor returned to the atmosphere by plant transpiration, the free oxygen released in photosynthesis, and the chemical energy that is eventually used as food.

The lower diagram illustrates schematically one of the ways that systems respond in time to an input. In this mode of systems behavior, the output does not reach full strength immediately upon application of an input. Time is required for a steady condition to be achieved. Some systems, such as the atmosphere, respond rapidly to new inputs; other systems, such as soils, may require hundreds of thousands of years to come to equilibrium. (Tom Lewis)

spheric pressure. The pressure variations are a consequence of place-to-place variations in solar energy input to the atmosphere. Solar energy itself is an output of atomic reactions within the sun. The material input of the beach system—sand—is an output of a complex erosional system on the land, and was probably delivered to the shore by a river. Finally, the output of the beach system—the changing form of the sand deposit—has important effects on the living organisms that inhabit the beach. This is but one simple example of the complexity of almost any open system.

Systems can store energy or materials for varying amounts of time. Rocks that are being stretched or squeezed in the earth's crust are storing energy. They are behaving like a compressed or stretched spring. Eventually, when the strain exceeds the strength of the rocks, they snap. This releases the stored energy in the form of an earthquake. The slow bending of rocks in western California has stored so much energy that a major destructive earthquake is expected sometime in the 1980s. Hurricanes are another example of the release of stored energy. In this case the summer build-up of both sensible and latent heat in the tropics finds release in powerful tropical storms that vent this energy into higher latitudes.

System Equilibrium

Many systems tend toward a stable condition in which continued inputs of energy and material produce no changes in the system or its output. Such a system is said to be in *equilibrium*. Short-term changes may occur, yet the long-term behavior is constant. For example, a flood may cause a stream channel to enlarge and deepen by eroding its bed and banks. But when the flood subsides, the stream channel returns to its previous state by depositing sediments that fill up the portions enlarged by erosion. Although there are short-term fluctuations, the channel remains about the same from year to year.

The time required for an equilibrium system to begin to change as a result of change in inputs is the *response time*. The time required for a system to reestablish equilibrium after a change is the *relaxation time*. These times vary greatly in different systems. A river channel responds in hours to changes in input and reestablishes equilibrium quickly, within hours or days. Other systems, such as hillslopes, may have response and relaxation times of years, centuries, or even longer.

Some types of disturbance do not permit a system to return to its original state. Rather, the system assumes a new form. Damming a river reduces the load of sediment it carries to the sea. This reduces the supply of sand to coastal beaches. In a few years or decades after a dam is built, coastal beaches may shrink or disappear altogether because their input has changed permanently.

Cycles of Materials

As long as the earth and its atmosphere do not experience a long-term change in temperature, we can assume that the earth's input and output of energy are in balance. But this energy is used and transformed in many ways between the time solar radiation, the major energy input, arrives on our planet and the time energy is returned to space in a different form. The intervening energy exchanges constitute the force driving most physical systems on the earth.

The materials composing these systems are fixed in amount. Except for rare meteorite falls and the escape of some gaseous molecules, the earth is composed of the same atoms that came together when the planet formed 4.6 billion years ago. Since the earth's materials are finite, they must be constantly recycled to permit the earth's systems to continue functioning.

The rate at which materials are recycled varies greatly from system to system. Water can be recycled from the sea surface to the atmosphere as vapor, then back to the oceans as precipitation, all in a few hours. However, if this same

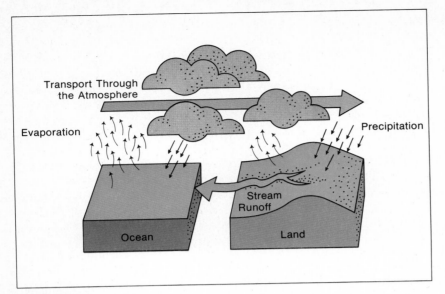

Figure 2.6
The hydrologic cycle is a key element in physical geography because water has important interactions with systems such as vegetation and landforms. The processes involved in the cycling of water include evaporation of water from the oceans, evaporation and transpiration from the continents, transport of water vapor through the atmosphere, and the return of water from the atmosphere to the surface as precipitation. The annual evaporation loss from the oceans exceeds the amount they gain from precipitation, but the deficit is made up by streams flowing from the continents to the oceans. (Tom Lewis)

water falls as snow in a polar region, it could become part of a glacier, where it might be held as ice for hundreds or even thousands of years before returning to the sea.

Other materials participate in even longer cycles. Oceanic salts have been stored for hundreds of millions of years in beds of rock salt extending from New York to Ohio and Michigan. Eventually, the slow erosion of the North American continent will return this salt to the oceans. The earth's crust is itself being recycled ever so slowly by geological processes, as we shall see in Chapter 11.

The earth's hydrosphere is a dynamic physical system that is dependent on the *hydrologic cycle* illustrated in Figure 2.6. Streams carry water endlessly from the continents to the oceans.

If there were no "return flow" back to the continents, the land masses would eventually be waterless. The hydrologic cycle in its simplest form begins with evaporation of sea water, which enters the atmosphere as water vapor. The atmospheric circulation transports much of this water vapor over the continents. Over the land, water vapor is condensed into clouds composed of water droplets. The droplets coalesce and fall as rain or snow. Some of the rainwater or melted snow runs off into streams that bring the water back to the oceans to complete the cycle.

Actually this is an oversimplification. Much precipitation also falls on the oceans themselves, producing a short subcycle within the larger cycle. And evaporation can take place on the continents—from various bodies of water and from plant leaves in the process of *evapotranspiration*. Most rainfall does not run off directly into streams but enters the soil and becomes soil moisture. Some of this percolates deeper to become groundwater, which sustains stream flow between rains. Nevertheless, the basic concept of the cycle is correct—involving circulation and changes of state of material, with periods of *residence* in different forms.

In the hydrologic cycle, water has several residence forms. About 97 percent of the earth's free water is salt water residing in the oceans. The next largest store of water is the ice in glaciers, particularly the polar ice sheets. Most of the remaining water is stored in porous rock or loose deposits below the land surface, forming the groundwater that is tapped by wells. Lakes, ponds, swamps, and rivers contain less than 1 percent of the total supply of free water. The atmosphere holds even less. If all the water present in the atmosphere at any moment were to descend to the earth, it would form a layer only about 2.5 centimeters, or 1 inch, in depth.

Another material cycle of great importance to life on the earth is the *carbon cycle* (see Figure 2.7, pp. 34–35). The most critical element in the carbon cycle is gaseous carbon dioxide (CO_2). Green plants require atmospheric carbon dioxide, along with water and solar energy, to manufacture carbohydrates in the process of photosynthesis. The carbon dioxide in the atmosphere is continually depleted by plant activity. But it is also continually replenished by plant decay and by animal respiration, which expels CO_2 as a waste product of the process by which animals oxidize nutrients to produce energy.

Only a small percentage of the earth's carbon is present as CO_2 in the atmosphere. Far larger amounts of carbon compounds are dissolved in the seas and stored in vegetation, limestone bedrock, and deposits of the fossil fuels (coal, oil, and natural gas). However, since the beginning of the Industrial Revolution in the nineteenth century, the burning of fossil fuels has increased the CO_2 content of the atmosphere by 10 to 15 percent (Figure 2.8, p. 35). There is concern

about this because CO_2 is a strong absorber of the thermal energy re-radiated from the earth's surface. The absorbed energy already appears to have raised the temperature of the lower atmosphere. It is estimated that a 10 percent increase in atmospheric CO_2 produces about a 0.3°C global temperature increase. Continuation of the present tendency toward global warming would eventually reduce the size of the polar ice caps. The removal of water from glacial storage would gradually raise the level of the seas, causing heavily populated coastal regions to become submerged.

Budgets of Energy and Materials

Inputs, outputs, storage, and balance can be treated as parts of the *budget* of a system. As the term suggests, the budget concept is similar to a financial account that balances income and savings against expenses.

An ordinary concrete swimming pool can be used to illustrate the budget concept. The pool gains water from the city water supply and from rainfall. The pool loses water by evaporation and by the splashing out of water by pool users. But, to account for all the water passing through the pool, we must understand all interactions in the system. If there are cracks in the concrete, leaks must be included on the output side of the budget. Similarly, any drainage of rainwater into the pool must be added to the input.

We can also develop an energy budget for the pool, in which energy inputs and outputs are reflected in the changing temperature of the pool water. Temperature changes reveal the energy budget in the same way that water-level changes reflect the material budget.

The accounting system for energy and materials sometimes reveals that things are occurring in the system that we do not understand or have failed to measure. For example, the global CO_2 budget shows that the increase in atmospheric CO_2 produced by the burning of fossil fuels over the past century is only about half the

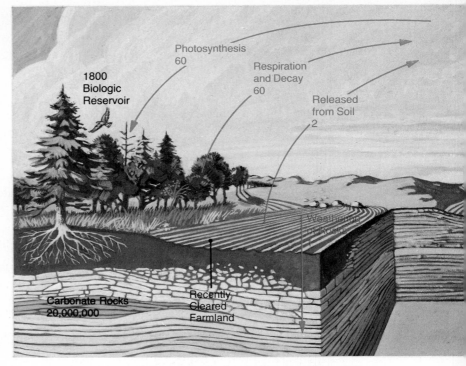

Figure 2.7
This diagram illustrates the large number of systems through which carbon can pass during its cycle. The numbers in parentheses are estimated values in billions of tons of carbon released or absorbed annually in each process or the total amount stored in each reservoir.

Transfer of CO_2 to and from the atmosphere is an essential part of the carbon cycle. CO_2 enters the atmosphere primarily from the respiration and decay of organisms, from the burning of fossil fuels, and from the CO_2 dissolved in the oceans. Photosynthesizing green plants and the oceans absorb CO_2 from the atmosphere. Some of the carbon absorbed by plants is locked into long-term storage as coal and peat. The largest reservoir of carbon

is limestone bedrock. Much of the calcium carbonate ($CaCO_3$) composing limestone is precipitated chemically from sea water, but some is composed of the shells and skeletal parts of tiny marine animals. The remains of marine organisms are also the source of petroleum and natural gas.

The transfer of CO_2 to and from organisms is believed to be nearly in balance over the year. The CO_2 content of the atmosphere, however, is increasing by several percent each decade. This increase is the result of man's industrial activities; it would be even greater if it were not for the CO_2 taken up by the oceans. (John Dawson after Gilbert and Plass, "Carbon Dioxide and Climate," *Scientific American*, 1959)

expected amount. The missing CO_2 is believed to have gone into solution in the oceans or been absorbed by vegetation on the land. The CO_2 budget shows that either ocean water or land vegetation has a higher CO_2 storage capacity than expected, somewhat reducing the harmful effect of fossil fuel use.

One of the tasks of physical geography is to identify the significant similarities and differences among the systems at work in the various regions of the world. Systems that are subject to similar inputs make similar responses. Thus the problems met, the lessons learned, and the solutions developed in one area may be applied to

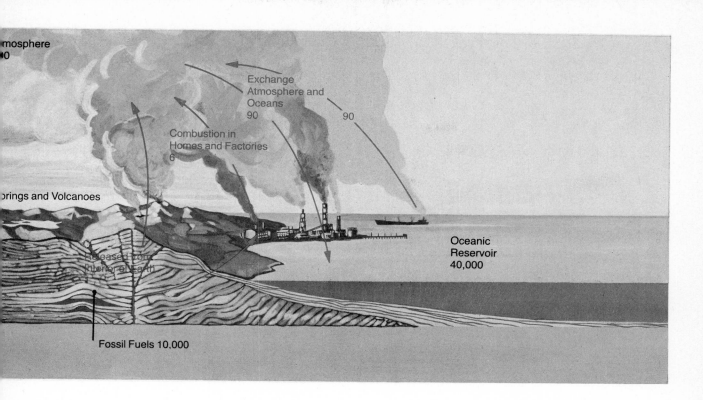

mosphere
0

Exchange
Atmosphere and
Oceans
90

90

Combustion in
Homes and Factories
6

prings and Volcanoes

Oceanic
Reservoir
40,000

eleased from
terior of earth

Fossil Fuels 10,000

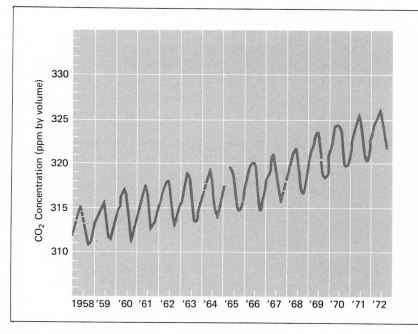

Figure 2.8
Mean monthly concentrations of CO_2 at Mauna Loa, Hawaii, at a site distant from any major local source of CO_2. The cyclical pattern is due primarily to the seasonal increases and decreases of vegetation in the northern hemisphere, but the long-term trend upward is obvious. (Vantage Art, Inc. after P. V. Hobbs et al., "Atmospheric Effects of Pollutants," *Science, 183* p. 910, March 8, 1974)

others of a similar type. This can help us avoid the costly errors that have already led to environmental degradation in many parts of our planetary home.

SUMMARY

Energy is the capacity to do work—that is, to cause change in the state or position of materials. The principal sources of energy for physical processes on the earth are solar radiation, gravitational force, and the earth's internal energy resulting from radioactive decay of unstable elements. Although energy cannot be created or destroyed, it can take many forms. Of most significance to environmental processes are solar radiant energy, heat energy, kinetic energy, gravitational (or potential) energy, and chemical energy. The earth's physical processes involve constant transformation of energy from one form to another. Some energy is lost as heat in every energy transformation.

Water is extremely important to the earth's physical systems because of its role in energy conversion, storage, and circulation. All changes of state of water between gaseous, liquid, and solid forms either absorb energy as latent heat or release energy as sensible heat. The atmospheric circulation transports energy globally as measurable sensible heat or as latent heat that is released when water vapor condenses to liquid water. The atmospheric circulation is impelled by pressure differences resulting from differential solar heating of the earth's surface. Water's properties allow it to store and release vast quantities of heat energy without undergoing large temperature changes. Thus water has a moderating effect on temperature changes at the earth's surface and in the atmosphere. Water also transmits work-producing kinetic energy in the form of flows of water and glacial ice and in water waves.

The earth is a complex of interacting physical systems. The largest-scale systems are the at-mosphere, hydrosphere, lithosphere, and biosphere. Most earth systems are open systems that receive inputs of energy and materials from other systems and have outputs to other systems. Most systems tend to move toward an equilibrium condition in which processes act and interact without causing long-term changes in the system itself. Physical systems can store energy or materials for varying lengths of time. The release of stored energy may cause violent events, such as earthquakes and hurricanes, but these are necessary to preserve system equilibrium.

The continued activity of the earth's physical systems requires energy and materials to be used over and over again, constantly changing their form. These cycles of energy and materials, involving inputs, transformations, storage, and outputs, can be evaluated in terms of budgets that must balance out over a period of time. The budget concept helps us to understand environmental processes better and can help us avoid errors in the management of the earth, our home.

APPLICATIONS

1. Diversion of "wasted" solar energy to controlled uses will cause decreases in solar energy input to other systems. Will this have noticeable consequences?
2. Latent heat is a type of energy that might be compared to gravitational energy. What are the similarities and differences between latent heat and gravitational energy?
3. How is national policy for energy production of concern to climatologists?
4. What is the source of the water you use each day? What artificial systems have been installed to bring this water to you? How has the simple hydrological cycle described in the text been altered or "short-

circuited" in your area? Has this produced any unexpected consequences?

5. What is the largest scale closed system you can think of? What is the smallest scale open system imaginable?

6. What sort of artificial disturbances of natural systems have occurred and are occurring in your area? What are the response times to these disturbances? Are the relaxation times known?

FURTHER READING

Asimov, Isaac. *Life and Energy.* New York: Bantam (1965), 378 pp. An account of energy principles and their applications to living organisms by one of the most adept of all writers on scientific topics.

Chorley, Richard J., ed. *Water, Earth, and Man.* London: Methuen (1969), 588 pp. This is a most useful collection of 38 selections, mostly by British authors, dealing with various aspects of the hydrologic cycle. The selections are at introductory and intermediate levels and complement a number of chapters in this text.

Energy and Power, A Scientific American Book. San Francisco: W. H. Freeman (1971), 144 pp. A collection of articles on energy printed originally in the periodical, *Scientific American.* Energy sources, flows, and transformations are all treated in a lively, highly readable manner, with many excellent illustrations.

Odum, Howard T. *Environment, Power, and Society.* New York: Wiley-Interscience (1971), 331 pp. This book attempts to show how natural and man-made systems are related in terms of energy flows and basic system structures. Odum uses the principles of ecology to offer solutions to current problems of global importance. A particular contribution is the symbolic language used to depict energy flows and transformations.

Weyl, Peter K. *Oceanography: An Introduction to the Marine Environment.* New York: John Wiley (1970), 535 pp. This book complements *Essentials of Physical Geography Today* very well. The chapters on oceanic salts and geochemical cycles are especially pertinent to Chapter 2 of this text.

CASE STUDY: **Understanding Nature**

The world around us is a wonderfully complex mosaic; its patterns have aroused curiosity from the very beginnings of humankind. What is the sky, the earth, the ocean? And how does each piece of the mosaic fit with all the others in the ever-changing natural world?

In our attempts to sort through the earth's myriad phenomena, we tend to construct mental *models* to help us understand things and events and how they are interconnected. With models we find a way of perceiving order in nature, of gaining deeper knowledge of why and how things happen as they do. Consider the model for "tree." One of us might think of a pine tree; another, an elm or a centuries-old redwood. Even though these are different trees, they nevertheless have enough similarities to be understood by all of us as belonging to the same category, and we share an understanding of that model.

Now consider the ecologist, whose model may be more complex: he views a "tree" as a marvelous microenvironment, where different creatures occupy different niches from roots to treetop, where intricate energy exchanges take place between the inhabitants and the surrounding environment. His model explains much more because he has probed a little deeper into a seemingly simple part of the world and discovered its larger dimensions.

All fields of inquiry claim many models that are useful in explaining the world. For example, there are innumerable ways to look at and understand the nature of one of the earth's most important and most common substances, water. Water covers three-fourths of the earth's surface; it is also beneath the surface, permeating the soil and rock, and in the atmosphere, circulating in the form of vapor.

Different types of models describe water's different properties. A chemist uses a molecular model to explain water's fundamental makeup of two hydrogen atoms and one oxygen atom. A physical scientist is interested in water's unique ability to form three states of matter. A physical geographer, however, is concerned with water as a transporter of heat energy through the atmosphere and as an agent of change on the earth's surface. So models have been developed that describe how water flows across the earth in streams, rivers, and glaciers, how it circulates in the atmosphere and oceans, how it affects climate, vegetation, and soil formation, and how it shapes the landscape.

The concepts of "cycle" and "system" and "budget" are important components of the models in physical geography. The hydrologic cycle mentioned earlier and described more fully in later chapters is a model to explain how all the earth's original supply of water is continually recycled. The same water is transported time and time again from the land and oceans into the air, and then back to earth again. The local water budget is a model to describe the availability of water for various processes and purposes in the environment. Water shapes the earth's surface features, and several of the models useful for analyzing flowing water are illustrated here.

Keep in mind that the models described in this book are the physical geographer's perceptions of the world—a world we can come to know much better through educated eyes.

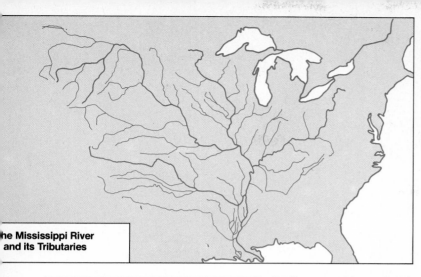

Wherever rainfall runs off the land, it erodes its own drainage system. This map, which is an example of a *graphic model*, shows the area drained by the Mississippi River system (see also Chapter 13). Graphs, photographs, and diagrams are types of graphic models used to aid our conceptualization of the physical world. (Doug Armstrong)

the Mississippi River and its Tributaries

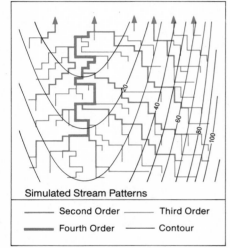

Simulated Stream Patterns

——— Second Order ——— Third Order
——— Fourth Order ——— Contour

Drainage patterns may also be simulated by computers, as this example of a *mathematical model* shows. This model is used to study the development of stream patterns and to predict changes that will occur in underground supplies of water as water is added or withdrawn. Computer models can also be used to simulate changes in variables such as precipitation or runoff to see the effect on a total system. In this illustration, the curved lines represent contours of elevation and the blue lines show the development of a drainage pattern on homogeneous material (see Chapter 13). The water flows down the slope in the direction of the arrow. Computer models are used in the oceanic and atmospheric systems as well and in other branches of science where complex systems with many variables are involved. (Doug Armstrong after A. Armstrong and John Thornes, *The Geographical Magazine*, © IPC Magazines, Ltd., 1972 by permission of New Science Publications)

An actual replica to scale of the Mississippi River system is an example of a *physical model*. It gives engineers answers to emergencies that arise in flood conditions. During the flood of 1973, this model was a valuable addition to computer models in calculating the rise in water level that would accompany the opening of a spillway. By using automatic instruments to make graphic recordings of water levels at critical points, inputs and outputs can be tested on the model, and the effects of proposed projects can be predicted. The portion shown here is the Atchafalaya River basin in southern Louisiana. (Wide World Photos)

Number 8 by Mark Rothko, 1952. (From the collection of Mr. and Mrs. Burton Tremaine, Meriden, Connecticut)

All life processes are driven by complex exchanges of visible and invisible forms of radiant energy from the sun. Desert heat and the cold of the polar zones are consequences of the earth-sun relation in space and of energy transfers on the earth.

CHAPTER 3
Energy and Temperature

Solar radiation is the principal source of energy for the natural processes that create diversity and change on the earth. However, if the earth continually received energy from the sun without returning an equal amount to space, the oceans would boil and the land would be scorched. Since the average temperature of the earth remains much the same from year to year, the earth must be returning as much energy to space as it receives from the sun.

Of course, not all locations on earth have equal energy gains and losses. Each year, tropical regions receive a greater amount of energy than they radiate back into space. Polar regions, on the other hand, annually lose more energy to space than they receive from the sun. We know that the tropical regions are not progressively heating up nor are the polar regions cooling off. This means that there must be a flow, or *flux*, of energy from areas of excess to areas of deficiency. The atmosphere and oceans circulate the energy that the earth receives, transporting warm air and water from the tropics to the poles while moving cool air and water back toward the equator. In this chapter we shall examine the gains and losses of energy that maintain the earth's temperature balance.

SOLAR RADIATION

Solar energy is transferred through space as *electromagnetic radiation*. Such radiation travels in waves and can be classified according to wavelength—the distance between similar points on successive waves. Radiation wavelengths are measured in units called *microns*, which are equal to one millionth of a meter. Figure 3.1 presents the *electromagnetic spectrum*, in which various types of electromagnetic radiation are identified according to wavelength. Visible light includes wavelengths from 0.4 to 0.7 microns. Our eyes sense the various wavelengths of visible light as different colors. When visible light is bent, or refracted, and then reflected by water droplets in the atmosphere, we see the various wavelengths as a rainbow. Solar radiation also includes wavelengths both longer and shorter than those of visible light. Ultraviolet radiation, X-rays, gamma rays, and cosmic rays are all emitted at wavelengths shorter than those of visible light, while infrared radiation and radio waves have longer wavelengths than those that are visible.

Every solid, liquid, and gas, whether warm or cold, emits electromagnetic radiation as a result of the motion of its molecules. The temperature of the radiating substance determines the wavelengths of emission. The hotter the object, the shorter the wavelength of the radiation emitted. When we turn on the heating element of an electric stove, the coil remains dull black while warming up. During this time it emits infrared radiation, which we can feel but not see. As it grows hotter, the wavelength of maximum emission becomes shorter, shifting over into the visible portion of the spectrum. The coil glows dull red, then bright red. If it could be heated fur-

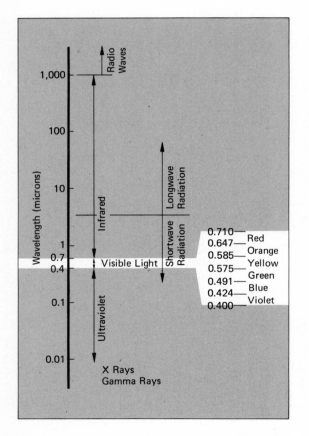

Figure 3.1
The electromagnetic spectrum is conventionally divided on the basis of wavelength. The divisions most important in physical geography are infrared radiation, visible light, and ultraviolet radiation. Visible light consists of wavelengths between approximately 0.7 and 0.4 micron. Infrared radiation consists of wavelengths intermediate between the wavelengths of radio waves and those of visible light, and ultraviolet light corresponds to wavelengths shorter than those of visible light. At extremely short wavelengths, the ultraviolet portion of the spectrum merges into the X-ray and gamma ray divisions.
Radiation with a wavelength longer than 4 microns is called longwave radiation; radiation in wavelengths shorter than 4 microns is called shortwave radiation. The distinction is useful in physical geography because solar radiant energy input to the earth is mostly at wavelengths shorter than 4 microns, while the energy radiated from the earth is at wavelengths longer than 4 microns (see Figure 3.2). (Doug Armstrong)

Figure 3.2
Because of the high surface temperature of the sun, energy is emitted primarily at wavelengths shorter than 4 microns, with much of the energy concentrated in the visible region of the spectrum. In contrast, the longwave radiation emitted by the earth is confined to wavelengths longer than 4 microns and has a broad peak at about 10 microns, because of the earth's comparatively low average surface temperature.

   The lower portion of the figure indicates the degree to which atmospheric gases, primarily carbon dioxide and water vapor, absorb electromagnetic energy near the earth's surface. Wavelength bands of strong absorption are shown in red, and bands of relative transparency are indicated by yellow. The lower atmosphere is relatively opaque to longwave radiation, so that much of the radiant energy emitted by the earth's surface is absorbed there. The atmosphere is relatively transparent to electromagnetic radiation in the band from 8.5 to 11 microns, and radiant energy in this band can escape to space if the sky is clear. (Doug Armstrong after G. M. Dobson, *Exploring the Atmosphere*, 1963, by permission of the Clarendon Press, Oxford)

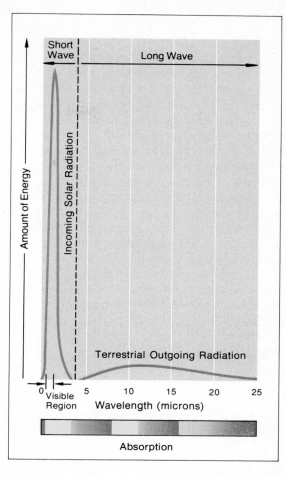

ther without melting, it would eventually glow yellow-white, like the sun.

At the earth's surface and in the lower atmosphere, temperatures normally range between $-40°$ and $+40°C$ ($-40°$ and $104°F$). Substances at such temperatures emit electromagnetic radiation with wavelengths of 4 microns or more. The surface of the sun, on the other hand, has a temperature of more than $6,000°C$ ($10,800°F$). At this temperature, most radiation is emitted at wavelengths of less than 4 microns. Hence we can think of most solar radiation as *shortwave radiation*, and all radiation emitted by terrestrial sources (the earth's surface and atmosphere) as *longwave radiation* (Figure 3.2).

One of the laws of physics is that a surface emits an amount of radiation per unit of time that is in direct proportion to the surface temperature. A very hot object, such as the sun, not only emits shorter wavelength radiation than a cooler object like the earth, but it also emits ra-

diation at a far greater rate. Accordingly, solar radiation is shortwave radiation emitted at a high rate, and any radiation emitted by the earth and its atmosphere must be longwave radiation emitted at a low rate.

SOLAR ENERGY INPUT TO THE EARTH

The sun radiates energy equally in all directions. The earth, which in relation to the sun is like a grain of sand to a football 100 yards away, intercepts only a tiny fraction of the total energy emitted by the sun. But this small fraction is an enormous quantity of energy, amounting to 2.6×10^{18} calories per minute (2.6×10 multiplied by itself 18 times). The solar energy intercepted by the earth in one minute is about equal to the total electrical energy artificially generated on earth in one year.

Not all this radiant energy reaches the earth's surface because, as we shall see later in this section, the earth's atmosphere modifies the solar radiation that strikes it. Nor do equal amounts of radiant energy strike all parts of the upper atmosphere. This is because the distribution of solar radiation reaching the top of the atmosphere is controlled by the length of the daylight period and the elevation of the sun above the horizon. Both of these factors vary with latitude and are determined by the earth-sun relationship.

The Earth–Sun Relationship

Like every planet in the solar system, the earth follows a fixed path, or *orbit*, around the sun. The orbit is slightly elliptical, so that the earth's distance from the sun varies a small amount through the year. However, this distance never varies more than 1.8 percent from the average distance of 149.5 million km (92.9 million miles). As a result, the monthly income of solar energy over the entire earth varies from the 12-month average by no more than 7 percent. The earth intercepts the greatest amount of solar radiation in early January, when it is closest to the sun (*perihelion*), and receives the least in July, when it is farthest from the sun (*aphelion*).

In addition to orbiting the sun, the earth rotates on its axis in a period that has been divided artificially into 24 hours. This introduces an additional daily cycle of solar energy income related to the times of sunrise and sunset. Because the earth is essentially spherical in shape, the sun illuminates only half of the globe at any time, as shown in Figure 3.3 (see pp. 46–47). The boundary line between the light and dark halves is the *terminator*, or *circle of illumination*.

The Earth's Tilted Axis

As the earth rotates on its axis, the circle of illumination appears to sweep around the earth from east to west in 24 hours. If the earth's axis of rotation were perpendicular to the plane of the earth's orbit, every place on the earth would have 12 hours of daylight and 12 hours of darkness each day of the year. In the middle and high latitudes the seasonal change from long daylight periods in summer to short daylight periods in winter would not occur. But the earth's axis tilts at an angle of $23\frac{1}{2}°$ from the perpendicular. This causes the duration of daylight and darkness to vary seasonally everywhere except at the equator, which is always cut exactly in half by the circle of illumination.

Figure 3.3 shows that as the earth circles around the sun, the earth's axis always remains tilted in the same direction. This phenomenon is the cause of the seasons and the variation in daylight periods from the equator to each of the poles. The inclination of the earth's axis also affects the elevation of the sun above the horizon. On December 21 or 22, for example, the sun appears to be directly overhead at noon at latitude $23\frac{1}{2}°$S. This latitude, called the *tropic of Capricorn*, marks the southernmost position at which an observer on the ground would see the midday sun directly overhead.

The day on which the noon sun is directly overhead at the tropic of Capricorn is known in the northern hemisphere as the *winter solstice*. In the northern hemisphere this is the day of the year that has the fewest hours of daylight, with the sun appearing low in the southern sky even at noon. In areas north of the Arctic Circle (latitude $66\frac{1}{2}°$N), the sun is not visible at all at the winter solstice. Areas situated within the Arctic Circle, such as Greenland and northern Scandinavia, experience 24 hours of darkness on this date. Moving south from the Arctic Circle, an increasing proportion of each parallel lies within the circle of illumination, so that the days become longer. Still, the period of darkness continues to be longer than the period of daylight until one arrives at the equator (0° latitude), where day and night are of equal duration at all times.

As one proceeds south of the equator, entering the summer hemisphere, proportionately more of each parallel lies within the circle of illumination, so that the daylight period becomes progressively longer than the period of darkness. At the Antarctic Circle (latitude 66½°S) and beyond it to the South Pole, the sun remains above the horizon for 24 hours, so that there is no night at all. Of course, while it is the winter solstice in the northern hemisphere, it is the summer solstice in the southern hemisphere. Table 3.1 indicates the varying lengths of the daylight period at different latitudes at the northern hemisphere winter solstice.

As the earth circles the sun, the tilt of the earth's axis causes the noon sun to be directly overhead at different locations at different times of the year (Figure 3.3). Six months after the northern hemisphere winter solstice, the noon sun is directly overhead north of the equator at latitude 23½°, known as the *tropic of Cancer*. The moment the vertical rays of the sun fall on the tropic of Cancer, about June 21, is the northern hemisphere *summer solstice*. At the summer solstice, the area north of the Arctic Circle has 24 hours of daylight, while in latitudes south of the Antarctic Circle the sun does not appear above the horizon.

Midway between the solstices—about March 21 and September 21—the noon sun is directly overhead at the equator. At these two times the circle of illumination cuts through the poles and coincides with the meridians of longitude. This causes the periods of daylight and darkness to be of equal duration everywhere. Thus these dates are known as the *equinoxes*—the *vernal equinox* in March, and the *autumnal equinox* in September, using the perspective of the northern hemisphere. The exact dates of the solstices and equinoxes vary slightly from year to year

Table 3.1
Length of Daylight During Winter Solstice*

LATITUDE	DAYLIGHT
90°N	0
80°N	0
70°N	0
60°N	5 hr 52 min
50°N	8 hr 4 min
40°N	9 hr 20 min
30°N	10 hr 12 min
20°N	10 hr 55 min
10°N	11 hr 32 min
0°	12 hr 7 min
10°S	12 hr 28 min
20°S	13 hr 5 min
30°S	13 hr 48 min
40°S	14 hr 40 min
50°S	15 hr 56 min
60°S	18 hr 8 min
70°S	24 hr
80°S	24 hr
90°S	24 hr

*Not counting twilight. Daylight includes time when at least the upper edge of the sun's disk is above the horizon.
Source: Adapted from List, Robert J. (ed.) 1951. *Smithsonian Meteorological Tables*, 6th rev. ed. Washington, D.C.: Smithsonian Institution Press.

because the astronomical relationships between the earth and sun do not coincide exactly with the days determined by the earth's rotation.

The tilt of the earth's axis not only produces solstices and equinoxes; it also affects the intensity of solar radiation. Imagine holding a flashlight close to a square of cardboard representing the earth and its atmosphere. With the flashlight perpendicular to the cardboard, the circle of light is small but bright. If the cardboard is tilted to make an angle with the beam of light, the illuminated area of the cardboard becomes larger but also dimmer. The same amount of light is emitted by the flashlight, but the light intensity received per unit area of the cardboard surface has decreased.

The geometric relationship between the ele-

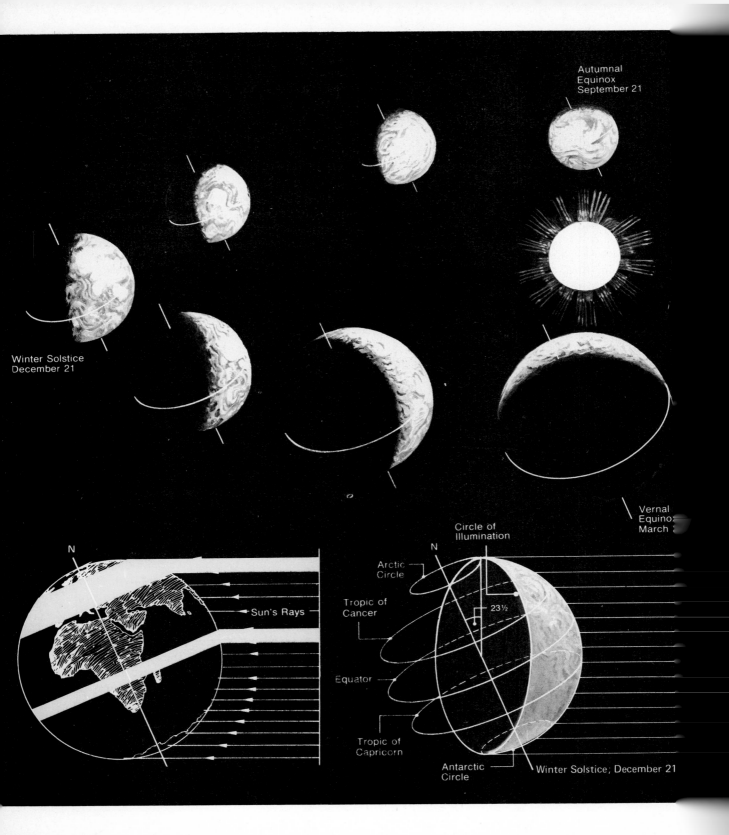

Autumnal
Equinox
September 21

Winter Solstice
December 21

Vernal
Equinox
March

N

→Sun's Rays

Circle of
Illumination

Arctic
Circle

N

Tropic of
Cancer

23½

Equator

Tropic of
Capricorn

Antarctic
Circle

Winter Solstice; December 21

Summer Solstice
June 21

Parallels

Meridians

N
90
80
70
60
50
40
30
20
10

Latitude

Longitude

W 90 80 70 60 50 40 30 20 10 0 E

Prime
Meridian

Figure 3.3

(top) The amount of solar energy that reaches a given location at the top of the atmosphere depends on the distance between the earth and the sun and on the orientation of the earth. The top half of the diagram shows the earth at twelve different times during the year. The maximum distance of the earth from the sun is 152 million km (94.4 million miles) and occurs early in July. The minimum distance is 147 million km (91.4 million miles) and occurs early in January. Because solar energy input to the earth varies with the earth-sun distance, the amount of solar radiation the earth receives is 1.07 times greater in early January than in July.

The earth's axis of rotation is 23 1/2° from the perpendicular to the plane of the earth's orbit. As the earth circles the sun, the orientation of the axis remains the same. At the winter solstice, the sun is overhead on the tropic of Capricorn in the southern hemisphere, and at the summer solstice, it is overhead on the tropic of Cancer in the northern hemisphere. At the vernal and autumnal equinoxes, the sun is overhead at the equator.

(bottom left) The amount of solar radiant energy that a unit area on the earth's surface intercepts depends on the angle between the sun's rays and the plane of the surface. As the sketch shows, the solar beam is spread over a wider area where it meets the surface obliquely, reducing the energy received per unit of area. A surface intercepts the greatest amount of radiant energy when the surface is perpendicular to solar rays.

(bottom center) The solar radiant energy input to a location on the earth during a given 24-hour period depends partly on the duration of daylight. At any single moment, one-half of the earth is illuminated by the sun. The circle of illumination is the boundary between the light and dark regions of the earth. Because of the tilt of the earth's axis, the duration of daylight is different at different locations. The diagram illustrates the situation at winter solstice.

(bottom right) The surface of the earth has been divided into two sets of intersecting grid lines to enable locations on the earth to be specified. The *parallels of latitude* are circles parallel to the plane of the equator. These circles connect points having the same angular distance north or south of the plane of the equator, the angle being formed at the center of the earth. The North Pole has latitude 90°N.

The second set of grid lines, the *meridians of longitude*, are circles drawn with the earth's axis as a diameter. One meridian, chosen to be the *prime meridian*, is designated as 0° longitude. In the United States, and in many other countries, the prime meridian is taken to be the meridian on which the astronomical observatory in Greenwich, England, is situated. The other meridians connect points having the same angular distance east or west of the plane of the prime meridian. Longitude can change from 0° to 180°.

For precision, a degree can be divided into 60 minutes of angular measure, and a minute can be divided into 60 seconds. The latitude and longitude of Washington, D.C., for example, can be written 38 degrees 54 minutes North, 77 degrees 2 minutes West, or in abbreviated form as 38°54′N, 77°02′W. (John Dawson)

vation of the sun above the horizon (beam angle) and the intensity of the solar beam at the earth's surface is shown in Figure 3.3. When the sun is directly overhead, the solar elevation is at the maximum of 90° and the intensity of the solar beam per unit of surface area is greatest. As solar elevation above the horizon decreases from the maximum, the intensity of the solar beam per unit of surface area decreases, reaching zero at sunset when the solar elevation is 0°.

The amount of solar energy received per unit of surface area thus depends primarily on the angle at which the incoming radiation strikes the surface. Winter sunbathers in Miami recognize this, propping themselves in lounge chairs in order to be perpendicular to the winter sun's rays. Vineyards are planted on steep south-facing slopes along the Rhine and Mosel rivers in western Germany for the same reason. When the sun is 60° above the horizon, a field on a 30° slope facing toward the sun receives nearly 15 percent more solar energy than a horizontal field of the same size.

Solar Radiation at the Top of the Atmosphere

The rate at which perpendicular rays of solar radiation strike the top of the earth's atmosphere is known as the *solar constant*. The average value over the year, based on satellite, rocket, and high mountain data, is about 1.94 calories per sq cm per minute. Scientists express amounts of solar radiation by a unit known as the *langley*. One langley is equal to 1 calorie per sq cm. The solar constant, therefore, can be expressed as 1.94 langleys per minute.

The amount of solar radiation striking an area at the top of the atmosphere during one day depends on the value of the solar constant, the true distance between the earth and the sun, the duration of sunlight, and minute-by-minute changes in the angle of the solar beam. Geographic variations in solar radiation received at the top of the atmosphere result from the last two factors, which are related to latitude.

Figure 3.4 (opposite)
This graph shows solar radiant energy input to the top of the atmosphere. The horizontal scale is marked off in months of the year. The vertical scale lists latitude north and south of the equator. The curved contour lines give the solar radiant energy input to the top of the atmosphere in units of langleys per 24 hours, based on a value of 1.94 langleys per minute for the solar constant. The shaded areas of the diagram poleward of latitude 66½° represent times of perpetual darkness when there is no solar radiant energy input.

The energy input into the northern and southern hemispheres is not perfectly symmetrical. At summer solstice in the northern hemisphere (June 21), locations at latitude 15°N receive about 900 langleys per 24 hours, but at summer solstice in the southern hemisphere (December 21), locations at latitude 15°S receive close to 1,000 langleys per 24 hours. The reason is that the earth is nearer the sun in January than in July, and the energy input to the earth is 7 percent higher in January than in July.

The orange curve in the graph represents the latitude where the solar elevation at noon is 90°. (Doug Armstrong after *Smithsonian Meteorological Tables*, edited by Robert J. List, 6th ed., 1971, by permission of the Smithsonian Institution)

The total daily solar radiation input to horizontal surfaces at the top of the atmosphere at different latitudes is shown in Figure 3.4. Each curve shows solar radiation input in langleys per 24 hours. For example, the energy input at the latitude of Philadelphia or Denver (about 40°N) on September 1 is about 800 langleys. This decreases to only about 325 langleys at the winter solstice. The shaded areas represent periods of continuous darkness in the polar regions.

Figure 3.4 shows that through the year the solar radiation input varies much more at the poles than at the equator. At the equator, the variation is from about 780 langleys per day at the summer solstice to about 880 during late February and early November. At the equator the hours of daylight and darkness are always about equal, and the solar altitude at noon is

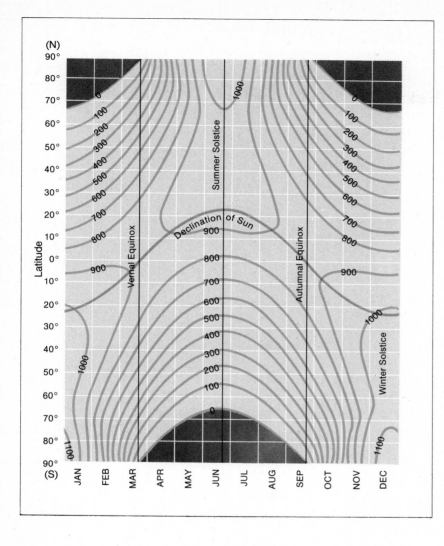

always high. The most extreme annual variation is at the South Pole, ranging from more than 1,100 langleys in mid-December, when the earth is at perihelion, to 0 from mid-March to early September. That is because the South Pole receives continuous sunlight between September and March, while between March and September the sun is always below the horizon. Averaged over a full year, however, locations at the equator receive nearly two and one-half times as much solar radiation as the South Pole.

Solar Radiation and the Atmosphere

Before solar radiation can reach the surface of the earth, it must pass through the earth's atmosphere, which is composed of gases, particles, and clouds that respond differently to solar radiation at various wavelengths. The atmosphere therefore exerts a strong influence on the amounts and types of solar radiation that reach the earth's surface.

Table 3.2

Principal Gases in the Earth's Lower Atmosphere

GAS	MOLECULAR FORMULA	NUMBER OF MOLECULES OF GAS PER MILLION MOLECULES OF AIR	PROPORTION (PERCENT)
Nitrogen	N_2	7.809×10^5	78.09
Oxygen	O_2	2.095×10^5	20.95
Water Vapor	H_2O	variable	variable
Argon	Ar	9.3×10^3	0.93
Carbon Dioxide	CO_2	330.0	0.03
Neon	Ne	18.0	1.8×10^{-3}
Helium	He	5.0	5.0×10^{-4}
Krypton	Kr	1.0	1.0×10^{-4}

Source: Adapted from List, Robert J. (ed.) 1951. *Smithsonian Meteorological Tables*, 6th rev. ed. Washington, D.C.: Smithsonian Institution Press.

The chemical makeup of the atmosphere is given in Table 3.2. In Chapter 1 we saw how events during the early history of the earth caused nitrogen, oxygen, and carbon dioxide to be introduced into the atmosphere. The inert gases neon, helium, and krypton are probably remnants from the earth's original atmosphere. Argon, a more abundant inert gas, seems to have been produced largely by the decay of radioactive potassium in the earth's crust.

The nitrogen, oxygen, and inert gases in the atmosphere are present in the same relative proportions everywhere. The amounts of other gases, such as carbon dioxide, water vapor, and ozone, vary with time, location, and altitude. Water vapor, which enters the atmosphere from the earth's surface through evaporation and plant transpiration, rarely appears above an altitude of 10 km (6 miles). Ozone occurs primarily at an altitude of about 25 km (15 miles), where it is produced by reactions between solar radiation and oxygen molecules.

One can think of the atmosphere as being divided into "spheres" related to temperature changes with altitude. In the *troposphere*—the portion of the atmosphere nearest the earth's surface—the temperature decreases with altitude. The average vertical temperature decrease in the troposphere is 6.5°C per kilometer (3.6°F per 1,000 ft). In the *stratosphere*, on the other hand, temperature increases with altitude. The level at which temperatures stop decreasing and begin to increase with altitude is known as the *tropopause*. Figure 3.5 shows these temperature/altitude relationships, as well as those in the still-higher *mesosphere* and *thermosphere*, which are composed of thinly scattered gaseous molecules and sub-atomic particles.

Solar radiation interacts with gas molecules in the atmosphere through the processes of *absorption* and *scattering*. In the process of absorption, a molecule takes up radiant energy and converts it to heat. Scattering refers to the process in which gas molecules, dust particles, and water droplets deflect incoming solar radiation from its original path. Because of scattering, the earth's surface receives solar radiation from the sky as well as directly from the sun. Shorter wavelengths, especially the blue portion of visible light, are scattered more effectively than longer wavelengths, causing the sky to appear blue. The water droplets that form clouds also reflect some solar radiation back out to space as well as absorbing and scattering incoming radiation.

Gas molecules absorb most strongly within

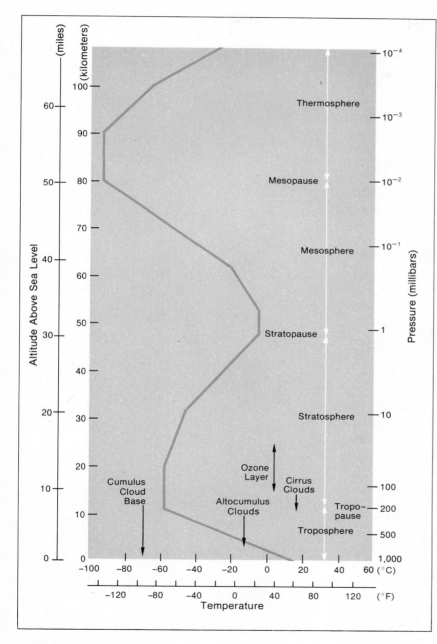

Figure 3.5
The atmosphere is conventionally divided into layers according to the variation of temperature with altitude. The *troposphere* decreases in temperature with increased altitude at the average rate of 6.5°C per km (3.6°F per 1,000 ft). The troposphere is warmest at the earth's surface, on the average, because it is heated primarily from below by the transfer of heat energy from the surface. The troposphere contains water vapor, and most clouds and weather phenomena are confined to this layer. Temperature increases with increased altitude in the *stratosphere*, largely because atmospheric gases such as ozone absorb a portion of the radiant energy incident from the sun. The stratosphere is nearly devoid of water vapor, so clouds seldom form there. Above the stratosphere, molecules of the gases forming the atmosphere become very widely dispersed and atmospheric pressure is negligible. The *thermosphere* is composed of scattered free electrons and other sub-atomic particles.

The temperature curve in the diagram corresponds to the values assumed for the 1962 United States standard atmosphere. A standard atmosphere is meant to represent the average state of the atmosphere, but the actual temperatures and pressures on a given day may differ from the standard values. (Doug Armstrong after the U.S. Air Force, 1965)

the longwave portion of the spectrum. Since most solar radiation is in the range of visible light, the majority passes through the atmosphere. More incoming radiation is reflected back to space by clouds than is absorbed by the atmosphere. As we shall see later in this chapter, nearly all the radiation absorbed by the earth's lower atmosphere is longwave radiation from the earth's surface, not incoming solar radiation.

Ultra-shortwave radiation, such as X-rays, gamma rays, and ultraviolet radiation, requires special mention because its energy can destroy the complex molecules required for life processes. Fortunately, little ultra-shortwave radiation actually reaches the earth's surface; most of it is absorbed in the upper atmosphere. Molecular oxygen in the outer layers of the earth's atmosphere and the ozone layer in the stratosphere strongly absorb ultra-shortwave radiation, causing these portions of the atmosphere to be heated to temperatures close to those at the earth's surface.

THE ENERGY BALANCE OF THE EARTH-ATMOSPHERE SYSTEM

Because the earth is neither heating up nor cooling off, the amount of energy the earth receives from the sun and the amount it radiates back into space must be in balance. To understand this we must examine the ways in which the earth and its atmosphere lose energy and thereby maintain an energy balance. It is important to keep in mind that there is both seasonal and geographical variation in energy exchanges within the earth-atmosphere system, but we shall be considering an average for the entire globe over a year. For this purpose, we shall assume, as in Figure 3.6, that the total solar radiation intercepted by the earth and its atmosphere over a year is equivalent to 100 units of energy.

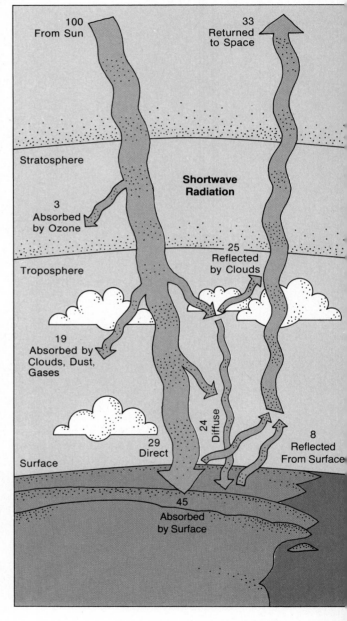

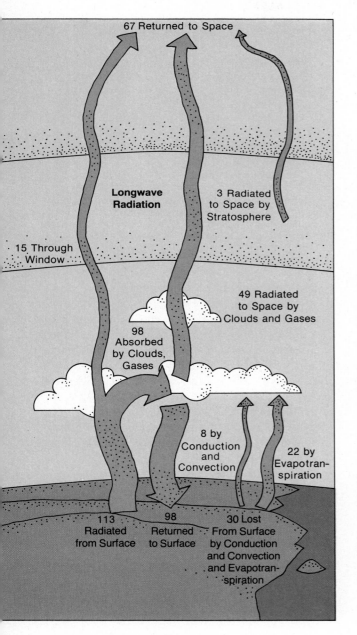

Figure 3.6 This diagram traces 100 units of shortwave solar radiation that arrive at the top of the earth's atmosphere and the interactions of this radiation with the atmosphere and the surface. The numbers represent average global values. The interactions of incoming shortwave radiation are shown on the left, and the interactions of outgoing longwave radiation are shown on the right. All the indicated interactions occur at a given location during daylight hours, but at night, when there is no solar radiant energy input, only the longwave interactions occur.

The earth as a whole is neither gaining nor losing radiant energy; for every 100 units of solar radiant energy received, an average of 33 + 67 = 100 units are returned to space. Similarly, energy is in balance at the earth's surface, on the average. The surface receives 45 + 98 = 143 units of energy and loses 113 + 30 = 143 units. Can you use the diagram to show that the average gain and loss of energy by the atmosphere are also in balance?

Note that the 98 units of longwave radiant energy returned to the surface by the atmosphere are important components in the energy balance at the surface. If it were not for the energy contributed by the atmosphere, the surface would cool to the temperature where the radiated energy was in balance with the absorbed energy. In the absence of an atmosphere, the average temperature of the earth would be approximately −20°C (−4°F). (Tom Lewis after Herbert Riehl, *Introduction to the Atmosphere*, © 1972 by McGraw-Hill Book Company)

Reflection and Albedo

Airline passengers are often startled by the dazzling brightness when their plane emerges from thick clouds after taking off under a heavy overcast. Thick clouds are very effective reflectors of shortwave radiation, throwing back as much as 90 percent of the solar energy that falls on their upper surfaces. This reflected radiation combines with the incoming solar beam to produce the extremely bright zone just above the clouds. The proportion of incoming solar radiation that an object reflects is its *albedo*.

Clouds and fresh snow reflect 55 to 95 percent of the incoming solar beam, and the albedo of a good mirror approaches 100 percent. Most land surfaces, except those covered by snow, have albedos between 10 and 30 percent. The albedo at a given location varies from season to season as snow appears and disappears and as bare fields become covered with crops (Table 3.3). It is particularly important to take albedo into account in analyzing the radiation balance, because in the process of reflection no radiation is absorbed; it is simply redirected—generally outward to space. The reflected energy is lost to the earth-atmosphere system.

Figure 3.6 shows that the albedo of the whole earth-atmosphere system is about 33 percent. About 25 percent of incoming radiation is reflected by clouds, and another 8 percent by the earth's surface. Absorption by atmospheric gases and dust accounts for an additional 22 percent. Of this, 3 percent is ultraviolet radiation absorbed by ozone in the stratosphere. Ultimately, 45 percent of the solar radiation is absorbed by the land and water areas of the earth. Figure 3.7 summarizes the average annual latitudinal distribution of solar radiation in the earth-atmosphere system.

Clouds have an important effect on the geographical distribution of solar radiation at the earth's surface. Figure 3.8 (see pp. 56–57) shows the average solar radiation input at the surface during winter and summer for the United States and for the entire earth. At the top of the atmosphere the amount of solar energy input varies only with day length and the angle of the sun's rays. But at the earth's surface differences in cloudiness can cause irregularities in the pattern of energy input. The southwestern deserts in the United States are farther north than Florida but receive more energy because their skies are generally clear. In North America the lowest energy inputs at the surface are in the east and northwest, where fog and clouds are common. The world maps show strong north-south differences in energy received at the surface in the winter hemisphere. The highest values occur over deserts in the summer hemisphere.

Table 3.3
Albedo of Various Surfaces

SURFACE	ALBEDO (PERCENT)
Fresh Snow	80–95
Dense Stratus Clouds	55–80
Ocean (Sun Near Horizon)	40
Ocean (Sun Halfway up Sky)	5
Bare Dark Soil	5–15
Bare Sandy Soil	25–45
Desert	25–30
Dry Steppe	20–30
Meadow	15–25
Tundra	15–20
Green Deciduous Forest	15–20
Green Fields of Crops	10–25
Coniferous Forest	10–15

Sources: List, Robert J. (ed.) 1951. *Smithsonian Meteorological Tables*. 6th rev. ed. Washington, D.C.: Smithsonian Institution Press. Budyko, M.I. 1974. *Climate and Life*. D. H. Miller, tr. New York: Academic Press, pp. 54–55.

Longwave Radiation: The Greenhouse Effect

Because the earth's surface is warm relative to its atmosphere and the space around it, having an average temperature of about 15°C (60°F), it emits longwave radiation upward to

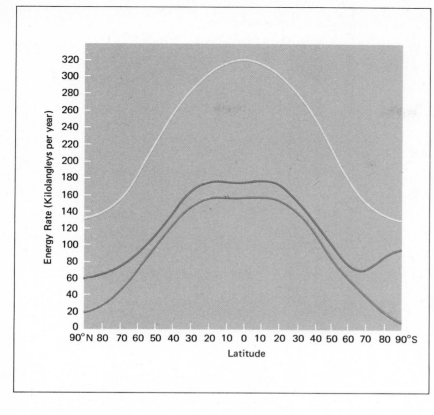

Figure 3.7
This graph shows the average annual distribution of solar radiation in the earth-atmosphere system by latitudes. The yellow curve represents incoming solar radiation at the top of the atmosphere, the red curve the incoming solar radiation just above the earth's surface, and the green curve the net solar radiation gain of the surface. In polar regions the combined effects of low solar altitudes, clouds, snow cover, and icecaps are evident. In the tropics seasonal cloudiness reduces the radiation gain at the surface. The highest gains are in the cloud-free subtropical desert regions. (Vantage Art, Inc. after William D. Sellers, *Physical Climatology*, © 1965, The University of Chicago Press)

the atmosphere and space. If the earth had no atmosphere, all this longwave radiation would be lost to space, and the surface would cool very rapidly after each sunset. This does not occur, however, because the atmosphere acts like a greenhouse, allowing shortwave solar radiation to pass through to the surface, but trapping the longwave radiation emitted by the earth itself. Thus the atmosphere acts as a blanket that keeps the earth warm.

Water vapor and carbon dioxide molecules are primarily responsible for the absorption of longwave radiation in the lower atmosphere. The absorbed radiation heats the lower atmosphere, especially in humid regions where there is abundant water vapor. The atmosphere, in turn, reradiates longwave radiation. Some of

this is directed upward and is lost to space, but much more is returned downward to the surface of the earth. On the average, such longwave radiation exchanges result in a net loss of longwave radiation at the surface and a net gain in the troposphere. Hence the surface is cooled and the troposphere is warmed by longwave radiation exchanges.

Surface cooling by longwave radiation is dramatically reduced by the presence of clouds. Clouds absorb most of the longwave radiation emitted from the surface, severely reducing losses to space. Like the atmospheric gases, clouds also emit longwave radiation, much of which is received by the ground. That is why cloudy nights tend to be much warmer than clear nights with starry skies.

ENERGY AND TEMPERATURE

Figure 3.8
The average radiation received at the surface in the United States during January and July and globally during December and June is shown in units of langleys per day.

During December and January, when the sun is overhead in the southern hemisphere, the solar radiation received in the northern hemisphere decreases rapidly with increased latitude. Southern Florida receives more than 300 langleys per day during January on the average, but northern Minnesota receives only 150 langleys per day. Solar radiation input is small at high latitudes during the winter because of the short duration of daylight and the low altitude of the sun in the sky. The solar radiation input to a given location is larger when skies are clear than when skies are cloudy; for this reason the arid southwestern United States receives more solar radiation than cloudy locations at the same latitude.

During the summer months (June to September in the northern hemisphere, December to March in the southern hemisphere), the solar radiation input exhibits little variation with latitude. Locations in northern Canada receive the same input as locations in Texas. The increased duration of daylight with increased latitude compensates for the lower altitude of the sun toward the pole. The variation in solar radiation input between different locations during the summer is caused primarily by differences in cloudiness. Where the isolines (lines connecting points of equal value) are dashed, the data are missing or incomplete. Data for Greenland are unavailable for June. (*The National Atlas of the United States of America*, 1970; and Löf, Duffie, and Smith, *World Distribution of Solar Radiation*, Report No. 21, University of Wisconsin, 1966)

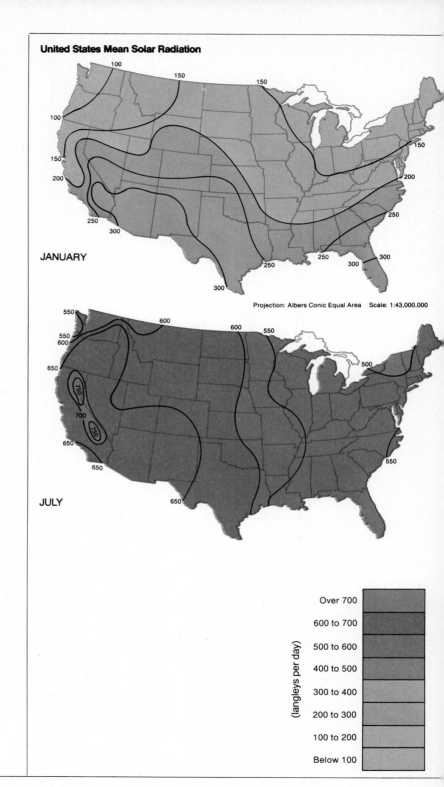

United States Mean Solar Radiation

JANUARY

Projection: Albers Conic Equal Area   Scale: 1:43,000,000

JULY

Over 700

600 to 700

500 to 600

400 to 500

300 to 400

200 to 300

100 to 200

Below 100

(langleys per day)

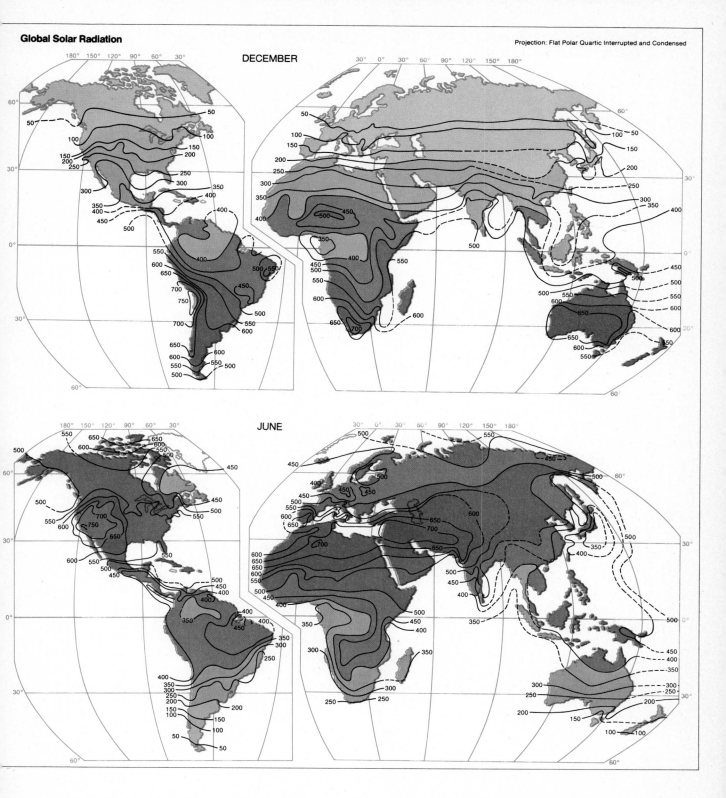

Global Solar Radiation

Projection: Flat Polar Quartic Interrupted and Condensed

DECEMBER

JUNE

Radiation Exchanges in the Earth-Atmosphere System

The right-hand side of Figure 3.6 traces the exchanges of longwave radiation within the energy balance of the earth-atmosphere system. The figure shows that an average of 113 units of longwave radiation is emitted by the earth's surface. Most of this (98 units) is absorbed by clouds, water vapor, and carbon dioxide in the troposphere, and only a small proportion (15 units) escapes to space. The troposphere, in turn, radiates the absorbed longwave radiation (98 units) back to the surface, and an additional 49 units upward to space. The warm stratosphere also loses some longwave radiation (3 units) to space.

The longwave radiation loss to space exceeds that received from the surface because it is supplemented by the shortwave radiation absorbed and converted to longwave radiation in the atmosphere itself (30 units of longwave plus 19 units of converted shortwave radiation make 49 units of energy that the atmosphere radiates into space as longwave radiation). With the radiation from the surface that escapes (15 units) and the loss from the stratosphere (3 units), the total longwave loss to space equals 67 units.

We have considered the solar energy intercepted by the earth and atmosphere as 100 units of shortwave radiation. As Figure 3.6 shows, returned to space are 33 units of reflected shortwave radiation and 67 units of longwave radiation emitted from the troposphere, the surface, and the stratosphere. Therefore, outgoing shortwave and longwave radiation (33 + 67 = 100) is equal to the incoming solar radiation (100 units), and the earth-atmosphere system as a whole is in balance.

Considered separately, however, the atmosphere and the earth's surface do not show balanced radiation budgets. Figure 3.6 shows that radiative losses in the troposphere exceed gains by 30 units. Similarly, the earth's surface gains 143 units and radiates away only 113 units of longwave radiation. Consequently, the surface experiences an annual net gain of 30 units—an amount just equal to the net loss in the troposphere. We know that the surface is not becoming progressively warmer, nor is the troposphere becoming cooler. Therefore, other energy exchanges must compensate the imbalance in the troposphere and at the earth's surface. These energy exchanges occur through conduction, convection, and evapotranspiration.

Conduction, Convection, and Evapotranspiration

When two objects at different temperatures are in contact, heat flows from the warmer to the colder object by the process of *conduction*. The same process plays a role in the transfer of heat energy at the surface of the earth. By midday, solar radiation causes the top layer of the ground to become warmer than the deeper layers. As a result heat is transferred deeper into the ground by conduction. At night, when radiation losses have cooled the surface layer, heat flows from the deeper layers back toward the surface.

Heat is also conducted upward from the earth's surface to the atmosphere when land and water surfaces are warmer than the air in contact with them. This helps use up some of the net radiation gain of the surface and offsets some of the net radiation loss of the troposphere.

The efficiency of the conduction process is increased by the process of *convection*—the vertical rise of parcels of low-density warm air. Convection exchanges warmer surface air with cooler air aloft. At times, however, warm air moves over cooler land or water surfaces. At such times, conduction causes heat to flow from the atmosphere downward to the surface. This also occurs at night when the surface cools by longwave radiation. Conduction in this direction chills the lower atmosphere and removes any possibility of upward convection. As shown in Figure 3.6, an average of 8 units of energy is transferred from the surface to the atmosphere by conduction and convection.

A most important energy exchange between the earth's surface and the troposphere involves the change of liquid water and solid ice to water vapor. Recall from Chapter 2 that every gram of water that is vaporized under average conditions transfers heat energy from the surface to the atmosphere. Evaporation from water surfaces of all types is a constant process that transfers heat to the atmosphere. Less obvious is the process by which plants absorb liquid water from the soil and emit water vapor to the atmosphere through pores in their leaves. This process is called *transpiration*. Since transpiration and evaporation involve similar energy exchanges and are difficult to measure separately, they are often considered jointly as *evapotranspiration*. Figure 3.6 shows that, on the average, 22 units of energy are transferred from the earth's surface to the troposphere by evapotranspiration. This energy is stored in the water vapor in the form of *latent heat*. It becomes *sensible heat*, capable of warming the surrounding air, only when the water vapor condenses back into water droplets or ice crystals elsewhere in the atmosphere.

Heating of the atmosphere by evapotranspiration, conduction, and convection offsets the net radiation gain of the earth's surface and the net radiation loss of the atmosphere, balancing the global energy budget. A final component, plant photosynthesis, also consumes solar energy, but utilizes too small a fraction to be of concern in an analysis of the energy balance of the earth-atmosphere system.

Latitudinal Differences in Solar Income and Radiational Losses

The energy balance of the earth-atmosphere system shown in Figure 3.6 represents an annual average for the system as a whole. If we look at specific locations on the earth, however, we discover that energy exchanges are not in

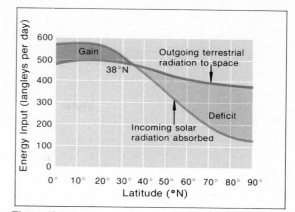

Figure 3.9
The graph shows average annual global values of absorbed solar radiant energy and emitted longwave radiation at each latitude in the northern hemisphere for the earth's surface and the atmosphere together. Between the equator and latitude 38°N, absorption exceeds emission, and there is a net input of energy to the earth. Poleward of 38°N, emission exceeds absorption, and there is a net loss of energy. The temperature in the lower latitudes therefore rises until the rate of heat flow toward the poles is sufficient to carry away the excess energy. (Doug Armstrong after F. K. Hare, *The Restless Atmosphere*, 1966, Hutchinson Publishing Group, Ltd.)

balance. Figure 3.9 shows that equatorial and tropical regions receive much more solar energy than they return to space through reflection and longwave radiation. Polar regions, on the other hand, lose more energy through reflection and longwave radiation than they gain from solar radiation.

These latitudinal imbalances in energy gains and losses are the cause of the atmospheric and oceanic circulations that are discussed in Chapter 4. Flows in the atmosphere and oceans transport warm air, latent heat in water vapor, and warm water into high latitudes, and colder air and water toward the equator. As Figure 3.10 (see p. 60) indicates, there is a strong seasonality to the poleward flux of energy. In the winter hemisphere only the zone extending about 20°

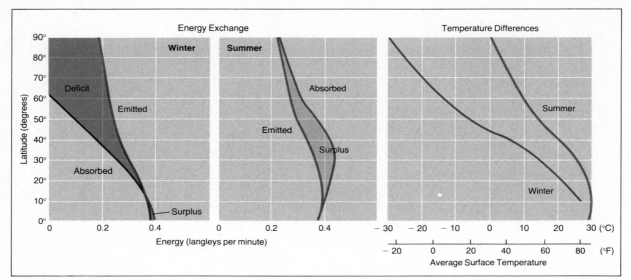

Figure 3.10
Energy exchange and temperature in the northern hemisphere. The diagram on the left shows the average rates at which the earth and the atmosphere absorb solar radiant energy and emit longwave radiation to space. During the winter, absorption exceeds emission only at latitudes near the equator, but during the summer the rate at which energy is absorbed exceeds the rate of emission at every latitude. Much of the excess absorbed energy warms the oceans and is stored there.

The diagram on the right shows the average surface temperature of the atmosphere at a height of approximately 2 meters above the surface. The temperature difference between low and high latitudes is greater during the winter than during the summer. The temperature differential helps to power atmospheric motion and a poleward flow of energy from the tropics. (Doug Armstrong after Herbert Riehl, *Introduction to the Atmosphere*, © 1972 by McGraw-Hill Book Company)

beyond the equator receives more energy from radiation than it loses, but in the summer hemisphere all regions gain more radiational energy than they lose. Because temperature differences between low and high latitudes are greater during the winter, the atmospheric circulation tends to be most vigorous then.

ENERGY BUDGETS AT THE EARTH'S SURFACE

The large-scale energy exchanges between the sun, the earth-atmosphere system, and space have been measured with increasing accuracy in recent years by a number of specialized satellites. Our knowledge of local energy budgets at the earth's surface is based on more routine measurements at various experimental stations. By interpreting these data according to the principles of energy exchange discussed earlier in this chapter, scientists are gaining a better understanding of how the energy balances at particular locations are affected by such factors as cloud forms, vegetation cover, local water bodies, and human activities.

Local Energy Budgets

Figure 3.11 is an example of a daily pattern of energy balance factors. During the daytime, the net radiation gain is usually used up by evapotranspiration, atmospheric heating, and heating of the soil. The portion used in plant photosynthesis is negligible.

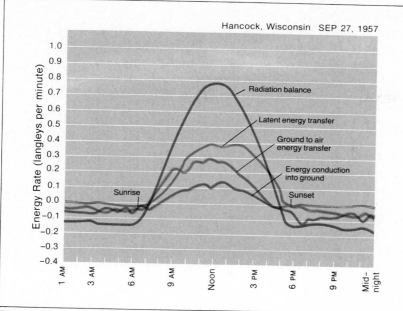

Figure 3.11

The local energy budget and its principal components during a 24-hour period are shown for Hancock, Wisconsin, a moist midlatitude location. The radiation balance is the difference between the rate at which solar radiant energy is absorbed at the ground and the net loss of longwave radiation in the exchange between the surface and the atmosphere. During most of the day, the radiation balance is positive, which indicates a net flow of radiant energy toward the surface of the ground. During the night the radiation balance is negative because the net flow of radiant energy is away from the surface.

Energy cannot accumulate at the surface, and the excess or deficiency of radiant energy flow to the surface can be accounted for as the sum of three components: the latent energy removed by the evaporation and transpiration of water; the heat energy flowing from the ground to the air by conduction and convection; and the heat energy that flows into the soil from the surface of the ground. At Hancock at midday, latent energy transfer is positive because of evaporation and plant transpiration. The small negative value of latent energy transfer before sunrise indicates the condensation of water vapor on the ground as dew. Heat transfer from the ground to the air is positive at midday, which indicates that heat from the ground is warming the air. Energy conduction into the ground is negative before sunrise and after sunset, when heat is conducted upward from the warmer ground to the cooler air above it. (Doug Armstrong after William D. Sellers, *Physical Climatology*, © 1965, The University of Chicago Press)

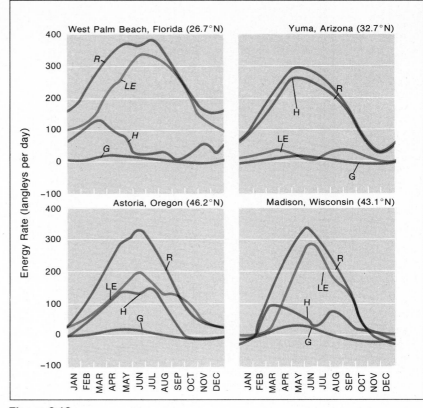

Figure 3.12
The graphs show the annual regime of the local energy budgets at four places in the United States. Madison is representative of humid midlatitude regions with low solar radiation income during winter, West Palm Beach of humid subtropical regions with moderate solar radiation income during winter, Astoria of humid and cloudy maritime climates with much winter cloudiness, and Yuma of hot desert regions with little cloudiness.

R is the radiation balance (solar energy received vs. longwave energy emitted). *LE* is the energy removed by evapotranspiration (latent heat transfer). *H* is the energy moved from the ground to the air by conduction and convection. *G* is energy conducted into the ground from the surface. The values of *R* are similar at all four locations during summer, but *R* is much smaller at Astoria and Madison during winter because of fewer hours of daylight and low solar altitudes; during winter in Madison, *R* is slightly negative.

LE is usually the largest component of *R* during the growing season. *LE* is zero at Madison during winter, when low temperatures inhibit evapotranspiration, and there is no net radiational gain. *LE* is low throughout the year at Yuma because of the absence of water for evapotranspiration, and most of the net radiational gain goes into heating the air. In each location, energy tends to be conducted from the surface into the soil during spring and back to the surface during fall. (Doug Armstrong, after William D. Sellers, *Physical Climatology*, © 1965, The University of Chicago Press)

In Figure 3.11, Hancock, Wisconsin, is selected to be representative of humid midlatitude locations. During the sunny daytime hours, more shortwave and longwave radiation is received at the surface than is reflected or radiated back to the atmosphere or into space. Following sunrise, the rate of incoming solar radiation rises rapidly, producing a net radiation gain that peaks near solar noon and returns to zero a few minutes before sunset. Thus, during the day, the energy budget at Hancock is said to be positive.

At night, however, more longwave radiation is lost to the atmosphere than is received from it, so that the energy budget is negative. Small amounts of energy are conducted downward from the warmer atmosphere and upward from the warmer subsoil to the cooling surface. Thus the lower atmosphere is heated by the surface during the day when the surface radiation budget is positive, and is cooled by the surface at night when the radiation budget is negative.

In Figure 3.11, evapotranspiration is also seen to be closely controlled by the net radiation. This is evident in the curve for latent energy loss due to vaporization of water. Evapotranspiration is primarily a daytime phenomenon; there is little or no vaporization of water at night. In a dry climatic region little soil moisture is available for evapotranspiration, and the net radiation gain goes almost entirely into sensible heat—the heating of the air and soil. On the average, equal energy inputs will cause dry regions to be warmer during daylight hours than areas with high rates of evapo transpiration.

Figure 3.12 presents annual energy budgets for four contrasting locations within the continental United States. Significant differences may be observed among these locations on the basis of latitude, yet it is apparent that other environmental factors, such as relative humidity, also cause the budgets to vary. The greatest contrast is between Madison, Wisconsin, and Yuma, Arizona. At Yuma, a desert location, very little of the net radiation gain can be used up by evapotranspiration, so that nearly all is available for heating the air and soil. As a consequence, Yuma is well-known as one of the hottest year-around locations in the United States. By contrast, at Madison and West Palm Beach, abundant moisture in the summer permits high evapotranspiration rates, so that less energy is available to heat the soil and air. Astoria is intermediate between these extremes.

Differential Heating and Cooling of Land and Water

Land surfaces and water bodies respond differently to equal energy inputs. This is because the *heat capacities* of land and water are very different. As we saw in Chapter 2, water can absorb much heat without increasing its temperature greatly, whereas land heats rapidly as it receives energy. Whereas the heat capacity of water is about 1, the heat capacity of most land surfaces is from 0.2 to 0.4. This means that for a given energy input a volume of ground will increase in temperature three to five times more than an equal volume of water.

On a summer day, the temperature of a soil surface exposed to the sun can easily reach 40°C (104°F). Dry beach sand and asphalt pavement may become too hot for bare feet. Most of the energy gain is concentrated near the surface, since conduction to greater depths is relatively slow. At night, the soil surface cools rapidly by longwave radiation and, in turn, cools the atmosphere immediately above by conduction. If there is little wind to mix the lowest layers of air, the air near the ground will become several degrees cooler than the air a few meters higher. Vertical profiles of air and ground temperatures during a fair winter day in southern New Jersey are shown in Figure 3.13 (see p. 64).

Surface temperatures of large water bodies change much less during the year than land surface temperatures. In addition to the greater heat capacity of water, the energy gain of a water surface is distributed to much greater depths. Solar radiation penetrates water to a

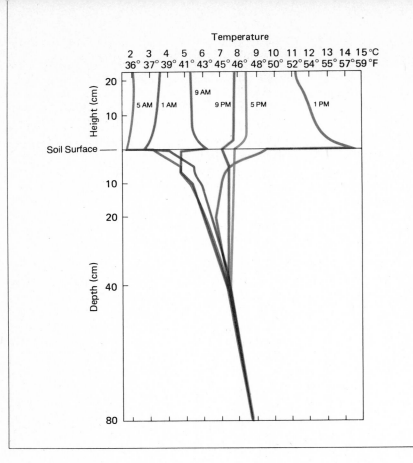

Temperature

Figure 3.13
This figure illustrates steep temperature gradients immediately above and below the ground surface during fair days with little wind. The data were taken at an experimental site near Seabrook, New Jersey, during December 5, 1956. Between 9:00 A.M. and 1:00 P.M., the warmest temperatures were within the first 5 cm above the surface, but during late afternoon, overnight, and early morning the coldest temperatures were immediately above the surface, which was losing heat by longwave radiational cooling. (Vantage Art, Inc. after J. R. Mather, *Climatology: Fundamentals and Applications*, © 1974 by McGraw-Hill Book Company)

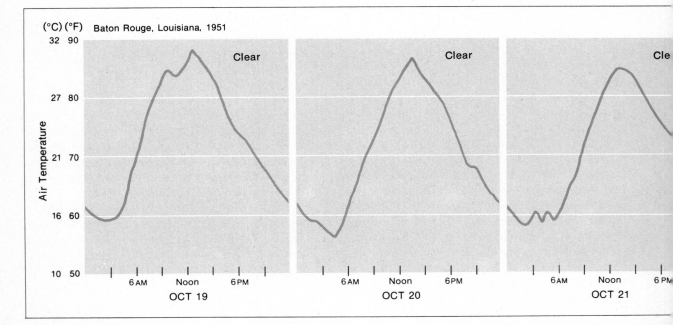

Baton Rouge, Louisiana, 1951

depth of tens of meters, and waves and currents also distribute heat energy downward.

Local Temperature Variations

The physical principles underlying the energy budget can help explain temperature variations on a local scale. Figure 3.14 shows air temperatures for a period of several days at Baton Rouge, Louisiana. Air temperature near the ground tends to reflect surface temperature. In the first three days, clear weather permitted ground and air temperatures to rise markedly during the day and to fall sharply at night. The lowest temperatures occurred just after sunrise, and the highest followed in mid-afternoon. Although the solar radiation input peaks near noon and then declines, cooling does not begin until late afternoon, when radiation losses begin to exceed gains. During cloudy weather, the temperature change throughout the day is much less. Clouds reduce incoming solar radiation during the day, and at night they trap outgoing longwave radiation and return much of it to the surface, smoothing out the daily temperature variation.

A large body of water also tends to reduce daily temperature ranges over adjacent land areas because the temperature of large water bodies changes little from day to day. Latent heat transfer between the water and the atmosphere helps maintain uniformly mild temperatures at coastal cities in comparison with cities in continental interiors.

Monthly temperature variations reflect the same influences as daily variations. Figure 3.15 (see p. 66) shows winter and summer temperatures at San Diego, California, a coastal city; Elko, Nevada, an interior city in an arid region; and Cleveland, Ohio, a midcontinent city with frequent cloud cover. The graphs show the *distribution* of temperatures for January and July, which is found by measuring the temperature every hour and plotting the number of hours each temperature occurred.

Because of the reduced energy input in winter, the January temperature distributions in all three cities fall well below the distributions for July. The January and July distributions for San Diego show the most overlap and the least separation between peaks because coastal temperatures tend to be uniform throughout the

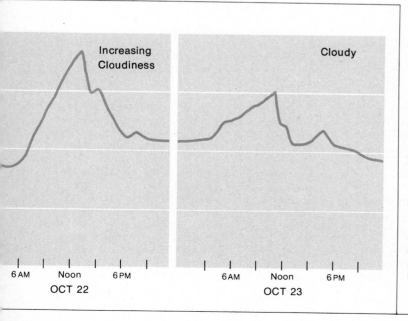

Increasing Cloudiness

Cloudy

6 AM   Noon   6 PM
OCT 22

6 AM   Noon   6 PM
OCT 23

Figure 3.14
The temperature records at Baton Rouge, Louisiana, for October 19 to 23, 1951, illustrate the daily variation in air temperature as the ground heats and cools. The temperature rises in the morning, when the energy budget becomes positive, and reaches a peak in the early afternoon. The radiation gain declines in the afternoon and becomes negative at night, causing the temperature to drop.

On October 19 to 21, skies were clear, but on October 22 and 23, skies were cloudy. On a cloudy day, the daily variation in temperature is less extreme than on a clear day, because clouds reduce solar energy input during the day and at night reduce surface heat loss caused by longwave radiation. (Doug Armstrong)

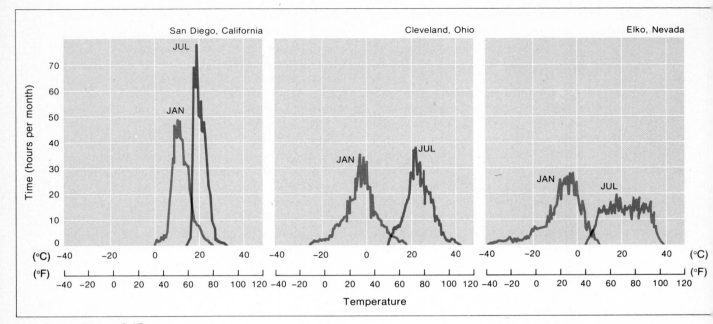

Figure 3.15
The average temperature distributions during January and July are shown for three cities in different climatic regions over the five-year period 1935 through 1939. Each point on a temperature distribution tells the number of hours during the given month that the specified temperature was recorded; in San Diego in July, for example, a temperature of 20°C (68°F) was recorded in 57 separate hours.

A narrow temperature distribution, which San Diego, a coastal city, shows, means that the temperature remains comparatively constant. At Elko in July, in contrast, the temperature distribution is broad, reflecting the hot days and cool nights of the cloudless desert. The distributions are moderately broad for Cleveland, a midcontinent city with a moist climate. (Doug Armstrong after Arnold Court, *Journal of Meteorology, 8,* 1951, American Meteorological Society)

Figure 3.16 (opposite)
Global temperature distributions are shown for January and July. Temperatures have been reduced to approximate sea level values by applying a correction for the decrease in temperature with increased altitude. In winter (January in the northern hemisphere, July in the southern hemisphere), temperatures generally depend on latitude and decrease regularly from the equator toward the poles. Note, however, the continental and maritime effects on temperature over North America and the North Atlantic during January. In summer temperatures depend less strongly on latitude and more on land and water contrasts (see Figure 3.17).

Note that blue tones on this map indicate below-freezing temperatures, and "warmer" colors indicate above-freezing temperatures. The extreme poles are not shown on the flat polar quartic map projections used in this book. (Andy Lucas and Laurie Curran after Glenn Trewartha, *Introduction to Climate,* © 1968, McGraw-Hill Book Company and *Goode's World Atlas* © 1970, Rand McNally & Company)

year. The monthly temperature distributions at Cleveland and Elko cover a greater range than those at San Diego because of the greater daily variation of temperature in the continental interior. The wide range of the July temperatures for Elko reflects the city's high elevation (5,000

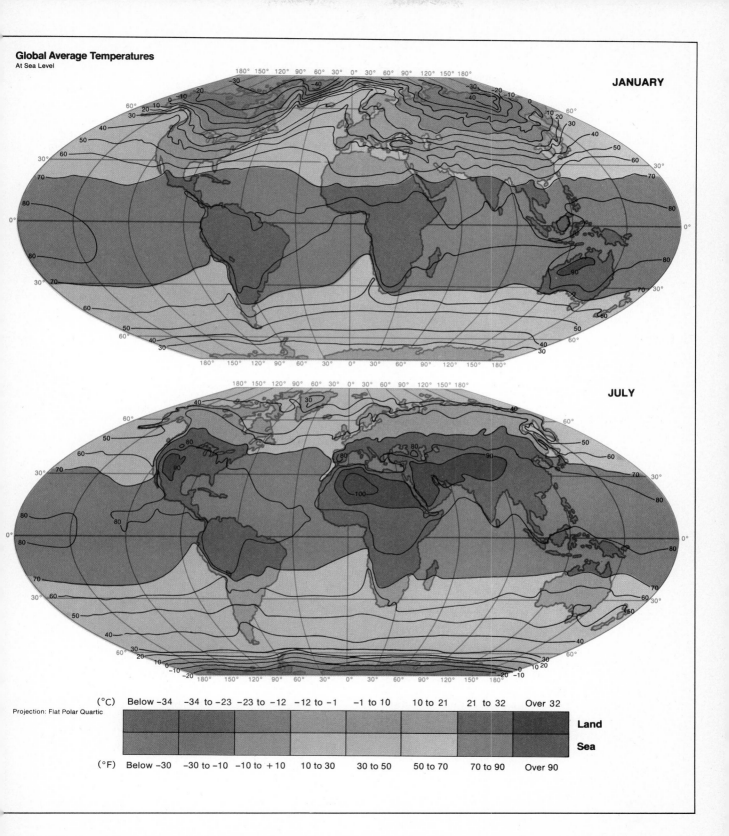

Global Average Temperatures
At Sea Level

JANUARY

JULY

Projection: Flat Polar Quartic

(°C)	Below −34	−34 to −23	−23 to −12	−12 to −1	−1 to 10	10 to 21	21 to 32	Over 32	
									Land
									Sea
(°F)	Below −30	−30 to −10	−10 to +10	10 to 30	30 to 50	50 to 70	70 to 90	Over 90	

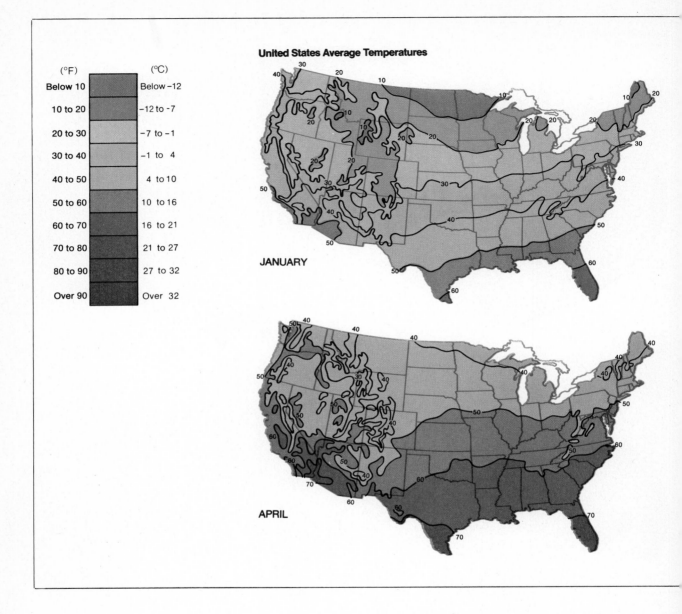

United States Average Temperatures

(°F)		(°C)
Below 10		Below -12
10 to 20		-12 to -7
20 to 30		-7 to -1
30 to 40		-1 to 4
40 to 50		4 to 10
50 to 60		10 to 16
60 to 70		16 to 21
70 to 80		21 to 27
80 to 90		27 to 32
Over 90		Over 32

JANUARY

APRIL

ft; 1,500 m) and desert location with minimal cloud cover, producing hot days and cool nights.

Average global temperatures for January and July are shown in Figure 3.16, and seasonal temperatures for the United States are depicted in Figure 3.17.

Modifications of Energy Budgets

Changes in any of a number of factors can alter local energy budgets. Cloudiness, atmospheric pollution, surface albedo, irrigation, vegetation—all influence the utilization of radi-

68

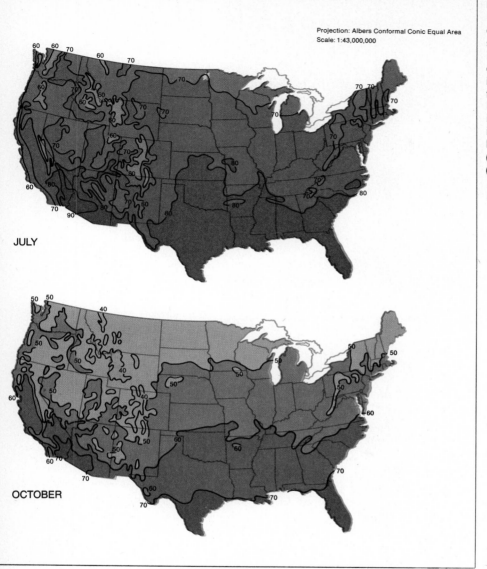

Figure 3.17
United States temperature distributions (in °F) are shown for selected months. In January, temperatures decrease regularly with increased latitude. In July, the variation with latitude is less marked because the input of solar radiant energy is relatively independent of latitude during the summer (see Figure 3.8, pp. 56-57). (Andy Lucas after *The National Atlas of the United States of America*, 1970)

JULY

OCTOBER

Projection: Albers Conformal Conic Equal Area
Scale: 1:43,000,000

ant energy. Human activities, both intentional and accidental, have modified local energy budgets and, if continued, could result in measurable changes in climate over large areas.

On the local scale, gardeners and orchard growers are concerned with the prevention of frost damage to plants. On clear nights the ground rapidly loses heat to the atmosphere by longwave radiation, and radiation back from the atmosphere is minimal. During early spring and late fall, such conditions can cause temperatures near the ground to drop below the freezing

ENERGY AND TEMPERATURE

Figure 3.18
The different colors in this *thermogram* of downtown New York City correspond to different temperatures, with red indicating the warmest temperature. The generation of heat within cities and the comparatively high heat capacities of building materials turn cities into *heat islands* that are several degrees warmer than the surrounding countryside. Washington, D.C., for example, is frost-free for a month longer than neighboring rural areas because of human modifications of the local climate. (Howard Sochurek)

point. Orchard growers often use large fans or even helicopters to disturb the cold air near the ground and mix it with warmer air above. These have largely replaced orchard heaters, or smudge pots, which produce a blanket of smoke to absorb and reradiate longwave radiation back to the ground. Radiant heaters are sometimes used to protect plants in courtyards and around patios. Often ornamental plants and small citrus trees are covered with plastic sheeting to reduce longwave radiation losses that cause frost damage.

Large urban areas produce their own distinctive climates by modifying the local radiation balance. Especially in winter, cities can be as much as 7° or 8°C (12° or 14°F) warmer than the surrounding countryside. Towering city buildings and canyon-like streets trap above-average amounts of radiant energy, even when the sun is low in the sky. Asphalt, concrete, and bricks store sensible heat that would be used for

evapotranspiration in a natural or agricultural environment. The burning of fossil fuels in homes, factories, and automobiles adds to the radiant heat input; and the dust, smoke particles, and pollutants in the air trap outgoing longwave radiation and return it to the surface. These effects combine to produce the phenomenon of the "urban heat island" (Figure 3.18).

Human modification of the energy balance did not begin with the Industrial Revolution. Ancient agricultural practices, such as tropical slash-and-burn agriculture (see Case Study following Chapter 9), midlatitude forest cutting, and overgrazing in subtropical latitudes, have altered surface albedo, water-holding capacity of soils, and evapotranspiration rates over vast areas. Removal of natural vegetation for agricultural purposes has allowed the wind to lift soil particles high into the troposphere, creating dust storms that darken the sky over areas of thousands of square kilometers.

The modern industrial age has intensified the human impact on energy budgets. Jet aircraft leave exhaust products and condensation trails high in the troposphere and even in the lower stratosphere. Jet condensation trails resemble some natural clouds and can significantly reduce shortwave radiation in regions of heavy air traffic. Perhaps more disturbing are findings that various man-made chemicals, such as the chlorofluorocarbons in spray-can products and the methyl chloroforms in solvents, may have the potential to destroy the ozone layer in the stratosphere that prevents dangerous amounts of ultraviolet radiation from reaching the earth's surface. At the surface, artificial ponds and reservoirs, paved areas, forest removal, and the drainage of marshes have all altered albedo and evapotranspiration patterns. Data from satellite surveys of albedo, land use and plant cover, and degree of cloudiness over the entire surface of the earth are required to assess the effects of these changes. The launching of the Earth Resources Technology Satellite, ERTS-1, in the summer of 1972 was an important first step in this program, which has continued through data transmissions from a series of orbiting satellites, such as LANDSAT and SEASAT.

SUMMARY

The sun provides the energy for most of the natural processes that affect the earth's surface. Solar energy is transmitted through space in the form of electromagnetic radiation, having a range of wavelengths. Solar radiation consists principally of ultraviolet, visible, and infrared radiation. Most solar radiation has a wavelength of less than 4 microns and is considered to be shortwave radiation.

The intensity of the solar beam at the top of the atmosphere is expressed as the solar constant. Daily amounts of solar radiation received at the top of the atmosphere depend on the number of daylight hours and the angle at which the sun's rays strike the atmosphere throughout the day. These factors are determined by earth-sun relationships, the most important being the distance of the earth from the sun and the inclination of the earth's axis of rotation with respect to its path around the sun. The inclination of the earth's axis is the cause of the seasons and controls the angle of the solar beam at various latitudes. Almost half of the solar radiation striking the atmosphere penetrates to the earth's surface. The rest is reflected back to space or is absorbed by the earth's atmosphere. Absorbed shortwave radiation is later re-emitted in the form of longwave radiation.

For its energy budget to remain in balance, the earth-atmosphere system must return the same amount of energy to space as it receives from the sun. This outgoing radiation consists of reflected shortwave radiation and longwave radiation from the earth and its atmosphere. Reflected shortwave radiation produces the earth's albedo. Water vapor and carbon dioxide molecules in the lower atmosphere absorb much of the longwave radiation emitted by the earth, heating the atmosphere from below. Longwave radiation re-emitted by the atmosphere is a substantial energy input to the earth's surface, preventing steep drops in nighttime temperatures and maintaining high average surface temperatures. The earth's surface, therefore, gains energy not only from the sun, but also from its own atmosphere.

The earth's surface loses energy by longwave radiation, by evapotranspiration and latent heat loss, by conduction and convection of heat into the atmosphere, and by conduction of heat from the surface down into water bodies and the soil. The daily variation of air temperature near the surface can be interpreted in terms of the local energy budget and the differential heating of land and water. Local energy budgets have been modified significantly by human activities, both deliberate and accidental. These activities and their consequences are now being monitored by orbiting artificial satellites.

APPLICATIONS

1. Using Figure 3.4, plot the annual regime of solar radiation at the top of the atmosphere at the latitude of your campus. Compare this with the regimes at latitudes 15 degrees to the north of your location and 15 degrees to the south of your location. Explain the differences in the three regimes.

2. Estimate how the landuse and landcover of your campus area have changed from pre-settlement days to the present. How would these landscape changes have affected the albedo of the campus area as a whole?

3. Over which regions of the earth would you expect the greatest seasonal changes in albedo?

4. Local daily temperature highs and lows are usually given on radio and TV and in newspapers: highs referring to late afternoon, and lows occurring in early morning hours. Keep track of the daily temperature range and the degree of cloud cover over a month-long period. How much do varying degrees of cloudiness affect the daily temperature range? Are other factors involved?

5. On a clear evening with no breeze, record the variations in temperature in different areas of your campus. You will need a thermometer that responds quickly to temperature changes (check with your instructor). Explain the pattern of temperature variations you measure.

6. On both a clear day and a cloudy day, make a record of hourly temperatures at an unshaded ground surface and compare these readings with hourly air temperatures 5 feet above the ground. Make the same experiment measuring temperatures on bare ground and grassy surfaces. Explain your results.

FURTHER READING

Gedzelman, Stanley David. *The Science and Wonders of the Atmosphere.* New York: John Wiley (1980), 535 pp. This beautifully illustrated book emphasizes introductory level explanations of physical processes. Chapters 3, 4, and 8 include very helpful examples of earth-sun relationships, radiation laws, and atmospheric optics.

Geiger, Rudolf. *The Climate Near the Ground.* (Tr. of 4th German ed.) Cambridge, Mass.: Harvard University Press (1965), 611 pp. This classic stresses the results of field studies on each of the continents, with particular emphasis on radiation and temperature. Geiger is sometimes called the "father of microclimatology."

Mather, John R. *Climatology: Fundamentals and Applications.* New York: McGraw-Hill (1974), 412 pp. Chapter 2 includes basic discussions of radiation and temperature, instrumentation and data, as well as their applications and limitations.

Miller, David H. "A Survey Course: The Energy and Mass Budget at the Surface of the Earth." Publ. No. 7, Comm. on College Geog., Assoc. of American Geog. (1968), 142 pp. This very useful monograph is organized into study units that proceed from energy exchange processes through local energy budgets to regional synthesis. Emphasis is placed on professional papers from almost every region of the globe.

Oke, T. R. *Boundary Layer Climates.* London: Methuen (1978), 372 pp. This recent text is a more advanced analysis of surface energy exchanges. Emphasis is placed on local and geographical differences, and the text is an excellent primer for students who want to do special studies.

Sellers, William D. *Physical Climatology.* Chicago: University of Chicago Press (1965), 272 pp. This work focuses almost entirely on fundamental analyses, in both descriptive and mathematical forms, of the radiation and energy budgets at the earth's surface.

Trewartha, Glenn T., and **Lyle H. Horn.** *An Introduction to Climate.* 5th ed. New York: McGraw-Hill (1980), 416 pp. Focus on geographical distributions of solar radiation and temperature is found in chapters 2 and 8–12.

CASE STUDY: **Solar Energy**

Suppose you leave your garden hose sprawling across a sun-baked patio or driveway on a hot summer day. When you come home at 5 in the afternoon, you rush out and pick up the hose to water some wilting plants. But you may do more harm than good. The water in the hose has been heated to more than 50°C (122°F) by absorption of solar radiation, and the hot water may be lethal to your tender plants.

At the same time, the soles of your feet will burn if you step barefooted out on the black-top pavement of the street. Black-top pavement and dark automobile tops can heat to more than 60°C (140°F). People have long been aware of these straightforward examples of solar heating, but, with few exceptions, there has been little attempt to harness this energy source. Only since the rapid rise in energy costs have people begun to pay serious attention to the potential uses of solar energy.

Solar systems for heating water and buildings represent the simplest and least expensive adaptations to solar energy. On a cold day the family cat will almost always take advantage of winter sun streaming through the glass panes of a window. The glass allows most of the solar radiation to enter the room, where it is absorbed by carpet and cat alike. On the other hand, most longwave radiation emitted from within the warm room is absorbed by the glass and reradiated back into the room—a simple application of the greenhouse effect discussed in Chapter 3.

Commercial systems for heating water by solar energy have been developed and are becoming common in Florida, California, and some foreign regions. These systems usually consist of a water-filled network of tubing coated with black paint and set into a sloping roof facing toward the south. The tubing is usually covered by glass and is set over dark metal sheets heavily insulated from the attic. The greenhouse effect is maximized. The heated water is then pumped to a storage tank, where it may be used directly, or it is circulated through a large volume of gravel or stone surrounding the tank, which serves for heat storage overnight and during cloudy days. This technology has also been adapted to heat air circulated through homes or buildings.

Direct solar heating normally supplements conventional heating systems. Initial costs are considerable, but operating costs are low. As the technology improves, costs should decrease, making solar heating increasingly competitive with conventional systems. Solar radiation is, of course, a continuously renewable source of energy, unlike gas, oil, or coal, and it is nonpolluting. It is especially adapted to Florida and the Southwest, where there is an abundance of solar radiation and an absence of large trees. In the more humid Northeast and the lower Mississippi River valley, the fossil energy saved by direct solar heating in the winter may be offset during hot humid summers by undesired solar heating of the structure. The trade-off will have to be compared against the much lower air-conditioning costs of one-story houses nestled within the shade of large trees. To take the greatest advantage of solar energy, the trees should be deciduous—giving shade in the summer, but losing their leaves during the winter—so that the winter sun may be absorbed by the building.

By use of multiple reflectors and mirrors, it

This experimental solar house at the University of Delaware has been designed so that solar heating and photovoltaic cells in the panels on the roof provide more than two-thirds of the energy needs of a family. (R. A. Muller)

is also possible to focus the solar radiation striking a number of receiving surfaces onto a single point, producing very high temperatures in solar furnaces. Probably the best known is the one at Odeillo in the French Pyrenees. This furnace can achieve temperatures higher than 3300°C (6000°F), and it is used for experiments requiring high temperatures in a few moments.

The ultimate objective of solar energy utilization is the conversion of sunlight into electricity that can be fed into regional power grids. The light meter of a camera includes photovoltaic cells that convert solar radiation into electrical current; the deflection of the meter needle is related to light impinging on the cells, resulting in an electrical current. Larger amounts of direct-current electricity can be generated by multiple cells of semiconductor materials, such as silicon cells, connected in series. This technology was originally developed for the satellite and space programs, where the solar cells have proven to be reliable and long lasting, but relatively expensive and inefficient. At the present time, they can convert to electrical current only about 10 to 13 percent of the solar energy they receive.

A number of more ambitious schemes to utilize solar energy have been developed in recent years. One such scheme involves

construction of giant power-station satellites consisting of banks of solar cells that would intercept solar radiation in space. The converted solar energy would then be beamed to receiving stations on the earth by means of microwave transmission. Another proposal calls for solar farms in the deserts of the American Southwest. The solar farms would cover thousands of acres of desert land, producing large amounts of electricity—with the environmental impact probably less than mining and drilling for fossil fuels.

Incidentally, most other energy sources are directly or indirectly related to solar energy. Wind and wave energy and hydroelectric power are all direct products of the earth's climate—all driven by variations in solar radiation over the surface of the earth. Wood products are the result of the photosynthetic conversion of solar radiation, carbon dioxide, and water; and the fossil fuels such as gas, oil, and coal also result from photosynthesis and represent solar radiation stored for millions of years within the earth's crust.

Direct solar heating and the conversion of solar radiation to electricity are especially appealing because total global energy consumption is far less than the solar energy received each day by the earth-atmosphere system. Pollution is minimal and the solar energy is a continuously renewable resource. But because solar radiation intensities are so low, the technologies of collection, concentration, and storage are more costly than fossil fuel utilization at the present time. However, there is no doubt that solar energy will be a growth industry in future years.

A furious storm at sea emphasizes the complex interactions of the atmospheric and oceanic systems. Winds drive the surface ocean currents, and the ocean temperatures influence the circulation patterns of the atmosphere.

Snow Storm—Steam Boat off a Harbour's Mouth Making Signals in Shallow Water and Going by the Land by William Turner. (The Tate Gallery, London)

General Atmospheric and Oceanic Circulations

A balloonist can determine only whether to rise higher or sink lower. But in 1979 and 1980 balloon trips were completed across the widths of both the Atlantic Ocean and North America. How can balloonists, with no control over their horizontal direction, decide that they are going to cross a continent, or an ocean? The answer is easy: the motion of the atmosphere does the deciding. In the middle latitudes all long-distance balloon trips go from west to east because the general motion of the atmosphere is in that direction. It is as impossible to go the other way as it is for a raft to drift up a river instead of down it. Similarly, a capped bottle that is thrown into the sea off the Florida coast will probably come ashore in Iceland, Norway, or Ireland, never in Brazil, or even nearby Cuba. To send a message to Cuba in a floating bottle, the best starting point would be Morocco; objects floating ashore in Brazil come from South Africa or Angola.

On a global scale, both the air and the seas move in paths that vary somewhat from season to season but are dependable from year to year. The large-scale semipermanent pattern of atmospheric flow, which drives the oceanic circulation, is one of the most important facts of physical geography and is known as the *general circulation.*

The general circulation of the atmosphere and the oceanic circulation that results from it are the principal mechanisms by which energy is transferred from equatorial and tropical regions having net solar energy gains to high-latitude regions having net energy losses to space. This flow of air carrying both latent and sensible heat is what maintains the earth's thermal balance. The general circulation also delivers to the continents water evaporated from the oceans. It determines local weather and is the foundation for the entire mosaic of global climates. This chapter explains the general circulation of the atmosphere and the resulting oceanic circulation.

FORCES CAUSING ATMOSPHERIC MOTION

The atmosphere is in ceaseless motion, whether churning with furious storms or drifting so slowly that hardly a breeze is felt. Although the details of atmospheric circulation are complex, the general features can be understood by looking at the forces involved.

Any force may be thought of as a push or a pull. According to the fundamental laws of motion developed in the seventeenth century by the English physicist Isaac Newton, a moving object's speed and direction of motion cannot change unless a force is made to act on the object. Horizontal motion is influenced only by forces pushing or pulling in the horizontal direction, and vertical motion is influenced only by forces acting in the vertical direction. Although the force of gravity pulls every parcel of air downward toward the earth, gravity has no di-

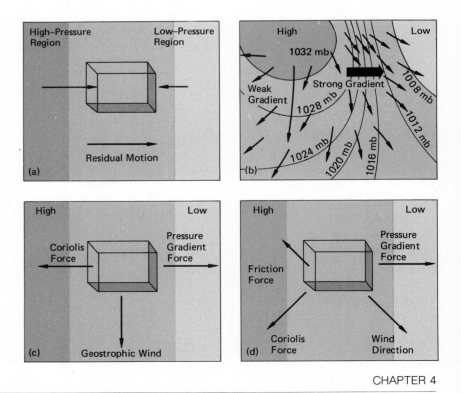

rect effect on the horizontal motion of air. The forces that act on a parcel of air moving horizontally in the atmosphere are the pressure gradient force, the Coriolis force, and friction.

The Pressure Gradient Force

Pressure is force per unit area. If the pressure on one side of a parcel of air is greater than the pressure on the other side, the parcel will be pushed toward the area of lower pressure (Figure 4.1a). Recall from Chapter 2 that differential surface heating on the earth produces horizontal differences of atmospheric pressure. The greater the difference in pressure, the greater the net push and the more rapid the resulting motion. This "push," caused by the horizontal difference in pressure across a surface, is called the *pressure gradient force.*

On a weather map, meteorologists represent atmospheric pressure with lines called *isobars.* Isobars connect locations on the map that have equal atmospheric pressures (see Figure 4.1b). The pressures shown do not correspond to measured values: they have been corrected to sea level values to compensate for the lower pressures measured at higher elevation weather stations. The pressure gradient force is at right angles to the isobars, and is strongest where the isobars are most closely spaced—that is, where the pressure decreases most rapidly.

The Coriolis Force

If it were not for the rotation of the earth, winds would simply follow the pressure gradient. But the earth's rotation complicates the motion of the atmosphere. Imagine two people tossing a ball back and forth on a merry-go-round that is rotating counterclockwise. One person is near the center and the other is at the rim, facing the center. If the person near the center aims the ball straight at the person on the rim, the ball will pass to the left of the catcher because the merry-go-round continues to rotate during the ball's flight. Although the ball actually moves in a straight line, the thrower sees it as curving to the right as though a sidewise force were acting to deflect it. A similar, but more complex, deflection occurs on the rotating spherical earth. This force is known as the *Coriolis force,* after Gaspard Coriolis, who

Figure 4.1 (opposite)
(a) The pressure gradient force pushes a parcel of air away from high pressure toward low pressure. If no other forces were present, the parcel would move toward the low-pressure region.

(b) Winds near the surface of the earth are partially determined by isobar patterns and spacing. Close spacing of isobars is associated with strong horizontal pressure gradients and strong winds.

(c) A parcel of air aloft moving horizontally on the rotating earth experiences a Coriolis force and a pressure gradient force. The Coriolis force acts at right angles to the direction of motion of the parcel, and will continue deflecting the parcel of air until it is moving parallel to pressure isobars. Then the pressure gradient force and the Coriolis force are equal, and no further deflection occurs. Air motion parallel to isobars is known as *geostrophic wind*; the speed of a geostrophic wind is highest where the pressure decreases most rapidly with distance. This diagram is drawn for the northern hemisphere. How would it appear for the southern hemisphere?

(d) Friction acts on air parcels moving near the ground. The diagram shows that because friction with the land surface reduces wind velocity, it also reduces the Coriolis force, which cannot come into balance with the pressure gradient force. As a result, air moves diagonally across isobars toward the area of lower pressure. (Vantage Art, Inc.)

first explained the phenomenon mathematically. Even though no real force is operating, the apparent deflection resulting from the earth's rotation affects all objects moving freely over the earth's surface. A correction for the Coriolis deflection must be added when aiming missiles or long-range artillery.

The Coriolis deflection acts at right angles to the direction of motion. In the northern hemisphere the apparent deflection is to the right, and in the southern hemisphere to the left, so that the pattern of deflection in the two hemispheres is symmetrical. Air flowing toward the equator is turned to the west in both hemispheres, and air moving toward the poles is turned to the east. Likewise, air moving to the west is turned toward the poles, and air moving

toward the east is turned toward the equator. Figure 4.2 shows how the deflection may be seen against the earth's imaginary grid of latitude and longitude lines. Figure 4.3 indicates that the Coriolis deflection is zero at the equator and increases toward the poles. The Coriolis deflection is also proportional to the speed of the moving mass, so that the deflection is greater with stronger winds.

Under ideal conditions the Coriolis force is equal to the pressure gradient force, causing air to flow parallel to pressure isobars (Figure 4.1c).

Figure 4.3
For an object such as an air mass moving horizontally on the earth's surface, the strength of the Coriolis deflecting force varies with latitude, as the graph shows.

At the North Pole the earth's surface rotates once in every 24 hours. At the equator the surface moves much faster—about 1600 km (1000 miles) per hour—but this motion does not cause rotation of the surface itself. A north-south line at the equator always points the same way, parallel to the earth's axis. It is not twisted or deflected by the earth's rotation. A north-south line close to the North Pole swings in a counterclockwise direction as the earth turns on its axis. This swing, or rotation, increases from the equator to the poles, causing increasing Coriolis deflection. (Doug Armstrong)

Figure 4.2
This figure illustrates how the earth's rotation and grid of parallels and meridians are related to the deflection of moving air to the right in the northern hemisphere. Air streams originating at 40°N, 90°W and moving from south to north and from west to east appear nine hours later as southwesterly and northwesterly flows. (Vantage Art, Inc. after Herbert Riehl, *Introduction to the Atmosphere*, © 1972, McGraw-Hill Book Company)

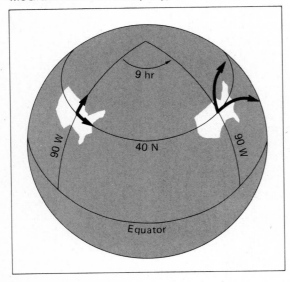

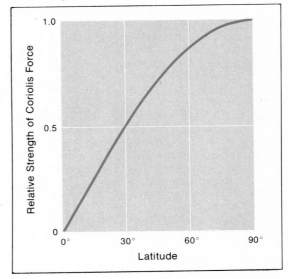

Meteorologists refer to air flow along isobars as *geostrophic wind*. The circulation of upper-level air, above the frictional effects of the earth's surface, approaches geostrophic flow. In the middle and higher latitudes of the northern hemisphere, the Coriolis deflection causes air to flow in a clockwise direction around high-pressure areas and in a counterclockwise direction around low-pressure areas (Figure 4.4). Utilizing these facts, the Dutch meteorologist Buys Ballot in 1857 described the relationship between pressure systems and the direction of flow aloft in simple terms. Buys Ballot's Law states that if one's back is to the geostrophic wind in the northern hemisphere, low pressure is to the left and high pressure is to the right. In the southern hemisphere, low pressure is to the right, high pressure to the left.

The geostrophic winds of the upper atmosphere are especially important to meteorologists concerned with weather forecasting, for they influence surface air movements, as will be seen later in this chapter. Upper air measurements of pressure, wind speed, and wind direction are taken twice a day at more than 100 weather stations across the continental United States. From these measurements, meteorologists construct weather maps that estimate pressure and wind at various levels of the atmosphere over North America. These projections are the basis for predicting weather as much as 72 hours in advance.

Figure 4.4
This figure shows the relationships between upper and surface winds for both northern and southern hemispheres. The Coriolis deflection is to the right in the northern hemisphere and to the left in the southern hemisphere, producing symmetrical patterns of deflection in the two hemispheres. In the northern hemisphere, the resultant geostrophic wind flows parallel to the isobars, with low pressure to the left, high pressure to the right. The surface flow converges toward low pressure and diverges away from high pressure because of frictional effects. (Vantage Art, Inc. after Arthur N. Strahler, *Introduction to Physical Geography*, 1st ed., 1951, John Wiley & Sons)

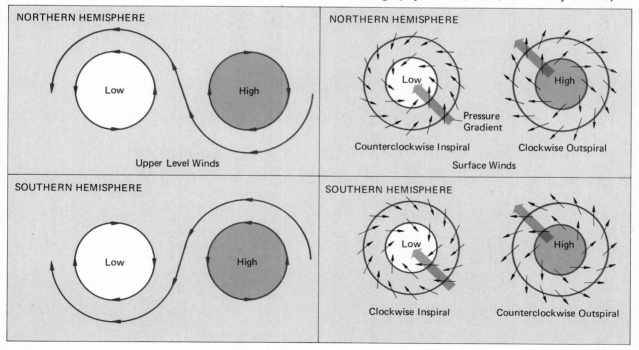

NORTHERN HEMISPHERE
Low   High
Upper Level Winds

SOUTHERN HEMISPHERE
Low   High

NORTHERN HEMISPHERE
Low   High
Pressure Gradient
Counterclockwise Inspiral   Clockwise Outspiral
Surface Winds

SOUTHERN HEMISPHERE
Low   High
Clockwise Inspiral   Counterclockwise Outspiral

The Force of Friction

If the energy sources that drive the atmospheric circulation were suddenly to disappear, friction between the atmosphere and the earth's surface, and within the atmosphere itself, would cause all atmospheric motion to slow and eventually cease. Scientists regard friction as a "force" that is opposed to the motion of masses on the earth or in the atmosphere. For a parcel of air to maintain its movement, the driving force must overcome frictional resistance.

The effect of friction is at its maximum at the earth's surface and decreases upward. Recall that the Coriolis force is proportional to the speed of the object. It follows that near the ground, where surface winds are slowed by frictional forces, the Coriolis effect is reduced, while the pressure gradient force is not affected. Consequently, instead of flowing parallel to pressure isobars, air near the surface responds to the pressure gradient force by flowing from higher to lower pressure at an angle across the isobars (Figure 4.1d). This causes surface air to spiral *into* centers of low pressure and to spiral *out of* centers of high pressure. In the northern hemisphere, the inward spiral is counterclockwise, the outward spiral clockwise. In the southern hemisphere, the reverse is true (see Figure 4.4). Relating Buys Ballot's Law to surface winds, if your back is to the wind in the northern hemisphere, and you rotate 45° to the right, low pressure will be on your left, high pressure on your right. In the southern hemisphere, rotate 45° to the left, and low pressure will be on the right, high pressure on the left (Figure 4.4).

THE GENERAL CIRCULATION OF THE ATMOSPHERE

The pressure gradient force, the Coriolis force, and the force of friction are the basic factors that determine atmospheric motion. But other large-scale mechanisms are also involved in the atmospheric circulation. The following sections describe a generalized model of the global circulation of the atmosphere. This model represents a sort of grand average for one year; the actual circulation at any moment may be somewhat different.

Pressure Patterns and Winds on a Uniform Earth

To understand the general circulation of the atmosphere, it is helpful to think first of the atmospheric circulation that would exist if the earth's surface were absolutely uniform. Let us start by also assuming that there is no Coriolis force. Under such conditions, a low-pressure belt would develop over the equatorial region, where there is excessive radiational heating. A high-pressure belt would develop over the polar regions, where there is excessive radiational cooling. The air would rise over the low latitudes and subside over the poles. The atmospheric circulation would consist of a surface flow of air from the high-pressure regions at the poles to the low-pressure zone at the equator. To complete the circulation loop there would be a return flow of air toward the poles in the upper atmosphere.

Such a vertical convective cell is called a *Hadley cell*, after the English meteorologist who proposed this model of circulation in 1735. The right side of Figure 4.5 shows a vertically exag-

Figure 4.5 (opposite)
This sketch is a highly diagrammatic representation of the general circulation over a homogeneous earth with a water surface. The Coriolis effect is included. Semipermanent pressure and wind belts are shown on the surface of the sphere, and a much enlarged vertical cross section is shown on the right hemisphere only. The text contains a description of the pressure and wind systems, and the text and figure should be analyzed together carefully. The polar front is shown only for the northern hemisphere, and it is discussed in more detail in later sections of this chapter. (Vantage Art, Inc.)

gerated cross-section of the Hadley cells north and south of the equator. Note that modern interpretations of the general circulation restrict the Hadley cells in each hemisphere to the lower latitudes between the equator and about 30° latitude. As the meridians of longitude converge poleward, so does poleward moving air. Thus cooling and convergence aloft force air downward in the subtropical latitudes.

If we add the Coriolis force to this circulation model, the circulation pattern is altered. Excessive radiational heating at the equator still produces rising air that flows out toward the poles aloft. Once away from the equator, however, the poleward flowing upper air in the northern hemisphere is turned to the right by the Coriolis force, producing a westerly (west-to-east) "jet stream." Leftward deflection of poleward moving upper-level air in the southern hemisphere creates a westerly jet stream there as well. The air aloft cools by longwave radiation to space and a portion sinks into the lower atmosphere at about 30° latitude in each hemisphere. This produces high-pressure zones in the subtropics. Much of this descending air flows back down the surface pressure gradient toward the equator. This surface flow is deflected by the Coriolis force, again toward the right in the northern hemisphere and toward the left in the southern hemisphere (Figure 4.5). In both hemispheres the deflected flow is turned toward the west, so that this flow is from northeast to southwest in

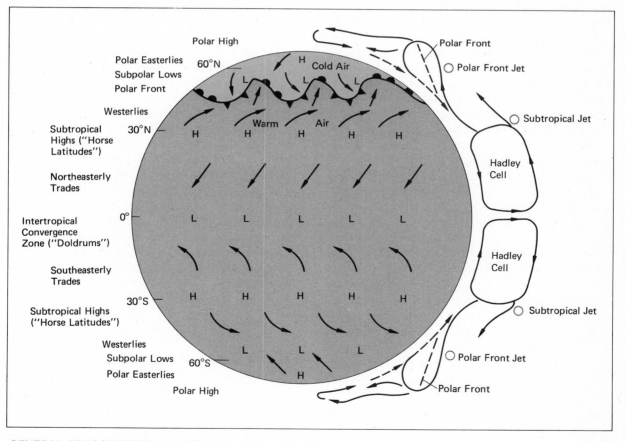

the northern hemisphere and from southeast to northwest in the southern hemisphere.

The surface pressure and wind systems described above exist in a general way and have traditional names dating from the era of wind-dependent sailing vessels. The belts of descending warm dry air, fair weather, and weak surface wind in the subtropics, called the Calms of Cancer and the Calms of Capricorn, were also known as the *horse latitudes*. Although other interpretations of this term have been made, it is generally believed that when sailing ships were becalmed here, and supplies of food and drinking water dwindled, horses were the first passengers to go overboard. Therefore, floating horse carcasses were sometimes sighted in this zone.

Winds are traditionally named according to the direction from which they blow. Between the horse latitudes and the equator in the northern and southern hemispheres lie zones characterized respectively by northeasterly and southeasterly *trade winds*. The "easterly trades" are among the earth's steadiest and most persistent winds, providing sea traders in sailing vessels with reliable westward routes over the oceans. Near the equator the trade winds converge into a low-pressure zone of generally light, variable winds and calms traditionally known as the *doldrums*. The doldrum belt is what present meteorologists refer to as the *intertropical convergence* (ITC).

Some of the subsiding air of the subtropical highs flows poleward near the surface (Figure 4.5). This flow is deflected toward the east to become the southwesterlies of the northern hemisphere and the northwesterlies of the southern hemisphere. Usually these winds are simply called *westerlies*. In the days of sailing vessels, these were the winds used for eastward passages across the oceans.

The surface air within the westerlies is normally of subtropical origin. When this warm "tropical" air meets colder "polar" air moving toward the equator, the tropical air flows up and over the denser polar air. The *polar front* is the boundary between the two types of air (Figure 4.5). As the warm air flows up over the wedge of colder air along the polar front, it is chilled by expansion and radiational cooling. This cooled air subsides over the polar regions, forming high-pressure centers known as the *polar highs*.

Surface air spreads from the polar highs in both hemispheres and undergoes westward Coriolis deflection to become the *polar easterlies*. The polar easterlies meet the westerlies in the zones of friction along the polar fronts. From time to time, particularly during winter, the polar front bulges toward the equator, allowing polar air to penetrate to subtropical latitudes. These bulges, known as *polar outbreaks*, are shown by the broken arrows in the cross-section in Figure 4.5.

Surface pressure and wind patterns on a uniform, rotating earth can be summarized as follows. There are three zones of low pressure and atmospheric convergence—the region of the polar front in each hemisphere and the intertropical convergence zone astride the equator. There are also four zones of high pressure and atmospheric divergence—the polar and subtropical highs in each hemisphere. In general, zones of low pressure and convergence are associated with clouds and precipitation, while fair weather prevails in zones of high pressure and divergence. Due to seasonal changes in the angle of the sun's rays (resulting from the tilt of the earth's axis), these zones shift poleward in summer and toward the equator in winter. The heat storage in oceans, however, delays poleward migration by one or two months, so that these belts do not reach their highest latitudes until late summer.

Observed Pressure and Wind Systems

The actual patterns of pressure and wind at the earth's surface are more complicated than the simple belts associated with the uniform surface model. The presence of continents and oceans considerably influences the circulation.

The specific heat of rock or soil is much smaller than that of water, so the continents become warmer than the oceans in summer and much cooler in winter. This differential heating and cooling of land and water masses affects surface pressure and winds, as thermal low-pressure cells develop over the land in the summer to be replaced by high-pressure cells in the winter. Mountain ranges also disturb the flow of the atmosphere, even at the atmosphere's upper levels.

The observed pattern of pressure distribution shown in Figure 4.6 (see pp. 86–87) consists of separate cells rather than continuous zones of high and low pressure. Nevertheless, these cells tend to be distributed along the same bands of latitude as the idealized pressure zones on a uniform earth. The average direction of surface winds is closely related to the pressure distribution. Surface winds spiral outward from high-pressure regions and inward toward low-pressure regions, indicating geostrophic flow modified by friction with the earth's surface.

In the southern hemisphere, there is almost continuous ocean between latitudes 45° and 65°S, so that the observed circulation, particularly in winter (July), tends to be similar to the model for the homogeneous surface. In the northern hemisphere, the presence of large landmasses produces well-defined pressure and wind cells. These cells migrate north during the summer and south during the winter, reflecting changes in the receipt of solar radiation.

In winter, the northern regions of North America and Eurasia become very cold in comparison with the adjacent oceans. This causes the polar high to be displaced in the direction of the equator, forming two high-pressure centers—the massive Siberian high over north-central Asia and the much smaller Yukon high over eastern Alaska and the Canadian Arctic. Two subpolar lows, the Icelandic and Aleutian lows, are situated over the oceans at similar latitudes.

During summer, when the continents become warmer than the oceans, the subtropical highs strengthen and expand over the cooler oceans, and low-pressure centers develop over the deserts of the southwestern United States and southern Asia (Figure 4.6).

The seasonal shifts in pressure patterns generate seasonal shifts in winds. The inset in Figure 4.6 shows how the average position of the intertropical convergence—the meeting place of the trade winds—shifts between January and July. Note that there is relatively little change in the position of the ITC over the Atlantic and eastern Pacific oceans. However, over land areas, especially those bordering the Indian Ocean, there is a large seasonal displacement. Associated with the extreme shift in the position of the ITC over eastern and southern Asia is a reversal in wind direction between winter and summer. This seasonal change in the direction of surface winds produces very strong seasonal changes in the prevailing weather. Half of the year the wind blows from the land to the sea; the other half it blows from the sea to the land. A wind system that reverses direction seasonally is known as a *monsoon*. Monsoons are discussed at greater length in Chapter 6 in connection with tropical weather types.

The Upper Atmosphere and Jet Stream

The circulation pattern of the upper atmosphere is much simpler than that near the surface. Poleward of the subtropical highs, the upper atmosphere flows from west to east in a vast circumpolar vortex. Since friction here is at a minimum, the Coriolis and pressure gradient forces are in balance. Therefore, air flow in the upper atmosphere approximates geostrophic conditions, with low pressure to the left and high pressure to the right in the northern hemisphere (Figure 4.7, p. 88). The strongest flows are concentrated in the relatively narrow *jet streams*.

High-velocity streams of air were first observed in the upper troposphere in the 1940s with the development of military aircraft that could fly at altitudes exceeding 10 km (6 miles).

Global Atmospheric Pressure and Prevailing Winds

Projection: Flat Polar Quartic

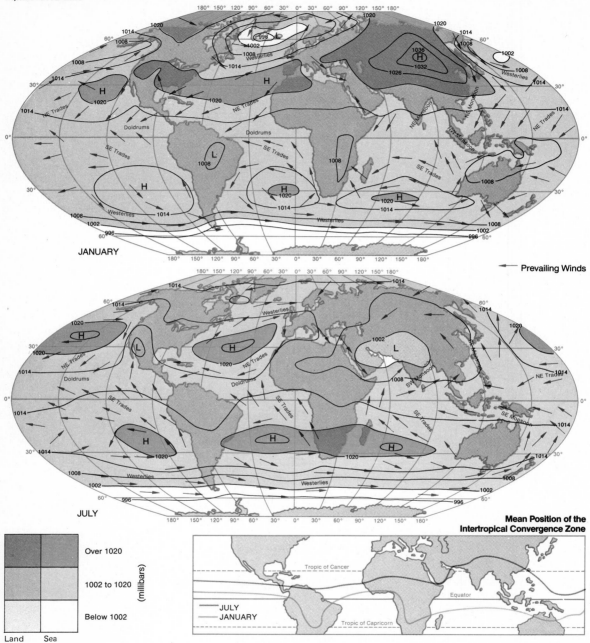

JANUARY

JULY

← Prevailing Winds

Mean Position of the Intertropical Convergence Zone

Over 1020

1002 to 1020

Below 1002

(millibars)

Land   Sea

Tropic of Cancer

Equator

Tropic of Capricorn

—— JULY
—— JANUARY

(Andy Lucas and Laurie Curran after Hermann Flohn, *Climate and Weather*, © 1969 by H. Flohn, used with permission of McGraw-Hill Book Company, and after W. Schwerdtfeger, "The Climate of the Antarctic," *World Survey of Climatology*, vol. 14, edited by S. Orvig, © 1970 by Elsevier Publishing Company)

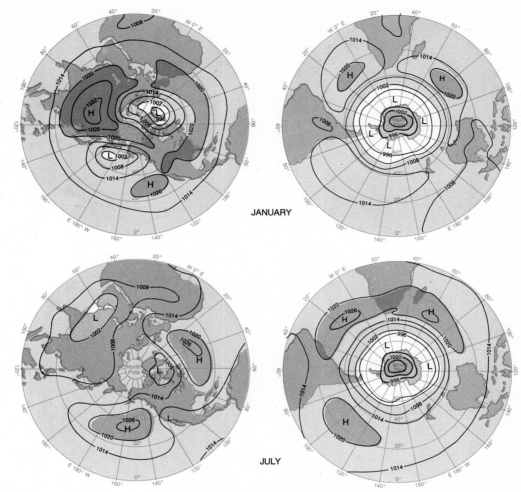

JANUARY

JULY

Figure 4.6

Global maps show the average direction of surface winds and the average atmospheric pressure at the surface for January and July. Pressure readings are in millibars. Near the Antarctic Circle, where there is open ocean around the entire globe, the isobars form a continuous belt. Elsewhere, the presence of landmasses causes the regions of high and low pressure to be broken into individual pressure cells. The cells tend to be distributed along bands of latitude that are analogous to the pressure belts

that would form on a uniform earth. The formation of pressure cells is affected by temperature differences between the oceans and the continents. (After the United States Weather Bureau)

The average distribution of surface winds closely follows the pressure patterns, with winds tending to flow outward from high-pressure regions and inward toward low-pressure regions. In many areas, the winds approach geostrophic flow modified by friction with the earth's surface.

In both the northern and southern hemispheres, the distribution of pressure cells

generally shifts to the north in July and to the south in January. Over the poles, the high-pressure cells are more diffuse and less numerous in summer than in winter. The resulting seasonal shift in wind patterns generates a seasonal variation in the weather of many regions. The winds on the east coast of central Africa, for example, change from the northeast trades in January to the southeast trades in July. The inset map shows how the mean position of the intertropical convergence zone, which is the meeting place of the trade winds, shifts between January and July.

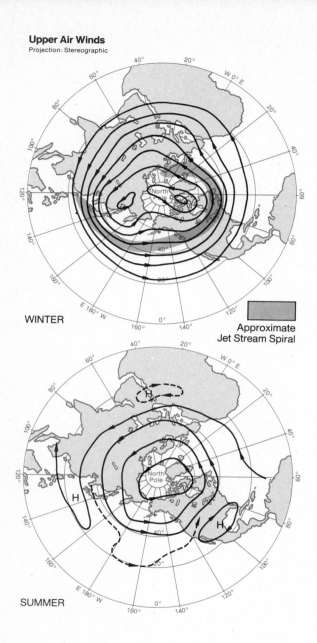

Upper Air Winds
Projection: Stereographic

WINTER

Approximate
Jet Stream Spiral

SUMMER

Figure 4.7
The maps show winter and summer wind patterns for the northern hemisphere in the upper atmosphere near an altitude of 5.5 km (3.4 miles), where the atmospheric pressure is approximately one-half the pressure at sea level. Wind directions and speeds in the upper atmosphere are usually determined by tracking freely drifting radio-equipped balloons that transmit meteorological data to the ground. In winter the upper air winds are strong and form a well-defined pattern of circulation with several undulations. The shaded region indicates the average location of fast-moving jet streams. Many winter storms in the northern latitudes appear to be generated with the help of jet streams. The flow of upper air winds tends to be weaker during summer than during winter. Dashed lines indicate particularly weak flows. (Andy Lucas after Herbert Riehl, *Introduction to the Atmosphere*, © 1972 by McGraw-Hill Book Company)

Figure 4.8 (opposite)
Circumpolar jet streams circle the earth at an altitude of 10 to 15 km (6 to 9 miles) and gain much of their energy from the temperature difference between warm tropical and cold polar air. (a) The jet streams appear in conjunction with large, slow-moving undulations, or waves, in the flow of the upper atmosphere. There are typically three to six waves in a complete pattern around the earth. The waves may undergo increasingly severe oscillations (b) (c), until cells of rotating warm and cold air are formed (d). The cells then die away, and the pattern once more resembles that shown in (a); the entire cycle is completed in four to eight weeks.

This mechanism results in the movement of warm air toward the poles and cold air toward the tropics and helps to maintain the poleward flow of energy from equatorial regions. Jet streams and upper air waves also appear to generate surface weather systems by causing convergence and ascent of air in cyclonic storms. (Doug Armstrong after Jerome Namias, "The Jet Stream," *Scientific American*, copyright © 1952 by Scientific American, Inc. All rights reserved.)

These so-called jet streams were found to occur at altitudes of 10 to 15 km. They are hundreds of kilometers wide and several kilometers thick. The wind speed along the core of a jet stream can exceed 300 km (200 miles) per hour for a distance of 1600 km (1000 miles) or more. Figure

4.5 shows that there are two jet streams in each hemisphere: the subtropical jets associated with the subtropical highs and the polar front jets that are normally above the polar fronts. The jet streams are utilized by eastward-bound airliners and are avoided as far as possible on westward flights. Assistance from the jet stream causes east-bound flights across the full width of the United States to require about an hour less time than west-bound flights.

The jet streams generally follow a sinuous path, as shown in Figure 4.8. There are typically three to six curves, or "waves," in each complete circuit of the jet stream around the earth. The waves usually remain within certain geographical limits, but at times they undergo severe oscillations. When the waves push outward for several weeks, the weather under displaced waves becomes warmer or colder and wetter or drier than normal, depending upon which way the air stream is curving. Areas of low pressure, known as *upper-level troughs*, are present where the jet stream bulges toward the equator. At the surface, cold air flows toward the tropics on the

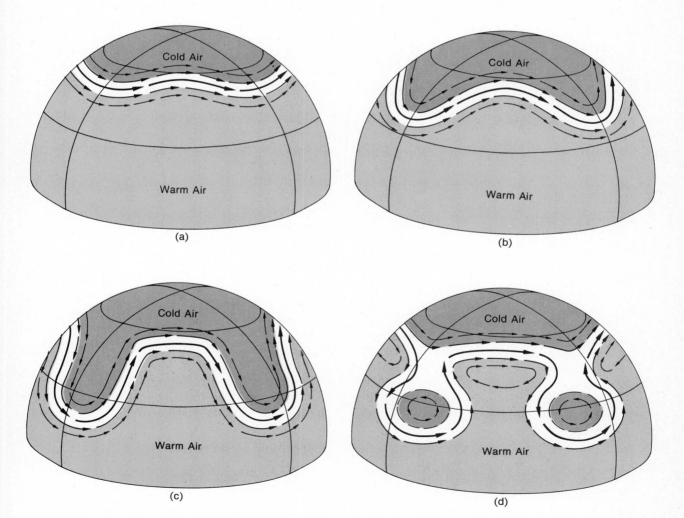

(a)

(b)

(c)

(d)

GENERAL ATMOSPHERIC AND OCEANIC CIRCULATIONS

western edge of each trough, and warm air flows poleward on the eastern edge. When no undulations are present, there is little exchange of energy between the tropics and the poles. The undulating pattern is necessary to maintain a net poleward flow of energy from low to high latitudes. Periodically, the oscillations become so extreme that waves are "pinched off" from the main stream, forming detached upper-level cells of low pressure (Figure 4.8d). These cells affect surface weather, but quickly die away, causing the pattern again to resemble that shown in Figure 4.8a. A wave cycle in which the path of the jet stream changes from straight to highly irregular to straight again generally requires one to two months.

The general circulation related to global patterns of pressure sets in motion *secondary circulations* that produce local weather conditions. In Chapter 6 we shall examine these secondary circulations, which directly influence our daily activities.

THE OCEANS

The phenomena of the oceans and atmosphere are so closely linked that many scientists prefer to speak of "the ocean-atmosphere system." Air moving over the sea surface sets water in motion, producing ocean currents. These currents (like air currents) help redistribute energy by carrying warm waters toward the poles and cool waters toward the tropics. The oceans, in turn, affect atmospheric circulation by feeding water and latent heat into the atmosphere through evaporation, and by functioning as reservoirs of heat that drives the atmospheric circulation and provides much of the energy for the hydrologic cycle.

The Oceans: Energy Banks

We have seen that the heat capacity of ocean water is two to three times that of continental surfaces and much greater than that of the air. The great heat capacity of the oceans makes them reservoirs of the energy that powers global weather. This energy is released into the atmosphere both by conduction and convection and by latent heat transfer during the evaporation process that moves enormous amounts of water vapor from warm sea surfaces to the atmosphere. While this water provides the earth's rainfall, it also helps power the atmospheric circulation by latent heat release during condensation and cloud formation.

Ocean temperatures are "conservative," changing very slowly through the seasons. But even small changes in ocean temperatures seem to be able to produce large changes in weather patterns. Oceanographers have recently discovered vast and persistent surface "pools" of ocean water that are 1° to 2°C warmer or cooler than the water around them. These ocean temperature abnormalities, or "anomalies," seem to produce changes in the average atmospheric circulation. For example, unusually warm water off the east coast of the United States during 1971 and 1972 was associated with very wet weather in the coastal states, and it probably helped provide energy for Hurricane Agnes, a tropical cyclone that was especially destructive between Virginia and Pennsylvania. Because of complex interactions with the upper air flow, the same anomaly may have played a role in freezes, droughts, and grain crop failures in the Soviet Union in 1972. A pool of colder than normal water in the central Pacific in 1976–1977 is believed to have caused changes in the general circulation, producing severe drought in the western United States along with the coldest winter in 60 years in the east, followed by a summer drought in the Midwest that equaled those of the Dust Bowl years of the 1930s.

Surface Currents

Wind blowing over the surface of the ocean exerts a push on the water, but the resulting motion of the water is only a fraction of the wind speed. This is due to the greater density and internal friction of water. Wind-driven ocean currents have speeds ranging from kilometers per hour to kilometers per day. Because friction causes current velocities to decrease rapidly with depth, strong wind-driven currents are confined to the upper hundred meters or so of the ocean. Deeper currents unrelated to wind stress have also been discovered, but their origin and pattern remain unclear.

A well-developed ocean current takes a long time to respond to changes in the wind, and even the waves produced by strong winds require many hours to reach their full development. Therefore, ocean currents reflect average wind conditions over periods of many months, and the circulation of the oceans is closely related to the general circulation of the atmosphere.

We have seen that if the earth were uniform, without any land and water contrasts, the winds would form well-defined belts. Assuming complete cover by water, wind-driven ocean currents would flow around the earth in a similar pattern. But the actual distribution of land and water means that only the ocean encircling Antarctica can circulate freely in a continuous belt. The Atlantic, Pacific, and Indian oceans are bounded by continents or island archipelagos on the east and west, blocking free flow of ocean water around the earth at most latitudes.

As a consequence, the surface currents of the oceans consist of closed circulation loops, or *gyres*, which correspond to the global wind patterns. The Atlantic and Pacific oceans ideally would include three gyres on either side of the equator, as represented in Figure 4.9. The trade winds drive the low-latitude east-to-west currents of the subtropical gyres, and the midlatitude westerly winds drive the higher latitude return flows from west to east. The actual oceanic circulation is shown in Figure 4.10. The strongest currents are on the perimeters of the gyres (near the coasts), with much less movement near the centers of the oceans. Gyres are clearly developed in the Atlantic and Pacific oceans,

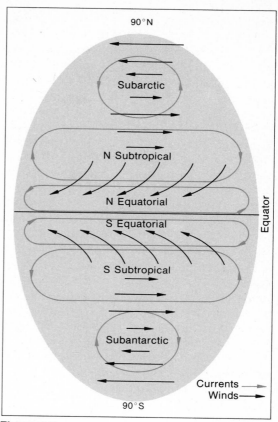

Figure 4.9
The idealized oceanic circulation in an ocean basin consists of loops, or gyres, which are driven by prevailing winds. (Doug Armstrong after P. Weyl, *Oceanography*, 1979, John Wiley & Sons)

but near Antarctica an eastward moving current flows around the entire earth; the idealized southernmost gyre in Figure 4.9 is replaced by the Antarctic continent.

Ocean currents play key roles in redistributing heat around the earth. A current of warm tropical water flows poleward at the western edge of each ocean basin, and cool waters drift toward the tropics along the eastern margins of the seas. Along the east coast of North America, the Gulf Stream is the warm, north-setting current. Its counterpart in the Pacific is the Kuroshio current off Japan. Both are fast, narrow currents, moving several kilometers per hour. The volume of water transported annually by the Gulf Stream alone is more than 30 times greater than the total amount of streamflow from all the continents.

The Gulf Stream merges into the slower-moving North Atlantic Drift near the Grand Banks off Newfoundland. The warm water of the North Atlantic Drift, which flows northeastward to Europe, causes the average winter temperatures of western Europe to be significantly higher than those of eastern North America, even though Europe lies farther poleward. Thus populous Denmark and Sweden are at the latitude of Hudson Bay in subarctic Canada.

Evaporation from such warm currents provides high-latitude maritime air masses with the latent heat and moisture supplies for the cyclonic storms that bring rain to Europe, eastern Asia, and western North America. Air moving over warm currents not only gains moisture, but is heated from below, producing steep lapse rates that cause air mass instability, favoring upward convection and heavy rainfall over the land.

Upwelling

The ocean currents shown in Figure 4.10 are driven by the general circulation. But the ocean currents do not exactly parallel the prevailing winds. The Coriolis effect tends to turn the currents a bit to the right in the northern hemisphere and to the left in the southern hemisphere. The surface water tends to drift off at an angle of 45° to the wind. Along the west coasts of continents, in particular, where winds blow equatorward around the eastern sides of the subtropical highs, the surface water moves away from the coasts. This is replaced by deeper water that rises to the surface along the coasts. This water is cold and produces several important effects.

Coastal *upwelling* of cool water occurs along the west coasts of all continents, but it is pro-

nounced off California, Peru, Morocco, Namibia, and Somalia. The resulting cold water chills the air above it, producing fog. This chilled air is denser than the warmer air above it. Such a temperature inversion prevents the vertical motion of air needed to build rain clouds. Therefore, coasts where upwelling occurs are foggy but receive very little rainfall. All have desert climates, either seasonally or the year around.

Upwelling is very important ecologically because it lifts nutrients up into sunlit surface waters, allowing the tiny marine plants and animals known as plankton to flourish and support a very rich marine food chain. This permits fishing industries to thrive along many desert coasts. The richest fishing grounds are off the Peruvian coast of South America. Here, however, disaster strikes every few years, when the waters become abnormally warm and upwelling seems to fail. Not being adapted to warm water, and lacking the nutrients supplied by upwelling, the plankton die off, and the entire food chain collapses. Fish die in great numbers, along with

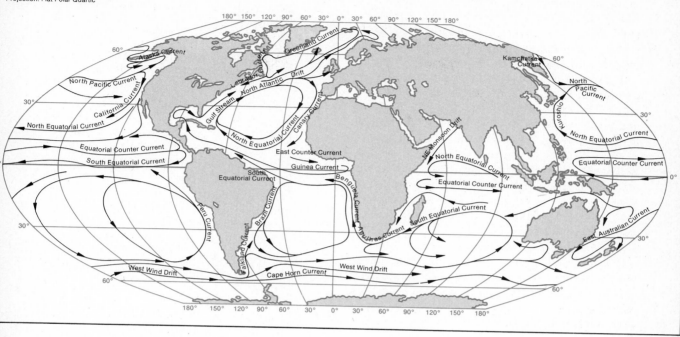

the birds that normally feed on them. Decomposing fish litter the beaches and the waters, giving off so much hydrogen sulfide that the white lead paint used on ships turns black. The phenomenon has been called the "Callao painter," after the port of Callao in Peru. Since this effect usually occurs around Christmas time, the southward drifting warm current is called El Niño, the Christ Child.

The cause of the disaster is periodic fluctuation in the strength of the west-blowing trade winds. The trade winds strengthen every few years, probably in response to sea surface temperature anomalies, and drive water into the western Pacific, actually raising the sea level there. When the trades slacken again, this warm water "sloshes back" eastward across the width of the Pacific, and a portion spreads southward along the Peruvian coast. This overruns the upwelling cold water and produces the disastrous El Niño phenomenon.

SUMMARY

The atmospheric circulation is a response to the energy surplus in low latitudes and the energy deficiency at high latitudes. Differential surface heating produces horizontal differences of atmospheric pressure. Air is pushed toward lower pressure but is deflected by the Coriolis force created by the earth's rotation. In the upper atmosphere the Coriolis force is equal and opposite to the pressure gradient force, so that air flows parallel to pressure isobars. Near the surface, however, the effect of friction causes air to flow obliquely across isobars toward lower pressure.

In the northern hemisphere a counterclockwise flow of surface air spirals into areas of low pressure, and a clockwise flow spirals out of high-pressure areas. These conditions are reversed in the southern hemisphere, producing a mirror-image of northern hemisphere condi-

GENERAL ATMOSPHERIC AND OCEANIC CIRCULATIONS

tions. On a uniform earth there would be several zones of high and low pressure related to the ascent and descent of air on a global scale. The effect of the continents and oceans is to break the system of pressure and winds into a series of cells that change seasonally in size and strength. The circulation of the upper atmosphere is dominated by the effects of the polar and subtropical jet streams found in both the northern and southern hemispheres. Strong undulations in the jet streams cause latitudinal interchanges of energy and affect surface weather.

The oceanic circulations are driven by the circulation of the atmosphere above and consist of gyres bounded by rapidly moving currents, with only slow currents at their centers. Warm currents moving poleward on the west sides of the oceans and cold currents moving from high to low latitudes on the east sides of the oceans help offset net radiation gains in equatorial latitudes and losses in subpolar and polar regions. Small changes in ocean temperatures appear to trigger major weather changes over land areas, accounting for unusual droughts, floods, and severe winters.

The cold currents on the east sides of the oceans are reinforced by the upwelling of cold water from greater ocean depths. This upwelling results from westward Coriolis deflection of equatorward moving currents. Areas of upwelling are unusually rich in marine life due to the nutrients brought up into sunlit surface waters. Massive die-offs of fish and marine birds occur when easterly trade winds weaken, permitting warm equatorial water to invade the area of upwelling.

APPLICATIONS

1. Suppose your residence is in a North American urban area, with a bakery to your west, a chocolate factory to the east, an oil refinery to the north, and a stockyard to the south. Describe the sequence of odors you would sense when a strong low pressure center passes from west to east along a line 100 km north of your location. Would the odors be the same if the low passed to the south of you?

2. By means of radio and TV weather reports and your own observations of cloud movements, keep a record of wind direction each day for the rest of the term. Do the wind conditions help you predict weather changes? How far in advance of weather changes do wind shifts occur?

3. Contrast your own data from the preceding exercise with the prevailing winds shown in Figure 4.6. What significant differences appear? How can these differences be explained?

4. Look at the existing pattern of ocean currents in Figure 4.10. Would it be possible to send a message in a floating bottle to every coastal location in the world? Could this always be done from another continent, or would some messages have to be sent from ships at sea?

5. The continents have not always occupied the positions they do today. About 250 million years ago all continents were united in one large elongated mass that formed a north-south strip extending from pole to pole. Draw a map of the earth's wind systems and ocean currents as they would have existed at that time. See the Appendix for information on map projections.

FURTHER READING

Gedzelman, Stanley David. *The Science and Wonders of the Atmosphere*. New York: John Wiley (1980), 535 pp. Relationships between pressure and wind are developed in Chapters 15 and 16 by drawing upon commonplace examples and simple explanatory equations.

Hare, F. Kenneth. *The Restless Atmosphere*. Rev. ed. London: Hutchinson (1956), 192 pp. This brief introductory classic contains outstanding regional and continental chapters that focus on the dynamics of the general circulation.

Lutgens, Frederick K., and **Edward J. Tarbuck**. *The Atmosphere: An Introduction to Meteorology*. Englewood Cliffs, N.J.: Prentice-Hall (1979), 413 pp. An introductory text with a strong geographical perspective. Chapters 6 and 7 are particularly helpful for pressure, winds, and the general circulation.

Neiburger, Morris, James Edinger, and **William Bonner**. *Understanding Our Atmospheric Environment*. San Francisco: W. H. Freeman (1973), 293 pp. Basic fundamentals are presented in a nonmathematical framework.

Riehl, Herbert. *Introduction to the Atmosphere*. 3rd ed. New York: McGraw-Hill (1978), 410 pp. This nonmathematical text emphasizes the upper air circulation and its relationships to weather and climate.

Young, Louise B. *Earth's Aura*. New York: Avon Books (1979), 305 pp. A highly readable account of the various phenomena of the atmosphere, written for the general public. Full of anecdotes; informative and accurate.

CASE STUDY: The Manila Galleons

The most tenuous lifeline ever to link a colonial outpost and its mother country was the galleon route between the Spanish colony established in the Philippines in 1571 and the empire (New Spain) founded earlier in Mexico by the Spanish conquistadors. The linkage in those days of sailing ships depended upon the prevailing winds and ocean currents, with the crossing from east to west normally made easy by the reliable trade winds, but with a much longer and more perilous west-to-east voyage from the colony back to the motherland. Each voyage was an adventure, not only because of the natural hazards involved but also because the lonely galleons moving eastward were laden with the riches that were the life blood of the colonial venture and a magnet for pirates at both ends of the lifeline.

The first documented east-to-west crossing of the Pacific was made in 1520–21 by the Portuguese adventurer Ferdinand Magellan, then in the employ of Spain. In 1571, after four unsuccessful expeditions to the Philippines, the Spanish finally established a foothold at Manila, a large native settlement and regional trade emporium. The objectives were to obtain gold, silk, and spices for Spain and to win converts for Christianity. The gold proved elusive, but silk, spices, exotic perfumes, Chinese porcelain, and other eastern treasures began to flow into a pipeline consisting each year of one to four lonely galleons on the great sea, following a storm-whipped northerly route back to New Spain and the port of Acapulco. After the Spanish opened the pipeline to luxury-hungry Mexico, Peru, and Europe, merchants from all parts of southern and eastern Asia and the East Indies converged on Manila, which quickly became one of the world's most cosmopolitan cities.

The economy of the Philippine colony depended utterly upon the famous Manila galleons dispatched east, loaded with luxury goods, each year from 1572 to 1811. In this 239-year period, the Spanish colonists in the Philippines were engaged in the export of merchandise brought to Manila from China, Japan, Persia, India, the Spice Islands (Moluccas), Siam, Cambodia, Java, Sumatra, and Borneo. The colonists were entitled to shares in the cargo space of each galleon and sought to fill their space with the most profitable goods available, including the world's finest silks, spices, and porcelain, along with Persian carpets, precious stones, and articles manufactured from jade, ivory, and rare woods. The potential profits were enormous. The most important annual events in the life of the community were the departure of the cargo ships, sent on their way ceremoniously with solemn priestly blessings and the prayers of the people, and the great fiesta that marked the return of the gold- and silver-laden galleons from New Spain, assuring prosperity for another year.

But the galleon trade was fraught with risk. The trip east was reckoned the most fearsome voyage in the world. Some thirty times in 239 years the galleons did not arrive at their destination. Each such event was not only a tragedy, but ushered in a year of economic ruin in the Philippines.

The east-bound vessels were usually overladen since every colonist attempted to put a cargo aboard. Such ships were easily swamped by typhoons in the low latitudes or by cyclonic storms on the east-bound track. Timing was important. The departure date was in June or July during the southwest monsoon, but before the onset of the typhoon season when cyclonic storms were shifted to their highest latitudes. But occasionally typhoons arrived early, and cyclonic storms departed from their usual track.

The departing galleons used the winds of the southwest monsoon to sail to higher latitudes and then counted on the westerlies to carry them across the northern Pacific to California. There they turned south in the California Current, following the coast to Acapulco and sometimes onward across the equator to Lima, Peru. The sea was not the only hazard. Pirates lurked at both ends of the route—Muslims from Borneo, Dutch and Portuguese in Philippine waters, and English, French, and Americans near Mexico. Moreover, disease often erupted on board to rage virtually unchecked. One treasure ship drifted past Acapulco with all of its crew dead. Sometimes ships were driven back to Manila by storms, or were turned back by sickness, loss of masts, or rebellion of the crew.

Accounts of the 7- or 8-month voyages make it clear that they were ordeals, "enough to destroy a man of steel," according to one passenger, who mentioned "hunger, thirst, sickness, cold, continual watching—besides the terrible shocks from side to side caused by the furious beating of the waves." Clouds of flies, gnats that bred in the biscuits, worms that squirmed in the food, and other vermin continually plagued the voyagers.

Nevertheless, the overall success of the Manila galleons was such that they caused an economic crisis in Spain itself. An increasing proportion of the wealth of Mexico and Peru was diverted away from Spain and to Manila in exchange for Oriental luxuries. The merchants of Spain despaired as the market for domestic textiles was captured by the finer Chinese goods. The Spanish kings' edicts restricting the size of cargoes and number of sailings from Manila were regularly violated. Still, Chinese goods arrived in Manila far faster than the annual voyages could transport them to Acapulco. Some fifty large seagoing junks crammed with trade goods were docking yearly in Manila, with only one to four galleons leaving for Mexico.

The eventual decline of the galleon trade resulted from changing social and economic conditions in the Philippines. In Manila the enterprise increasingly became less of a community effort. Speculation in cargo shares became rampant, destabilizing the trading system. Ironically, despite the intensity of economic activity, the Philippine colony produced nothing itself and required an annual subsidy from the treasury of New Spain. Official interest in financial independence and development of the islands' own resources finally began to develop. The last Manila galleon sailed east in 1811, with the final return trip in 1815, ending a remarkable era in which daring men in flimsy vessels for the first time used the atmospheric circulation to link alien civilizations on opposite sides of the world.

A Manila Galleon. (From Zoilo M. Galang, ed., *Encyclopedia of the Philippines*, Manila: Philippine Education Co., Inc., 1936)

Water enters the atmosphere through evaporation, is transported as water vapor, eventually condenses into clouds, and falls again to the earth as rain, snow, sleet, or hail. Condensation and precipitation are the results of a number of complex physical interactions that involve energy and moisture.

Seascape Study with Rain Clouds by John Constable, *c.* 1824–1828. (Royal Academy of Arts, London)

CHAPTER 5
Moisture and Precipitation Processes

During the afternoon of June 9, 1972, clouds gathered over the Black Hills of South Dakota. Over large areas more than 30 cm (12 in.) of rain—equal to a full year's precipitation—poured down in a 6-hour period. Mountain streams became rampaging torrents, overflowing their banks and sweeping away cars and even houses. Automobiles and buildings blocked the spillway of a dam just upstream from Rapid City, causing water levels behind the dam to rise an additional 3 m (10 ft). Late in the evening, the dam burst, and a flood wave swept through the city, causing enormous destruction and taking more than 200 lives. The total damage exceeded $120 million.

This type of disaster was repeated north of Denver, Colorado, on the night of July 31, 1976, when a similar deluge of rain fell in a 1- to 3-hour period, causing floodwaters to rush through the narrow canyons of the Big Thompson and Cache La Poudre rivers in the foothills of the Rocky Mountains. Within a few hours, the raging waters destroyed nearly all traces of human habitation in the canyons and drowned 144 persons. The amount of water discharged through some sections of the Big Thompson

Figure 5.1
Because precipitation occurs only when certain conditions are present, the amount of moisture received by a given region may vary markedly from year to year. The coming of rain, or the lack of it, is a central concern of farmers each year.

(left) People in India rejoice in the coming of the summer monsoon rains, which supply the country with much of its annual precipitation. (Brian Brake/Rapho/Photo Researchers, Inc.)

(opposite) This farm in Oklahoma was abandoned in the 1930s when a succession of drought years made farming impossible. The dry soil was transported by the wind, forming drifts of sand and turning the region into a dust bowl for many years. (The Bettmann Archive)

Canyon was nearly seven times greater than the maximum recorded in the previous 29 years.

How is it possible for so much rainfall to be poured over small areas in so short a time? The water that fell on both locations was evaporated from the Gulf of Mexico, thousands of kilometers away. Moving masses of warm, wet air carried the water vapor northward across the Great Plains. Along the way, some of the molecules of water vapor condensed to form droplets of water only microns in diameter—much too small to fall to the ground as rain. Finally, however, the moisture-laden air was forced to rise over the steep slopes of abruptly rising mountain ranges, where towering thunderclouds formed and water droplets coalesced to form raindrops that fell in a deluge.

The movement of water from the earth's surface to the atmosphere and then back again to the surface constitutes the atmospheric part of the hydrologic cycle that was described in Chapter 2. The transfer of water between the surface and the atmosphere occurs by means of three related processes. *Evaporation* represents the transfer of liquid water at the surface to water vapor in the atmosphere. *Condensation* is the conversion of water vapor to water droplets or ice crystals in the form of fog or clouds, or dew at the surface. *Precipitation* involves the coalescence of water droplets or transformation of ice crystals into raindrops, snowflakes, or hailstones large enough to fall to the ground. During the condensation process the latent heat absorbed during evaporation is released as sensible heat. Thus the hydrologic cycle can also be thought of as part of the energy balance of the earth-atmosphere system.

Under average conditions, nearly half of the earth's surface is covered by clouds. But only a fraction of the clouds produce significant rain-

fall. Precipitation is the result of complex atmospheric interactions, and the conditions that lead to it are so restricted that rain, snowfall, or hail should be considered an unusual rather than a commonplace event. This chapter describes how the atmosphere is continuously supplied with water vapor and how condensation, cloud formation, and precipitation occur.

TRANSFER OF WATER TO THE ATMOSPHERE

When water evaporates, water molecules are detached from the surface of the liquid and enter the air as water vapor, a dry gas. Molecules in liquids are in continuous motion but are kept from separating by attractive forces. The higher the temperature of the liquid, the greater the

energy and motion of its molecules. Eventually, some molecules gain enough energy to break away from the liquid and enter the air. Because the escaping molecules carry kinetic energy with them, evaporation tends to cool the surface from which the molecules are removed. The evaporation of water molecules on your skin is what makes you feel a chill when you step out of a swimming pool, even when the air is warmer than the water. The sun supplies most of the energy needed to vaporize water—590 calories for every gram of water transformed from the liquid to the vapor phase under average conditions.

Evaporation from Water Surfaces

If a jar filled with water is left open in a room, all of the water will eventually evaporate. Heat stored in the room's air provides all the energy required for vaporization of the water. Even if the jar is sealed, some water will evaporate into the air above the liquid. Eventually, however, a condition of equilibrium is reached, and the water level in the jar remains constant. The air in the jar is then said to be *saturated*; it contains as much water vapor as it can hold at its particular temperature.

Water vapor exerts pressure, just like any other gas. The amount of pressure exerted by molecules of water vapor in the air is called the *vapor pressure*. When the air is saturated, the vapor pressure is at its maximum, but this maximum varies with the temperature. In other words, *saturation vapor pressure* depends on the temperature of the air. As the air temperature rises, the saturation vapor pressure increases rapidly. Figure 5.2 shows that at the temperatures found near the earth's surface, the saturation vapor pressure nearly doubles for each 10°C increase in temperature. Hence, warm air has a much greater capacity to hold water vapor than does cold air.

Although the rate of evaporation from water surfaces depends especially on the energy supply, it is also affected by the degree of saturation of the air above the water surface. Under normal atmospheric conditions there is variation in vapor pressure from the water surface through the air above. This is known as the *vapor pressure gradient*—the rate at which vapor pressure varies with distance. How much evaporation takes place depends on how "steep" the gradient is (how rapidly the vapor pressure decreases upward from the water surface).

The steepness of the vapor pressure gradient is determined mainly by the turbulence in the lower atmosphere. When there is no wind, a layer of nearly saturated air forms immediately above water surfaces. The vertical change in vapor pressure within this layer is small, so that the vapor pressure gradient is low, or "weak." Under such conditions the rate of evaporation is low. During windy conditions, however, water vapor is mixed through a much deeper layer of the atmosphere. This permits a steeper vapor

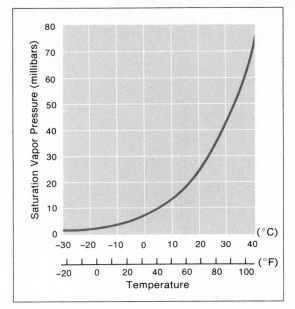

Figure 5.2
The curve shows that the saturation vapor pressure of air increases rapidly with increased temperature. The data extend to temperatures lower than 0°C (32°F), the normal freezing point of water, because small droplets of water can remain liquid at temperatures as low as −40°C (−40°F); this is known as "supercooled" water. (Doug Armstrong after *Smithsonian Meteorological Tables*, edited by Robert J. List, 6th ed., 1971, by permission of The Smithsonian Institution)

pressure gradient, allowing evaporation rates to approach maximum values. To maintain high rates of evaporation, energy for vaporization must be supplied by warm water, warm air immediately above the surface, or absorbed solar radiation. Evaporation, then, depends on both the energy supply and the vapor pressure gradient.

There are few regular measurements of evaporation from the world's oceans or large lakes. The global distribution of evaporation has to be estimated indirectly by means of energy-budget calculations. In general, average annual evaporation is related to latitude and the corresponding solar radiation gains and longwave radiation losses, as shown in Figure 3.9 (p. 59).

Evaporation rates also vary from place to place due to such factors as winds, cloudiness, water temperature, and water vapor content of the atmosphere. The highest annual evaporation rates occur over the subtropical oceans, where the water is warm, the air aloft is dry, and the weather is usually fair. Mean annual evaporation ranges from more than 100 cm (40 in.) over equatorial and sub-tropical oceans to less than 10 cm (4 in.) over polar oceans.

Evapotranspiration from Land Surfaces

Where plants cover the land, only a small part of the water vapor entering the atmosphere comes from direct evaporation of water in the soil. Instead, most is released through small openings in plant leaves called *stomata* (Figure 5.3). After plants absorb soil water through their

Figure 5.3
This vastly enlarged cross section shows the functional parts of a typical plant leaf. The leaves of green plants possess openings known as *stomata* in the bottom layer of protective epidermis. When the leaf is exposed to light, photosynthesis occurs in the chloroplasts, and the stomata are open to allow the entry of carbon dioxide and the exit of oxygen and water vapor. Soil moisture absorbed by the roots is transported to the leaves through the vascular bundles in the veins. The transpiration of water vapor is an essential process in green plants, and most of the water vapor entering the atmosphere in heavily vegetated regions is from plant transpiration rather than from evaporation.

The guard cells around the stomata close the openings if the plant lacks moisture and begins to wilt. In plants adapted to hot, dry climates, the guard cells keep the stomata closed during the hottest part of the day, when water loss by transpiration would be greatest. In some plant species adapted to dry climates, the covering, or cuticle, is particularly thick and waxy to prevent loss of moisture through cell walls. (John Dawson)

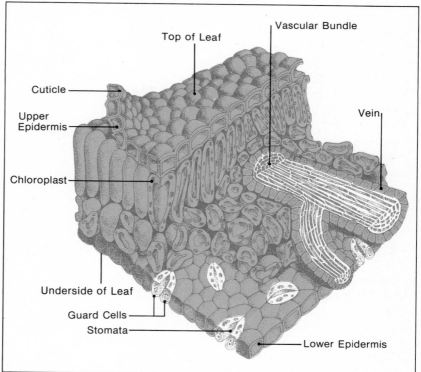

roots, water and dissolved nutrients are transported up the stem of the plant, and excess water is "transpired" through the stomata as vapor. During the day, when a plant is photosynthesizing, the stomata are open to allow the entry of carbon dioxide; at the same time, water vapor escapes by *transpiration*. At night, the stomata are closed and there is little transpiration. The energy requirement for transpiration is the same as that for evaporation: 590 calories per gram of water vaporized at normal temperatures. Hence transpiration is likewise a cooling process.

Water loss from land surfaces is usually regarded as the combination of transpiration from vegetation and evaporation from soil. As we saw in Chapter 3, the two processes together are known as *evapotranspiration*. In areas covered by vegetation, water loss by transpiration is usually two to three times greater than water loss by evaporation directly from the soil. The clearing of forests noticeably increases both stream flow and groundwater recharge, and forest regrowth clearly reduces both surface and subsurface water supplies. The ratio between transpiration and direct evaporation can vary considerably, depending upon degree and type of vegetation cover, making it very difficult to evaluate annual patterns of evapotranspiration from the continents. Estimates of evaporation from the oceans are far less difficult. The effects of evapotranspiration on local water budgets are treated in detail in Chapter 7.

MOISTURE IN THE ATMOSPHERE

There is always some water vapor in the air. At a given temperature, the air may contain any amount of water vapor up to the maximum value at which saturation occurs. Figure 5.4 shows that air at 10°C (50°F) can contain up to 10 grams of water vapor per cubic meter. Ten grams, the weight of two American nickels, may

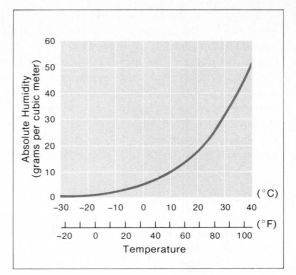

Figure 5.4
The curve shows the maximum amount of water vapor that can be contained in a cubic meter of air at a given temperature. If the air is not saturated, the absolute humidity lies below the curve. The absolute humidity of saturated air rises rapidly with increased temperature, so that warm air is able to contain more water vapor per unit of volume than cool air can contain. The shape of the absolute humidity curve is similar to the saturation vapor pressure curve shown in Figure 5.2, p.102. (Doug Armstrong after *Smithsonian Meteoroligical Tables*, edited by Robert J. List, 6th ed., 1971, by permission of the Smithsonian Institution)

not seem like much water, but rain clouds that contain 10 grams of water vapor per cubic meter have the potential to produce significant rainfall on the ground below. The water vapor content of the atmosphere is referred to as *humidity*. Humidity can be described by several different measures, each one useful for specific purposes.

Measures of Humidity

The most direct measure of the air's moisture content is the *absolute humidity*, which is the weight of water vapor in a given volume of air (Figure 5.4). Absolute humidity is usually expressed in grams of water vapor per cubic meter of air. In the atmosphere, however, warming air expands and cooling air contracts. Even when there is no change in the water vapor content, the absolute humidity changes when the temperature changes. The use of absolute humidity, therefore, tends to be restricted to controlled scientific experiments.

Relative humidity is the ratio between the water vapor in the air and the amount the air could hold if it were saturated. The relative humidity is always expressed as a percentage and does not indicate the specific amount of water vapor present in the air. For example, air at 40°C (104°F) with a relative humidity of 50 percent contains about 25 grams of moisture per cubic meter, but cool air at 20°C (68°F) and 80 percent relative humidity contains only about 14 grams of moisture per cubic meter. The cooler air in this example has a lower water vapor content but a higher relative humidity than the hot air because the cooler air is closer to saturation. Figure 5.4 shows that air can become saturated merely by cooling without any change in water vapor content. As we shall see, cooling of air to its saturation point is the most important factor in moisture condensation and precipitation.

As a measure of atmospheric moisture, relative humidity has the same disadvantage as absolute humidity: whenever the temperature changes, the humidity measure also changes, even if there is no change in water vapor content. Figure 5.5 shows that temperature changes in a 24-hour period usually cause relative humidity to be highest at night and in the early morning and lowest in mid-afternoon.

Several other measures are used to describe the water vapor content of air. The *specific humidity* is the weight of water vapor per kilogram of air. The *mixing ratio* is the weight of the water vapor present per kilogram of dry air (air minus its water vapor). Both of these measures are useful because they change only when the amount of water vapor itself changes. They are used especially in studies of air masses, which are discussed in Chapter 6.

A final and increasingly important method of expressing the moisture content of air is by means of the *dew point* temperature—the temperature at which saturation occurs and moisture condensation begins. The higher the dew point temperature, the greater the moisture content of the air. The dew point is discussed more fully in the section on condensation processes.

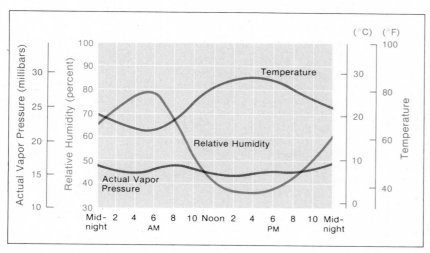

Figure 5.5
During fair weather, the vapor pressure of the air near the ground remains almost constant through the day. However, the relative humidity changes markedly through the day as the air temperature changes, because the saturation vapor pressure of air is strongly dependent upon temperature. (Doug Armstrong adapted from *Hydrology Handbook*, 1949, published by the American Society of Civil Engineers)

Distribution of Water Vapor in the Atmosphere

Only a tiny fraction of the earth's water supply is stored in the atmosphere at any one time. If all the atmospheric water vapor fell to the earth as rain, it would produce a layer of water only about 2.5 cm (1 in.) deep. Heavy rains are possible only because there is a constant recycling of liquid water and water vapor within the hydrologic cycle. In the atmosphere, water that is lost by precipitation at one place is at the same time being replaced by evapotranspiration at another place.

The transfer of moisture to the atmosphere by evapotranspiration proceeds most rapidly when the ground is warm and moist on a fair summer day in which there is a large net gain of radiation. Convection carries water vapor aloft quickly, so that a steep vapor pressure gradient (rapid pressure change with height) is maintained near the ground (see Figure 5.6). At night, however, the net radiation is negative and there is little or no energy available for

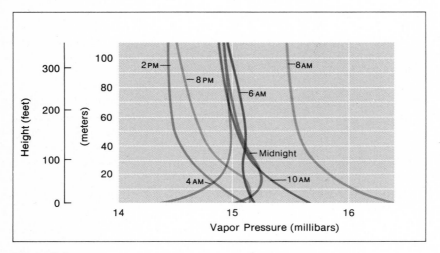

Figure 5.6
Each curve on this graph shows how the vapor pressure in the atmosphere near the ground varies with height throughout a clear summer day. At 8:00 A.M. the vapor pressure decreases rapidly with increased height, indicating a flow of vapor from the surface into the atmosphere. The vapor pressure decreases from 8:00 A.M. to 2:00 P.M. because the air becomes heated during the morning and early afternoon, causing the onset of convection and consequently a more efficient removal of moist air from near the surface. Much of the water vapor remains in the lowest 50 meters of the atmosphere, forming a humid blanket. Water vapor continues to flow upward from the surface during the day while plants are transpiring. By 8:00 P.M. the vapor pressure is nearly constant through the lowest 15 meters of height, indicating that little water vapor is leaving the surface. Later at night, the vapor pressure immediately above the surface increases with increased height, indicating a flow of water vapor from the atmosphere toward the ground. By 6:00 A.M., after sunrise, the flow is once again from the surface to the atmosphere. (Doug Armstrong after R. Geiger, *The Climate Near the Ground*, © 1965, Harvard University Press)

evapotranspiration. At the same time, the air immediately above the surface usually is cooled to its dew point. Then water vapor condenses on the ground, roof tops, and automobiles as dew or frost.

The water vapor content of the atmosphere varies horizontally as well as vertically. Average atmospheric storage of water over North America ranges from about 5 cm (2 in.) near the Gulf of Mexico in the summer to less than 0.8 cm (0.3 in.) over central Canada during the winter. This pattern of water vapor distribution varies constantly, however. Daily rainfalls of as much as the 60 cm (24 in.) that fell near Houston, Texas, in tropical storm Claudette during July 1979 indicate that atmospheric circulations are capable of concentrating immense quantities of water vapor in a very short period of time.

THE CONDENSATION PROCESS

Condensation is the process by which water vapor in the atmosphere changes phase to tiny liquid droplets or ice crystals. At the earth's surface, condensation on cool or cold objects produces dew or frost. When water vapor condenses in the atmosphere, the result is a mass of water droplets or ice crystals known as a cloud. Fog is simply a cloud at the earth's surface.

For condensation to occur, the water vapor content of at least one layer of the atmosphere must reach saturation. Saturation can be produced by adding water vapor to the atmosphere, by cooling the atmosphere, or by a combination of added moisture and cooling. Most cloud formation and precipitation result from atmospheric cooling.

Condensation Nuclei

When moist air is cooled to saturation, condensation does not occur automatically. Fine particles, called *condensation nuclei*, must be

present to act as centers for condensation and for the growth of water droplets. In moist air that is completely free of particles of any size, condensation takes place only when the vapor pressure is three or four times the saturation value. Air containing more than its normal saturation amount of water vapor is said to be *supersaturated*. However, since air is usually well supplied with microscopic particles, condensation generally begins almost as soon as the air is cooled to the saturation point.

Figure 5.7 shows that condensation nuclei are

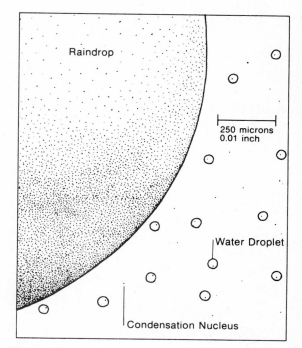

Figure 5.7
This drawing shows the relative sizes of condensation nuclei, water droplets, and a raindrop. Most condensation nuclei are less than 1 micron in radius; water droplets are usually less than 20 microns in radius; and raindrops have radii of 1,000 microns or greater. Each water droplet forms in saturated air by condensation of water vapor on a condensation nucleus, and other processes cause millions of water droplets to coalesce and form a raindrop. (John Dawson)

Figure 5.8
Dew forms on plants and other natural and man-made surfaces when the surrounding air cools to the dew point. (David Cavagnaro)

extemely small. Few exceed a radius of 1 micron (0.001 mm), and most have radii smaller than 0.1 micron (0.0001 mm). Giant condensation nuclei, with radii greater than 1 micron, number only about 1 per cubic cm. Principal sources of condensation nuclei are industrial pollutants and smoke, forest fires, salt crystals from sea spray, pollen, and dust particles. Natural processes and industrial and agricultural activities renew the supply of atmospheric particles as fast as they fall to the surface or are washed out by rain. Hence nuclei for condensation are almost always available when the air becomes saturated.

The water droplets that form clouds usually grow to a radius of 10 or 20 microns before the addition of water molecules ceases. In a typical cloud there may be a million water droplets per cubic meter of volume, and the average water content of the cloud may be about 1 gram per cubic meter of air.

A cloud droplet with a radius of only 10 mi-

crons has a settling velocity (rate of fall) of less than 1 centimeter per second, and it would normally evaporate in the drier air below the cloud. Under average conditions, droplets do not grow larger than 10 or 20 microns in radius, nor can they easily combine in larger drops. Studies have shown that when such small droplets approach one another in a cloud, they are forced apart by the air between them. Thus most clouds do not contain water drops large enough to fall as rain. We shall see later in the chapter what processes are necessary to produce precipitation.

Dew and Frost

When a volume of air cools with no change in water vapor content, its relative humidity increases. The temperature at which the relative humidity becomes 100 percent is the *dew point temperature*. When the air cools to the dew point, water vapor begins to condense into small drops of liquid water. In humid summer weather the outside of a glass of ice water becomes covered with beads of moisture because the moist air near the glass has been chilled to the dew point temperature. The same process accounts for the formation of dew itself. Longwave radiation cools plants and the soil after sunset, and the air in contact with them is also cooled by conduction. If the air cools to the dew point, its water vapor condenses on the ground and exposed plant surfaces, forming dew (see Figure 5.8). In urban areas, dew formation is perhaps more apparent on automobiles parked outside overnight in the spring and fall.

When the dew point is at or below the freezing point of water, it becomes the *frost point*, with water vapor condensing as ice crystals, and frost appearing on exposed surfaces. Because condensation liberates latent heat, dew and frost formation keeps plant and soil surfaces from cooling as much as they would without this release of heat.

Fog

Fog occurs when a thick layer of moist air near ground level is cooled to its dew point. This can occur in several ways, so that different types of fog are recognized.

Radiation fog occurs at night when the ground and the air immediately above lose heat by longwave radiation. If the ground and air are moist, the lowest 10 meters or more of the moist air may be cooled to the dew point, resulting in condensation in the form of dense fog. Moist air and damp ground are particularly susceptible to radiation fog on clear nights, which promote longwave radiation losses to space. But even moderate breezes will prevent the formation of radiation fog by mixing the colder air near the ground with warmer air above.

Advection refers to the horizontal movement of air across the earth's surface. When warm, moist air passes over snow or cold ocean water, the air may be cooled to its dew point, resulting in *advection fog*. Warm air moving over cold ocean water off the coasts of California and New England frequently produces advection fog. San Francisco is famed for its summer fog, which forms over cold coastal waters and is drawn inland in the afternoon by thermal convection over the land (Figure 5.9). As Figure 5.10 (see p. 110) shows, fog is most common along coastlines.

When advection fog occurs over snow, there may be a marked increase in snowmelt. Normally, snow melts slowly because it has a high albedo and reflects much incoming solar radi-

Figure 5.9
This photograph shows San Francisco Bay and Oakland, California, shrouded in fog, with Mount Diablo in the distance to the northeast. Low-lying advection fog frequently occurs in the coastal region around San Francisco as moist air moving eastward over the Pacific Ocean cools to its dew point. (David Cavagnaro)

MOISTURE AND PRECIPITATION PROCESSES

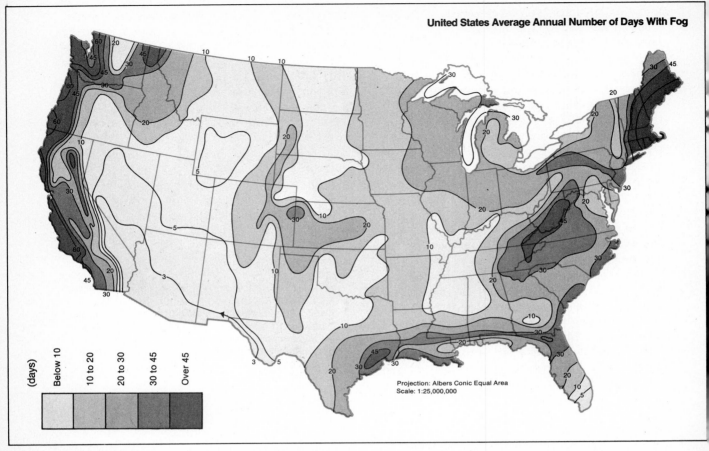

Projection: Albers Conic Equal Area
Scale: 1:25,000,000

(days)

Below 10 | 10 to 20 | 20 to 30 | 30 to 45 | Over 45

Figure 5.10
This map shows the average annual number of days of fog in the conterminous United States. Fog occurs most frequently along the Pacific Coast and along the coast of New England, where moist air is cooled by cold ocean waters. Fog also occurs frequently in the Appalachian Mountains of the east central United States. It seldom occurs in the warm, dry air of the western deserts. (Andy Lucas and Laurie Curran adapted from Arnold Court and Richard Gerston, *Geographical Review, 56*, © 1966, American Geographical Society of New York)

ation back into space. However, the advection of warm, moist air over snow-covered areas causes condensation on the snow surface. Every gram of water vapor that condenses releases enough latent heat to melt more than 7 grams of snow. Thus when warm, moist air from the Gulf of Mexico moves northward over snow-covered landscapes in the American Midwest and Northeast, rapid melting of snow can occur even on a cool, overcast day. Quite reasonably, skiers hate fog!

Orographic fog is associated with mountain areas, such as the Appalachians. When moist air is forced up mountain slopes, it can be cooled to the dew point, resulting in the formation of fog and clouds. Orographic cooling is a major cause of condensation and cloud formation, and will be discussed in more detail later in this chapter.

Fog of any type can be very hazardous to human activities. Throughout history the vast majority of ship collisions have occurred in fog. Many other accidents on the water, including frequent collisions between boats and bridge piers, are attributable to fog. On one day in September 1923, seven destroyers of the U.S. fleet were wrecked by running aground in dense fog, one on the heels of another, near Point Conception on the southern California coast. Such accidents gain even more importance as an increasing number of the ships at sea are tankers laden with enormous quantities of oil that can produce disastrous pollution of coastal waters.

On the land massive chain accidents involving dozens to more than a hundred vehicles have occurred on fog-shrouded highways in California and New Jersey. Airports are usually located to avoid fog, but many continue to be affected by this hazard, causing frequent and inconvenient diversions of flights to alternate points. Airports serving Boston, New York, Los Angeles, San Francisco, Seattle, London, Zurich, and many other important centers have frequent fog problems.

Cooling of surface air by radiation or advection is restricted to a shallow layer of the atmosphere near the surface. The resulting fog and low clouds are simply not thick enough for the processes producing rainfall, snow, or hail to be effective. The air may be filled with a mist of small droplets, sometimes falling as light drizzle, but significant precipitation can occur only when a much thicker mass of air is cooled throughout its depth by being lifted to higher elevations.

Clouds

Clouds are condensation forms that develop above the ground, usually by the lifting and chilling of moist air. Although cloud types seem almost infinite in number, they can be categorized fairly easily. Clouds occur in three general types and are recognized as falling into three elevation classes (see Figure 5.11, pp. 112–113). At the highest altitudes are *cirrus* clouds composed of ice crystals that form thin filaments,

wispy puffs, or translucent veils. A second general cloud type is the horizontal *stratus* cloud, which forms an extensive continuous sheet. The third general form is the *cumulus* type, which has vertical rather than horizontal development, appearing puffed up, billowy, and sometimes forming awesome white towers in the sky. Cumulus clouds indicate rising currents of air, whereas stratus types develop in air that lacks strong vertical currents. These simple cloud types can be combined: thus *cirrocumulus* clouds are high puffs; *cirrostratus* clouds are high thin veils; *stratocumulus* clouds are layers of coalescing puffy clouds.

Cloud types are also designated by altitude: as high clouds (cirrus types); middle-altitude clouds, including *altostratus* and *altocumulus* types; and low clouds (stratus, stratocumulus). The term *nimbo*, or *nimbus*, means that the clouds are capable of producing abundant precipitation. Low-level cloud blankets that produce heavy rains are *nimbostratus* clouds. Towering cumulus clouds that produce torrential downpours, often including hail, are *cumulonimbus* clouds. The cumulonimbus type may reach to an elevation of 12,000 to 15,000 m (40,000 to 50,000 ft) and frequently flares out at the top in the shape of a blacksmith's anvil. Meteorologists have more sophisticated names to distinguish the great variety of clouds that exist, but the names used here are the ones most widely recognized.

THE PRECIPITATION PROCESS

A water drop must have a radius of at least 100 microns in order to fall from a low cloud. Thousands of 10-micron droplets must combine to produce one 100-micron drop, and millions of 10-micron droplets are needed to form an average raindrop with a radius of 0.1 cm (1,000 microns). The processes that cause water droplets to coalesce to drops large enough to fall from clouds have been studied in detail since the 1940s. Scientists have described two separate

Figure 5.11

(a) Fair weather cumulus clouds over Baton Rouge, Louisiana. Each cloud forms in a rising thermal with flat and even bases at the altitude where the rising thermals are cooled to the dew point and saturation. These clouds tend to be short-lived, and they usually evaporate into the surrounding drier air within an hour or so. (R. A. Muller)

(b) An altocumulus, or "mackerel sky," cloud over Boulder, Colorado, takes the form of a layer of patchy cloud puffs. It forms at moderate altitudes and often heralds the approach of a warm front and accompanying precipitation. (National Center for Atmospheric Research, N.C.A.R., Boulder, Colorado)

(c) Cirrus clouds are feathery wisps of ice crystals that form at high altitudes. They are frequently the first indication in the sky of the approach of a midlatitude cyclone system from the west or south. (R. A. Muller)

(d) Cumulonimbus clouds, or "thunderheads," rise to great heights and are sometimes associated with heavy rain, flash floods, hail, and tornadoes. This view shows massive thunderstorms organized in an orographic situation with warm, moist air being forced up and over the Rocky Mountains at Boulder, Colorado. (R. A. Muller)

(e) The leading edge of a squall line passing over Baton Rouge. The squall line was just ahead of a vigorous cold front, and the dark clouds are indicative of severe turbulence and violent winds. (R. A. Muller)

(f) Shallow radiation fog near sunrise in autumn in southeastern Wisconsin. (R. A. Muller)

(g) "Man-made" cumulus over petrochemical plants in the Los Angeles basin. These small clouds are associated with heat and water vapor added to the lower atmosphere as by-products of industrial processes. (R. A. Muller)

processes that are believed to be responsible for precipitation. These are known as the *coalescence model* and the *Bergeron ice crystal process*.

Formation of Rain and Snow

The generation of rain by the coalescence process depends on the occurrence of oversized water droplets that are larger than 20 or 30 microns in radius. A larger droplet falls just a bit faster than the small droplets, and it grows by colliding with and sweeping up smaller droplets in its path. Rising currents of air carry the drops upward faster than they can fall out of the cloud, allowing them more time to grow in size. A droplet requires about half an hour to grow to raindrop size by coalescence, and the rain clouds must be at least 1 km (0.6 mile) thick for the growing drops to remain in the cloud long enough to become raindrops. Thinner clouds limit the growth of drops by coalescence but may produce *drizzle*, a form of precipitation that consists of very tiny drops that "float" rather than fall to the surface. Wet pavements caused by drizzle can be very hazardous for motorists, but drizzle never produces significant quantities of precipitation.

Sea salt particles 1 micron or more in radius make particularly effective condensation nuclei for oversize droplets. That is because of the *hygroscopic* nature of salt—its natural tendency to absorb water vapor. Even far inland, salt particles blown off the ocean play an important role. Nor is it necessary for the salt particles, or other giant condensation nuclei, to be present in large quantities. Only about one giant condensation nucleus is needed for each cubic meter of cloud.

Figure 5.12 (below and opposite)
The principal models that successfully explain precipitation from clouds are the coalescence model and the Bergeron ice crystal model; the main steps of each process are outlined in this figure.

The coalescence process of precipitation is dominant in the tropics, where many clouds are too warm to contain supercooled water droplets and ice crystals. The Bergeron process dominates at higher latitudes, where the upper portions of many clouds are considerably colder than the normal freezing point of water, even in summer.

The type of precipitation that falls from a cloud depends on the mechanism involved and on the variations of temperature in the atmosphere. Precipitation is formed as snow in the Bergeron process, and if the atmosphere beneath the cloud is below freezing, the precipitation reaches the ground as snow. If the snow falls into warm air, however, the precipitation arrives at the ground in the form of rain. The coalescence process cannot produce snow, but either process can produce sleet or ice pellets if raindrops freeze while falling through a layer of cold air near the ground. (John Dawson)

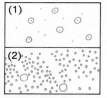

Coalescence Model

(1) Condensation of vapor onto nuclei and growth of droplets from vapor

(2) Growth of raindrops by coalescence with droplets

Bergeron Ice Crystal Model

(1) Condensation of cool vapor onto nuclei and growth of supercooled droplets

(2) Freezing of supercooled droplets onto ice-forming nuclei

(3) Growth of ice crystals at the expense of water droplets

(4) Further growth of ice crystals by coalescence with ice crystals and supercooled droplets

The coalescence process is most effective in thick vertical clouds, which are common in the tropics, but less so in higher latitudes, where some process other than coalescence is needed to produce significant quantities of precipitation. In the middle latitudes the tops of many rain clouds are in altitudes where the temperature is well below freezing. In 1933 Tor Bergeron, a Swedish meteorologist, proposed a process of droplet growth to explain rainfall from cold clouds that were not always thick.

Although puddles of water on the ground freeze at 0°C (32°F), tiny water droplets in the atmosphere can remain liquid at temperatures

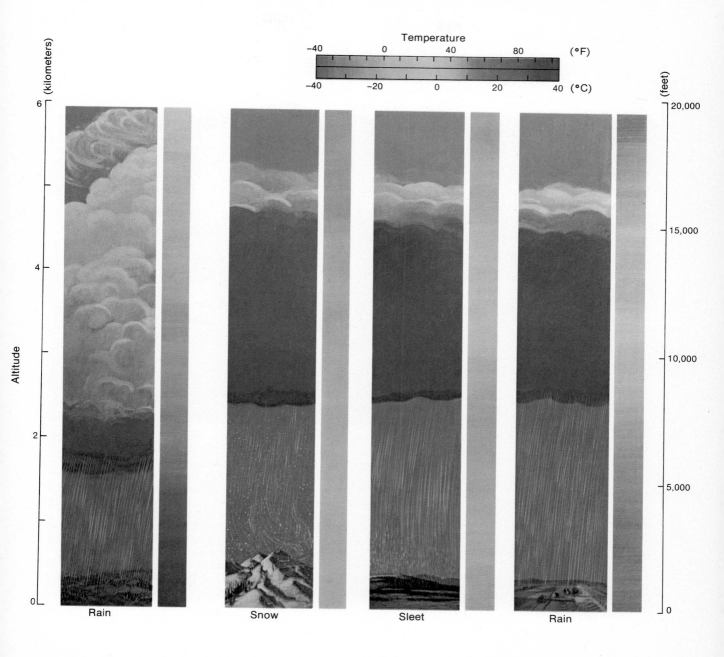

Rain   Snow   Sleet   Rain

down to $-40°C$ ($-40°F$). Because of this *super-cooling*, cloud droplets often remain liquid even when the upper parts of a cloud are far below $0°C$. Particles in the air, known as *ice-forming nuclei*, promote the freezing of supercooled droplets into ice crystals, usually after the tem-

perature has dropped below about $-15°C$ ($5°F$).

Once ice crystals have formed, a vapor pressure gradient is established between the ice crystals and nearby supercooled water droplets. The ice crystals grow at the same time that the droplets are losing water by evaporation. The

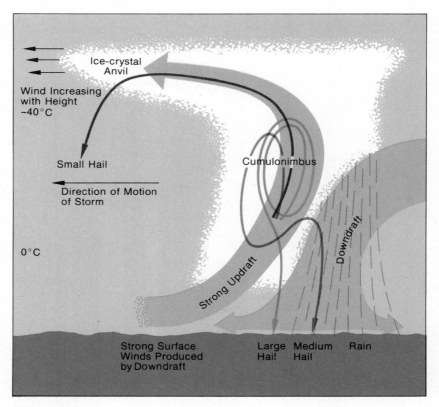

Figure 5.13
This cross section of an intense thunderstorm illustrates the cyclic processes involved in the formation of hailstones. Powerful currents of air in the central updraft carry falling raindrops and ice crystals upward again. Because of the forward motion of the storm, ice particles may be swept up in the updraft a number of times. On each passage through the cloud, the particle gains a new layer of ice, thus becoming a hailstone. Clouds with strong updrafts are able to support hailstones through many cycles until they become too heavy for the air currents to support. Particles are also blown away from the central column by strong winds aloft, giving the cloud its characteristic anvil shape. (Doug Armstrong after Hermann Flohn, *Climate and Weather,* © 1969 by H. Flohn, used by permission of McGraw-Hill Book Company)

ice crystals also grow by collision with other supercooled water droplets. Most summer rain in middle latitudes therefore begins as ice and snow in the upper parts of towering clouds. In winter, the ice crystals fall as snow if the air is not warm enough to melt the snowflakes. The precipitation processes are illustrated in Figure 5.12.

Formation of Hail

Hailstones are spherical balls of ice that fall from towering summer thunderheads (cumulonimbus clouds) that have strong updrafts. They range from tiny grains to spheres the size of a grapefruit. When cut open, hailstones show an onionlike layering of clear and opaque shells of ice. This layering suggests that hailstones grow by being cycled through a cloud many times (Figure 5.13).

It is generally believed that strong updrafts in a thundercloud carry small ice particles upward through the cloud's central column. If the particles rise through a section containing many supercooled water droplets, a wet layer will form around each ice particle. This water gradually freezes into a shell of clear ice. If the particles then rise into a region having only a few small supercooled droplets, the collisions result in instant freezing, producing a layer of ice that is cloudy due to pockets of trapped air. Near the top of the thunderhead, the ice particles emerge from the central updraft and fall through the outer areas of the cloud. As they near the base of the advancing cloud, however, they may be swept back into the central updraft.

With each cycle through the cloud, the particles gain new layers of ice. After a number of cycles, they grow too large to be supported by the updrafts, and they fall to the ground as hail. Large hail is associated with violent thunderstorms in cumulonimbus clouds containing updrafts with speeds of 20 to 30 meters per second (45 to 60 miles per hour). Hail is most common east of the Rocky Mountains on the high plains from Alberta to Kansas, where it can destroy late spring or early summer wheat crops in a few moments.

ATMOSPHERIC COOLING AND PRECIPITATION

To generate significant rainfall by either the coalescence or ice crystal model, large masses of air must be cooled throughout their depth. This can be accomplished only by lifting the air to higher elevations. Vertical displacement of large masses of air, which can produce heavy rainfall, snow, or hail from clouds, can be brought about in several ways. The remainder of this chapter discusses the phenomena that trigger precipitation from clouds.

Adiabatic Cooling

Net radiation gains at the earth's surface may either heat the surface or cause evapotranspiration. Largely because albedo and evapotranspiration vary from place to place, some locations heat up much more than others. Figure 5.14 (see p. 118) shows how air above local "hot spots" becomes organized into rising bubbles of warm air. Soaring birds and glider pilots use these rising currents, called *thermals*, to gain or maintain altitude. Circling within the updraft, they are carried upward until the air cools by expansion and loses its buoyant tendency.

The temperature decrease in rising air results from a process known as *adiabatic cooling*. The term "adiabatic" refers to the fact that the process occurs without energy gains or losses by the air. As a parcel of air rises to higher levels, the surrounding air is increasingly less dense. The air in the rising parcel expands in response to the decrease in the density and pressure of the atmosphere. Because the molecules of the expanding air become more widely separated, they collide less often, so that the sensible temperature of the rising air decreases.

The rate of adiabatic cooling of rising air in

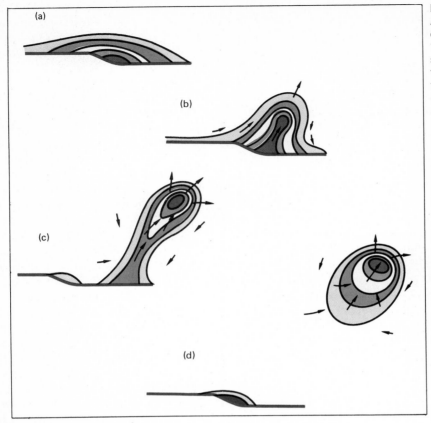

Figure 5.14
A thermal is a rising bubble of air emanating from a local "hot spot" on the earth's surface where the temperature exceeds that of the surrounding area. This differential heating is associated with variations in albedo, orientation of slopes to the sun, and the availability of water for evapotranspiration. The rising thermal initially exchanges little heat with the surrounding air. Each concentric circle, or isotherm, represents an increase of 0.1°C; thus the outer circle is 0.1°C warmer than the general atmosphere, the next circle is 0.2°C warmer, and so on. The highest temperature occurs at the center of the thermal. (Vantage Art, Inc. after Herbert Riehl, *Introduction to the Atmosphere*, © 1972 by McGraw-Hill Book Company)

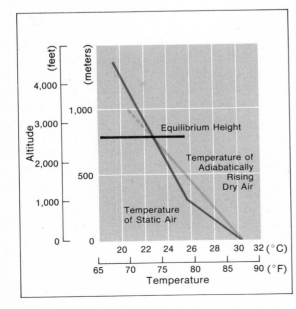

Figure 5.15
When a parcel of dry air rises, it expands in volume in order to equal the pressure of the surrounding atmosphere. The expansion causes the temperature of the parcel to fall. The figure compares the temperature of a typical parcel of rising dry air with the temperature of the air in the surrounding static atmosphere. The temperature of the parcel and the temperature of the static air both decrease with increased altitude, but in general the temperature of the rising parcel falls more rapidly, as indicated in the figure. The parcel experiences a net upward force, causing it to continue rising only as long as the temperature of the parcel exceeds the temperature of the static air. The equilibrium height of the parcel is near the height where the temperatures are equal. There is no upward convection past this level, but lifting and cooling at the adiabatic rate can continue along frontal surfaces and over mountain barriers. (Doug Armstrong)

which condensation is not occurring is known as the *dry adiabatic rate*. Its value is 10°C for every kilometer of increasing altitude (5.5°F per 1,000 ft), as illustrated in Figure 5.15.

The process can also be reversed. When a parcel of cool air descends, it becomes compressed. As a result, the molecules in the parcel collide more frequently and the temperature increases. This is called *adiabatic heating*. A descending parcel of cool air, subject to adiabatic heating, increases in temperature at the dry adiabatic rate.

Rising air cools at the dry adiabatic rate only until it reaches its dew point temperature and becomes saturated. After saturation, it may continue to rise, but it cools at a reduced rate called the *moist adiabatic rate*. The rate of cooling decreases at saturation because the water vapor in the parcel begins to condense, releasing latent heat. The added heat increases the buoyancy of the parcel and enables it to rise to greater altitudes. This mechanism permits towering clouds to develop with potential for high-intensity precipitation.

Not all the water vapor in the air parcel condenses at once. Condensation continues as the parcel rises, and at any given altitude the parcel contains just enough water vapor to maintain saturation. When the rising air cools to the same temperature as the surrounding atmosphere, its upward movement by convection ceases. The moist adiabatic rate varies with the moisture content of the rising air. In warm, moist air it may be 5°C per km (2.8°F per 1,000 ft), but at very low temperatures it approaches the dry adiabatic rate of 10°C per km.

Thermal Convection

On a fair day, puffy clouds form within thermals at the altitude where saturation is reached and condensation begins. Figure 5.16a (see p. 120) shows that the thermal continues to rise until the temperature of the rising air equals that of the surrounding atmosphere. This *equilibrium point* marks the greatest height to which the cloud can grow. The vertical movement of thermals responsible for fair-weather cumulus cloud formation is known as *thermal convection*.

Since the atmosphere is heated principally from below by longwave radiation from the earth's surface, air temperature normally decreases upward. A global average rate of temperature decrease for all weather conditions is about 6.5°C per km (3.6°F per 1,000 ft), which is known as the *average lapse rate*. Locally, however, the rate of change of temperature with altitude may deviate considerably from the average. The actual vertical temperature change measured at any location, known as the *environmental lapse rate*, has important effects on thermal convection and cloud formation.

The environmental lapse rate in Figure 5.16b shows warmer air aloft over cooler air at the surface. This *temperature inversion* is a common result of nighttime radiational cooling during fair weather. If dry air near the surface were forced to rise by flowing against the slopes of a mountain range, it would cool at the dry adiabatic rate and would soon become cooler and denser than the air above the inversion. Being cooler than the surrounding air, the lifted air would tend to sink downward rather than rise through the inversion. Therefore a temperature inversion restricts upward motion in the atmosphere. Meteorologists refer to environmental lapse rates that discourage vertical movement of air as *stable lapse rates*. Temperature inversions are an extreme type of atmospheric stability, in which there is no opportunity for thermal convection or the development of thick clouds and precipitation. By preventing the upward movement of air an inversion also acts as a lid that traps pollutants emitted by automobiles and industries and causes them to become concentrated in a layer of "smog."

On clear days the ground is warmed by solar radiation, and the lower atmosphere is heated by conduction and thermal convection. This usually eliminates any overnight temperature inversion at the surface. Air can rise, and ther-

Figure 5.16
(a) When the land surface heats, it warms the lower atmosphere, causing the environmental lapse rate (red line) to bend to the right (higher temperature) at the surface. A rising thermal or parcel cools at the dry adiabatic rate (yellow line) until condensation begins. The base of the cloud at about 0.7 km marks the altitude at which the thermal cools to the dew point, initiating condensation. The parcel then cools at the moist adiabatic rate, which is less than the dry rate. The release of the latent heat of condensation enables the parcel to rise to much higher altitudes before it reaches equilibrium with the surrounding static air. Condensation continues as the parcel rises, and at any given altitude the parcel contains just enough water vapor to keep the air in it saturated.

(b) This shallow inversion, with warmer air above cooler air at the surface, is characteristic of radiational cooling at night. If a parcel were forced to rise, it would come quickly into equilibrium with the surrounding static air. Inversions are associated with atmospheric stability.

(c) With daytime heating, the inversion is much higher aloft, permitting thermals to reach much higher altitudes. The equilibrium level is the limit of thermal convection. (Doug Armstrong)

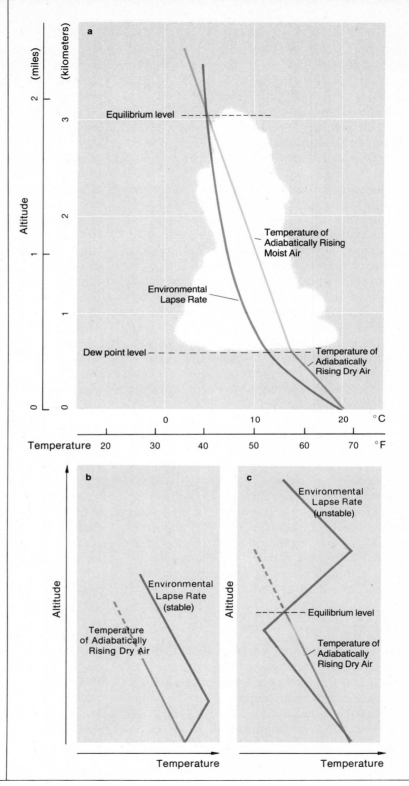

mals become active (Figure 5.16a). If the environmental lapse rate is steep (temperatures decreasing rapidly with height), rising thermals continue to be warmer and less dense than the surrounding air and can continue moving upward buoyantly. The rise of this air to great altitudes permits the development of towering cumulonimbus clouds that are associated with thunderstorms (see Figures 5.16a and 5.11d). Meteorologists refer to the steep environmental lapse rates leading to the development of thermal convection as *unstable lapse rates.*

During summer, when rapid heating of the ground creates steep lapse rates and conditions of atmospheric instability, rain-producing cumulus-type clouds follow a regular pattern of development. Early in the day, the sky is usually clear. After a few hours scattered wisps of cloud begin to appear at an altitude of about 1 kilometer (3000 ft). These are formed by condensation in convective updrafts. The wisps increase in size and eventually become large cumulus clouds with bases at the altitude at which the rising air has cooled to its dew point. Cumulus clouds indicate the tops of rising thermals. Between the clouds air sinks to replace the air that rises in the thermals. For this reason, cumulus clouds never cover the sky completely, and precipitation from them takes the form of scattered showers and thunderstorms that typically break out in the afternoon and persist into early evening. Although an individual shower may produce more than 2.5 cm (1 in.) of rain locally, thermal convection rarely results in widespread rains.

Orographic Lifting

When moisture-laden air is forced to rise over a mountain barrier, the air cools, and condensation and precipitation often result. This process, which may exert a strong local effect on precipitation, is known as *orographic lifting.*

To the frequent disappointment of vacationers, clouds, fog, and precipitation are characteristic of most mountainous regions. In western

North America, the air normally flows from west to east, so the western, or windward, slopes of mountain ranges are often cloudy and wet. When the air descends the eastern slopes, however, the sinking air warms at the dry adiabatic rate. In the warmer descending air, water droplets evaporate and the clouds dissipate. The eastern slopes and adjacent lowlands are usually sunny and dry (Figure 5.17) and are said to lie in the "rainshadow" of the mountains. Other mountain areas are subject to different patterns of orographic lifting. In the eastern United States, moist air may advance against the Appalachian mountain system from either the east or the west, so that the mountains have no dry side.

A spectacular example of the effects of orographic lifting is found on the island of Kauai in the Hawaiian chain. The average annual rainfall on the windward side of Mount Waialeale is 1,170 cm (460 in.), but it is only 51 cm (20 in.) on the lee side, just 25 km (15 miles) away. This extraordinary rainfall record is produced by orographic clouds maintained by persistent moisture-laden winds from the northeast. It was orographic lifting of unusually moist unstable air, producing enormous cumulonimbus clouds, that caused the Rapid City and Big Thompson Canyon disasters noted at the beginning of the chapter. The truly unusual aspect of these storms was that the rain-producing clouds remained over the same areas for several hours, rather than drifting onward and spreading their effects, as normally happens.

Frontal Lifting and Convergence

The lifting of large masses of moist air, causing widespread clouds and precipitation, does not require mountains. In the middle and higher latitudes, the atmospheric circulation often brings together large masses of warm and cold air. Largely because of density differences, the warm and cold air masses do not mix but remain separated by a boundary zone known as

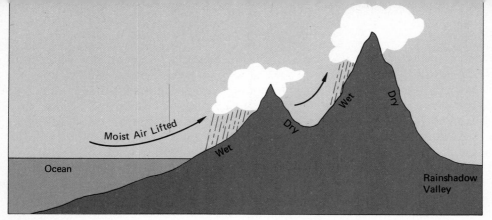

Figure 5.17
Mountains exert a strong local influence on precipitation because they force
moisture-laden air to altitudes where condensation can occur. The mountain ranges
along the Pacific coast of the United States intercept moisture transported from the
Pacific Ocean by prevailing westerly winds, so that coastal ranges are moist and
inland regions are dry, as shown in Figure 5.19b. (John Dawson)

Figure 5.18
(a) When a large mass of warm air moves into a region occupied by
colder air, the less dense warm air flows up and over the surface cold
air. This atmospheric cross section shows that the warm front slopes
gently upward through the lower atmosphere. Characteristic clouds and
precipitation from the warm air tend to be widespread, with precipitation
usually at light to moderate intensities.

(b) When a cold air mass moves into an area of warmer air, the warm
air is pushed upward more sharply. As a result, the slope of the cold
front is steeper, and the precipitation associated with it tends to be more
intense and localized. (Vantage Art, Inc. after Horace R. Byers, *General
Meteorology*, 3rd ed., 1959, McGraw-Hill Book Company)

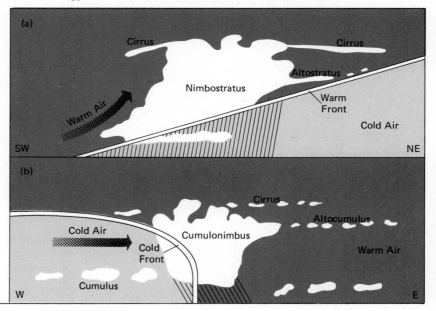

a *front*. Fronts, which are discussed more fully in Chapter 6, are zones of rapid transitions in the temperature and humidity characteristics of adjacent air masses. Where the air masses are converging, the warmer, less dense air either rides up and over the cooler air or is wedged upward by invading colder air. As it rises, the warm air cools to its dew point, clouds form, and precipitation is likely to occur.

Idealized cross sections of fronts are shown in Figure 5.18. When warm air is advancing horizontally over a surface previously occupied by colder air, a *warm front* exists (Figure 5.18a). The advancing warm air slides over the retreating cold air. The warm air rises only about 1 km in a horizontal distance of about 100 km; thus the ascent is gentle, unlike airflow over a mountain barrier. High cirrus clouds come first, followed by lower altostratus and nimbostratus clouds that produce widespread but gentle precipitation.

When cold air is advancing horizontally over a surface previously occupied by warmer air, a *cold front* is present. Figure 5.18b illustrates typical cloud distribution along a cold front, where advancing cold air pushes warm air aloft. The slope of a cold front is about twice as steep as that of a warm front, rising about 1 km over a horizontal distance of some 50 km. This more abrupt uplift causes cold front precipitation to be more intense but shorter in duration than warm front precipitation. In the eastern United States, cold fronts usually arrive in the form of a "squall line" of cumulonimbus thunderheads that produce violent rain. The thunderheads pass by rather quickly to be followed by clearing skies with scattered fair-weather cumulus clouds.

Surface air can also rise and cool by *convergence*. In areas of low atmospheric pressure, designated as "lows" on weather maps, surface air spirals toward the low-pressure center like the water moving toward a bathtub drain. As air converges into the low-pressure area, it is sucked upward and escapes outward aloft. This convergence and ascent frequently results in cloud formation and precipitation, even though no air mass fronts are present. The specific weather associated with atmospheric pressure systems, air masses, and fronts, and the resulting global precipitation patterns, are the subject of the next chapter.

GLOBAL PRECIPITATION DISTRIBUTION

A knowledge of the atmospheric and oceanic circulations and their interactions allows us to predict the distribution of precipitation over the earth—a most important element of global climates. The processes producing thick clouds and precipitation are most often associated with air mass convergence in low-pressure systems, causing the uplift of unstable moist maritime air. Where air is subsiding and diverging from surface high-pressure centers, cloud formation and precipitation are unlikely. It follows, therefore, that in equatorial and tropical latitudes precipitation should mirror the pattern of convergence of moist tropical air. In higher latitudes precipitation should be concentrated in zones of frontal interactions. Coasts washed by warm ocean currents will be zones of heavy precipitation as will the windward sides of mountain ranges. Cold water coasts and the leeward sides of mountain ranges will be dry.

The map of average annual precipitation over the continents, shown in Figure 5.19a, bears out these expectations. Precipitation is especially high in three general locations: the equatorial region where the intertropical convergence zone is present most of the year; the monsoon regions of India and southeast Asia; and the region of the polar front in the mid-latitudes. Wet spots also appear wherever moist air is regularly forced up and over mountain ranges, as in the Pacific Northwest from northern California to southeastern Alaska, in the Andes of southern Chile, and on the south side of the Himalaya Mountains.

Figure 5.19

(a) This map of average annual precipitation illustrates the complex distributions associated with the positions of continents, ocean basins, and mountain barriers. In spite of the complexity, however, global patterns of high and low precipitation are evident, and these should be related to the global pressure and wind systems shown in Figure 4.6, pp. 86–87. Particularly evident are the subtropical deserts in both hemispheres, separated by rainfall zones associated with the intertropical convergence, and succeeded on their poleward sides by rainfall zones associated with the polar fronts.

(b) The pattern of average annual precipitation over North America displays some broad regularities. The East tends to be moist because of atmospheric moisture from the Gulf of Mexico and the adjacent Atlantic. Precipitation decreases westward from the Mississippi River valley. Moist air from the Gulf of Mexico seldom flows westward, and moist air from the Pacific releases most of its moisture as orographic rainfall on coastal mountain slopes.

(c) The Great Plains are dry in winter because of the influx of cold dry air from the Canadian Arctic. The Far West receives most of its precipitation in winter, when midlatitude cyclones sweep the coast; a strong subtropical high over the eastern North Pacific inhibits precipitation during summer. The East receives precipitation during both winter and summer. (After *Goode's World Atlas*, 1970)

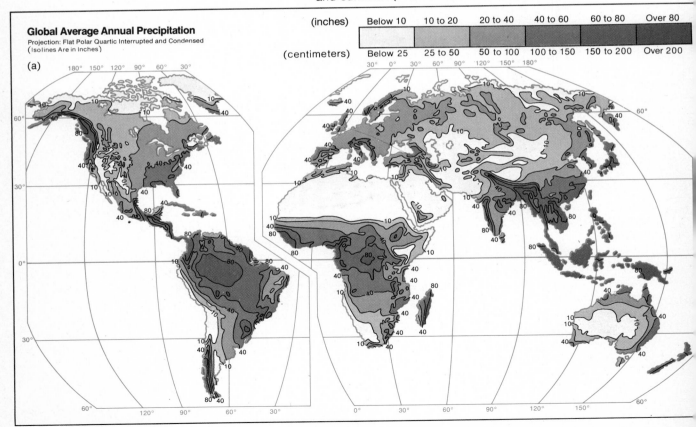

Global Average Annual Precipitation
Projection: Flat Polar Quartic Interrupted and Condensed
(Isolines Are in Inches)

(a)

(inches)	Below 10	10 to 20	20 to 40	40 to 60	60 to 80	Over 80
(centimeters)	Below 25	25 to 50	50 to 100	100 to 150	150 to 200	Over 200

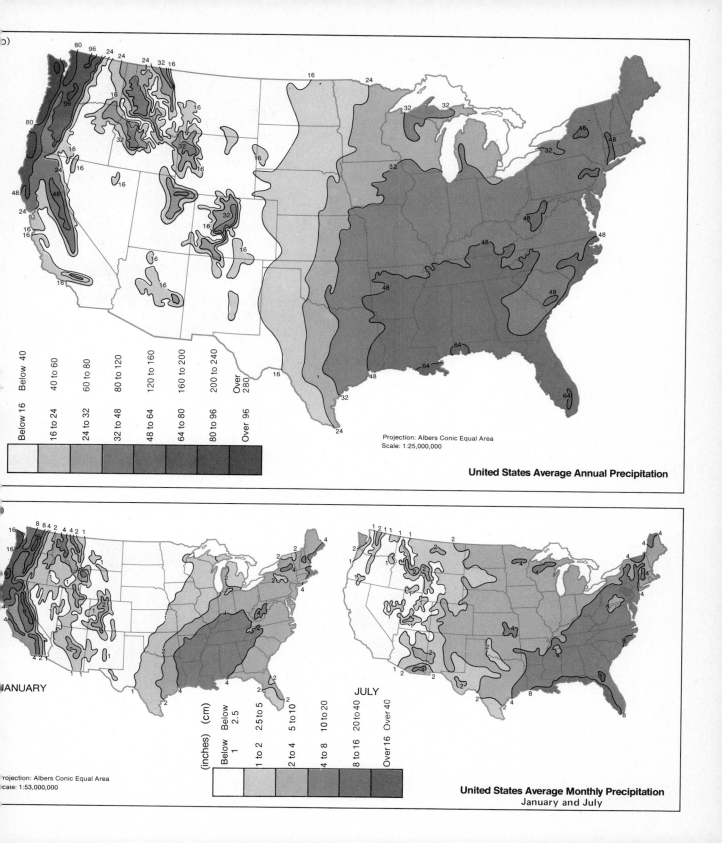

United States Average Annual Precipitation

Projection: Albers Conic Equal Area
Scale: 1:25,000,000

(inches) (cm)

Below 16	Below 40
16 to 24	40 to 60
24 to 32	60 to 80
32 to 48	80 to 120
48 to 64	120 to 160
64 to 80	160 to 200
80 to 96	200 to 240
Over 96	Over 280

JANUARY

JULY

Projection: Albers Conic Equal Area
Scale: 1:53,000,000

(inches) (cm)

Below 1	Below 2.5
1 to 2	2.5 to 5
2 to 4	5 to 10
4 to 8	10 to 20
8 to 16	20 to 40
Over 16	Over 40

United States Average Monthly Precipitation
January and July

In contrast, precipitation is low over most subtropical regions, from the Sahara Desert through Arabia and Iran to Pakistan; over the interiors of Asia and North America, which are far from oceanic moisture sources; and over the polar regions where the capacity of the atmosphere for water vapor is low.

SUMMARY

Moisture enters the atmosphere by evaporation from water surfaces and by evapotranspiration from the land. Condensation is the process by which water vapor in saturated air changes to water droplets or ice crystals, and precipitation refers to the growth of water droplets and ice crystals to sizes large enough to fall to the surface. Evapotranspiration is a cooling process in which latent heat is removed from the local environment and stored in water vapor, to be released as sensible heat elsewhere in the atmosphere when condensation occurs.

The moisture content of air can be expressed in several different ways, including absolute, relative, and specific humidity, the mixing ratio, and the dew point temperature. The capacity of the atmosphere to hold water vapor declines as temperature decreases, and condensation normally occurs when air is cooled to its dew point, when it is saturated with water vapor. In the atmosphere, condensation, which forms clouds and fog, is not automatic at dew point temperatures but requires microscopic condensation nuclei.

Condensation at the land surface creates dew and frost. Condensation within a thick layer of air at the surface produces fog. The chilling required to cause this condensation occurs in various ways, producing radiation, advection, and orographic fog, any of which can be a hazard to humans. Condensation in the atmosphere above the land surface produces clouds of various types. Clouds are classified in terms of their form and the level at which they occur. High, middle, and low level clouds can be of the cirrus (high only), cumulus, or stratus types, or combination forms of these. Cumulus clouds reflect vertical currents of air, whereas stratus types indicate an absence of strong vertical motion.

Precipitation occurs only when water droplets become raindrops by coalescence in warm clouds, or when ice crystals in cold clouds grow by transfer of water from nearby supercooled droplets. Hailstones are formed in clouds with strong updrafts and are built up layer by layer, by being recycled through the cloud several times.

To generate significant rainfall, large masses of air must be cooled, which requires the air to be lifted to higher elevations. This lifting and cooling can be accomplished in several ways: by thermal convection and adiabatic cooling, resulting in cumulus clouds; by orographic lifting, in which air is forced to rise over mountain barriers; by frontal lifting, in which warm, moist air rises to flow over cooler surface air or is forced upward by invading cold air; and by convergence, in which air flowing into a low-pressure system ascends aloft. Warm ocean waters increase air mass instability, while cold ocean currents create the stable atmospheric conditions that allow deserts to be present next to the sea. The global distribution of precipitation is largely controlled by the general atmospheric circulation, but is also strongly affected by the patterns of the continents, ocean basins, and the form and elevation of the land surface.

APPLICATIONS

1. From Figure 5.19, identify the regions of the earth where the annual precipitation averages less than 25 cm (10 in.). Which of these regions appears to be associated with the subtropical highs, rainshadows, continental interiors, and cold ocean currents? Explain the reason for each of the following deserts: the Gobi, Atacama,

Thar, Namib, Taklamakan, Kalahari, Sahara, Simpson, Sonoran, and Great Basin.

2. Maintain a diary of cloud types for several weeks. Does there appear to be a relationship between cloud types, wind directions, and air temperature and humidity changes? What about precipitation?

3. Fill a small plastic container with water, and for the rest of the term measure the daily amount of evaporation from the container. Do this by weighing the container on a laboratory balance each day. Compute the depth of evaporation by remembering that 1 g of water is equal to 1 cm^3 of water. How do the rates of evaporation vary from day to day? What is the relationship to the weather?

4. In your neighborhood, make a survey of the occurrence of dew after sunset. On what types of surfaces does dew appear first? Does dew play any role in the local ecological system in your region?

5. Are there fog-prone locations in your area? If so, how are these fogs related to the types discussed in the text? Do you see any relationship between fog and the amount of dew formed on particular nights?

6. Why is it unlikely that there will ever be complete agreement about plans for large-scale weather modification schemes? Are there any locations where it seems possible that a small-scale weather modification of some type would not provoke a controversy?

FURTHER READING

Barry, Roger G., and **Richard J. Chorley.** *Atmosphere, Weather, and Climate.* 3rd ed. London: Methuen (1976), 432 pp. This paperback is written largely from the perspective of the British Isles. Concepts are illustrated frequently by examples from the professional literature. Chapter 2 includes a particularly useful development of evaporation.

Battan, Louis J. *Harvesting the Clouds: Advances in Weather Modification.* Garden City, N.Y.: Anchor Books (1969), 148 pp. This is an especially readable account of techniques and objectives of weather modification.

————. *Fundamentals of Meteorology.* Englewood Cliffs, N.J.: Prentice-Hall (1979), 321 pp. Chapter 8 is especially helpful for its discussion of the growth of cloud droplets and the formation of rain and snow.

Chagnon, Stanley A., Jr. "The La Porte Anomaly— Fact or Fiction?" *Bull. Amer. Meteor. Soc.,* Vol. 49 (1968): 4-11. This analysis of an unusual 40-year precipitation record southeast of Chicago has proved somewhat provocative.

————. **Floyd A. Huff,** and **Richard G. Semonin.** "METROMEX: An Investigation of Inadvertent Weather Modification." *Bull. Amer. Meteor. Soc.,* Vol. 52 (1971): 958-967. This is an early summary of research on the effects of the St. Louis metropolitan region on precipitation processes and patterns.

Holzman, B. G., and **H. C. S. Thom.** "The La Porte Precipitation Anomaly." *Bull. Amer. Meteor. Soc.,* Vol. 51 (1970): 335-337. The authors present a contrasting view of the original 1968 analysis of Chagnon.

Neiburger, Morris, James Edinger, and **William Bonner.** *Understanding Our Atmospheric Environment.* San Francisco: W. H. Freeman (1973), 293 pp. This is an excellent source for basic meteorological concepts in a nonmathematical framework. Chapters 8, 9, and 15 are especially pertinent.

Scorer, Richard, and **Harry Wexler.** *Cloud Studies in Colour.* Oxford: Pergamon Press (1967), 44 pp. A most interesting set of 122 color plates, mostly from the British Isles, is well coordinated with descriptions and analyses of associated weather events.

CASE STUDY: Weather Modification— Cloud Seeding

Most regions of the earth experience occasional departures from their regular weather patterns: they become wetter, drier, warmer, or colder than usual for periods of weeks, months, and sometimes even years. The winter of 1976–1977, for example, was very dry from the northern Great Plains westward across the Rocky Mountains to the Pacific coast. Farmers in western Iowa, ski resort operators in Colorado, and water-resource managers in northern California wished that precipitation could be brought under effective control. At the same time, weary snow-removal crews in Buffalo, New York, wanted an end to the unusually persistent snow squalls that swept off Lake Erie, eventually bringing normal activities to a standstill during mid-February.

Unusually dry weather in the Soviet Union during 1972 forced the Russians to buy large quantities of American grain, and these purchases produced serious economic consequences within the United States. The effects of the severe winters of 1976–1977 and 1977–1978 across the northern United States persisted for several years.

Scientists have learned to exert some control over the weather through the process of cloud seeding to produce rain or snowfall. The key to modern rainmaking is the realization that many clouds possess all the conditions necessary for precipitation except a natural way to trigger the growth of cloud droplets into raindrops.

In 1946, Vincent Schaefer of the General Electric Research Laboratories discovered that he could change droplets to ice crystals by dropping dry ice (solid carbon dioxide below $-40°C$) into a chamber containing supercooled water droplets. Field tests by Schaefer and Irving Langmuir showed that a few pounds of crushed dry ice dropped from an airplane into a supercooled cloud could produce light precipitation. Precipitation is usually produced if the cloud is thick enough, cold enough, and long-lived—in other words, if the cloud already has the necessary conditions for producing rain.

Tiny crystals in the smoke from burning silver iodide also make excellent ice-forming nuclei. Supercooled droplets will form ice crystals on silver iodide at temperatures as high as $-4°C$ (24.8°F). An effective number of nuclei can be produced by burning only a few ounces of silver iodide in burners mounted on aircraft. Hence, commercial rainmakers are able to seed updrafts and the most promising cloud systems directly.

But the overall effectiveness of weather modification is difficult to judge. Experiments in seeding large cumulus clouds in southern Florida have produced dramatic changes in cloud structures and large but local increases in rainfall. Other experiments with cumulus cloud systems suggest that precipitation over large regions may not have been substantially increased. In the dry and mountainous West, where supercooled clouds are forced over mountain barriers, favorable results seem to depend on careful selection of weather situations. Analyses suggest that winter-spring snowfall in the mountains can be increased from 10 to 30 percent. Whether increased snowfall in the mountains means less precipitation over

(above)This photograph shows the first conclusive field test of cloud seeding, carried out in 1946 by Irving Langmuir and Vincent Schaefer. The supercooled stratus clouds were seeded by dry ice pellets dropped from an airplane flying around an oval track. In less than an hour, the seeded area cleared because of the induced precipitation. (General Electric Company)

(left) In this early attempt at rainmaking, electricity from a hand-cranked generator was pumped into clouds in the mistaken belief that the electricity of a lightning flash was somehow responsible for rain. (The Bettmann Archive, Inc.)

downwind dry areas farther east is not known.

Rainmaking is not the only objective of weather modification. Cloud seeding has also been applied in efforts to modify conditions that lead to fog, hail, and hurricanes.

Fog is especially dangerous over airports and along highways. If fog droplets are supercooled, seeding with dry ice can dissipate the fog as snow crystals, and some airports are able to use this technique. Fogs in most areas are warm, however, and no inexpensive way of clearing a warm fog has been developed. Heating the air to a temperature above the dew point will clear a warm fog, but the energy required makes the method too costly for general application.

Hailstorms cause so much damage to crops that efforts to disrupt the formation of hail in turbulent clouds are believed worthwhile. Seeding with silver iodide seems to inhibit further growth of hailstones or to cause the cloud to produce only rain. Because hailstorms tend to be concentrated in certain regions it has been feasible to set up hail suppression facilities in particularly susceptible areas.

Some efforts have also been directed toward modifying the intensity of hurricanes, but the erratic movements and growth of these storms make it difficult to determine whether a seeding experiment has reduced their winds. Hurricane Debbie was seeded by Project Stormfury planes on August 18 and 20, 1969. Maximum winds decreased 31 percent on the first day, increased on the second day, and decreased again (by 15 percent) after seeding on the third day. Much more research is needed, however, before seeding of hurricanes can become routine.

The panorama of the skies often changes from
hour to hour as migrating systems of high and
low atmospheric pressure generate flows of
moist and dry air that bring us our daily weather.
Air masses, fronts, local winds, and cyclonic
storms are all aspects of the secondary
circulations of the earth's atmosphere that
directly affect our lives.

Wheatfields by Jacob Isaacksz Van Ruisdael (The Metropolitan Museum of Art, New York)

CHAPTER 6

Secondary Atmospheric Circulations

The day-to-day changes in the weather expected by people in much of the United States and Canada are not normal in all parts of the world. In the tropics people do not encounter the successions of fair and stormy or warmer and colder days experienced across much of North America. Yet even in the supposedly monotonous tropics, dramatic changes of weather do occur. Over large areas monsoon conditions cause extreme contrasts between wet and dry seasons. In the late summer, hurricanes and typhoons batter many areas with destructive winds and torrential rains, and even under the prevailing calmer conditions, groups of wetter than normal days are noticeable.

Whatever the geographical area—the midlatitudes, the tropics, or the poles—changes in weather are not isolated events. They are linked to the larger scale phenomena of the general circulation described in Chapter 4. A summer shower that covers only a few square kilometers is a local weather event. But the local weather is controlled by larger scale weather systems, or *secondary circulations*, that move through the middle and low latitudes. Such features endure for days and cover areas of thousands of square kilometers.

MIDLATITUDE SECONDARY CIRCULATIONS

Most midlatitude weather is dominated by the interaction of large air masses of unlike characteristics. In the northern hemisphere, especially during winter, warm, moist air moving poleward out of the subtropical highs meets cool, dry air flowing equatorward out of the Siberian and Yukon highs. Where the air masses meet, a front is formed. The interacting air flows frequently become organized into vast, spiraling eddies that constitute the "storms" of the midlatitudes. These usually migrate from west to east and may be as much as 1,000 km (600 miles) in diameter. Such migrating secondary circulations and their associated air masses and fronts produce the changeable weather of the middle latitudes.

Air Masses

An *air mass* is a large, nearly uniform segment of the atmosphere that moves as a unit, usually in association with secondary high or low pressure systems. Air masses retain the temperature and moisture characteristics of their source region even after traveling thousands of kilometers. Air mass types are therefore identified according to their source region: *polar* (*P*) or *tropical* (*T*), *continental* (*c*) or *maritime* (*m*). The capital letter indicates air mass temperature characteristics; the lower case letter suggests relative moistness. Thus, four general air mass types are commonly recognized: *cP*, *mP*, *cT*, and *mT*. Some classifications also include arctic (*A*) and equatorial (*E*) air masses. Air masses are also designated according to their stability at the surface and aloft: *K* and *W*, meaning colder and warmer, respectively, than the surface beneath, and *u* and *s*, meaning unstable or stable aloft. An *mTKu* air mass has maximum instability: it is colder than the surface, therefore subject to heating from below, which produces a steep lapse rate; and it is unstable aloft as well.

Extremely cold winter weather in the United States is associated with *continental polar* (*cP*) or arctic (*A*) air masses that form over the snow-covered plains of northern Canada. Arctic air masses actually originate closer to the polar region and are colder in the summer than those air masses called polar. In the following discussion of North American air masses we shall regard all cold air masses as being the polar type.

During the long subarctic winter, the sun remains below the horizon much of the time and frequent clear skies promote intense radiational cooling of the snow-covered land surface. After a few days these conditions create a nearly homogeneous cold, dry *cP* air mass that extends horizontally for more than 1,000 km (600 miles).

Eventually, the *cP* air spreads out of the source region and moves toward lower latitudes. The air is modified only slightly as it sweeps southward, traveling more than 4,000 km (2,500 miles) across the Great Plains and down the Mississippi Valley to the Gulf of Mexico. During midwinter, even at New Orleans the mean temperature of the *cP* air is close to 0°C (32°F), with occasional temperatures lower than −5°C (23°F). However, when the *cP* air moves out over the warm waters of the Gulf of Mexico, it gains moisture and is warmed rapidly. Modification is so quick that a "new" air mass is produced within 48 hours or so. This warm, moist air mass is designated *maritime tropical* (*mT*), just like air masses that originate over the tropical oceans. Most heavy rains in the midlatitudes are produced by condensation of moisture within poleward-moving *mT* air masses.

The mountains of western North America prevent most *cP* air from spreading to the Pacific coast. Instead, the west coast is usually under the influence of *maritime polar* (*mP*) air that has moved eastward from the seas off Japan and Alaska. This air is never as cold as polar air that originates over the land, but it is much wetter than *cP* air. That is why the Pacific coast receives heavy rain and wet snow in the winter, when polar outbreaks are common, but seldom experiences freezing temperatures. Similar *mP* air from the North Atlantic occasionally invades the northeastern United States, bringing freezing rain in the fall and winter. Occasionally in the summer the Pacific coast receives an outbreak of *continental tropical (cT)* air. This air is hot and extremely dry, coming from the deserts of Mexico and the southwestern United States. Globally, the most important source of *cT* air is the Sahara Desert of North Africa. Saharan *cT* air affects all the Mediterranean coastlands, occasionally crosses the Alps, and even penetrates as far as Great Britain and Scandinavia.

Air masses move over the earth's surface according to predictable patterns. Figure 6.1 shows typical tracks of air masses in the United States in winter and summer. The movement of these air masses largely determines weather conditions throughout North America. But the weather also depends on surface characteristics. For example, over the large land areas of the Great Plains and Mississippi Valley, *cP* air is generally associated with weather that is cold and windy, but clear. However, over the Great Lakes the lower layers of *cP* air receive heat and moisture from large expanses of water. The result is spectacular snow squalls on the eastern and southeastern shores of the Great Lakes.

Continental air masses normally produce very little precipitation. Instead, they tend to gain moisture from evapotranspiration. The opposite is true of maritime air masses; once formed, they lose more moisture by precipitation than they receive back by evapotranspiration. They are the sources of rain and heavy snowfall around the world.

Fronts

A *front* is an interface between two air masses that differ in temperature or humidity, or both. The most distinct fronts are those separating air masses whose properties contrast

Figure 6.1
The air masses that enter a region can strongly influence weather conditions there by replacing the existing air with air of different temperature and humidity. Maritime air is humid, and continental air tends to be dry. The maps in this figure depict the average movements of air masses over the conterminous United States during winter and summer. During winter much of the northern half of the United States is invaded by cold polar air, whereas during summer the northward flow of tropical air dominates the weather in most regions. (Calvin Woo after Dieter H. Brunnschweiler, *Geographica Helvetica*, vol. 12, 1957)

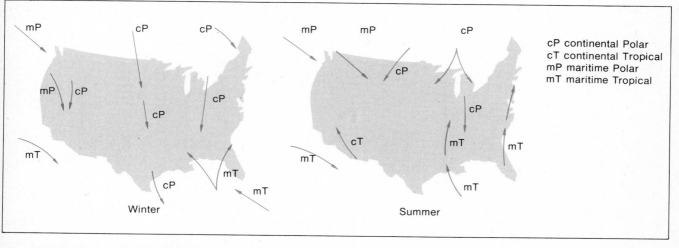

cP continental Polar
cT continental Tropical
mP maritime Polar
mT maritime Tropical

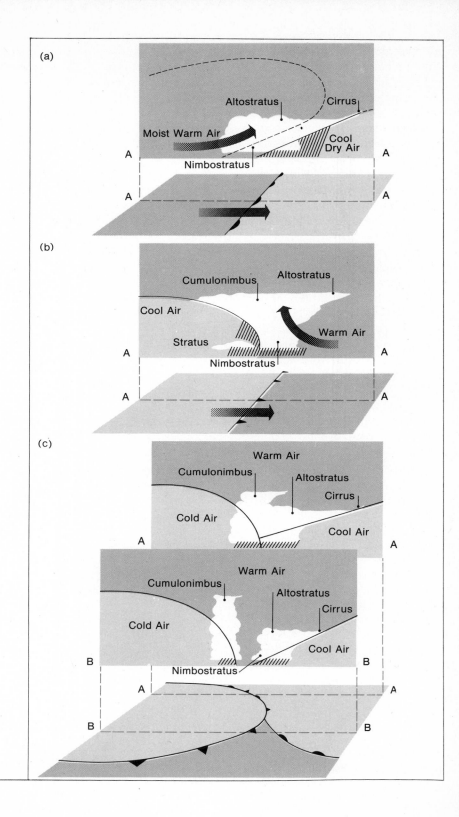

Figure 6.2 (opposite)
(a) A warm front is formed when a mass of warm air encounters a denser mass of cool air. The diagram shows a vertical cross section of a warm front, together with the surface symbol as it would appear on a weather map. As the uplifted warm air becomes cooler, condensation and cloud formation may occur. The vertical section shows that the resulting precipitation arrives at a location ahead of the surface warm front itself.

(b) A cold front develops where a mass of cool air pushes into a region of warmer air. The vertical cross section of the front shows the warm air ahead of the front being forced upward over the incoming cool air. Cloud formation and precipitation may occur as the uplifted warm air cools. The vertical section shows that precipitation normally arrives at a location just ahead of the surface front.

(c) An occluded front involves three air masses of different temperatures, and it combines some of the features of both warm and cold fronts. The vertical cross sections show that the mass of incoming cold air forces warm air to rise above both regions of cool air. In the forward vertical section, the occlusion is not complete, and the warm air is in contact with the ground in close proximity to a cold front and a warm front. In the rear vertical section, the occlusion has formed, and the warm air is lifted completely above the ground. Precipitation may occur on both sides of an occluded front, as the diagrams indicate.
(Calvin Woo adapted from Hermann Flohn, *Climate and Weather*, © 1969 by H. Flohn, used with permission of McGraw-Hill Book Company)

most sharply—in particular, the cold, dry *cP* and warm, moist *mT* types. Such fronts are common in eastern North America. On the Pacific coast most fronts merely separate successive *mP* air masses of varying temperature and humidity, although *mT* and *cT* air masses are occasionally involved, especially in southern California. On rare occasions *cP* air penetrates into central California and the Pacific Northwest, with disastrous consequences to ornamental vegetation and winter crops in these areas.

When warm air moves into a region previ-

ously occupied by a colder air mass, the forward edge of the warm air mass is designated as a *warm front*. Figures 5.18 (see page 123) and 6.2a show how a warm front slopes forward from the surface as the lighter warm air slides over the denser cold air it is replacing. The cool air is shown retreating toward the right, corresponding on the map to the east or northeast, the usual direction in which a warm front advances in the middle latitudes. At an altitude of one kilometer, the warm air may be several hundred kilometers ahead of the front at the surface.

Weather conditions in the vicinity of a warm front depend on the properties of the air masses as well as the nature of the land surface. Nevertheless, there is a characteristic sequence of weather conditions associated with a warm front. Condensation and cloud formation begin where the warm, moist air rises over the cooler air and cools adiabatically to its dew point. Because the warm front slopes so gently, it is usually detected through the appearance of high cirrus clouds a day or more before the surface front arrives (Figures 5.18 [page 122] and 6.2a). As the surface front approaches, sheet-like stratus clouds become thicker and lower. Just ahead of the surface front is a broad band of precipitation. Warm fronts move slowly and often produce steady rains that may last a day or more.

A *cold front* develops when cold air advances into a region occupied by warm air. Figure 6.2b shows that the advancing cold air pushes under the warm air, forcing the warm air to rise abruptly, so that the frontal slope is much less gradual than that of a warm front. There is rapid development of towering cumulonimbus clouds, with precipitation occurring just ahead of the surface front. The passage of a cold front in the summer is usually associated with the sudden appearance of a line of thunderstorms and a rapid drop in temperature (Figure 6.3). The zone of precipitation along a cold front may pass by in only an hour or two.

More complex *occluded fronts* consist of three or more air masses, with one front overtaking another so that the air mass between loses con-

Figure 6.3
Instability of the air in this squall line over western Texas is indicated by the development of cumulus cloud turrets overtopped by giant cumulonimbus thunderheads. (T. M. Oberlander)

tact with the ground. This creates a broad band of heavy precipitation falling from the occluded (lifted) air mass (Figure 6.2c). Occluded fronts will be treated below in connection with cyclonic storms.

All fronts are segments of the polar front, which is a component of the general circulation. As such, they are associated with moving secondary circulations, including midlatitude cyclones and anticyclones.

Cyclones and Anticyclones

Almost any daily weather map for a midlatitude region shows centers of high and low pressure. The spiraling flows of air into moving centers of low pressure, called "lows," are *cyclones*. The diverging flows around high-pressure regions, referred to simply as "highs," are *anticyclones*. The horizontal differences in pressure within large highs or lows usually amount to less than 10 millibars over a distance of 100 km (60 miles), only a small fraction of the normal sea level atmospheric pressure of 1,013 millibars.

As shown in Figure 6.4, a traveling cyclone consists of converging surface air that ascends and diverges in the upper atmosphere. As air rises in a cyclone, it cools adiabatically, resulting in cloud formation and perhaps rainfall. A traveling anticyclone, on the other hand, consists of air subsiding from aloft that diverges at the surface. The descending air in an anticyclone is heated adiabatically, reducing its humidity and producing clear skies.

Midlatitude cyclones develop along the polar front where warm and cold air come into contact. In North America cyclones most often form just east of the Rocky Mountains and migrate toward the east and northeast, eventually passing out into the Atlantic Ocean. Most precipitation in the central and eastern United States and Canada is associated with the ascent of warm, moist *mT* air over the cold *cP* air within the cyclonic circulation. Along the Pacific coast most cyclonic disturbances involve the interaction of *mP* air masses of differing character. Having originated far away off the coast of Japan or Alaska, these cyclonic storms commonly arrive at the west coast in the form of mature occluded fronts.

The evolution of midlatitude cyclones usually follows a characteristic pattern. Cyclones begin along a stationary segment of the polar front. Figure 6.5a shows a stationary front with a northeast-southwest orientation. Such a front may remain over the same region for as much as 24 to 48 hours before a cyclone begins to develop, possibly due to an upper air disturbance, such as a wave in the jet stream.

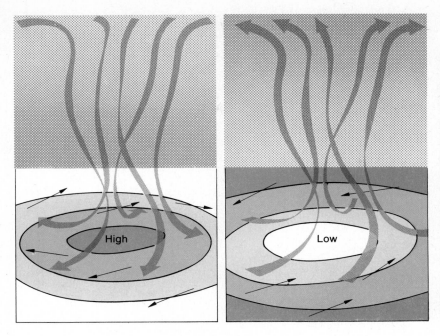

Figure 6.4
In a region of high pressure (left), air descends from the upper atmosphere and diverges outward along the surface of the ground. The air streaming outward from highs therefore tends to be dry. In a region of low pressure (right), air converges inward along the surface of the ground and ascends to the upper atmosphere. If the air streaming into a low is warm and moist, condensation and cloud formation occur as the air rises and cools. The directions of circulation are shown for the northern hemisphere. (Calvin Woo)

The isobars in Figure 6.5a (see p. 138) show that pressure increases away from the front on both sides. On the polar side of the front, cold air flows from the northeast; on the tropical side, warm air flows from the southwest. As is normal where warm and cold air are in contact, there is often warm air aloft over the cold air at the surface. This produces a narrow band of clouds on the polar side of the front, with some light precipitation.

The cyclone begins as a small wave disturbance caused by a local drop in pressure, as shown in Figure 6.5b. As pressure falls even more, air begins to converge toward the center of the low in a counterclockwise circulation. Ahead of the center, the warm air advances into the former domain of the cold air, forming a warm front. The cold air begins to sweep southeastward behind the center, forming a cold front. A broad apron of warm-front drizzle develops ahead of the center, and a narrow band of showers breaks out along the cold front.

Many small waves may move toward the northeast along the polar front, dissipating in 6 to 12 hours. A few waves, however, grow into well-developed cyclonic circulations. One of the difficult tasks of the weather forecaster is to predict the development of a significant midlatitude cyclone from a small wave disturbance.

Figure 6.5c shows the cyclone at a more advanced stage. The central pressure of the cyclone may fall to less than 1,000 millibars (29.5 in.), and the converging counterclockwise circulation may expand to a diameter exceeding 1,000 km (600 miles). In eastern North America the cyclonic system normally moves in a northeasterly direction at a speed between 25 and 50 km (15 to 30 miles) per hour. Steady and substantial rains usually occur ahead of the surface position of the warm front, and occasionally severe thunderstorms and even tornadoes are associated with the cold front (see Case Study, pages 150–151).

Figure 6.5d shows the cyclone in its mature phase. The cold front moves more rapidly toward the east than the cyclonic system does,

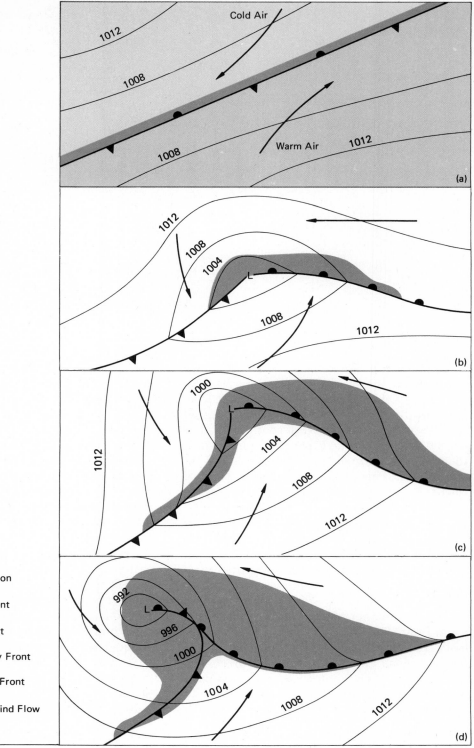

Cold Air

1012

1008

1008

Warm Air

1012

(a)

1012

1008

1004

L

1008

1012

(b)

1000

1012

L

1004

1008

1012

(c)

992

L

996

1000

1004

1008

1012

(d)

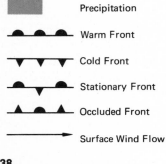

Precipitation

Warm Front

Cold Front

Stationary Front

Occluded Front

Surface Wind Flow

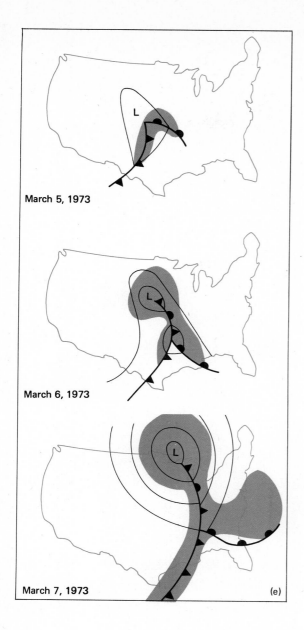

March 5, 1973

March 6, 1973

March 7, 1973

(e)

Figure 6.5 (opposite and left)
Cyclones in the midlatitudes usually follow characteristic patterns of evolution along the polar front. This figure should be studied in conjunction with Figure 6.2, which shows other properties of fronts.

(a) The process begins along stationary segments of the polar front where cold and warm air stream in opposite directions.

(b) The stationary front tends to be unstable, and a bulge, or wave, usually develops within one or two days. The waves move along the polar front toward the northeast, and most dissipate within six to twelve hours.

(c) A few waves, however, grow into well-developed circulations with diameters of 1,000 km (600 miles) or more. In the northern hemisphere the counterclockwise converging circulation of the cyclone includes warm and cold fronts that separate polar and tropical air masses.

(d) The cold front eventually overtakes the warm front, lifting the warm air away from the surface and forming an occlusion. The occlusion eliminates the surface air temperature differences that provide energy for the system, and the circulation then weakens and finally dissipates.

(e) This sequence of development can be traced on the three map sketches for March 5, 6, and 7, 1973. As a cyclone occludes, a new wave disturbance may form farther back along the trailing stationary front. Over the oceans, the polar front often supports a cyclone family of three to five members in various stages of development. Most late fall, winter, and spring precipitation over North America is associated with midlatitude cyclones and associated fronts. (Vantage Art, Inc.)

Figure 6.6 (left)
This weather satellite image of the eastern Pacific shows the typical comma-shaped cloud pattern produced by an occluded front that has moved into British Columbia and is approaching Washington and Oregon. The low pressure center is marked by the cloud spiral south of Alaska. The cloud band at the bottom of the view is the result of convection in the intertropical convergence zone. (National Oceanic and Atmospheric Administration)

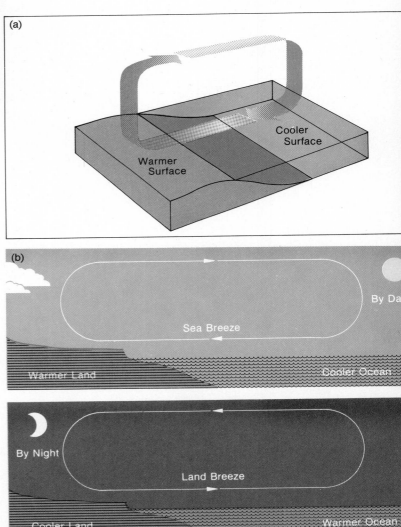

(a)

Cooler Surface

Warmer Surface

(b)

By Da

Sea Breeze

Warmer Land

Cooler Ocean

By Night

Land Breeze

Cooler Land

Warmer Ocean

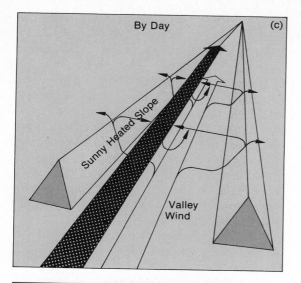

By Day (c)

Sunny Heated Slope

Valley Wind

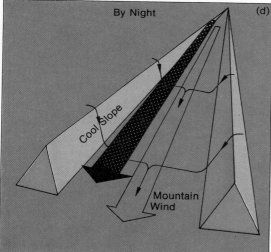

By Night (d)

Cool Slope

Mountain Wind

(e)

Figure 6.7 (opposite and left)
Small-scale circulations of air can significantly modify local weather conditions.

(a) This diagram shows a small convective cell generated near the boundary between a warm and a cool region, such as a land and a water area.

(b) Land and sea breezes develop because of the difference in temperature between the ocean and the land. During the day, the land heats more rapidly than the ocean, and a surface *sea breeze* develops from the ocean toward the land. At night, the land cools rapidly, and the surface flow of air is a *land breeze* from the land to the ocean.

(c) Local winds develop because of the temperature differences between valleys and mountain slopes. During the day, the slopes are warm, and as air rises up the slopes, a *surface valley* wind flows up the valley to replace the ascending air.

(d) At night, cool air descends the slopes as a *mountain wind* and flows down the valley. (Calvin Woo)

(e) This view shows cloud formation resulting from convective rise of air above heated mountain slopes adjacent to Lake Brienz, Switzerland. (T. M. Oberlander)

SECONDARY ATMOSPHERIC CIRCULATIONS

141

so the cold front eventually overtakes the warm front. This lifts the warm air away from the surface, forming an occlusion. Occluded fronts have the properties of both warm and cold fronts (Figure 6.6). Heavy precipitation usually falls from the moist warm air that is lifted during the occlusion. Thus precipitation occurs on both sides of the surface front. The intensity and size of cyclonic storms are often greatest at the beginning of the occluded phase. Occlusions are especially characteristic of cyclonic systems over northern California and the Pacific Northwest, where the Coast Range, Cascade Mountains and Sierra Nevada are often inundated by heavy rains (and snows at higher elevations) during winter and spring.

In its final stage, the energy supply of the cyclone undergoes a significant alteration. The developing occlusion eventually eliminates the surface air temperature differences that power the circulation. Warm, moist air from the warm front no longer feeds into the center of the system. It is not long before the low-pressure center begins to fill with cooler, drier air, and the cyclone itself loses its energy. The life cycle of a midlatitude cyclone ranges from 24 hours to as much as five days, so that it is possible for a particularly long-lived cyclone to travel across most of the United States.

The passage of midlatitude cyclones and associated fronts accounts for most of the late fall, winter, and spring precipitation over North America. Cyclones that develop along the eastern slopes of the Rocky Mountains from Alberta to Colorado and along the Gulf and Atlantic coasts migrate along the western, southern, and eastern margins of polar outbreaks, following paths that lead them toward the northeast and eventually to the region of the Icelandic low. Cyclones that reach the Pacific coast of North America in an occluded stage have formed along the eastern coast of Asia.

Jet streams play a part in the formation of midlatitude cyclones. The presence of waves in a jet stream aids in the ascent of surface air in the low-pressure center of a cyclone. Cyclones intensify when the surface flow of air coincides with favorable conditions aloft; when air flow aloft is unfavorable, development of the cyclone seems to be suppressed. That is why weather forecasters constantly monitor the upper air.

In North America during winter, cyclonic storms and anticyclones bringing clear skies often cross the continent in rapid succession, producing very changeable weather. In some winter weeks as many as three pairs of cyclones and anticyclones move down the St. Lawrence River valley toward the Atlantic Ocean or invade the Pacific coast from the west.

The configuration of coastlines and the land surface significantly affect local circulations and weather. Under anticyclonic conditions, *land and sea breezes* and *mountain and valley winds* dominate the weather in coast and highland regions. Both result from daily cycles of heating and cooling of different surface types, as seen in Figure 6.7.

TROPICAL SECONDARY CIRCULATIONS

Much less is known about tropical weather patterns than about the weather of the middle latitudes. There are far fewer observing stations in the tropics than in the middle latitudes, and weather data from tropical ocean areas are especially sparse. The Coriolis force is small near the equator, so equatorial winds are not geostrophic. Rotating cyclones and anticyclones do not form near the equator, and air mass contrasts and frontal activity are absent. In the continental tropics, the principal influence on the weather is the daily cycle of heating and cooling of comparatively homogeneous humid air. The temperature variation from day to night in the tropics often exceeds the variation in average monthly temperatures through the year.

The tropics may lack variety in day-to-day weather, but many areas within the tropics have strong seasonal weather contrasts and exhibit weather phenomena found nowhere else.

Easterly Waves

Poleward of the band of persistent clouds and precipitation that marks the intertropical convergence, weak troughs of low pressure occasionally form in the trade wind zone and drift slowly westward (Figure 6.8). These *easterly waves* extend roughly north and south for a distance of several hundred kilometers. They form most often over the seas during the high-sun season. In the tropics, a warm layer of subsiding air usually overlies the surface layer, which produces a weak temperature inversion and prevents the surface air from rising to higher altitudes. This subsidence inversion is temporarily destroyed by an easterly wave, resulting in weather disturbances that vary from mild to quite violent.

Ahead of a wave, to the west of the trough, the surface winds diverge. In this area warm, dry air descends from the upper atmosphere, and the weather is fair. Behind the wave, to the east of the trough, the winds converge. Here, the inversion is broken, moist air can ascend to great heights, and severe thunderstorms may be generated. Each year a few easterly waves increase in intensity and develop into tropical cyclonic storms.

Tropical Cyclones

A dangerous interruption of the monotony of tropical weather is the late summer appearance of intense cyclonic disturbances. If their wind speeds exceed 120 km (75 miles) per hour, such disturbances are classified as *tropical cyclones*. In the Caribbean area and on North American

Figure 6.8
The kink in the isobars denotes the presence of an easterly wave. Easterly waves move toward the west with the trade winds. Ahead of an easterly wave, to the west of the trough, warm, dry air descends from the upper atmosphere, and the weather is fair. Behind an easterly wave, to the east of the trough, moist air from the surface ascends to great heights, and severe thunderstorms may be generated. Each year a few easterly waves increase in intensity and develop into the severe tropical cyclonic storms known as hurricanes. (Andy Lucas after Hermann Flohn, *Climate and Weather*, © 1969 by H. Flohn, used with permission of McGraw-Hill Book Company)

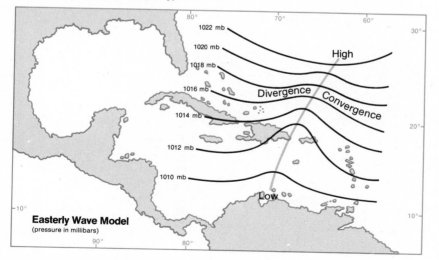

Easterly Wave Model
(pressure in millibars)

coasts these are called *hurricanes*; in the western Pacific they are known as *typhoons*; in the Indian Ocean and Australia they are simply called *cyclones*.

A tropical cyclone is an unusually compact and intense low-pressure center, much smaller in diameter than midlatitude cyclones, but having a pressure gradient that can exceed 30 millibars per 100 km. The winds that whirl around the center of a tropical cyclone commonly have velocities greater than 200 km (120 miles) per hour, accompanied by driving rain and severe thunderstorms.

Hurricanes begin as a rotating tropical storm generated from a strongly developed easterly wave. Why some storms die out and others continue to build to hurricane strength is not known, although upper air flows may play a role. The characteristic structure of a fully formed tropical cyclone consists of a relatively cloudless central "eye" surrounded by a rapidly rotating wall of towering clouds, as illustrated in Figure 6.9. The height of the wall clouds is typically 10 to 15 km (6 to 9 miles), while the eye may be several tens of kilometers in diameter. Dry air from the upper atmosphere descends in the eye and is heated adiabatically, keeping the eye relatively free of clouds. Moist air spirals upward around the eye, and massive condensation

Figure 6.9
This cross section of a hurricane (vertical scale much exaggerated) shows the central column, or eye, and the swirling cloud bands that give a hurricane its characteristic appearance as seen from above. A tropical cyclone in the northern hemisphere has an intense counterclockwise rotation. Wind speeds near the center may exceed 200 km per hour (120 miles per hour). The central eye is a region of low atmospheric pressure, and the dry air descending into the eye from above makes weather in the eye clear and calm. (Tom O'Mary after *The Atlas of the Earth*, p. 30, © 1971, Mitchell-Beazley, Ltd.)

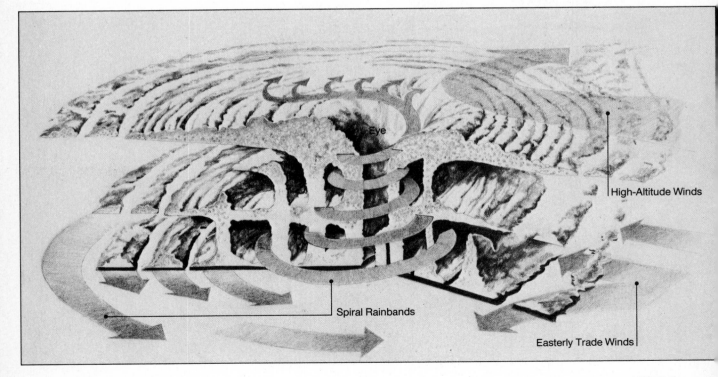

Eye

High-Altitude Winds

Spiral Rainbands

Easterly Trade Winds

CHAPTER 6

produces the cloud walls and releases enormous amounts of latent heat.

Because of the weakness of the Coriolis force at low latitudes, tropical cyclones rarely form within 5° of the equator. Nor do they usually form at latitudes higher than 30° (Figure 6.10). The warm ocean surfaces between these two latitudes provide the necessary conditions for the formation of tropical cyclones. Unlike mid-latitude cyclones, which are powered primarily by the temperature differences between dissimilar air masses, the energy source for tropical cyclones is the latent heat released by massive condensation of water vapor. Water vapor is most abundant in warm tropical air over oceans with surface temperatures greater than 27°C (81°F). The oceans reach their maximum temperature in the late summer or fall, making this the hurricane season. Once a tropical cyclone begins to travel over land or cold water, it is cut off from the water vapor that is its source of energy, and its strength diminishes.

On a densely populated low-lying coastline, a direct hit by a strong hurricane is devastating. When gale-force hurricane winds drive coastal waters up onto low shores, the sea may rise sev-

Figure 6.10
This map shows the tracks of some devastating North Atlantic hurricanes for the years 1954 through 1980. The path of an individual hurricane tends to be erratic, but as the map indicates, many of the hurricanes generated in the western Atlantic follow the same general course westward and northward. Some hurricanes turn northward early and skirt the east coast of the United States; others enter the Gulf of Mexico before heading north. Hurricanes lose strength over land because of friction and lack of water vapor, but some hurricanes, such as Hazel, 1954, and Frederic, 1979, have traveled long distances across the United States. A hurricane traveling over land brings heavy rains that cause rivers to overflow. (Andy Lucas after *The National Atlas of the United States of America*, 1970)

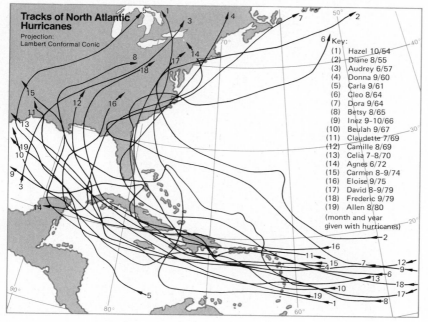

Tracks of North Atlantic Hurricanes
Projection: Lambert Conformal Conic

Key:
(1) Hazel 10/54
(2) Diane 8/55
(3) Audrey 6/57
(4) Donna 9/60
(5) Carla 9/61
(6) Cleo 8/64
(7) Dora 9/64
(8) Betsy 8/65
(9) Inez 9-10/66
(10) Beulah 9/67
(11) Claudette 7/69
(12) Camille 8/69
(13) Celia 7-8/70
(14) Agnes 6/72
(15) Carmen 8-9/74
(16) Eloise 9/75
(17) David 8-9/79
(18) Frederic 9/79
(19) Allen 8/80
(month and year given with hurricanes)

Figure 6.11
(a) This photograph of hurricane Gladys, 1968, was taken from the Apollo 7 spacecraft. The open central eye and the counterclockwise circulation of the hurricane are visible. (NASA)

(b) This satellite image of weather conditions on August 30, 1975 shows three hurricanes in various stages of development. All are moving westward in the zone of easterly winds. Hurricane Caroline, which originated in the western Atlantic, is moving onshore from the Gulf of Mexico, and caused destruction on the mainland. Hurricane Katrina is seen near its point of origin off the west coast of Mexico, and Hurricane Jewell is dissipating farther to the northwest. An occluded midlatitude cyclonic storm can be seen decaying over British Columbia and Washington. (National Oceanic and Atmospheric Administration)

(a)

(b)

eral meters above the normal high tide, surging tens of kilometers inland. The resulting destruction and loss of life can be enormous. Some 250,000 people were drowned by such an event in Bangladesh in 1970. A rainfall of more than 25 cm (10 in.) is not unusual as a hurricane passes by, and, if the hurricane moves over land, heavy rains soon bring rivers to the flood stage. The number of hurricanes generated in the Atlantic and the Caribbean varies from three or four to a dozen per year, largely depending on the sea surface temperature and the behavior of the subtropical jet stream.

While it is difficult to predict the course of a hurricane, its distinctive cloud formations are easy to see on weather satellite images (Figure 6.11b). Therefore, the moving storm can be tracked in time to give warning to threatened areas. Improved warning and communications systems have steadily reduced the loss of life resulting from hurricanes in the United States, even though those in peril often underestimate the danger and do not always take the warnings seriously. In less developed areas of the world, lack of adequate communications leaves great numbers of people unaware of the hazard, even after it is detected. When the cyclone in the Bay of Bengal struck Bangladesh in 1970, vast numbers received no warning, which accounts for the catastrophic loss of life.

Although every hurricane appears to be a potential disaster, these powerful storms are vital to the earth's heat balance. They are a means of transporting energy from areas of excess to areas of deficiency. If there were fewer hurricanes each year, those that occurred would have to be even larger and more violent to perform their vital function of spilling energy poleward.

Tropical Monsoons

We have already seen that monsoons are secondary circulations that involve a reversal in the direction of surface winds. This wind regime results from seasonal pressure changes on a continental scale. The most extreme case is in Asia.

During the winter, the Asian interior is cold, and an almost steady outflow of dry cP air from the Siberian high brings clear weather to most of the continent. The dry northwest monsoon affects China and Japan, and the somewhat warmer northeast monsoon prevails in India and Pakistan. But during the summer, the heating of southern Asia produces a thermal low, easily seen in Figure 4.6 (pp 86–87). This causes the wind direction to reverse, as warm, humid mT air sweeps over the continent from the Indian Ocean and the southwestern Pacific. The southwest winds of the Indian summer monsoon soak up moisture as they cross the warm Indian Ocean. Summer precipitation from this air accounts for 70 percent of India's annual rainfall. The mountains of India trigger very heavy orographic precipitation, producing annual totals of more than 1,000 cm (400 in.) in some locations—85 to 90 percent of which falls between May and September.

The Indian monsoon also seems to be related to seasonal changes in the circulation of the upper air in the tropics. At the start of the summer monsoon, the westerly subtropical jet stream located above the south slope of the Himalaya Mountains shifts to the north of Tibet. At the same time, another jet stream blowing from the east appears over northern India. This easterly jet accentuates the ascent of air in the low-pressure center seen in Figure 4.6, and draws the ITC into India, far north of its usual position.

Weaker monsoon circulations also occur in China and Japan, southeast Asia, northern Australia, and the Guinea coast of Africa. There is a slight monsoon effect over the Mississippi Valley, with frequent outbreaks of polar air (called "northers") in winter, and waves of humid, sultry air invading from the Gulf of Mexico during summer.

SUMMARY

The general circulation of the atmosphere produces smaller scale secondary circulations that are the controls of day-to-day weather in the middle and low latitudes. This weather is associated with moving high and low pressure centers and accompanying air masses and zones of air mass interaction.

Air takes on the heat and moisture properties of the underlying surface when it occupies a homogeneous region for days or weeks at a time. Most weather can be explained in terms of the interactions between cP, mT, mP, and cT air masses having differing temperatures and humidities. Continental and maritime air masses differ greatly in humidity, while polar and tropical types present strong temperature contrasts.

Precipitation in the midlatitudes is produced mainly by frontal activity in cyclonic storms that move from west to east along the polar front. Cyclonic storms are generated by pressure drops along stable fronts that separate air masses from dissimilar source regions. If a low-pressure center develops along a stable front, cyclonic circulation is initiated, in which air behind a cold front overtakes warmer air advancing in a warm front. Warm air slides up over cooler air along the warm front, and cold air pushes under warm air along the cold front. Eventually the cold front overtakes the warm front, and the warm air loses contact with the ground, forming an occluded front. The development of cyclonic storms is assisted by the flow of air in the upper atmosphere.

Air mass contrasts and fronts rarely exist in tropical and equatorial regions. Convective showers and thunderstorms develop almost daily in the unstable air of the intertropical convergence zone. Weather disturbances in the tropics begin as easterly waves that produce clusters of rainy days. Heating of the oceans during the summer causes easterly waves to intensify into tropical cyclones, also known as hurricanes and typhoons, that are enormously destructive. These powerful storms are energized by latent heat released by moisture condensation and cloud formation in extremely humid air. Tropical cyclones are part of the mechanism by which net energy gains in the low latitudes are drained off to compensate for net energy losses in high latitudes.

Monsoon circulations are another tropical weather phenomenon of importance. The monsoon effect is a seasonal reversal of winds and associated weather types related to changes in atmospheric pressure over land and water areas. The most extreme example is seen in India, which is cloud-free in winter as dry air streams southward toward the ITC from the high-pressure center in the interior of Asia. Heavy rainfall follows in the summer as moist air from the Indian Ocean is drawn into a low-pressure center that develops over southern Asia.

APPLICATIONS

1. Keep a daily log of the air mass types present in your area, using your own judgement and recollection of past extreme conditions to establish the actual air mass types. How abrupt are changes in air masses? Are some air masses transitional between the basic types discussed in the text? What is the nature of the weather during periods when air mass types change in your area?

2. On the basis of the midlatitude cyclone model described in Figure 6.5, determine the most likely progression of cloud types and weather events when a cyclonic storm approaches your area from the west in winter and passes to the north and northeast. What would change if the cyclonic disturbance passed to the south?

3. How far toward the equator do midlatitude cyclones penetrate? Does Hawaii experience such storms? Are there places that experience both midlatitude cyclones and tropical cyclones? If so, would both be possible at the same time of year? If the NOAA periodical publication *Environmental Satellite Imagery* is available on your campus, check the images to answer the above question.

4. What was the heaviest rainfall ever received in your area? What unusual meteorological conditions occurred or combined to produce it?

5. Study the hurricane history of a segment of the North American coastline. Annual summaries of hurricane tracks are published in the periodical, *Weatherwise* (February issue), as well as by NOAA, and newspaper files are excellent sources for day-by-day accounts of weather events. As an alternative do a similar study of tornadoes, or memorable midlatitude cyclones in some region—such as those that paralyze urban areas for days because of snow removal problems.

6. What would be the long-range effect of a workable program to stop the growth of every storm that threatened to be destructive or costly?

7. What have been the greatest floods in the nation's history? Were they produced by similar meteorological events?

FURTHER READING

Anthes, Richard A., et al. *The Atmosphere*. 2nd ed. Columbus, Ohio: Merrill (1978), 442 pp. This introductory text has an interesting historical perspective. Other unusual features include snyoptic and seasonal analyses of midlatitude weather.

Eagleman, Joe R. *Meteorology: The Atmosphere in Action*. New York: D. Van Nostrand (1980), 384 pp. This very interesting new text emphasizes aspects of the circulation, and interrelationships to the upper air flow. Part 3 is especially helpful.

Hidore, John J. *Workbook of Weather Maps*. 3rd ed. Dubuque, Iowa: Brown (1976), 81 pp. This paperbound compilation of official U.S. weather maps for eleven series of weather events includes the record polar outbreaks of January and February 1962 and Hurricane Agnes in June 1972.

Hughes, Patrick. *American Weather Stories*. Washington, D.C.: U.S. Dept. of Commerce, NOAA (1976), 114 pp. Brief accounts of weather and climate events that had major impacts on our climate.

Lehr, Paul E., R. Will Burnett, and Herbert S. Zim. *Weather*. New York: Golden Press (1975), 160 pp. This excellent paperback is one of the well-known Golden Nature Guide series. The color diagrams of fronts and midlatitude cyclones are especially informative.

Muller, Robert A. "Snowbelts of the Great Lakes." *Weatherwise*, vol. 19, no. 6 (1966): 248–255. Focus is on weather events that produce persistent and deep snowfalls over small areas of the Northeast. Anyone interested in weather and its consequences should enjoy each issue of this journal.

Neiburger, Morris, James Edinger, and William Bonner. *Understanding Our Atmospheric Environment*. San Francisco: Freeman (1973), 293 pp. Basic fundamentals are presented in a nonmathematical framework. Includes a chapter on modern forecasting techniques.

Simpson, Robert H., and Herbert Riehl. *The Hurricane and Its Impact*. Baton Rouge: LSU Press (1981), 420 pp. An excellent new book by two of the outstanding American experts on tropical storms. In addition to meteorological processes, this book emphasizes what can happen to people and buildings along the coastline.

Stewart, George R. *Storm*. New York: Random House (1941), 349 pp. This fictional classic follows the evolution of a midlatitude cyclone, named Maria, across the Pacific and then over the United States. Much of the perspective is through the eyes of forecasters at San Francisco and managers and workers of transportation and communications networks. This novel is a must for students interested in interactions between weather and human endeavors. (Reprinted in paperback in 1974 by Ballantine Books, New York.)

Weems, John Edward. *A Weekend in September*. New York: Henry Holt (1957), 180 pp. This is a careful journalistic account of the 1900 hurricane that swept Galveston, Texas, with a great loss of life.

CASE STUDY: **The Tornado Hazard**

Of all the earth's winds, tornadoes are the most feared. Along the track of a severe tornado, destruction is nearly total, while just a few meters away there may be no damage at all. Tornadoes are especially frightening because of their sudden appearance. At night or in densely forested country, it is difficult to see the funnel cloud of an approaching tornado; the sound, frequently compared to the roar of an express train, is sometimes the first warning. The National Weather Service has an effective program of tornado watches for alerting the public to areas where tornadoes may break out, but there is no way to predict the precise time and place of occurrence.

A *tornado* is a narrow vortex of rapidly whirling air. It is almost always associated with severe thunderstorm activity. The rotating vortex extends downward from a cumulonimbus cloud, and it becomes visible when water vapor condenses, producing the familiar *funnel cloud*. Dust and debris swept up from the ground create a much darker and more ominous-looking funnel. Occasionally, several funnels may dangle from the same cloud, and many funnel clouds aloft never reach the ground. As the mother cloud moves on, the funnel is often retarded at the surface by friction, so that it becomes tilted or crooked. Tornado funnels over coastal waters and seas are called *waterspouts*.

Tornadoes usually advance at a speed of 32 to 48 km/hour (20 to 30 mph). Thus a tornado will normally pass a given point in a matter of seconds. The tornado and its track along the surface are usually only a few meters wide, although occasionally extending up to several hundred meters. Some tornadoes skip across the landscape, leaving a broken track of destruction. Most tornado tracks are 5 to 10 km (3 to 6 miles) in length, but a few are much longer; on May 26, 1917, a single tornado tracked more than 400 km (250 miles) across Illinois and Indiana in a little less than eight hours.

Although tornadoes are short-lived, they are extremely violent. The winds of the tornado vortex have been estimated to reach speeds of up to 600 km (370 miles)/hour. Within the funnel cloud, atmospheric pressure is as much as 50 millibars lower than the adjacent air, and pressure drops of up to 100 millibars have been estimated from damage patterns. As a tornado passes over a building, the strong winds of the vortex rip at the exterior, and the abrupt pressure drop causes the building literally to explode, with the roof and walls blown out. The debris is then caught by the rotating winds and strewn along the tornado's path; each piece of debris becomes a flying missile of destruction. Cellars, interior halls, and bathrooms offer greater safety than rooms with outside walls. When severe tornadoes pass over modern slab homes, sometimes only the plumbing fixtures, such as bathtubs and toilets, remain. House trailers are especially vulnerable to tornado winds.

In the United States, weather situations favorable for tornado development occur most frequently over the Great Plains, the Mississippi Valley, and the Southeast. The region extending from the Texas Panhandle northeastward across Oklahoma and eastern Kansas is sometimes called "tornado alley." Tornadoes are very infrequent west of the Rocky Mountains and across northern New

England and the upper Great Lakes region. The mean annual number of tornadoes reported has increased significantly in recent decades. Since 1970 the annual average has been about 900. The recent increase is attributed mainly to improved observations and detection.

Tornadoes break out most commonly along thunderstorm or squall lines ahead of cold fronts, where temperature contrasts are large and instability great. In the United States instability is usually associated with a deep layer of warm, dry *cT* air from the Southwest above warm, moist *mT* air from the Gulf of Mexico. Tornadoes usually occur in the warm sector, and track from the southwest toward the northeast.

The seasonal distribution of tornado outbreaks tends to follow the geographical distribution of fronts separating *mT, cP,* and *cT* air masses, midlatitude cyclones, and

A close view of the tornado which swept across sections of Dallas, Texas, on April 2, 1957. (ESSA Weather Bureau)

maximum instability through a deep layer of the troposphere. In the United States, therefore, tornadoes are most frequent in the South during late winter and early spring, with the hazard migrating northward into Kansas and Missouri, and finally to Iowa and Nebraska by June, when there are few tornadoes near the Gulf Coast. Many tornadoes occur during late afternoon, when instability tends to be greatest, but unfortunately occurrences remain relatively common at night, when visual detection is difficult.

(top) This map shows the number of tornadoes that were reported between 1955 and 1967 in the United States. The main features of the map include "tornado alley" over the Great Plains and the very small number of tornadoes westward from the Rocky Mountains.

(bottom) This map shows the direction and path lengths of tornadoes that occurred in Kansas between 1950 and 1970. Most tornadoes are dragged along within thunderstorms by the upper winds from the southwest that steer the thunderstorm cells. (Joe R. Eagleman, Vincent U. Muirhead, and Nicholas Willems, *Thunderstorms, Tornadoes, and Building Damage*, 1975, Lexington Books)

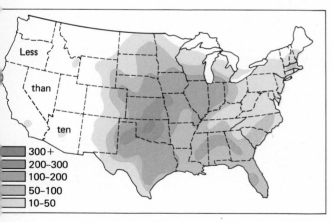

Less than ten

300+
200–300
100–200
50–100
10–50

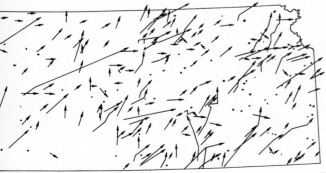

The sun's radiant energy provides the power for the atmospheric circulation to deliver energy and moisture to the earth's surface. On the land the water interacts with the rocks, soils, and vegetation. In time, most of it returns to the atmosphere in a never-ending cycle of renewal.

Rainy Season in the Tropics by Frederick E. Church, 1866. (Fine Arts Museum of San Francisco)

The Hydrologic Cycle and the Local Water Budget

The processes by which water in various states moves from the ocean to the atmosphere to the land and back to the ocean are part of the *hydrologic cycle*. The oceans are vital to this cycle since water evaporated from them is the source of rainfall that supports life on the land. The oceans are also enormous reservoirs of heat energy, and their temperature variations influence the general circulation of the atmosphere, which powers the hydrologic cycle.

On the continents the hydrologic cycle is

much more complicated than over ocean areas, where we need think only of the processes of evaporation and precipitation. Consider, for example, the state of Washington. Inland from the gently rolling hills surrounding Puget Sound and the city of Seattle, the land rises abruptly to a spine of high mountains, the Cascade Range that divides the state. To the west of the Cascades, rain falls much of the year, brought by storm clouds that sweep in from the North Pacific. As much as 2,500 cm (1,000 in.) of snow can accumulate in a year on major peaks like Mt. Rainier, and water is almost always present in the streams that rush westward toward the coast. But the land to the east of the Cascades is a region of dry grasslands, cut off from the coastal weather by the mountains. Here are the sun-drenched fields and orchards of the Yakima and Wenatchee valleys and the Grand Coulee area. In this region irrigation is necessary to produce crops. The contrasts on either side of the Cascades and many similar mountain systems illustrate how different and complex the hydrologic regime can be on the land, and how human activities are adjusted to accommodate these differences.

If we add up all the ways in which water is used, we find that the average daily consumption in the United States is equal to 6 cubic meters (1,500 gal) per person—not including water used for the generation of hydroelectricity. The public uses only about 10 percent directly for such things as cooking, sanitation, and lawns and gardens. Industry uses another 10 percent. The remaining 80 percent is divided about equally between agriculture and thermal power plants that use water for cooling.

If all this water were truly consumed and permanently removed from the hydrosphere, disastrous water shortages would develop in short order. But water is a renewable resource. Nearly all the water used on farms, in homes, and by industry takes some path back into the hydrologic cycle. Most of the water diverted from rivers or pumped from wells to be used in irrigation is returned to the atmosphere through evapo-

transpiration in the fields. The water used in cities and in suburban homes is usually channeled into sewage systems to return to surface streams and finally to the ocean.

Throughout history people have labored to ensure reliable water supplies. As early as five thousand years ago, the floodwaters of the Nile River in Egypt were channeled through long canals to vast basins surrounded by clay dikes. Aqueducts carried to ancient Rome a water supply that was ample even by modern standards. Today similar aqueducts carry water hundreds of kilometers to the Los Angeles basin and agricultural areas in the California desert (see Case Study following this chapter). Water projects of amazing scale, involving gigantic reservoirs and transfers over thousands of kilometers, have been proposed, but costs and environmental issues have so far prevented their development.

The need for fresh water grows ever more pressing as populations increase and expand into areas where natural supplies are scarce. Because of this growing need to manage available water, there is increasing emphasis on the study and analysis of water on and below the earth's surface—a field known as *hydrology.*

In this chapter we shall first review the global water budget before examining in detail the pathways of water on the continents. A key concept in this chapter is use of the local water budget as a tool for evaluating the availability of water at a particular place.

THE GLOBAL WATER BUDGET

As we saw in Chapter 2, the water of the earth's hydrosphere is stored in several different conditions. Figure 7.1 shows that the oceans, which cover about 70 percent of the earth's surface, contain nearly 98 percent of the total water supply. Since ocean water contains dissolved minerals, primarily salt, it is unfit for consumption by humans, land animals, and even land plants. But when water evaporates from the

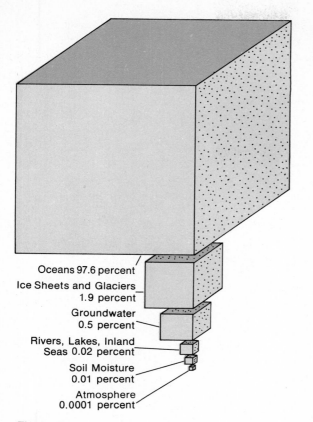

Oceans 97.6 percent

Ice Sheets and Glaciers 1.9 percent

Groundwater 0.5 percent

Rivers, Lakes, Inland Seas 0.02 percent

Soil Moisture 0.01 percent

Atmosphere 0.0001 percent

Figure 7.1
The volumes of the cubes show the relative amounts of free water in storage on the earth. Nearly 98 percent of the water is stored in the oceans, which contain an estimated volume of 1.3 billion km³ (0.3 billion mi³) of water. Glaciers contain the largest store of fresh water, but the turnover is slow. Most of the readily available fresh water is stored in porous rock beds as groundwater. The amount of water stored in the atmosphere is relatively small, but because it is actively transported and released, it plays a key role in the hydrologic cycle. (Tom Lewis after R. L. Nace, *Water, Earth, and Man*, edited by R.J. Chorley, 1969, Methuen & Co., Ltd., Publishers)

oceans, the dissolved salts are left behind. Hence, water vapor evaporated from the oceans is a source of liquid water that is fresh and essentially "pure."

About three-fourths of the earth's nonsaline fresh water is stored as glacial ice, mainly in the ice sheets covering Antarctica and Greenland. These ice sheets receive an annual snowfall equivalent to only about 10 cm (4 in.) of liquid water. Glacially stored water returns to the oceans centuries or even thousands of years later, when icebergs break away from coastal glaciers and melt in the sea. The ice sheets, then, represent long-term storage of fresh water.

The next largest reservoir, accounting for only one-half of 1 percent of the total water supply, is groundwater. Much of this is unusable because of mineral contamination or problems of extraction. Surface water supplies are an even smaller part of the total, as is moisture stored in the soil.

The amount of water stored in the atmosphere at any one time is even less. As noted previously, all the water in the atmosphere at any moment would form a layer only 2.5 cm (1 in.) deep over the earth. Over an entire year, however, the atmosphere transports and recycles enough water to cover the earth with a layer about 95 cm (37 in.) deep.

For the global water budget to remain in balance, the amount of water that leaves the atmosphere as precipitation must return to the atmosphere through evapotranspiration. Figure 7.2 (see pp. 156-157) shows how the global water budget is kept in balance. If we consider the oceans alone, we find that precipitation into them is less than evaporation from them. Over the continents precipitation exceeds evapotranspiration. But ocean levels are not falling, and the continents are not becoming flooded. These imbalances are offset because water continues to be exchanged between the oceans and continents. The land sheds its excess precipitation by contributing moisture to continental air masses that move out over the oceans, and by the flow of rivers to the sea.

Transport of Water Vapor from the Oceans (94)

(12)

Storage as Ice and Snow

Precipitation over the Continents (106)

Evapotranspiration from the Continents (69)

Interception by Plants

Temporary Surface Storage

Surface Runoff

Infiltration

Soil Moisture Storage

Percolation

Groundwater Storage

Storage in Rivers and Lakes

Groundwater Runoff to Streams

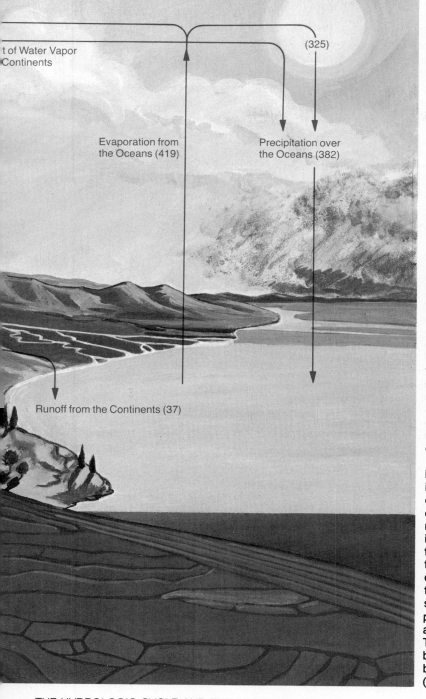

t of Water Vapor
Continents

(325)

Evaporation from
the Oceans (419)

Precipitation over
the Oceans (382)

Runoff from the Continents (37)

Figure 7.2
The movement of water through the hydrologic cycle involves numerous interactions and storage processes. Each year about 419,000 cu km of water evaporate from the oceans into maritime air masses, and evapotranspiration from the continents into continental air masses amounts to an additional 69,000 cu km (values in parentheses in the figure are water volumes in 1,000 cu km). This total volume of 488,000 cu km is equivalent to a mean annual precipitation over the globe of about 95 cm (37 in.).

Precipitation back to the oceans amounts to only 382,000 cu km, with 325,000 cu km originating from maritime air masses and 57,000 cu km supplied by continental air masses that move over ocean areas. Over the continents, precipitation (106,000 cu km) is greater than evapotranspiration (69,000 cu km); the difference (37,000 cu km), represents runoff in streams and rivers that eventually returns to the oceans. Note that precipitation over the continents is supplied largely by water vapor from the oceans.

When precipitation falls on the land, a portion of the moisture is intercepted by vegetation and evaporates from temporary storage on leaves. The moisture that reaches the ground either infiltrates the soil, runs off across the surface, or evaporates from temporary storage in pockets and depressions. Some of the water that infiltrates the soil is stored as soil moisture, and a portion percolates deeper into the ground and enters groundwater storage. The flow of streams is maintained both by direct surface runoff and by groundwater contributions. (John Dawson)

WATER ON THE LAND

What happens to precipitation that falls on the continents? Depending on the characteristics of the land surface, this water can take several different pathways back to the atmosphere or to the seas to complete the hydrologic cycle.

Interception, Throughfall, and Stemflow

Not all the precipitation that falls reaches the soil. In urban environments, rain strikes roofs, building walls, and areas paved with concrete and asphalt. Most of this water runs off quickly into gutters and subsurface storm drains that empty into streams, lakes, or the ocean. The little that remains in puddles eventually evaporates, returning directly to the atmosphere.

Beyond urban areas, except where it is arid, plant growth covers the surface much of the year and prevents some rain from reaching the soil. When rainfall commences, leaves catch and store much of the water; Figure 7.3 shows how leaves *intercept* raindrops. If the rain is heavy or lasts very long, the capacity of the leaves to retain water is exceeded, and water begins to drip down to the soil as *throughfall*. Trees become less effective shelters as rainfall continues. Water also reaches the soil as *stemflow* by trickling along branches and down the trunks of trees. This change in the route of falling water diverts an above-average amount to areas at the bases of plants.

After the rain ends and leaves stop dripping, some water remains on the leaves and evaporates back into the atmosphere. Because of in-

Figure 7.3
Some of the precipitation that is intercepted by vegetation and held in temporary storage by leaves returns to the atmosphere by evaporation and does not reach the soil. (David Cavagnaro)

terception, the amount of water reaching the soil is less than the total precipitation, and the timing and distribution of water arrival at the surface are changed.

Infiltration and Soil Moisture Storage

A few days after a rainfall the surface soil has usually dried out, but you can find moist soil by digging down a few centimeters. Water enters, or *infiltrates*, soil from the surface downward. The maximum rate at which the soil can absorb water is the soil's *infiltration capacity*. Infiltration capacity depends upon the soil's porosity, permeability, surface condition, and moisture content.

Soil consists of fine grains of mineral and organic matter of various sizes (Chapter 10). Water moves through the spaces, or *pores*, between grains. Clay soils have much smaller grains and less pore space than sandy soils. If the pores are large and interconnected, as in sandy soils, the soil is *permeable*. Gravity pulls water downward through permeable soils; water also rises upward toward the surface through capillary action.

The infiltration rate depends upon the condition of the surface layer. If the soil has been compacted by vehicles, animals, or human traffic along a path, the soil may not be able to absorb moisture. Bare soil can also be compacted by rain itself. A cover of vegetation absorbs the impact of raindrops and helps maintain higher infiltration rates. In most situations, heavy grazing by sheep or cattle reduces the vegetation cover and significantly lowers infiltration rates.

If water arrives at the surface at a rate less than the infiltration capacity, all the water will be absorbed. Once the upper layers of soil are saturated, additional water cannot enter until water already in the soil begins to drain to lower levels. As Figure 7.4a (see p. 160) illustrates, dry porous loam (a mixture of coarse and fine soil particles) can absorb water at an initial rate of over 6 cm (2.4 in.) per hour. As the soil becomes wet, infiltration rapidly decreases to a slow but constant rate.

As the water in saturated soils drains, or *percolates*, downward, molecular attraction causes some water to cling in the pores between soil particles. This water is held most effectively in small pores, as in clay soils. In the large pores of sandy soils, molecular attraction is too weak to support the greater volumes of water against the pull of gravity; this water drains away over several days. The maximum amount of moisture that can remain stored in the soil after percolation is known as the *field capacity* of the soil. Moisture stored in the soil eventually returns to the atmosphere, either by direct evaporation from the soil or by transpiration through the leaves of vegetation.

Vegetation withdraws soil moisture for transpiration throughout its rooting zone. However, some water is unavailable for transpiration because it is held too tightly to the soil particles by molecular attraction. Most plants begin to wilt when the soil moisture is reduced to this level, which is called the *wilting point*. If soil moisture is not replenished by rains or irrigation, the plants will die. Too much soil moisture is also harmful to plants, with the exception of swamp and marsh vegetation. Soil pores must contain air as well as water because roots require oxygen and space to eliminate carbon dioxide waste.

Groundwater

Below the surfaces of the continents lies a vast and accessible reservoir of fresh water. A well-driller will strike water within 100 m (330 ft) of the surface just about everywhere—even in desert areas where there is little surface water. The top of this zone in which all openings are saturated with water is called the *water table*, and the water in the fully saturated zone below the water table is *groundwater*. This water is held in underground deposits of sand and gravel and porous types of rock.

The water table rises and falls in response to

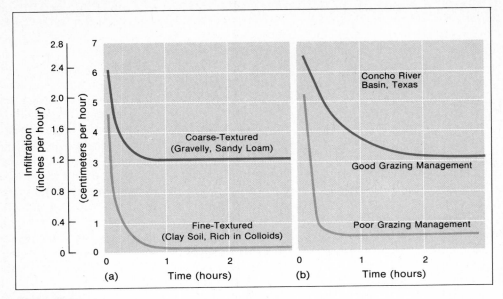

Figure 7.4
The infiltration rate of water into a soil depends on such factors as the texture, porosity, permeability, and moisture content of the soil, and on the condition of the surface layer. The graphs show the infiltration rates of various soils measured from the time at which water is added to their surfaces. The infiltration rate falls sharply at first in all cases as the top layer of soil becomes well moistened; then a constant rate of infiltration is attained.

(a) Water infiltrates a coarse, permeable soil more easily than it does a dense clay soil, in which the water passages are small.

(b) The surface of well-managed grazing land retains an open texture and has a high infiltration rate. The infiltration rate is lower on poorly managed land because overgrazing exposes the soil, which allows the bare surface to become compacted by raindrops and animal hooves. (Doug Armstrong after *Yearbook of Agriculture*, U.S. Department of Agriculture, 1955, and E. E. Foster, *Rainfall and Runoff*, © 1949, Macmillan Co.)

precipitation, evapotranspiration, and groundwater flow. It represents the minimum depth to which a well must be drilled for a reliable supply of water. Deposits from which groundwater may be obtained either naturally (in springs) or arti-ficially (in wells) are *aquifers*. Subsurface aquifers contain about 30 times the amount of fresh water in the streams, lakes, and swamps of the earth.

Under natural conditions the quality of groundwater is usually good. The major exceptions are in arid and coastal regions, where aquifers may be contaminated by dissolved salts. Generally, the porous rock of an aquifer filters the water and removes suspended particles and harmful bacteria. But it is possible for urban and industrial contaminants and agricultural pesticides to seep into aquifers. This is a particular problem where geological material is very permeable, as in areas underlain by sand or gravel deposited by streams or past glaciers, or by limestone that contains interconnecting cavities caused by the dissolving action of groundwater. From New York's Long Island to California's Central Valley, wells have been declared unsafe because of recent contamination by pesticides and industrial wastes.

The problem of maintaining groundwater quality is a growing one. This is important because it is more practical for cities to utilize

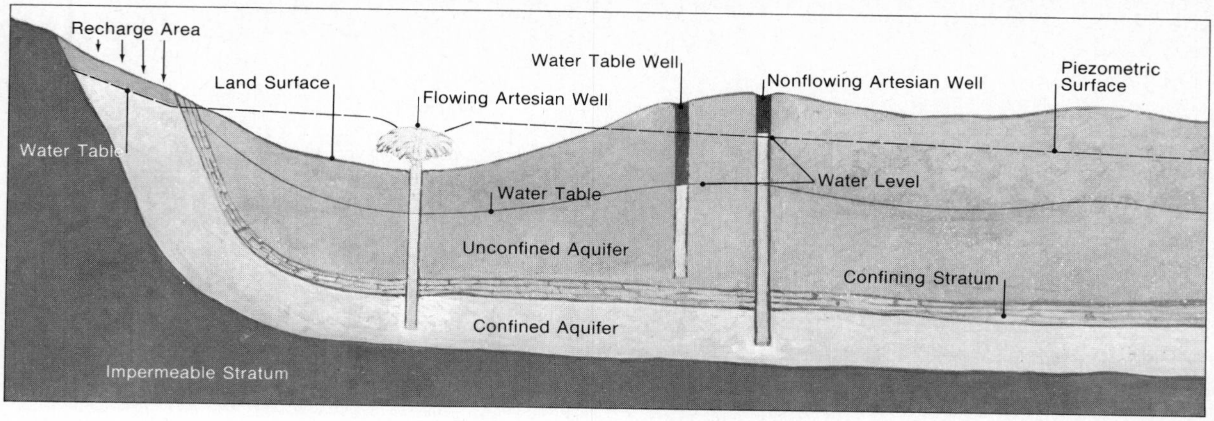
Recharge Area
Land Surface
Water Table Well
Flowing Artesian Well
Nonflowing Artesian Well
Piezometric Surface
Water Table
Water Table
Water Level
Unconfined Aquifer
Confining Stratum
Confined Aquifer
Impermeable Stratum

Figure 7.5
This diagram shows the principal features of aquifers in schematic form. If the rock above an aquifer is permeable enough to allow the vertical movement of water, the aquifer is said to be unconfined. The water table, or the water level in an unconfined aquifer, is the level to which water will rise in a well sunk into the aquifer. If the rock above an aquifer is impermeable, the aquifer is confined and must be replenished from a recharge area that is permeable to water from above. When a well is sunk into a confined aquifer, the level to which the water rises in the well is called the *piezometric surface*. The piezometric surface can be a considerable height above a confined aquifer, particularly in the lower portion of a sloping aquifer, where the water pressure is high. The piezometric surface shown here slopes to the right, toward a region where the aquifers drain slowly into surface streams. (John Dawson after Raphael G. Kazmann, *Modern Hydrology*, 2nd ed., © 1972 by R. G. Kazmann, used by permission of Harper & Row)

comparatively pure groundwater than to purify badly polluted river water. In the midlatitudes the temperature of groundwater from depths of 10 to 20 m (30 to 60 ft) is usually only 1° to 2°C higher than the average annual temperature. The relatively constant cool temperature makes groundwater very attractive for urban and industrial users.

A variety of subsurface materials are porous and permeable enough to act as useful aquifers. Most of the aquifers used in North America are beds of sand and gravel. Some were deposited as outwash from glaciers during the Ice Ages, and others are the result of much earlier deposition. Individual sand and gravel aquifers are commonly 50 m (160 ft) thick and often cover several thousand square kilometers. Among solid rocks, sandstones, limestones, and lava beds form the best aquifers because they have interconnected openings that collect and transmit water.

All the water stored in subsurface aquifers comes originally from precipitation. After field capacity is reached, additional rainfall that does not flow directly over the surface percolates through the ground to the water table. In arid regions, water may also seep downward from stream beds and lake bottoms. In humid areas, by contrast, groundwater seeps *out* into streams and lakes.

The surface region from which water drains into an aquifer is called the *recharge area.* Groundwater usually moves laterally in an aquifer at rates varying from meters per year to kilometers per day. When an extensive aquifer slopes gently for a long distance, the principal recharge area may be hundreds of kilometers distant from the wells where the water is extracted (Figure 7.5).

Aquifers may be either *unconfined* or *con-*

fined. In an unconfined aquifer the water is not under pressure and will not rise above the level of the water table unless it is pumped to the surface. In confined aquifers the water-bearing layer is covered by a layer of impermeable material. Confined aquifers may lie far below the level of the water table. If the *confining layer* slopes downward, the difference in elevation between the upper and lower portions of an aquifer can result in a considerable difference in water pressure. If wells are drilled into a lower section of the confined aquifer, where the water pressure is high, the water in the wells will rise considerably above the level of the aquifer, as is shown in Figure 7.5. The elevation to which water will rise in such wells is known as the *piezometric surface.* The piezometric surface occasionally lies above the land surface. This creates flowing *artesian wells,* from which water gushes out at the surface with no necessity for pumping. Natural *artesian springs* occur where fractures in rocks permit water to escape to the surface from a confined subsurface aquifer.

Water can also collect on top of impermeable layers that lie above the normal water table. This creates *perched water tables.* These sometimes feed springs along canyon walls. Perched water tables are common in areas of layered sedimentary rocks (Chapter 11) and in thick masses of sediment deposited by streams and glaciers.

Approximately 50 percent of all the groundwater extracted in the United States is used for irrigation in Texas, Arizona, and California. Groundwater also supports extensive agricultural development and livestock industries in the Great Plains east of the Rocky Mountains and in eastern Australia, North Africa, Arabia, Iran, and other arid regions. Groundwater must be carefully managed so that the amount of water pumped out does not exceed the net flow into the aquifer. Some aquifers were recharged under climatic conditions that no longer prevail and are receiving little or no input of water today. This is especially true in desert areas. But in many areas the groundwater is being "mined"—the rate of pumping greatly exceeds the natural recharge so that water tables and

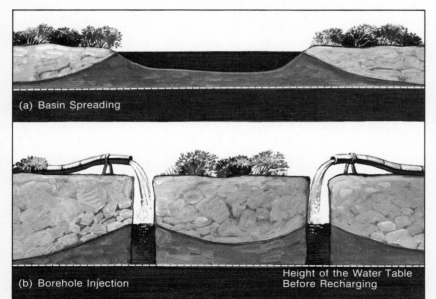

(a) Basin Spreading

(b) Borehole Injection

Height of the Water Table Before Recharging

Figure 7.6
This diagram illustrates two methods that have been used to recharge aquifers and raise the level of the water table in regions where supplies of groundwater have been depleted by excessive withdrawal.

(a) Water pumped into shallow surface depressions in the natural recharge area of an aquifer seeps through the permeable rock into the aquifer.

(b) Water pumped into boreholes sunk into an aquifer seeps into the permeable rock to recharge the aquifer. (John Dawson)

piezometric surfaces are falling. In such areas pumped wells have to be deepened, and former artesian wells have to be pumped. In some areas groundwater withdrawal has caused the land surface to subside by as much as 8 m (25 ft). This disrupts irrigation canals and has caused damage to streets and buildings in such scattered locations as Venice (Italy), Mexico City, Shanghai (China), Tokyo (Japan), and Phoenix (Arizona).

In some areas groundwater is being recharged artificially (Figure 7.6). This is common in southern California, using water brought some 600 km (400 miles) by aqueduct. Artificial recharge is also used in coastal regions where excessive pumping of groundwater has lowered the water table, allowing saline sea water to seep under the land, contaminating aquifers. This problem has appeared on Long Island, in southern California, in Israel, and in many other coastal areas.

Runoff and Streamflow

If rain falls at a rate greater than the infiltration capacity of the soil, water begins to collect on the surface. Surface irregularities and vegetation store some of the water for evaporation or later infiltration, but water also begins to trickle across the ground or move in sheets under the influence of gravity. This *surface runoff* increases from small rivulets at the beginning of a heavy rainstorm to a steady torrent when no more storage opportunities are available. Toward the end of a prolonged rain, nearly all the rainfall may become surface runoff.

Virtually all land surfaces not covered by ice or loose sand are laced with networks of small erosional channels created and maintained by surface runoff. The initial streams carry water to a smaller number of larger streams that, in turn, feed major rivers (Chapter 13). The runoff carried to the sea by rivers represents precipitation that did not go into subsurface storage or return to the atmosphere by evapotranspiration.

Hydrologists are concerned with the variable flow of rivers over time, especially the effect of precipitation on runoff. *Runoff*, used technically, is a measure of the average depth of water that flows from a drainage basin during a specified amount of time. Let us imagine that an intense rainstorm lasting an hour dropped 2.5 cm (1 in.) of rain on an impervious parking lot equipped with storm drains. The storm drains are efficient, and 15 minutes after the storm an average depth of only 0.05 cm of water, concentrated in a few depressions and cracks, is left on the parking lot. Later this water will return to the atmo-

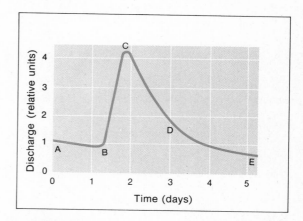

Figure 7.7
This graph shows the effect of an upstream rainstorm on the amount of water carried by a stream, measured from the time of the storm. From *A* to *B*, water from the rain has not reached the stream and the discharge, or rate of flow, measures the base flow supplied by groundwater. From *B* to *C*, the discharge rises rapidly as direct surface runoff from the upstream drainage basin reaches the stream. From *C* to *D*, the discharge falls slowly as the last of the surface runoff, including the runoff retarded by the vegetation cover, makes its contribution. From *D* to *E*, the discharge again primarily measures groundwater supplies, which have been newly recharged by the rain. The discharge of the stream gradually decreases as groundwater inflow to the stream diminishes. (Doug Armstrong after R. C. Ward, *Principles of Hydrology*, © 1967, McGraw-Hill Book Co. (U.K.) Ltd., used with permission)

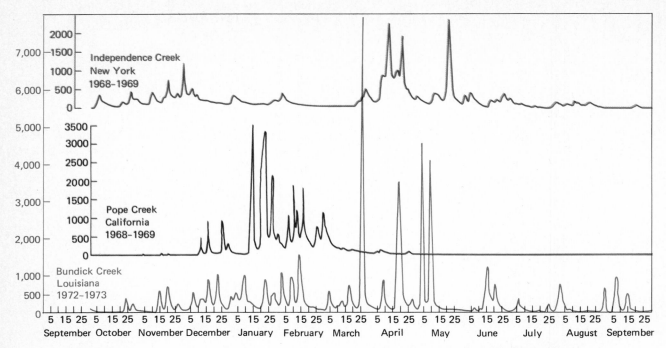

Figure 7.8

Hydrographs for three small streams in different climate regions, each draining an area between 194 and 310 sq km (75 and 120 sq miles). The hydrographs are organized by water years, which begin October 1 and end September 30. The water year is a useful calendar for water-resource management because streamflow tends to be lowest in late September, with minimum groundwater outflow to support base flow. Typically, storm runoff is superimposed as spikes, or peaks, on the base flow contributed by groundwater. Storm runoff tends to rise quickly and to recede more slowly.

Bundick Creek is located in the warm and humid climate of southwestern Louisiana. Groundwater provides for some base flow all year, but streamflow is largest on the average during winter and spring. A considerable proportion of the annual flow is produced in just a few days by the very large spring flood flows, and the maximum daily discharge of 7,980 cfs on March 25, 1973, was the highest daily flow in 17 years of measurements.

Independence Creek is located on the western flanks of the Adirondack Mountains in northern New York. This watershed is representative of climates where persistent low temperatures during winter allow most of the precipitation to accumulate as snowpack. Much of the annual flow occurs during spring as the snow melts, but there is usually a secondary discharge peak in autumn before the winter snows begin to accumulate.

Pope Creek drains a low mountainous region of the coastal ranges of northern California west of Sacramento. The climate is hot and dry during summer with no precipitation. Groundwater contributes to streamflow only during winter and spring, and normally there is no water in the creek from June through October. (Vantage Art, Inc.)

sphere by evaporation. The runoff from the parking lot amounts to 2.45 cm, or 98 percent of the rainfall. At the same time, we know that the runoff from an equal area of nearby parkland would be far smaller, if there were any runoff at all, for much of the rainfall there would be stored in the soil, eventually returning to the atmosphere by evapotranspiration.

Streamflow, on the other hand, represents the volume of water flowing down a stream channel during a short time period. It is usually expressed as *stream discharge* in cubic feet per second (cfs) or cubic meters per second (Chapter 13). A plot of stream discharge fluctuations is called a *stream hydrograph*; Figure 7.7 shows a hypothetical example for a stream of intermediate size in a humid region. The stream continues to flow between rains due to an almost steady input of groundwater that seeps into the channel. This minimum flow between storms is known as the *base flow*. During or shortly after each rainstorm, the water table rises and surface runoff reaches the stream. Figure 7.7 shows the characteristic rapid rise and slower recession associated with storm runoff, groundwater outflow, and recharged groundwater supplies. Daily fluctuations in discharge for three small streams in different climatic regions are shown and explained in Figure 7.8. The peaks rising from the base flow are caused by surface runoff during storms.

THE LOCAL WATER BUDGET: AN ACCOUNTING SCHEME FOR WATER

Water is not always available in the desired amounts just when and where we need it. The supply of moisture is so variable that a special technique is needed to estimate moisture availability. This is why climatologists and hydrologists make use of the *water budget* concept. The water budget is the local version of the hydrologic cycle. It takes into account four principal components of water distribution: precipitation, soil moisture, evapotranspiration, and runoff. When appropriate, snow accumulation and snowmelt may also be included. Of all these components, evapotranspiration is the most difficult to estimate accurately because of its dependence on complex meteorological and biological factors.

Potential and Actual Evapotranspiration

Measurements of actual evapotranspiration are limited to detailed studies of small field plots using expensive instrumentation. To overcome the difficulties of actual measurement, and to allow water budget components to be studied over large regions, the American climatologist C. Warren Thornthwaite introduced the concept of *potential evapotranspiration* (*PE*) and developed formulas for estimating *PE* under varying conditions. Potential evapotranspiration, the key to the water budget, is worth examining in detail.

Potential evapotranspiration is the rate at which water would be lost to the atmosphere from a land surface completely covered by growing vegetation that has been supplied with all the soil moisture it can use. *PE* is normally expressed as the depth of liquid water that is converted to vapor in a given time. A typical value for *PE* during a summer month in the eastern United States is about 15 cm (6 in). During a winter month in the same region, *PE* is usually less than 2.5 cm (1 in).

Evapotranspiration normally proceeds at the potential rate as long as moisture is readily available in the soil. After a number of days without rain, however, soil moisture becomes partly exhausted, and it is increasingly difficult for plants to extract the remaining moisture. The actual rate of evapotranspiration will then fall below the potential rate. For this reason Thornthwaite distinguished between *PE* and *actual evapotranspiration* (*AE*). During wet seasons, the *AE* will be the same as the *PE*, but in prolonged dry periods, the *AE* is less than the *PE*. The term "actual evapotranspiration" is somewhat misleading; in Thornthwaite's water budget analysis, the *AE* is an estimation, rather than an "actual" measurement.

Solar radiation is the principal factor that determines *PE*. In fact, one way to think about potential evapotranspiration is in terms of solar energy input and utilization. Both evaporation and transpiration require the expenditure of about 590 calories per gram of water. An estimate of *PE*, therefore, would require knowing the duration of daylight, the amount of cloud cover, and the albedo (or reflectability) of the plant cover.

Potential evapotranspiration from a field, or even an entire landscape, is almost independent of the type of plant cover. Imagine looking down from an airplane on a forest or field during the middle of the growing season. The plants usually present a nearly uniform cover of overlapping green leaves. The albedo of almost all green plants is from 15 to 20 percent. Therefore,

an acre of forest and an acre of soybeans would absorb about the same amount of solar radiation on a clear day.

Thornthwaite's estimation of monthly *PE* depends mostly on two factors: (1) average monthly air temperature, and (2) latitude, which controls the length of the daylight period from month to month. Data on temperature and length of daylight are available for most places on earth. Together, these two factors give an indication of solar energy input, which is itself measured in only a few places. Estimated average annual *PE* variations over the United States, calculated from Thornthwaite's formula,

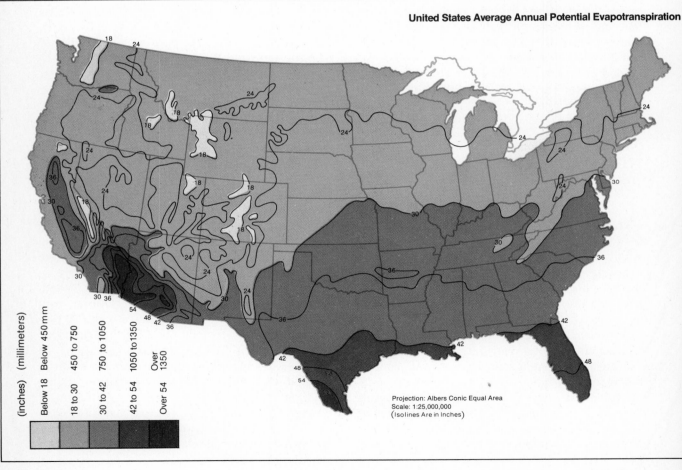

United States Average Annual Potential Evapotranspiration

(inches)	(millimeters)
Below 18	Below 450 mm
18 to 30	450 to 750
30 to 42	750 to 1050
42 to 54	1050 to 1350
Over 54	Over 1350

Projection: Albers Conic Equal Area
Scale: 1:25,000,000
(Isolines Are in Inches)

are shown in Figure 7.9. Thornthwaite's formula is widely used for regional climatic resource analyses, but much more complex formulas that require more data have been devised for detailed local studies.

Potential evapotranspiration at a particular location can be measured in a device called an *evapotranspirometer*, which is an open tank about 60 cm (2 ft) in diameter and 90 cm (3 ft) deep that is sunk flush with the ground surface. The tank is filled with soil and usually planted with a cover of grass. If the grass is watered adequately, soil moisture is always available for evapotranspiration. Once the soil is wetted to field capacity, a further weight increase represents water added to the tank by precipitation or irrigation; a weight loss, on the other hand, represents evapotranspiration and percolation measured as it collects in a false bottom. The various measurements comprise a water budget of the tank. As long as there is no water shortage in the tank, AE is equal to PE.

Regular operation of an evapotranspirometer is expensive and requires skilled personnel. Therefore measurements have been limited to experimental studies, mostly in dry regions where it is important to know how much water is enough for a particular type of vegetation.

Figure 7.9 (opposite)
This map shows the average annual potential evapotranspiration, or *PE*, for the coterminous United States, calculated from Thornthwaite's formula. The Southwest, Texas, and Florida have high values of *PE* because of greater solar radiation income and very warm weather. *PE* is much lower farther north because of decreased solar radiation income in winter and much cool and cloudy weather. Across the high mountain areas of the West, where temperatures are low, the *PE* is also low. There is also a strong seasonal regime of *PE* throughout the United States: *PE* is low in winter and high in summer, although this seasonality is least marked along the West Coast and across the South. (Andy Lucas and Laurie Curran adapted from *Geographical Review*, vol. 38, © 1948, American Geographical Society of New York)

Calculating the Water Budget

The local water budget simply represents a systematic accounting of the input, output, and storage of water at a location (see Figure 7.10). The computation is essentially a comparison between *PE* and *AE*. If *PE* exceeds *AE*, there is a

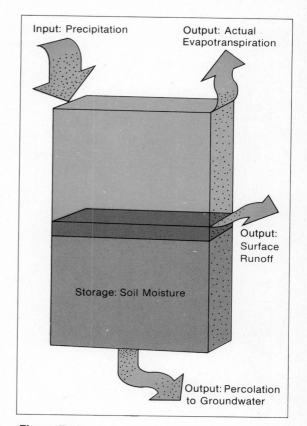

Figure 7.10
This schematic diagram illustrates the principal components of the local water budget. The input to the system of soil and vegetation is the amount of moisture supplied by precipitation. A large portion of the input returns to the atmosphere by evaporation and plant transpiration. Some of the moisture input is stored in the soil. However, the storage capacity of the soil is limited; when the storage is full, a moisture surplus becomes available to supply surface runoff and groundwater recharge. (Tom Lewis)

water deficit. If *AE* equals *PE*, there may be a water surplus that can produce soil moisture storage or surface runoff. In the computation, the incoming precipitation (*P*) during a given time period is allocated to evapotranspiration (*AE*), soil moisture recharge (ΔST), and surplus (*S*). Surplus represents surface runoff and groundwater recharge, and eventually finds its way to streams and rivers. If *PE* exceeds *AE*, there is no soil moisture recharge and no runoff; there is not enough water available to satisfy the climatic demand for it.

A Local Water Budget: Baton Rouge

To illustrate the local water budget, consider the example of Baton Rouge, Louisiana, in 1962. In an average year, Baton Rouge receives more than 125 cm (50 in) of rainfall. Although Baton Rouge experiences one of the highest average annual rainfall totals for cities within the 50 states, a water budget analysis shows that water deficits existed and that crops in the Baton Rouge area needed irrigation water during several months of 1962.

To illustrate the 1962 water budget at Baton Rouge, water income, output, and storage are considered month by month in Figure 7.11. Weekly or daily periods could also be used. The

Figure 7.11
This is the local water budget for Baton Rouge, Louisiana, during 1962. The water budget equation is $P = AE + S \pm \Delta ST$. For 1962 the equation works out: $52.9 = 39.3 + 15.7 - 2.1$. Similarly, the energy budget equation is $PE = AE + D$, and for 1962 this is $41.9 = 39.3 + 2.6$. Each equation balances, so we can be quite confident that we have not made any calculation errors. (Doug Armstrong)

	JAN	FEB	MAR	APR	MAY	JUN	JUL	AUG	SEP	OCT	NOV	DEC	Total
1. Precipitation (*P*)	6.4	0.7	3.3	9.7	1.6	11.4	2.0	4.5	4.3	5.2	0.9	2.9	52.9
2. Potential Evapotranspiration (*PE*)	0.4	1.7	1.3	2.6	5.3	6.1	7.4	6.8	5.2	3.4	1.1	0.6	41.9
3. Precipitation minus Potential Evaporation (*P-PE*)	6.0	−1.0	2.0	7.1	−3.7	5.3	−5.4	−2.3	−0.9	1.8	−0.2	2.3	11.0
4. Change in Stored Soil Moisture (ΔST)	0	−1.0	1.0	0	−3.7	3.7	−5.4	−0.6	0	1.8	−0.2	2.3	−2.1
5. Total Available Soil Moisture (*ST*)	6.0	5.0	6.0	6.0	2.3	6.0	0.6	0	0	1.8	1.6	3.9	——
6. Actual Evapotranspiration (*AE*)	0.4	1.7	1.3	2.6	5.3	6.1	7.4	5.1	4.3	3.4	1.1	0.6	39.3
7. Deficit (*D*)	0	0	0	0	0	0	0	1.7	0.9	0	0	0	2.6
8. Surplus (*S*)	6.0	0	1.0	7.1	0	1.6	0	0	0	0	0	0	15.7

Figure 7.12
Average water budget for Baton Rouge, Louisiana, based on standard climatological data for 1941–1970.

(a) This average budget is based on 30 years of temperature and precipitation data. Average monthly precipitation is much less variable than monthly precipitation in individual years, such as 1962 in Figure 7.11. In this particular average water budget, *AE* is always equal to *PE*, and there is no deficit (*D*). The distinction between the components of an average water budget and the budget of an individual year (Figure 7.11) should be kept clear.

(b) This graph of the average water budget for Baton Rouge is an example of standardized graphs that have appeared in research publications. In graphical form it shows the seasonal regimes of *P, PE, AE,* and *S,* as well as soil moisture withdrawal and recharge. (Vantage Art, Inc.)

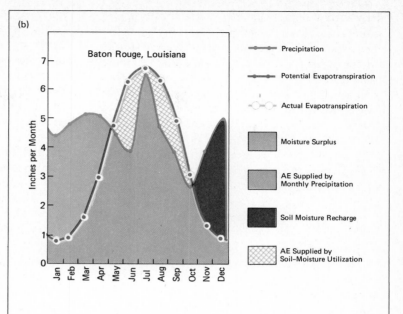

(b) Baton Rouge, Louisiana

Legend:
- Precipitation
- Potential Evapotranspiration
- Actual Evapotranspiration
- Moisture Surplus
- AE Supplied by Monthly Precipitation
- Soil Moisture Recharge
- AE Supplied by Soil-Moisture Utilization

(a)	Jan	Feb	Mar	Apr	May	Jun	Jul	Aug	Sep	Oct	Nov	Dec	Year
1. Precipitation (*P*)	4.4	4.8	5.1	5.1	4.4	3.8	6.5	4.7	3.8	2.6	3.8	5.0	54.0
2. Potential Evapotranspiration (*PE*)	0.7	0.9	1.6	3.0	4.7	6.2	6.7	6.3	4.9	2.8	1.2	0.8	39.8
3. Precipitation minus Potential Evapotranspiration	3.7	3.9	3.5	2.1	−0.3	−2.4	−0.2	−1.6	−1.1	−0.2	2.6	4.2	14.2
4. Change in Stored Soil Moisture (Δ*ST*)	0	0	0	0	−0.3	−2.4	−0.2	−1.6	−1.1	−0.2	2.6	3.2	0
5. Total Available Soil Moisture (*ST*)	6.0	6.0	6.0	6.0	5.7	3.3	3.1	1.5	0.4	0.2	2.8	6.0	—
6. Actual Evapotranspiration (*AE*)	0.7	0.9	1.6	3.0	4.7	6.2	6.7	6.3	4.9	2.8	1.2	0.8	39.8
7. Deficit (*D*)	0	0	0	0	0	0	0	0	0	0	0	0	0
8. Surplus (*S*)	3.7	3.9	3.5	2.1	0	0	0	0	0	0	0	1.0	14.2

THE HYDROLOGIC CYCLE AND THE LOCAL WATER BUDGET

first row in Figure 7.11 shows monthly precipitation in inches. The second row shows monthly *PE* estimated from Thornthwaite's formula. Since *PE* tends to follow the seasonal regime of temperature, it is much smaller in winter than summer. The third row shows the difference between precipitation and *PE*. When *P − PE* is positive, we have a "wet" month, and when *P − PE* is negative, we have a "dry" month. For 1962, six months were wet, and six were dry.

Soil moisture is the storage component within the local water budget. Because 1961 had been a wet year, 1962 began with the soil moisture storage in Baton Rouge at its full capacity of 6.0 in. This storage capacity figure is representative of the Baton Rouge area; on a global basis, soil moisture storage capacity ranges from about 2 to 12 in. Our simple water budget will assume that plants use all the water they need from the soil until there is no more moisture in storage. Rows 4 and 5 show the change in soil moisture storage during each month and the amount of soil moisture storage at the end of each month.

During wet months, *AE* is always equal to *PE*. During dry months, plants will draw on soil moisture as needed (note the storage change in July). *AE* represents water passing through the plant system, and the amount of *AE* is related in a general way to the building of plant tissue. According to the table, *AE* was less than *PE* during August and September. This indicates a *deficit* (*D*), shown in row 7. The deficit represents the amount of additional water that the plants could have used. Therefore, it is an index of irrigation needed to maintain crops at their full growth potential.

Any excess water that remains after soil moisture is brought to capacity is *surplus* (*S*). During 1962, surpluses occurred during only four months: January, March, April, and June. Surplus water is available for surface runoff and groundwater recharge. It is this water that enters streams and changes the land surface by erosion and sediment deposition.

Figure 7.12 shows both tabular and graphic representations of the average annual water budget over a 30-year period for Baton Rouge. The long-term average monthly precipitation is much less variable than the actual monthly precipitation in individual years. This is because a wet July in one year compensates for a dry July in another year, and so on. Long-term average monthly precipitation in Baton Rouge, therefore, is never low, and in the average water budget for Baton Rouge there are no deficits.

The seasonal and annual variations in energy and moisture regimes revealed by water budgets are extremely important to the activities of plants, wildlife, and humans. A more complex water budget model was used in Figure 7.13 to illustrate this variability. This figure shows monthly *PE*, *AE*, *D*, and *S* for Baton Rouge between 1960 and 1967. Deficits occurred each summer, but they ranged from small deficits with little environmental consequence to large ones during 1962, 1963, and 1965. Surpluses, on

Figure 7.13 (opposite)
Calculated monthly water budget components for Baton Rouge between 1960 and 1967. This graph emphasizes seasonal consequences of the variability of precipitation, which itself is not shown.

A more realistic, but at the same time more complex, model of soil moisture storage and depletion was used for these calculations. In the water budgets for Figures 7.11 and 7.12, it was assumed that soil moisture was "equally available" to meet the climatic demand of *PE* during months when *P − PE* was negative. In the model used for this figure, it was assumed that the soil moisture would become "decreasingly available" as it was depleted; in other words, the vegetation would not be able to withdraw all the needed soil moisture even though the soil still contained some available moisture. The decreasing availability model of soil moisture depletion requires special tables or equations.

The water budget graphics in the following chapters are based on the decreasing availability model. Note especially that this model shows that small deficits recur each year at Baton Rouge. (Vantage Art, Inc., after Muller, 1976)

the other hand, were extremely variable from year to year (compare the large surplus during 1961-1962 with the small one for 1962-1963). This type of variability, which is due to changes in the upper atmospheric circulation, has far-reaching impact on the environment, as well as on economic activities.

Problems in Application of the Water Budget Model

The water budget model presented here is a simple representation of a very complex natural system. Better estimates of *PE* are possible, but they require exact measurements of each of the radiation components discussed in Chapter 3. For simplicity, the model assumes that soil moisture is always equally available to the vegetation, regardless of the moisture content of the soil. In general, plant scientists report that plants are less able to utilize soil moisture as the soil begins to dry. The model also assumes that all rainfall infiltrates soil having any storage opportunity; allowances are not made for intense rainfalls that exceed infiltration capacities or for variations in specific soil types.

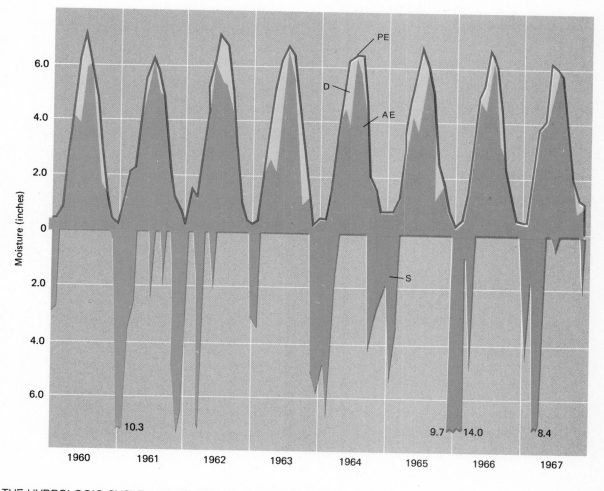

Figure 7.14
This graph compares the stored soil moisture estimated by the daily water budget method to the measured amount. The excellent agreement indicates the suitability of the local water budget approach to agricultural problems.

The amount of stored soil moisture attains its maximum in late winter and spring, when precipitation is heavy and transpiration is small due to lack of plant cover. Soil moisture decreases in April and May and reaches a minimum during summer, when precipitation is low and high temperatures and dense plant cover cause evapotranspiration to be high. The local peaks in the soil moisture are caused by rapid recharge during rainstorms. Can you explain why the average level of soil moisture increases in late autumn? (Doug Armstrong after C. W. Thornthwaite and J. W. Mather, *Publications in Climatology*, vol. 8, 1955)

For irrigation and flood forecasting, the monthly unit is too long, and a daily water budget is necessary. Modern computer technology permits analyses of daily water budgets for research and environmental monitoring. Most of the simplifications discussed in the preceding paragraph can be taken into account by the computer models. Figure 7.14 is a comparison of a water balance estimate of the daily regime of soil moisture during an entire year and the actual measurement of soil moisture changes. It can be seen that the water budget estimate produced by the computer closely approximates reality.

A further problem with the water budget model is that precipitation data are not always representative of weather and climate condi-

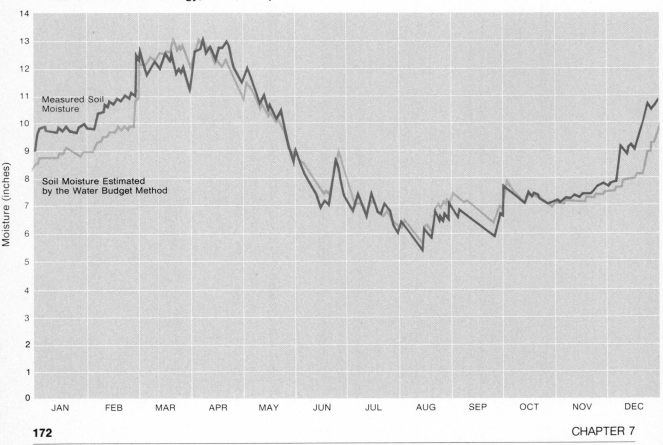

tions in the region. Precipitation is normally measured by noting the accumulation of water in a standard rain gauge exposed under standardized conditions. The gauges, however, do not catch a fully representative sample on windy days or when the precipitation falls as snow. In the United States, there is an average of only one official rain gauge for every 500 sq km (200 sq miles); all the official rain gauges in use could fit between the 40-yard lines of one football field! Because storm rainfall can be very localized, the rain gauge network often does not provide accurate precipitation data for sizeable areas.

Despite these problems, the water budget model is a valuable tool for a wide range of objectives. In its more complex form, it can be used for research in fields ranging from agriculture to flood hydrology. Different water budget components can be related to particular objectives. For example, actual evapotranspiration (AE) is related to plant growth and the yields of many crops; the deficit (D) is an index of irrigation needs; and the surplus (S) represents water potentially available for water-resource projects, sewage disposal, and industrial uses. Calculations of surplus have revealed that streamflow, or water yield, is affected by changes in land use and vegetation cover. Urban areas and some croplands produce the largest water yields per unit of precipitation. Forested regions, on the other hand, contribute the smallest proportions of precipitation to streamflow but sustain higher base flows between rainfalls.

SUMMARY

The movements of water from place to place and its transformations into liquid, solid, and vapor phases are all part of the hydrologic cycle. The oceans are the largest reservoir of water on the earth, and the evaporation of water from the tropical oceans and warm currents is the beginning of the hydrologic cycle. The cycle comes full circle when streamflow returns surface runoff and groundwater outflow to the sea.

The amount of water gained by the atmosphere as a whole over a year must equal the amount that falls from the atmosphere as precipitation. When the oceans or continents are considered separately, however, imbalances appear. The oceans lose more water by evaporation than they gain by precipitation. The continents gain more by precipitation than they lose by evapotranspiration. Runoff to the sea from streams and rivers and the horizontal transport of moisture in the atmosphere compensate these imbalances.

In unpaved land areas a small proportion of the precipitation is intercepted by vegetation and returned to the atmosphere by evaporation. The remainder reaches the ground by throughfall and stemflow. Water infiltration into the soil is determined by soil porosity, permeability, and moisture content. When the ground is saturated, surplus water either runs off over the surface or percolates downward to the water table, entering storage as groundwater. Groundwater outflow supports the flow of surface streams between rainfalls. Groundwater is stored below the water table in unconfined and confined aquifers. Chemical contamination of groundwater by human activities and the removal of groundwater faster than it is recharged are increasingly important problems.

Surface runoff and streamflow are two different measures: runoff refers to the average depth of water removed from an area of the earth's surface; streamflow is the amount of water passing through a stream channel per unit of time.

The availability of water as soil moisture or as surplus, producing runoff, can be estimated by using the local water budget model. PE represents the maximum rate of evaporation and transpiration from a vegetation-covered landscape with no water shortage. The water budget method compares PE with precipitation and soil moisture storage for varying time periods and permits the estimation of water surpluses or deficits on a daily, monthly, or annual basis.

The water budget model is a useful tool in a variety of situations ranging from agricultural planning to flood forecasting and water-resource development.

APPLICATIONS

1. Where are there stream gauging stations in your area? What do they look like? How do they work?
2. Using graph paper, plot daily or monthly stream discharge for local gauged streams in your area for the wettest and driest years on record. Stream discharge data are available from U.S. Geological Survey Water Resource offices in each state and from state departments of water resources, and are published in annual bulletins covering both states and major drainage basins. Remember that stream discharges represent mostly the surplus or "leftover" water within local water budgets.
3. Observe the ground during a heavy rainfall. Do you see runoff occurring on "natural" surfaces? On surfaces modified by human activity? Go to an area of bare ground with no vegetative cover during a heavy rain—a large roadcut or a spoil heap will do. What do you see there? How does the timing of rainfall, as distinct from the quantity, affect the amount of moisture absorbed by the ground?
4. How do the water supply agencies in your region obtain their water? Are there times when supplies are critically short? Has there been a water supply emergency in your area? What measures were taken to overcome it?
5. Produce an "artificial rainfall device" by perforating the bottom of a 46-ounce juice can with holes approximately 1 mm in diameter and 1 in. apart. Pouring a measured amount of water into the perforated can, observe how rapidly this is absorbed by different types of ground—wet, dry, sandy, clayey, grass-covered, forest-covered. Over vegetation note how much water is intercepted and retained by plant parts. Where is it easiest and hardest to produce runoff?
6. Compute a continuous monthly water budget using data from the nearest weather station, or some other location of interest. The computation procedure can be found in John R. Mather, *The Climatic Water Budget in Environmental Analysis* (see Further Reading). Then prepare a graph illustrating the variability of water budget components such as AE, D, and S, as in Figure 7.13. As an alternative, maintain a daily water budget, using temperature and precipitation data given each day by radio, TV, or newspapers.

FURTHER READING

Chorley, Richard J., ed. *Water, Earth, and Man.* London: Methuen (1969), 588 pp. This unusual book is a well-organized collection of essays on various aspects of the hydrologic cycle and their applications in hydrology, geomorphology, and socio-economic geography. Many of the contributors are British, giving the book a perspective from the British Isles.

Dunne, Thomas, and **Luna B. Leopold.** *Water in Environmental Planning.* San Francisco: W. H. Freeman (1978), 818 pp. This is the most thorough treatment of all aspects of water in the environment, by two of the leading authorities in the fields of hydrology and water management. Very well illustrated with maps, graphs, and photographs.

Kazmann, Raphael G. *Modern Hydrology.* 2nd ed. New York: Harper & Row (1972), 365 pp. Emphasis in this work is on critical evaluation of the hydrologic variables for water-resource management.

Leopold, Luna B. *Water: A Primer.* San Francisco: W. H. Freeman (1974), 172 pp. This excellent little book stresses the interrelationships between hydrologic principles and environmental responses.

Mather, John R. *The Climatic Water Budget in Environmental Analysis.* Lexington, Massachusetts: Lexington Books (1978), 239 pp. This recent book is a most useful summary of ideas and applications of the water budget, including appendices for calculations.

Thornthwaite, C. W., and **John R. Mather.** "The Water Balance." *Publ. in Climatology*, Vol. 8 (1955):1–86. This is a summary of the early water budget research and applications by Thornthwaite and his colleagues at the Laboratory of Climatology.

————. "Instructions and Tables for Computing Potential Evapotranspiration and the Water Balance." *Publ. in Climatology*, Vol. 10, 3(1957):185–311. This instruction manual provides explanations and tables for calculation of daily and monthly water budgets using the decreasing availability model of soil moisture depletion.

Ward, R. C. *Principles of Hydrology.* 2nd ed. London: McGraw-Hill (1975), 367 pp. Various components of the hydrologic cycle and the water budget are treated systematically. This text incorporates a geographic perspective, and it also includes an extensive bibliography organized by topic.

CASE STUDY: The California Water Plan

The management of fresh water is often a controversial subject. According to one view, there is plenty of water on the earth—it simply needs to be redistributed. Others argue that major economic development should not be extended into areas having a shortage of water, but should be concentrated where water is naturally available.

A case to consider is southern California, which has become greener and more productive by the importation of water from remote areas. In the early 1900s it became clear that the growth of Los Angeles—then a small coastal town—would be stimulated by irrigating larger areas. For a source of water, Los Angeles turned to the Owens Valley, 400 km (250 miles) to the north, at the eastern foot of the Sierra Nevada. The farmers of the Owens Valley had hoped for their own irrigation system, but it was argued that the climate there was too severe to support the profitable citrus groves that thrive around Los Angeles, where year-round agriculture is possible given adequate water.

The Los Angeles Aqueduct from the Owens Valley was completed in 1913. Soon more water was needed, however, and aqueducts were built to carry water across the Mojave Desert from the Colorado River to Los Angeles and San Diego. Today more than a hundred cities in southern California rely on water from the Colorado. This supply has recently been supplemented by water from northern California in the world's longest artificial water transfer. Rather than continuing to rely on local and regional initiative, the government of California has taken over water resources development for the entire state.

The California State Water Project is founded on the recognition that most of the state's population, voting power, and irrigated land are in the south, whereas most of the water is in the north. The State Water Project includes aqueducts, canals, dams, reservoirs, and power stations needed to transport water from the north to the south. The project begins at the massive Oroville Dam on the Feather River north of Sacramento. The current source of water for the present stage of the plan is the Sacramento River delta, formed by a confluence of streams from the west slope of the Sierra Nevada that carry over 40 percent of the state's runoff to San Francisco Bay through the only opening in the coastal range. A future project, bitterly opposed by residents of the delta area, is to route Sacramento River water around the delta, rather than through it. This could allow salt water intrusion into the delta, which would destroy delta farmlands and wildlife habitats. The south end of the water transfer system is Perris Lake, an artificial reservoir located southeast of Los Angeles. Never before has one of man's projects sent so much water flowing so far. At one point in its 1000 km (600 mi) journey southward, the water is lifted nearly 600 m (2000 ft) through the Tehachapi Mountains—a record lift for so much water.

In addition to supplying water for southern regions, the aqueducts provide irrigation for the dusty, windblown San Joaquin Valley, which is currently the major beneficiary of the project. Eventually, a million acres of formerly unproductive land will be made available for agriculture.

The water development program, which has spanned some 30 years, has instigated an angry sectional feud between the moist

This simplified map shows the conduits of the California State Water Project. The project extends nearly the length of California, bringing water from the north to the more densely populated south. (Steve Harrison and Louis Neiheisel)

either urban or agricultural areas. However, under the 1928 Boulder Canyon Act much of the water now taken from the Colorado River by California will soon have to be relinquished to Arizona. If the water used for irrigation in California were cut to 80 percent, the amount available for other uses would double from 10 to 20 percent. However, what effect would reducing irrigation have on food supplies for the entire country? California is now the leading supplier of dozens of agricultural commodities, almost all of them grown by irrigation.

The allocation of water resources is a matter of economics as well as of technology. Water demand is, in reality, often a demand for water at a sufficiently low cost to make irrigation profitable. The emphasis on higher-priced specialty crops, such as fruit, nuts, vegetables, and cotton, in California agriculture reflects the need for profitable crops to offset the cost of irrigation. However, vast amounts of water are also used to grow alfalfa, a low-value crop that is a heavy consumer of water. The California water project has been criticized for using public funds to subsidize and increase the value of privately held farmland.

Although more aqueducts, canals, and pumping plants are planned for the 1980s, it is uncertain when and if they will be completed. The pace of the water plan is influenced by projected water needs, the development of new technology, ecological impacts, and political considerations.

"north" and the dry "south," but the more numerous votes in the South carried the issue. Environmentalists claim that the project has upset the natural balances of streams, estuaries, vegetation, and wildlife and that it will destroy the ecology and economy of the Sacramento delta. It is also argued that providing more water to Los Angeles will promote population growth, more congestion, and associated problems.

Many of the questions and controversies over the water plan center on whether the water is really needed. Ninety percent of the water used in southern California is for irrigation; there is at present abundant water for domestic and industrial use. There is currently no attempt to conserve water in

Farmlands in the Sacramento River delta. Smoke is from intentional burning of crop residues. (T. M. Oberlander)

Dynamic processes in the atmosphere cause the weather patterns that give each region of the earth a distinctive climate. Climate can be viewed in two ways: as the delivery of energy and moisture by the atmosphere, or as the interaction of energy and moisture at the earth's surface with systems of soils, vegetation, and landforms.

Sky Above Clouds II by Georgia O'Keeffe. (Collection of Mrs. Potter Palmer, Lake Forest, Illinois; with permission of Georgia O'Keeffe)

Global Systems of Climate

An area's *climate* is the average condition of its weather over a period of years. Climate is generally described in terms of seasonal or annual temperature, precipitation, wind, and degree of cloudiness. A more comprehensive view of climate also includes such surface conditions as evapotranspiration, surface runoff, and soil moisture availability. The climate of a region is fundamental to the development of the region's soils, vegetation, landforms, and agricultural possibilities. Understanding the global distribution of the earth's varying climates is the key to understanding the environments of different places and the human activities in them.

From time to time we hear of proposals to change the climate artificially in some part of the world. For example, one such scheme is intended to increase precipitation in the arid southwestern part of the United States so that moist woodlands and agricultural land could exist where only cactus and desert shrubs grow today. The idea in this case is to change the climate by building a gigantic dam across the Bering Strait between Siberia and Alaska. The dam would have to be 100 km (60 miles) long, far surpassing any dam existing on the earth.

How could a dam at the Arctic Circle affect the climate in a desert 5,000 km (3,000 miles) away? If such a dam could be built, it would stop the circulation of cold Arctic waters into the Pacific Ocean. This would cause the Pacific Ocean to become warmer. We have seen (Chapter 4) that unusual cooling in the Pacific Ocean seems to trigger drought conditions in the western United States. Accordingly, advocates of the Bering Strait dam anticipate that *warming* Pacific Ocean water would *increase* rainfall in the American Southwest. Russian proponents of the plan hope that it would indirectly improve the cold, dry climates over much of Siberia as well. But warming the Pacific would almost certainly create a longer hurricane season in the tropics and perhaps bring hurricanes into higher latitudes. The dam would make the polar region cooler, which could produce stronger outbursts of polar air and more violent cyclonic storms along the polar front.

It is impossible to predict all the consequences of such a comprehensive scheme to modify climate. Changing the temperature of the Pacific Ocean would change the temperature of the air over the ocean and would affect the general circulation of the atmosphere. Because the general circulation controls the climates of the earth, the effects of the dam would be far-reaching. Beneficial effects in the American Southwest would quite likely be outweighed by harmful effects elsewhere.

The point of this rather far-fetched example is that the earth's climates are not independent phenomena—they are determined by the general atmospheric circulation. To change them, the circulation system as a whole must be changed, causing climates everywhere to be affected. It is much more sensible to learn how existing climates are produced and how to take advantage of what they have to offer.

This chapter discusses regional climates. Two quite different methods of climatic classification are described, and the local water budget is used to show why climate is the active factor controlling other environmental systems.

MEASURING CLIMATE

The type of climate a region experiences can be measured by various indexes and on varying time and space scales. These measures are also important in climatic classification schemes, as we shall see later in the chapter.

Climatic Indexes

Climatic indexes are numerical measures used to distinguish one type of climate from another. The choice of an appropriate index of climate depends on its intended use. Someone interested in constructing tall buildings might wish to consider wind speed and direction to assess the possibilities of wind damage. If the concern is transportation, interest would center on the frequency of fog and icing conditions. An irrigation engineer would be concerned with the amount of evapotranspiration from cropland. To assess a region's potential for air pollution, wind regimes and the timing and duration of temperature inversions must be known.

Classifications of climates by physical geographers generally employ temperature and precipitation as indexes. These are easily measured and clearly influence the distribution of vegetation and water supplies and the potential for different types of agriculture. Wind, temperature lapse rates, evapotranspiration, and other atmospheric phenomena are not commonly used as climatic indexes because data on such variables are not widely available.

Time and Space Scales of Climate

The general controls of climates are factors we have considered in previous chapters. These include *latitude*, which influences temperature and day length; proximity to warm and cold *ocean currents*, which affect temperature and air mass stability; proximity to *moisture sources* and *rain-producing mechanisms*, including the polar front and the intertropical convergence zone; *topography*, which causes ascent of air and

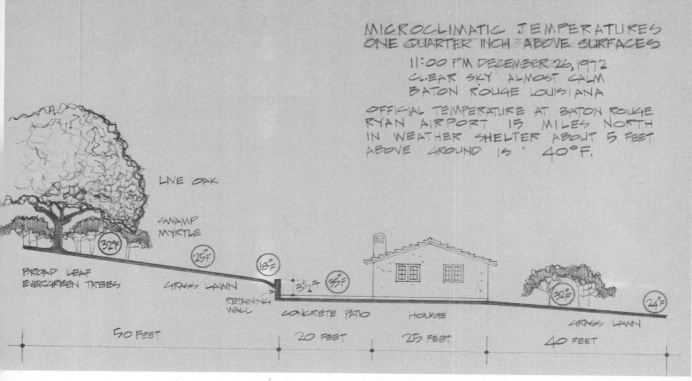

MICROCLIMATIC TEMPERATURES
ONE QUARTER INCH ABOVE SURFACES

11:00 PM DECEMBER 26, 1972
CLEAR SKY ALMOST CALM
BATON ROUGE LOUISIANA

OFFICIAL TEMPERATURE AT BATON ROUGE
RYAN AIRPORT 15 MILES NORTH
IN WEATHER SHELTER ABOUT 5 FEET
ABOVE GROUND IS 40°F.

LIVE OAK

SWAMP MYRTLE

BROAD LEAF EVERGREEN TREES

32°F   25°F   18°F   3½F   35°F   32°F   24°F

GRASS LAWN

RETAINING WALL

CONCRETE PATIO   HOUSE   GRASS LAWN

50 FEET   20 FEET   25 FEET   40 FEET

Figure 8.1
The nighttime temperatures in this Louisiana backyard show a large degree of variation from place to place because of such factors as local movements of air, differences in the amount of heat stored in the ground and buildings, and differences in cooling rates. Climate measured on such a small scale is called the *microclimate* of the area. Such details are lost when the climates of large regions are classified into broad categories. (Ron Wiseman after Muller, 1973)

high rainfall or descent of air and rainshadows; and the *general circulation*, which determines the sources of air masses and the general directions in which they move.

The movement of air masses can change local weather conditions significantly within a few hours. Weather conditions can also vary markedly over short distances. Even if temperature and precipitation are accepted as the indexes of climate, geographers must still decide when and how often to analyze these factors. The following example shows how important the time scale—the "when"—of measurement is.

The average annual temperatures at Aberdeen, Scotland, and at Chicago, Illinois, differ by less than 2°C (3°F). The average annual precipitation at both locations is the same (84 cm, or 33 in.). Despite similarities in annual averages, the climates at the two locations are quite different. The monthly averages give a much more accurate indication of the climate in each city. During January, Chicago averages 8°C (14°F) colder than Aberdeen. During July, Chicago averages 9°C (16°F) warmer than Aberdeen. The climate at Chicago is therefore considerably more extreme than the climate at Aberdeen. It should be evident from this that the seasonal distributions of temperature and precipitation, as well as the average annual values, are important in characterizing the climate of a location.

The "where" of measurement—the size of the climatic region—is also important. Again, the best areal scale depends on the intended use of the information. On a world map most of the west coast of the United States is considered to be in a single climatic region. But someone concerned with water resources on the Pacific coast would need to see a much more detailed picture, showing the gradation from the abundant rainfall in the Pacific Northwest to the general dryness of southern California. Even a single residential lot has several microclimates, with a wide range of temperatures over a small area. Figure 8.1 shows temperature readings made at 11 P.M. at various points in a backyard located in Baton Rouge, Louisiana. The microclimates of the yard are too small to appear on even a city-wide climate survey, but climatic conditions in different parts of the yard are important to someone trying to protect subtropical vegetation from frost damage.

This chapter focuses on global climatic types. Excessive detail is not desirable at this scale. The system of global climatic classification discussed later in this chapter assigns New York City and Nashville, Tennessee, to the same climatic region. On a nationwide scale, the climates of the two cities are obviously different, but on a global scale the climates of these two regions are more like each other than they are like climates of tropical rainforests or Asian steppes, which is the type of distinction possible on world maps.

CLIMATE DISTRIBUTION: THE DELIVERY OF ENERGY AND MOISTURE

The availability of energy and moisture in different regions of the earth is a result of the global distribution of solar radiant energy and the general circulation of the atmosphere. To understand how solar energy and the general circulation control the gross features of the earth's climates, we can analyze the climatic regions of an idealized hypothetical continent, on which there are no surface features to impose distorting effects.

Distribution of Climatic Regions on a Hypothetical Continent

The hypothetical continent shown in Figure 8.2 is flat and featureless with no interior mountains, seas, or gulfs. Nevertheless, it embodies many of the features of the actual continents: it is surrounded by oceans; it is broad at the north and extends to high latitudes, like North America and Eurasia; and it tapers toward the south and ends near latitude 60°S, like South America.

The distribution of temperature on the hypothetical continent is directly related to the pattern of incoming solar radiation. Tropical and subtropical regions receive the greatest annual amounts of solar energy and are warmest year-around (see Figure 3.16, page 67). Regions at higher latitudes receive less solar energy and have lower average temperatures.

The annual temperature range varies according to latitude and distance from the oceans. Because the tropical regions receive a nearly uniform seasonal input of solar energy, they maintain comparatively constant temperatures all year. But because of the tilt of the earth's axis, the higher latitudes receive a large amount of solar energy in the summer and a small amount in the winter. As a result, the difference between average summer and winter tempera-

Figure 8.2 (opposite)
The climatic regions represented on this hypothetical continent are determined by the input of solar radiant energy and the delivery of air masses and moisture by the atmospheric systems. The regions are depicted with overlapping boundaries to suggest that climate changes gradually from one location to another across the surface of the earth. The distribution of the climates should be studied in conjunction with the text and with the general circulation maps in Figure 4.6, pp. 86–87. (Doug Armstrong)

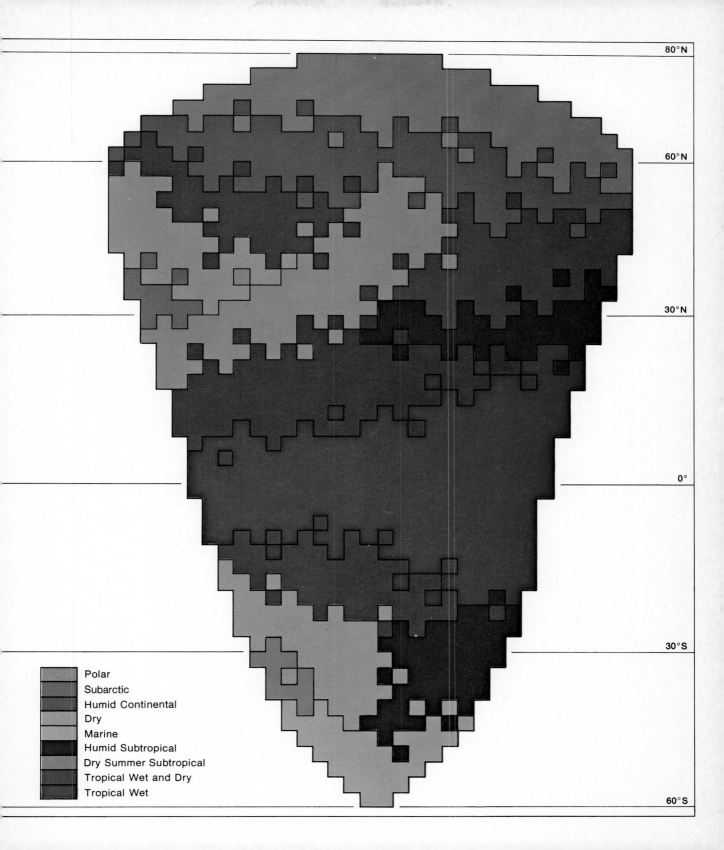

	Polar
	Subarctic
	Humid Continental
	Dry
	Marine
	Humid Subtropical
	Dry Summer Subtropical
	Tropical Wet and Dry
	Tropical Wet

80°N

60°N

30°N

0°

30°S

60°S

tures on the hypothetical continent is greater the farther regions are from the equator. The annual temperature range also increases with distance from the moderating effect of the sea. This phenomenon is known as *continentality*. The continental interiors heat strongly in the summer because of the low heat capacity of land areas and become cold in the winter as a consequence of rapid heat loss by longwave radiation.

The distribution of atmospheric moisture and precipitation is controlled by air masses, wind patterns, and pressure systems. Air masses reaching the hypothetical continent after crossing the ocean would be moisture laden as a result of evaporation. This humid, maritime air would become a source of precipitation if it were cooled sufficiently to cause condensation. Most cooling takes place when the air is forced to rise by convergence in a low-pressure center or by convection due to heating at the continental surface.

On the hypothetical continent, the equatorial trough of low pressure, or intertropical convergence zone, occupies the region near the equator, as shown in Figure 4.6, pp. 86–87. The easterly trade winds that converge into this trough are usually moist from evaporation of ocean water. As the converging air rises, cloud formation and rainfall occur. Rainfall is frequent year-around near the equator; thus the climate of the equatorial region is classified as *tropical wet*. Because the trade winds sweep onshore and across the eastern side of the hypothetical continent between about 15°N and S, the eastern side receives more moisture than the western side in equatorial latitudes. This is the climate of the equatorial portion of the Amazon Basin in Brazil.

The tropical wet zone astride the equator is succeeded both north and south by *tropical wet and dry* climatic regions. These zones receive rain in the summer high-sun season as a result of the presence of the intertropical convergence zone with its rising moist air. Drought occurs in the winter low-sun season when the ITC moves into the opposite hemisphere and is replaced by the subsiding air and temperature inversion of the subtropical high-pressure cells. Here winter is a time of drought rather than cold. This is the climate of the African Sahel.

Near 30°N and S on the hypothetical continent are *dry* climates that are under the permanent influence of the subtropical zone of high pressure, where dry air from the upper atmosphere subsides almost to the surface. As it subsides, the air warms adiabatically, resulting in a persistent temperature inversion and stable environmental lapse rates at middle altitudes of the atmosphere. Despite intensive surface heating, the dry air and stability aloft combine to inhibit precipitation processes. Thus subtropical deserts occur on the poleward sides of the tropical wet and dry climates, as in the North African Sahara Desert and the South African Kalahari.

In the interior of the hypothetical continent in both hemispheres, dry regions extend poleward of the subtropical deserts. These midlatitude dry climates are not directly associated with the subtropical zones of high pressure but occur because the regions are far from oceanic moisture sources. Weak midlatitude cyclones are unable to produce much rainfall from the relatively dry air masses in the continental interiors. The midlatitude dry regions would be much cooler during winter than their subtropical counterparts. Such cold deserts are best developed in inner Asia, the outstanding example being the Gobi Desert of Mongolia.

Along the western margins of the hypothetical continent in the subtropical latitudes, the delivery of moisture is governed mostly by the subtropical high-pressure cells centered over the ocean to the west. High pressure shifts to lower latitudes in the winter, allowing midlatitude cyclones to bring moist maritime air and frequent precipitation. In the summer the high-pressure cells move poleward and stabilize the atmosphere over the subtropical region so that it remains quite dry. As a consequence, this climate is known as the *dry summer subtropical* type. Cold ocean currents offshore further strengthen the inversion and stability. This is the climate of coastal central and southern California. While the dry summer subtropical climatic region has a wet winter and a very dry summer, the *marine* climatic region farther poleward is beyond the influence of the subtropical high-pressure cells and receives moisture throughout the year, as in the case of western Washington State and coastal British Columbia.

On the eastern side of the hypothetical continent at subtropical latitudes, precipitation occurs year-around, producing a *humid subtropical* climate. Atmospheric subsidence is less well-developed over the west sides of the subtropical highs than over the east sides. The oceanic circulation brings warm water to the east coasts at these latitudes, in contrast to the cold water along west coasts. Maritime tropical air is swept around the western margins of the subtropical highs and over the coastlines toward the interior. In the summer this air has a steep lapse rate due to heating by the land surface. Summer showers and thunderstorms recur frequently in the unstable, warm, humid air. Precipitation is also heavy during winter, when midlatitude cyclones occur along the polar front. The southeastern United States epitomizes this type of climate.

Like the real northern hemisphere continents, the hypothetical continent is broad in the middle latitudes of the northern hemisphere. Here we find a region of *humid continental* climate. Because of the minimal moderating effect of the far-off oceans, these interior climatic regions are characterized by very large annual temperature ranges, or continentality. Radiational heating of the land produces hot summers, and radiational cooling leads to cold winters. The prevailing westerly wind carries the continentality effect from the interior to the east coast. In the southern hemisphere, the continent is narrow and the moderating effect of the oceans is greater, so there is no development of continental climate. In the area of humid continental climate, most precipitation would occur

in the summer in association with midlatitude cyclones. During winter and spring, stormy periods with heavy snowfall would be interspersed between outbreaks of cold but fair weather, as in the North American Plains east of the Rocky Mountains.

Poleward of the humid continental climates in the northern hemisphere is the *subarctic* climatic region, which extends from coast to coast. In winter the subarctic climatic region is dominated by the polar high-pressure cell displaced southward over the colder continent. The winter air is always cold and snowfall would be light except near the coasts. Continentality again produces a large annual temperature range. Most of the precipitation would fall during the mild summer in association with midlatitude cyclones. Most of northern Canada and the majority of the Soviet Union have such a climate.

Beyond the subarctic climate region on the hypothetical continent is the region of *polar* climate. The land poleward of latitude 60°N would be dominated by the polar high-pressure cell much of the year and would also be influenced by the nearby cold and often ice-covered ocean. Winters would be very cold with meager snowfall. Most precipitation would occur during the short summer, when temperatures would be above freezing over all but extensive ice-covered areas. Only Antarctica and the Arctic Ocean and its fringes experience a true polar climate.

Distribution of Climatic Regions on the Earth

Many climatic regions on earth follow the pattern of the hypothetical continent (Figure 8.3, pp.186-187). Near the equator are tropical wet regions, with tropical wet and dry climates immediately to the north and south. Subtropical deserts sprawl across North Africa, the Arabian peninsula, Iran, and Pakistan. Smaller desert areas occur in the southwestern United States and western Mexico. Africa and Australia show the expected dry regions near latitude 30°S. The dry summer subtropical climates are

Figure 8.3
This global map classifies climates according to solar radiation input and atmospheric delivery of energy and moisture to the earth's surface. Compare it with the distribution of climates on the hypothetical continent shown in Figure 8.2. The landmasses of the high northern latitudes possess polar and subarctic climates like the hypothetical continent because of the low input of solar radiant energy. Like the hypothetical continent, equatorial Africa, South America, and Southeast Asia possess tropical wet climate regions flanked by tropical wet and dry climate regions. North America, Africa, and Australia exhibit dry climate regions near latitude 30°. The effect of mountains on climate distributions is apparent in the way the tropical wet region of South America does not cross the Andes to the west coast, which is a desert over much of its length.

The distribution of continents and oceans on the earth affects the movements of air masses. The southeastern United States, for example, possesses a humid subtropical climate instead of a humid continental climate because of the northward movement of moist tropical air from the Gulf of Mexico and the Caribbean. Parts of western Europe and the west coast of North America possess marine climates. In Europe the marine climate gradually merges into a humid continental climate region, but in North America the west winds from the Pacific Ocean are blocked by mountains. The region immediately inland from the mountains is desert. Central and eastern North America have a humid continental climate because of cold winds from the north in winter and warm winds from the south in summer. (Andy Lucas and Laurie Curran after Vernon C. Finch and Glenn Trewartha, *Physical Elements of Geography*, 1949, McGraw-Hill Book Co.)

Major Global Climatic Regions

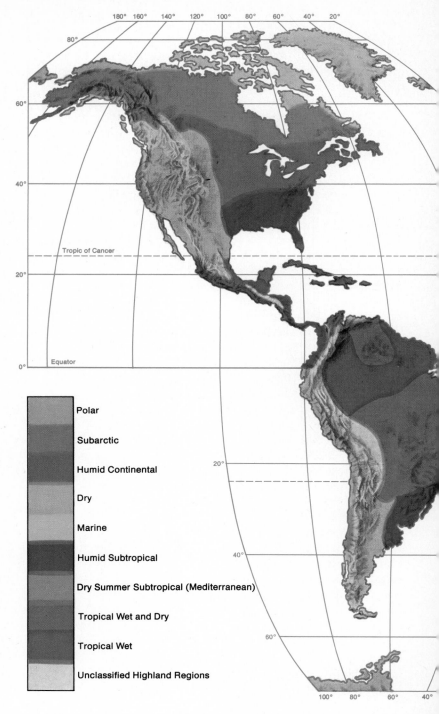

Polar

Subarctic

Humid Continental

Dry

Marine

Humid Subtropical

Dry Summer Subtropical (Mediterranean)

Tropical Wet and Dry

Tropical Wet

Unclassified Highland Regions

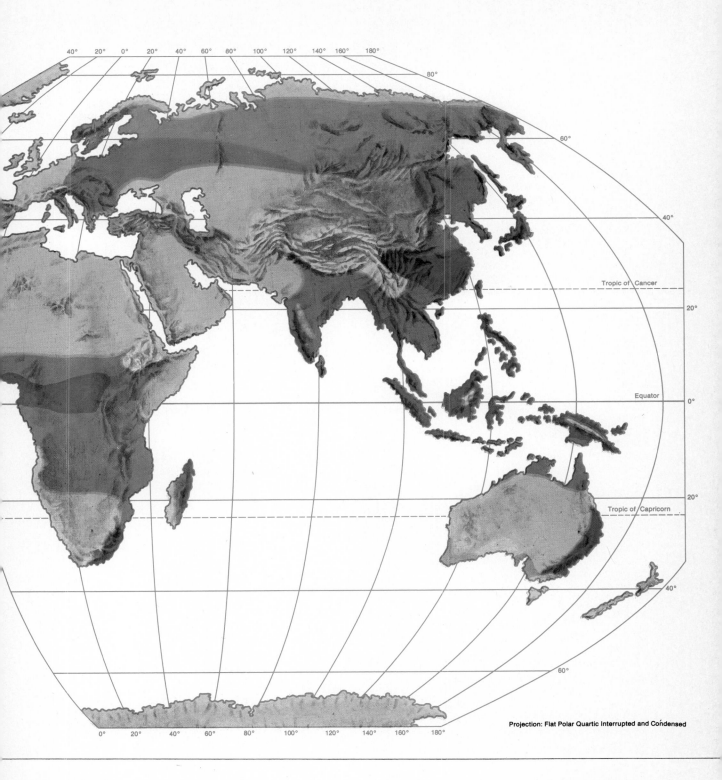

80°

60°

40°

Tropic of Cancer

20°

Equator 0°

20°

Tropic of Capricorn

40°

80°

Projection: Flat Polar Quartic Interrupted and Condensed

Figure 8.4

(a) England has a marine climate because of moist westerly winds from the Atlantic Ocean. The climate is moist throughout the year, with moderate temperatures. This early morning mist in Devon, England, formed in a layer of cool air near the wet ground. (G. R. Roberts, Nelson, N. Z.)

(b) Orographic lifting of persistent easterly trade winds creates a cloud canopy over the Hawaiian island of Kauai. These clouds produce frequent precipitation that makes the summit area in the background of this view one of the wettest places on earth. (Dan Budnik/Woodfin Camp & Associates)

(c) This snow-covered cornfield in Iowa exemplifies the seasonal temperature variations of a humid continental climate region. Snow is frequent in Iowa during the cold winter months, but sunshine and high temperatures make the summer an excellent growing season. Moisture is available throughout the year. (Erich Hartmann/Magnum)

(d) Precipitation in arid climates seldom exceeds a few centimeters per year. This oasis in the Sahara depends on groundwater to supply the needs of humans and animals. (Marc & Evelyne Bernheim/Woodfin Camp & Associates)

(e) The dry summer subtropical climate of the coastal regions of central California is moist during the winter and fosters the growth of vegetation. During the rainless summer season, the vegetation dries out and brush fires are common. The fires in this view burned for several days in August 1977 in hills near Oakland. (T. M. Oberlander)

(f) The climate near San Francisco, California, is moderate and supports the growth of subtropical vegetation. A rare occurrence of freezing temperatures during the winter of 1972 caused dieback of these normally green eucalyptus trees in the Berkeley Hills, to the east of San Francisco. (T. M. Oberlander)

(g) In subarctic climate regions, snow and ice may persist through the year. The photograph shows icebergs breaking off, or *calving*, from a glacier in northern Iceland. (Earl Dibble/Photo Researchers, Inc.)

located on the western sides of the continents in five expectable locations: California and Oregon; central Chile; Portugal, Spain, Morocco, and the northern Mediterranean coastlands; the Cape Town region of South Africa; and small areas of western and southern Australia. The midlatitude west coasts of North America, South America, and Europe show the expected marine climates, and the interiors of the United States and Europe are humid continental climatic regions. Subarctic climates occupy northern Asia and North America, and polar climates occur in the polar regions.

The climatic regions of the earth differ from those of the hypothetical continent only as a result of variations in the sizes, shapes, and topographic features of the land areas. The mountain ranges in western North and South America prevent the west coast marine climates from extending as far inland as they do in Europe. The regions to the east of the mountains are dry because moist air from the Pacific is blocked off or dried out in its passage across the mountains. The Gulf of Mexico provides a fortunate source of warm moist air that keeps these dry regions from reaching farther eastward. Similarly, the Andes Mountains of South America block the easterly trades, preventing moisture from reaching the Pacific coast of South America between the equator and latitude 30°S. Thus the coasts of Peru and northern Chile are almost completely rainless.

The tropical wet and dry climate reaches unusually far northward in Asia because of the monsoon effect (Chapter 6). As the vast continent of Eurasia heats in the summer, the resulting low pressure pulls the ITC far north of its average summer latitude, causing heavy rain over India and Southeast Asia. Drought follows when the ITC moves back into the tropics. But the causes of the monsoons are complex and include the high-pressure cell over Siberia in winter, low pressure over the Indus Valley in summer, the presence of the Himalaya Mountains, and sudden changes in the location of the jet stream over Asia.

Figure 8.3 correctly suggests that a resident of Seattle would find familiar weather in such far-off places as the British Isles and New Zealand, while an Alabaman or Georgian would find the weather in South China or eastern Australia more like that at home than the weather only a long day's drive to the north or west. The repetition of similar climates and associated environments in widely separated locations on the earth is one of the most important facts to be learned in the study of physical geography.

CLIMATIC TYPES BASED ON ENERGY AND MOISTURE INTERACTIONS

In describing the distribution of climatic types, we have ignored the interaction of energy and moisture with physical systems on the earth's surface. Because these systems share the energy and moisture that the atmosphere delivers, it is useful to classify climate on the basis of its relationship to other aspects of the environment.

To be most useful, a classification scheme should make enough distinctions to account for a system's principal features, but not so many that general understanding is lost in a welter of detail. No two places on earth have exactly the same climate, so any classification scheme inevitably submerges individual details. Because climatic characteristics do not change abruptly, the boundaries of climatic categories should be interpreted as zones of transition rather than as sharp divisions. There are many climate classification systems, but we shall look at the two that are most often utilized. They are instructive because their objectives are quite different.

The Köppen System of Climate Classification

The most widely used system of climatic classification was developed and refined in the early decades of this century by Wladimir Köppen, a German botanist and climatologist. Köppen's system was influenced by the work of nineteenth-century plant geographers who had mapped the world's vegetation on the basis of extensive field studies.

Köppen's classification is related to vegetation types that he thought to be a response to climate; it recognizes five general climatic types, designated *A*, *B*, *C*, *D*, and *E*. The types *A*, *C*, and *D* represent the moist climates in which precipitation exceeds evapotranspiration, so that there is an annual water surplus. These areas support forests under natural conditions. The dry *B* climates are those in which precipitation is exceeded by evapotranspiration, so that there is an annual water deficit. The *E* climates are polar types, in which low temperatures may reduce potential evapotranspiration to zero. The *B* climates are subdivided into the *BW*, or arid (desert) climatic type (*W* from German *Wüste*, or desert), and the *BS*, or semiarid (steppe) climatic type. The *E* climatic type includes the *ET*, or tundra climate, and the *EF*, frost or perpetual ice climate.

Köppen's definition of the *ET* climatic type illustrates his use of vegetation as an indicator of climate. In the *ET* climate the average temperature of the warmest month falls between 0 and 10°C (32 and 50°F). The 10°C (50°F) limit for the warmest month corresponds approximately to the poleward boundary of tree growth. By choosing this limit, Köppen ensured that regions with the *ET* climate would be essentially treeless. Still, the 0°C (32°F) lower limit ensures that temperatures above freezing do occur, so that some vegetative growth is possible, which is not true in the *EF* climate.

Köppen established specific temperature and precipitation amounts and ratios to distinguish between his principal climatic types and their subdivisions. Figure 8.5 is a schematic diagram of his classification; the specific criteria are given in Appendix III. The *A* climatic types are hot and moist, the *C* types are warm and moist, and the *D* types are cool and moist. The *B* cli-

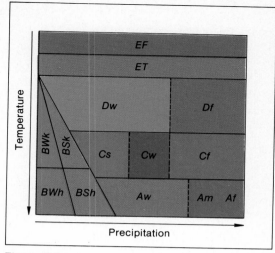

Figure 8.5
This schematic diagram represents the principal climatic regions of the Köppen classification in terms of average annual temperature and precipitation. The *E* climates, near the top of the figure, are the coldest, and the *A* climates, near the bottom, are the warmest. Reading from left to right in the direction of increasing precipitation, the *BW* climates are the driest and the *Af* climates are the wettest. The precipitation scale in particular is highly schematic. In the *Cs* (dry summer) type, for example, a smaller annual precipitation can support forest vegetation because most of the rain falls during winter when *PE* is low. The precipitation scale shows general trends in each temperature zone, but it does not indicate that the annual precipitation in *Df* regions is much less than in *Af* regions. (Vantage Art, Inc.)

matic types include a wide span of temperature and moisture values. As Figure 8.5 indicates, the boundary of the *B* climatic types is not an absolute value, but is a ratio between temperature and precipitation. The boundary between the dry and moist climates is intended to fall where precipitation and potential evapotranspiration are equal. Since potential evapotranspiration increases with temperature, the amount of precipitation required to equal the potential evapotranspiration must also increase with temperature.

Figure 8.6

This map depicts the broad global distribution of climates according to the Köppen system of climatic classification. The definite boundary lines shown between the principal climatic regions are assigned by the Köppen system; in actuality, climatic regions shade gradually into one another. The map of climates according to Köppen differs in an important respect from the map of climates presented earlier in this chapter (Figure 8.3, pp. 186–187). The earlier map shows a classification based on the delivery of energy and moisture. The Köppen system of classification takes into account very general relationships among seasonal temperatures, precipitation, and vegetation. Although Köppen's estimates of evapotranspiration are far less accurate than more recent formulations, his classification of global climates is useful for a relatively simple global regionalization. (See Appendix II for a more detailed explanation of Köppen's criteria for climate classification.) (Andy Lucas and Laurie Curran after Köppen-Geiger-Pohl map [1953], Justes Perthes, and Köppen-Geiger in *Erdkunde*, volume 8; and Glenn T. Trewartha, *An Introduction to Climate*, fourth edition. New York: McGraw-Hill, 1968)

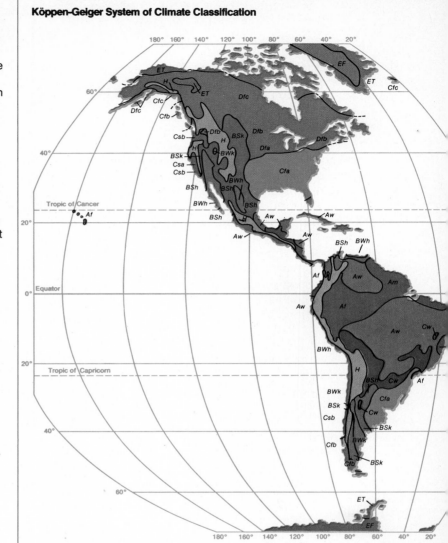

Köppen-Geiger System of Climate Classification

The Köppen system was an outstanding achievement for its time, and it still provides a useful introduction to world climatic patterns (Figure 8.6). The system allows additional symbols to be used to describe special features of climatic regions (see Appendix III). The symbol *f*, for example, means that precipitation adequate to offset evapotranspiration falls in every month of the year. Therefore, the symbol *Af* specifies a tropical wet climate, as in the equatorial rainforests of South America. The symbol *Cf* describes the warm, moist climate of the eastern United States. The symbols *w* and *s* signify winter and summer dry seasons, respec-

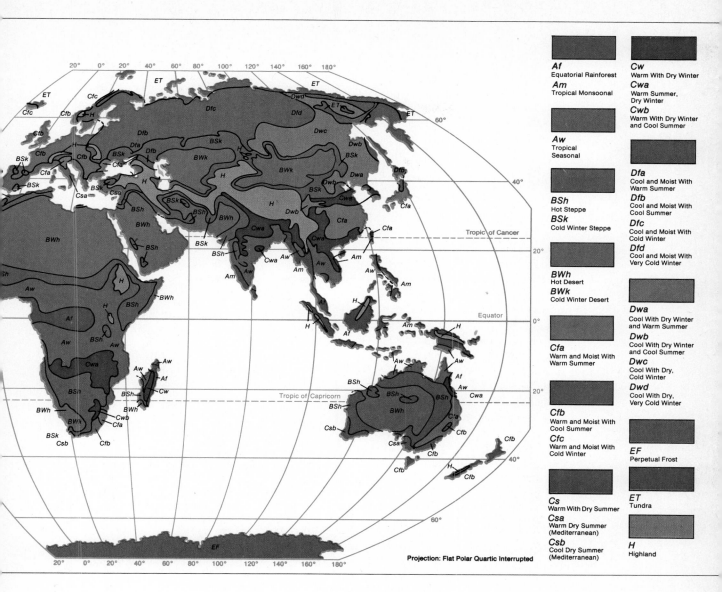

Af Equatorial Rainforest
Am Tropical Monsoonal
Aw Tropical Seasonal

BSh Hot Steppe
BSk Cold Winter Steppe

BWh Hot Desert
BWk Cold Winter Desert

Cfa Warm and Moist With Warm Summer
Cfb Warm and Moist With Cool Summer
Cfc Warm and Moist With Cold Winter

Cs Warm With Dry Summer
Csa Warm Dry Summer (Mediterranean)
Csb Cool Dry Summer (Mediterranean)

Cw Warm With Dry Winter
Cwa Warm Summer, Dry Winter
Cwb Warm With Dry Winter and Cool Summer

Dfa Cool and Moist With Warm Summer
Dfb Cool and Moist With Cool Summer
Dfc Cool and Moist With Cold Winter
Dfd Cool and Moist With Very Cold Winter

Dwa Cool With Dry Winter and Warm Summer
Dwb Cool With Dry Winter and Cool Summer
Dwc Cool With Dry, Cold Winter
Dwd Cool With Dry, Very Cold Winter

EF Perpetual Frost

ET Tundra

H Highland

Projection: Flat Polar Quartic Interrupted

tively. Thus the tropical wet and dry climate is an *Aw* type, while the dry summer subtropical climate is the *Cs* type. Additional symbols are added to characterize special conditions, such as a monsoon climate or frequent fog.

The Köppen system has been criticized because in many regions its boundaries do not truly coincide with the boundaries of the supposedly related vegetation regions. Köppen's assumptions concerning the relationship between climate and vegetation appear to have been oversimplified and often cannot be demonstrated. His assumption that forests stop where precipitation is equaled by potential evapotran-

$$\text{Moisture Index} = 100\,\frac{P - PE}{PE}$$

$$\text{Moisture Index} = 100\,\frac{S - D}{PE}$$

$P =$ Precipitation

$PE =$ Potential Evapotranspiration

$S =$ Surplus

$D =$ Deficit

Figure 8.7
The value of Thornthwaite's moisture index can be calculated from either of these two equations. The value of the moisture index is positive at places where mean annual *P* is greater than *PE*, and it is negative where *P* is less than *PE*. The moisture index is also positive at places where *S* is greater than *D*. Even though mean annual *P* is greater than *PE*, moisture deficits can occur where there is a strong seasonality of precipitation, such as in dry summer subtropical regions. (Vantage Art, Inc.)

spiration is incorrect. On the contrary, precipitation exceeds potential evapotranspiration at most forest boundaries. Because his empirical criterion for the *B* climatic type was not very precise, Köppen sometimes placed the boundary of the *B* climate hundreds of kilometers from the true boundary between the dry and moist realms. Finally, the criteria for subdivision are too inconsistent for some scientists to accept.

Nevertheless, the Köppen system is comprehensive and rather flexible. Most of the climatic types can be related to solar energy input and the general atmospheric circulation. The seasonal regimes of temperature and precipitation are clearly summarized. The terminology is recognized by most geographers and environmental scientists, and the system has been used in recent years primarily to introduce the regional aspects of global climates.

The Thornthwaite System of Climate Classification

In an attempt to overcome the shortcomings of the Köppen system, C. Warren Thornthwaite in 1948 devised a method for classifying cli-

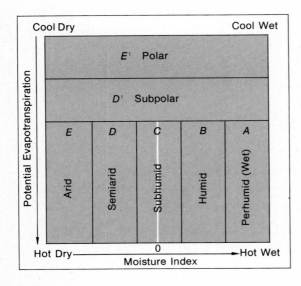

Figure 8.8
This schematic diagram represents the principal climatic regions in the Thornthwaite classification system. Potential evapotranspiration is used as a measure of moisture utilization by environmental systems on the earth's surface. The climates run from cool and dry at the upper left corner of the diagram to hot and wet at the lower right. The climates are divided according to orderly ranges of values for potential evapotranspiration and moisture index. In the higher latitudes where energy is in short supply, the emphasis is on energy availability. In lower and middle latitudes where large amounts of energy in terms of radiation and heat are normally available, the emphasis is on moisture availability. Five principal moisture regions, *A* through *E*, are recognized for middle and lower latitudes. The subhumid climatic region, *C*, is centered where the moisture index equals 0. (Doug Armstrong after D. Carter and J. Mather, Publications in Climatology, vol. 19, no. 4, 1966)

Figure 8.9

This map of the coterminous United States shows the principal moisture realms as classified according to the Thornthwaite moisture index. Precipitation exceeds potential evapotranspiration where the index is positive, and it is less than potential evapotranspiration where the index is negative. The line where the moisture index is 0 extends from western Minnesota south to the Gulf of Mexico. To the east of this line there is a surplus of moisture, and to the west there is a deficiency, under average conditions. The moisture index of a region bears a close relation to the form of its vegetation. Forests, for example, can maintain themselves only where the moisture index is positive and has a value of at least 20 or so. Only grasslands and desert vegetation can survive where the moisture index is negative. (Andy Lucas and Laurie Curran after C. W. Thornthwaite and J. Mather, "The Water Balance," *Publications in Climatology*, vol. 8, 1955)

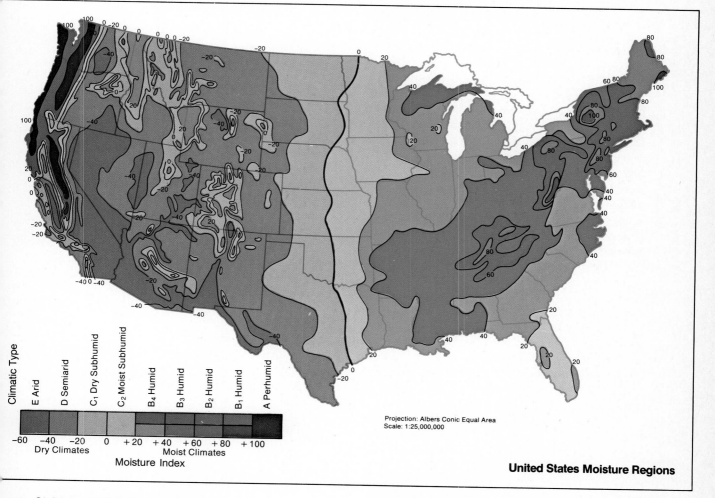

Climatic Type

E Arid | D Semiarid | C₁ Dry Subhumid | C₂ Moist Subhumid | B₄ Humid | B₃ Humid | B₂ Humid | B₁ Humid | A Perhumid

−60 −40 −20 0 +20 +40 +60 +80 +100
Dry Climates   Moist Climates
Moisture Index

Projection: Albers Conic Equal Area
Scale: 1:25,000,000

United States Moisture Regions

mates according to water budget evaluations of energy and moisture. In Thornthwaite's classification system, energy is specified by potential evapotranspiration (*PE*), and moisture is expressed by a *moisture index*, which is defined in Figure 8.7.

Thornthwaite's moisture index depends on the difference between precipitation and calculated values of potential evapotranspiration. The moisture index has the value −100 when there is no precipitation, and may exceed +100 where rainfall far exceeds potential evapotranspiration. At the boundary between the dry and moist realms, the moisture index is zero.

Figure 8.8 shows schematically how climatic types are defined in Thornthwaite's classification system. Each climatic region corresponds to a range of energy and moisture values that are calculated according to uniform criteria.

One of the principal differences between the Köppen and Thornthwaite classification systems is in the way the dry and moist realms are divided. Like Köppen, Thornthwaite divided the non-polar climates into five general types: two in the moist realm, two in the dry realm, and one, the *subhumid*, that occupies the transitional zone where precipitation and potential evapotranspiration are about equal. When the Köppen and Thornthwaite climatic classifications are applied to the United States, the difference between them emerges clearly. Consider the North American grasslands east of the Rocky Mountains, which are so distinctive that they appear to occupy a separate climatic region between the moister areas to the east and the drier regions to the west. When Köppen's criteria are used, the dry grasslands do not emerge as a distinct climatic region. However, the dry grasslands correspond closely to the subhumid climatic type of Thornthwaite's system, as shown in Figure 8.9.

Perhaps because its computations are time-consuming, the Thornthwaite system is not often used to define climatic regions on a global or continental scale. It does not give a graphic portrayal of specific temperature and precipitation amounts and regimes as does the Köppen system. Still, Thornthwaite's concept of the moisture index based on month-by-month water budget calculations (Chapter 7) represents an improvement over Köppen's temperature and precipitation relationships. World maps have been constructed that show individual parameters in water budgets: potential evapotranspiration, the moisture index, annual water surplus, and annual water deficit. These are employed by planners and engineers in the design of water projects and agricultural strategies, and other scientists use them to analyze climatic effects on land use, vegetation, soils, and landforms.

WATER BUDGET CLIMATES: INTERACTIONS WITH ENVIRONMENTAL SYSTEMS

A useful tool for understanding climate in terms of energy and moisture is the local water budget discussed in Chapter 7. The individual components of the water budget that can be used as indexes of climate include potential evapotranspiration, precipitation, soil moisture, actual evapotranspiration, water surplus, and water deficit. By analyzing these components, it is possible to understand how climate affects various systems in the environment. In particular, studies of water budget components are useful for making detailed comparisons of environments from place to place, as well as in revealing variations in environmental conditions through time.

To illustrate, we can consider the average local water budgets at two locations: Cloverdale, California, and Manhattan, Kansas. Both locations have a total annual potential evapotranspiration of nearly 80 cm (31.5 in.), so their energy endowments are essentially equal.

On the basis of annual precipitation, Cloverdale (100 cm, or 39.4 in.) appears to be more humid than Manhattan (80 cm, or 31.5 in.). But,

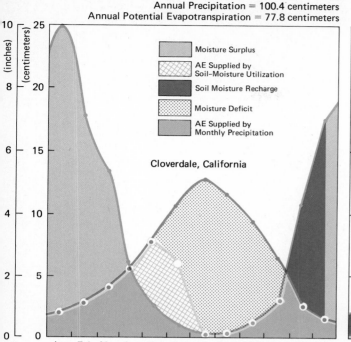

Annual Precipitation = 100.4 centimeters
Annual Potential Evapotranspiration = 77.8 centimeters

Moisture Surplus

AE Supplied by
Soil–Moisture Utilization

Soil Moisture Recharge

Moisture Deficit

AE Supplied by
Monthly Precipitation

Cloverdale, California

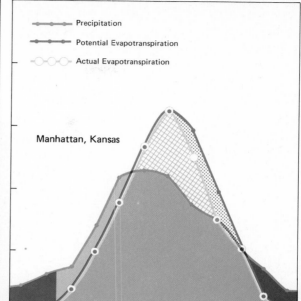

Annual Precipitation = 80.0 centimeters
Annual Potential Evapotranspiration = 78.3 centimeters

Precipitation

Potential Evapotranspiration

Actual Evapotranspiration

Manhattan, Kansas

Figure 8.10
Cloverdale, California, and Manhattan, Kansas, receive similar amounts of precipitation annually and have nearly the same total annual potential evapotranspiration. The climates at these locations appear to be similar on the basis of the annual inputs of moisture and energy. But their local water budgets indicate that the timing of precipitation compared to the demand for moisture at each location makes the climates quite different. At Cloverdale, heavy precipitation and a low moisture demand by vegetation during the winter generate large moisture surpluses. Deficits occur during the dry summer. Thus Cloverdale is moist for runoff and dry for vegetation. At Manhattan, the timing of precipitation through the year is in close accord with the demand by vegetation, making surpluses and deficits small. Manhattan is therefore dry for runoff and moist for vegetation. (Vantage Art, Inc. after Douglas B. Carter, Theodore H. Schmudde, and David M. Sharpe, "The Interface: As A Working Environment: A Purpose for Physical Geography," *Tech. Paper no. 1*, Comm. on College Geog., Association of American Geographers, 1972)

as noted in Chapter 7, the utilization of moisture at a place depends on the timing of precipitation. Figure 8.10 shows the water budgets for the two locations. At Cloverdale, large deficits of soil moisture are generated in the summer months when potential evapotranspiration is high and precipitation is low. Cloverdale's vegetation has evolved adaptations to survive a very dry growing season. In the winter and spring, however, heavy precipitation results in a large surplus of water. Therefore, Cloverdale's climate supports strong seasonal streamflow.

At Manhattan, the water budget shows that precipitation amount and timing tend to mirror the variation of potential evapotranspiration through the year. Therefore, only small deficits and surpluses are generated at Manhattan. There is moisture for vegetation throughout the growing season, but little to support streamflow at any time.

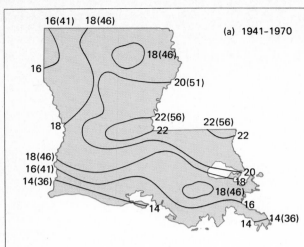

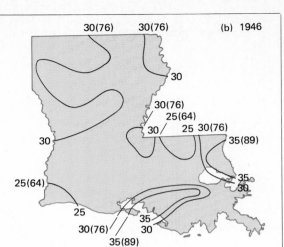

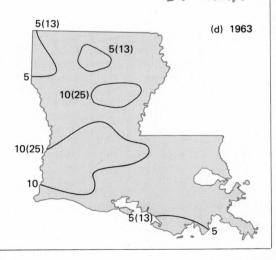

Figure 8.11
These maps illustrate in inches and centimeters the spatial and temporal variability of the moisture surplus for Louisiana during the winter-spring season (November through May).

(a) The mean surplus over a 30-year period shows considerable variability across the state.

(b) During 1946 a wet winter-spring season occurred everywhere.

(c) Extreme spatial variability existed in the winter-spring of 1953, when record floods occurred across south-central Louisiana.

(d) Very small surpluses were recorded during the extremely dry winter-spring of 1963. (Vantage Art, Inc. after Robert A. Muller and Philip Larimore, Jr., "Atlas of Seasonal Water Budget Components of Louisiana," *Publications in Climatology*, *28*, no. 1, 1975)

Figure 8.12

Components of the local water budget can be used as climatic indexes. These graphs emphasize variability of the climate at seven locations across the United States for the 30-year period between 1941 and 1970. The annual values of the components are plotted with the largest value to the left and the smallest value to the right.

(a) Annual moisture surpluses are shown here. The largest annual surplus over the 30 years at Blue Canyon, California, amounted to more than 229 cm (90 in.); the smallest surplus was only about 64 cm (25 in.). Over the 30-year period, the average surplus was about 140 cm (55 in.). On the basis of the cumulative frequency curve, we can see that the surplus amounted to 157 cm (62 in.) or more during 30 percent of the years. Similarly, during 90 percent of the years, the surplus was equal to or greater than 76 cm (30 in.); we can transpose the perspective and also say that during 10 percent of the years the surplus amounted to less than 76 cm (30 in.). The largest surpluses were at Blue Canyon; Morgan City, Louisiana, ranked second of the seven.

(b) The ratio of actual to potential evapotranspiration is a measure of the degree to which precipitation and soil moisture meet the climatic demand for water. Most crops and forest vegetation do poorly when the ratio is low over the growing season. For this index, Morgan City ranked first and Blue Canyon only fifth! (Doug Armstrong after R. Muller, "Frequency Analyses of the Ratio of Actual to Potential Evapotranspiration for the Study of Climate and Vegetation Relationships," Proceedings of the Association of American Geographers, vol. 3, 1971)

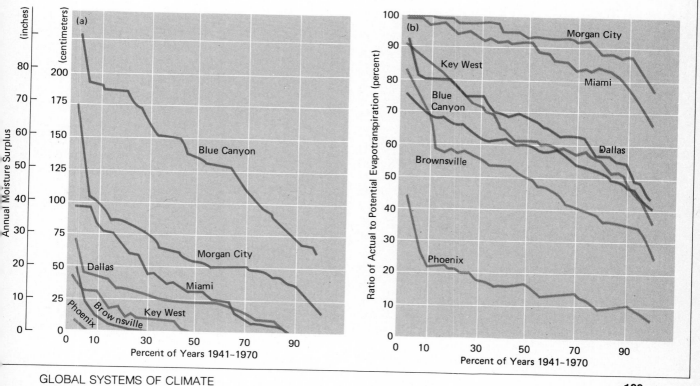

The examples of Cloverdale and Manhattan show that when one considers the relationship of climate to other environmental systems, simply knowing the total amounts of energy and moisture delivered is not enough. The distribution of energy and moisture through time must also be taken into account. A given location offers particular problems and possibilities that are revealed by different components of the water budget.

The surplus in the water budget is a good index of the variability in space and time that can exist within a given climate. Figure 8.11a shows the average winter-spring surplus across Louisiana for a 30-year period, 1941–1970. Although Louisiana has the highest average precipitation of the 50 states, the average seasonal surplus shown on the map ranges from more than 55 cm (20 in.) to less than 35 cm (14 in.) along the Gulf Coast. Even in this wet state, the spatial variability in the water supply is considerable.

Variability from one year to the next, which depends on small changes in the upper atmospheric circulation, can be greater than variation from place to place. The surpluses across Louisiana for three winter-spring seasons are shown in Figure 8.11b–d. These examples represent a season that is "wet" everywhere, a season with extreme spatial variability, and, finally, a "dry" season everywhere. Clearly the range between extreme years is very large. This exerts stress on vegetation and animal life; however, it is clear that the region's native plants and animals are adapted to survive such extremes, and themselves express the nature of the prevailing climate, including its year-to-year variation.

In 1972, the year of "the great freeze" in central California, it was apparent that native evergreen vegetation was unaffected by several successive days of highly unusual subfreezing temperatures. Imported plants, including whole forests of non-native eucalyptus trees, were either temporarily or permanently damaged because their evolution had taken place in other regions where such weather never occurred. Despite their infrequent occurrence, past "freezes"

in central California have eliminated such plants from the local flora.

Figure 8.12a shows the annual moisture surplus generated at seven locations in the United States. The data are presented in terms of frequency distributions over a 30-year period. This shows the degree of variability that can occur at each place. At Blue Canyon, California, a moisture surplus greater than 135 cm (53 in.) was generated in 50 percent of the years. But in individual years the surplus ranged from over 229 cm (90 in.) to under 64 cm (25 in.). In 70 percent of the years an annual surplus greater than 110 cm (43 in.) occurred at Blue Canyon. From the standpoint of moisture surplus, Blue Canyon should be classified as exceptionally humid; Morgan City, Louisiana, as humid; and Phoenix, Arizona, as arid.

In Figure 8.12b the same seven locations are arranged according to the ratio of actual to potential evapotranspiration—meaning water supply relative to water demand. Vegetation that is native to places with high water supplies relative to demands will not do well without irrigation in dry climates, such as that of Brownsville, Texas, where actual evapotranspiration can rarely supply more than 60 percent of the water demand. On this climatic scale Morgan City and Miami are best supplied with moisture, Blue Canyon is only intermediate, and Phoenix is again lowest. If the stations were plotted in terms of annual water deficit, Phoenix would rank first. Depending upon which water budget components are selected, locations change their relative positions along the climatic scale.

SUMMARY

Climate is the average condition of the atmosphere near an area of the earth's surface over a period of years. Its most common measures are temperature and precipitation. Climates show significant spatial and temporal variations, whether they are viewed on the scale of a continent, a region, or even within the confines of a suburban backyard. The range of possible climates can be divided into a small number of specific types in order to classify geographical areas according to their dominant climatic characteristics. Different schemes of climatic analysis and classification are useful for different purposes.

The geographical distribution of climatic types resulting from the delivery of energy and moisture by the sun and the general atmospheric circulation can be demonstrated on a hypothetical continent with a smooth surface and a regular coastline. The influences of mountain ranges and continental shapes are clear on a global map of actual climatic patterns resulting from the general circulation. The repetition of specific climatic types in widely separated parts of the earth is one of the most important facts of physical geography and affects the form and distribution of vegetation, soils, landforms, and human activities.

The climatic classifications of Köppen and Thornthwaite represent progressive attempts to evaluate energy-moisture interactions at the earth's surface. The Köppen system has the advantage of specifying the nature of annual temperature and moisture regimes, but has little application beyond displaying the geography of climatic types. The Thornthwaite system, based on the local water budget, analyzes interactions between energy and moisture in terms of precipitation, evapotranspiration, soil moisture, runoff, and water deficit. This system is a useful tool for understanding the ways in which climate controls various environmental systems. It is also helpful in the analysis of the spatial and temporal variability within climates.

By evaluating water budget components, we can see that areas having similar annual inputs of moisture and energy may, in fact, have quite different climates. This may be reflected in such phenomena as streamflow, vegetation cover, soil characteristics, and erosional processes and resulting landforms.

APPLICATIONS

1. Along portions of the coasts of North and South America, North and South Africa, Portugal, and Australia there are climates that are generally regarded as "Mediterranean" in type. However, the climates of these coasts do not exactly duplicate the climates of the Mediterranean coastlands themselves. What is the difference?

2. The largest high latitude land area that was not covered by an ice sheet during the Ice Age was Siberia, which now has the coldest winters of any area outside of Antarctica. How do you explain this contradiction?

3. Suppose the Gulf of Mexico and the Caribbean Sea did not exist and the North American coastline extended continuously from Florida through the Bahamas, Puerto Rico, and the Lesser Antilles and Trinidad to Venezuela. How would this affect the climates of North America? What would happen to the geographic location of Thornthwaite's zero moisture index that separates dry and moist climates? Draw a map showing hypothetical climatic regions on such a continent, using the Köppen system of classification.

4. The Mississippi River valley interrupts an otherwise continuous belt of hilly terrain and low mountains extending from Oklahoma to New England. In some places these mountains were much higher in the past. If they were today as lofty as the Rockies are, how would the climates of North America be affected?

5. Climatic variability from year to year can have significant impact on environmental systems and economic prosperity. Using Appendix II and National Weather Service data on monthly temperature and precipitation for some place of interest, determine the Köppen climatic type for each of five successive years. What portions of the earth would be most likely to

fall into different Köppen climates from year to year?

6. Imagine the effect on the earth's climates if the axis of the earth's rotation were perpendicular to the plane of the earth's orbit around the sun, rather than being inclined at the present angle of 23½ degrees. What differences in climatic types would result? Would the earth be more or less habitable? You might be assisted by Figure 3.4.

FURTHER READING

Carter, Douglas B., and **John R. Mather.** "Climatic Classification for Environmental Biology." *Publ. in Climatology,* Vol. 19, No. 4 (1966): 305–395. The evolutionary sequence of the several Thornthwaite climatic classifications and some of their relationships to vegetation distribution are presented in this technical paper. Thousands of average water-budget tables from thousands of places around the world are included in other issues of this specialized journal.

———, **Theodore H. Schmudde,** and **David M. Sharpe.** "The Interface as a Working Environment: A Purpose for Physical Geography." *Tech. Paper No. 7,* Comm. on College Geog. Assoc. of American Geog. (1972), 52 pp. This monograph stresses the relationship between water budget components and environment responses.

Critchfield, Howard J. *General Climatology.* 3rd ed. Englewood Cliffs, N. J.: Prentice-Hall (1974), 446 pp. An introductory text that emphasizes the interrelationships of climate to global patterns of vegetation and soils, as well as to a wide range of economic activities.

Hare, F. Kenneth. *The Restless Atmosphere.* Rev. ed. London: Hutchinson (1956), 192 pp. This brief introductory classic contains outstanding regional and continental chapters that focus on the dynamics of the general circulation.

Roberts, Walter O., and **Henry Lansford.** *The Climate Mandate.* San Francisco: W. H. Freeman (1979), 197 pp. This little paperback provides an outstanding up-to-date perspective on climatic variation and on some of its causes and consequences.

Rumney, George R. *Climatology and the World's Climates.* London: Macmillan (1968), 656 pp. The main portion of this book consists of descriptions of climates of natural regions (biomes) and the vegetative responses to biome climates. The treatment is quite detailed and deals with regional variations from typical conditions.

Thornthwaite, C. Warren. "An Approach Toward a Rational Classification of Climate," *Geographical Review,* Vol. 38, No. 1 (1948): 55–94. Thornthwaite first set out the potential evapotranspiration and water budget systems in this classic paper.

Trewartha, Glenn T., and **Lyle H. Horn.** *An Introduction to Climate.* 5th ed. New York: McGraw-Hill (1980), 416 pp. Temperature and precipitation properties of climatic regions over the globe are stressed in this introductory text.

Wilcock, Arthur A. "Köppen After Fifty Years." *Annals Assoc. of American Geog.,* Vol. 58, No. 1 (1968): 12–28. This article presents a concise review of Köppen's development of his climatic classification and the subsequent modifications by other climatologists.

CASE STUDY: How Much Do Climates Change?

Although the weather seems to be relatively the same from year to year, one summer may be hot and dry, and another cool and wet. Climatic fluctuation on this scale is very costly to technological civilizations, especially in terms of energy consumption and agricultural production. An early season freeze on Labor Day weekend in 1974 devastated crops across much of the upper Midwest. The record-breaking heat in Texas, the Plains states, and the South during the early summer of 1980 increased energy use, seared pastures, and withered crops. Fluctuations in water budgets are especially significant. Above normal deficits can greatly reduce crop yields (the "dust bowl" years of the 1930s are a dramatic example), and along the Mississippi, alternations in the surplus have resulted in both floods and low water problems (see Case Study following Chapter 13).

Only in recent decades have scientists begun to unravel the detailed outline of the earth's climatic history for the past million years, using studies of tree rings, fossil plant pollen, ancient soils, lake and deep-sea sediments, mountain glaciers, and cores from ice caps on Greenland and Antarctica. The figure on page 206 illustrates the major trends of global climate based on these analyses. Perhaps the most astounding features are the glacial-interglacial cycles that last about 100,000 years, with shorter interglacials separated by long glacial climates, as shown in graph a. Glacial conditions appear to develop slowly and irregularly, with interglacial climates evolving suddenly. Recent studies indicate that climatic variability on time scales of 100,000 and 20,000 years may be related to systematic variations of the earth's orbit about the sun, which produce seasonal and latitudinal changes in the distribution of incoming solar radiation. Graph b shows the rapid warming trend that began about 15,000 years ago. During the very sudden and brief Younger Dryas cold interval about 10,500 years ago, much forest in Europe was destroyed. Graph b also shows that the mildest post-Pleistocene temperatures occurred about 5,000 to 7,000 years ago.

Temperature variations in eastern Europe over the past 1,000 years are shown in graph c; the temperature reconstruction is based on historical records of the positions of the snouts of alpine glaciers and of the wine production at various monasteries. The lower temperatures associated with the "Little Ice Age" between 1400 and 1850 resulted in social and economic dislocations. The thermal maximum of the 1940s represents the highest average temperatures of the past 1,000 years. Graph d shows the cooling that began about 1945, but recent data suggest that the global cooling may have leveled off since the mid-1970s.

Much shorter-term climatic variability has direct impact on environmental processes and economic activities. The next figure shows monthly temperature and precipitation variability for southeastern Louisiana from 1911 through June 1977. Each month's deviation from its respective average for the 30-year period 1931–1960 is shown in graphs a and c. Graphs b and d show short-term variations in terms of six-month "running averages."

There are two outstanding features of

these graphs for southeastern Louisiana. One is the downward step of temperature beginning in the fall of 1957 and continuing, with the exception of several winters, to the present. The other is the tendency for clusters of months or of years to be warmer or colder, or wetter or drier, than normal. Especially obvious are the dry periods of 1915–1918, 1933–1938, and 1951–1955, all associated with droughts on the Great Plains; other intense dry periods include 1924–1925 and 1962–1963. Climatic variability on these time scales significantly affects agricultural yields and water-resource management.

Explaining why climates change is extremely difficult because many interactions of the general circulation are not understood in detail. Furthermore, slight climatic changes can produce disproportionate effects in the environment. If the average global temperature were to drop a few degrees, ice and snow would cover a greater proportion of the earth's surface. Because of its high albedo, the ice would reflect more radiant energy back to space, which would cause further cooling and further spread of ice and snow.

A number of different explanations have been proposed to account for climatic variability over geologic time, but none is entirely satisfactory. Glacial climates appear to have been associated with continental drift and plate tectonic processes (see Chapter 11), which caused continents to move into polar regions, allowing ice sheets to develop. In addition to Pleistocene glaciation, there is evidence of at least three widely separated glacial periods over the last billion years. Within these glacial periods, it is now believed that systematic variations in the earth's orbit account for the 100,000-year glacial-interglacial cycle. Furthermore, mountain uplift is believed to be connected with colder temperatures and increased snowfall over uplands. Stratospheric dust from intense volcanic activity has been linked

to periods of somewhat diminished solar radiation income and lower temperatures at the earth's surface. The early summer freezes and snows in northern New England in 1816 followed the massive eruption of the volcano Tambora in the Dutch East Indies in 1815, and several years of spectacular red sunsets followed the eruption of Mt. Agung in Indonesia in the early 1960s. More recently, cool summer weather caused by the 1980 ash eruptions of Mount St. Helens produced a record wheat crop in eastern Washington.

Numerous attempts have been made to relate sunspot activity to climatic variation, especially drought periods in the Great Plains. Although there is some correlation between sunspots and dry periods on a time scale of several years, no cause-and-effect relationship has been demonstrated. Climatic variability on shorter time scales, such as those shown for southeastern Louisiana, are obviously related to changing patterns of the circumpolar vortex of westerlies in the upper atmosphere. Meteorologists and geophysicists continue to debate the causes of these circulation changes and the eventual return to more "normal" patterns. Precise forecasting of the circulation patterns a season or more in advance remains an elusive objective at present.

In predicting the climates of the near future, most concern centers on the effect of human activities, particularly the use of the atmosphere as a dumping ground. The atmosphere is a sensitive controller of climate; cloudiness and dust in the air directly affect the amount of energy reaching the earth, and carbon dioxide in the air affects the amount of energy that is radiated away. Large-scale industry and agriculture add smoke and dust particles to the air, increasing the albedo of the atmosphere and decreasing the amount of solar radiation received at the surface, and possibly lowering surface temperatures as well.

The interconnections of the earth's

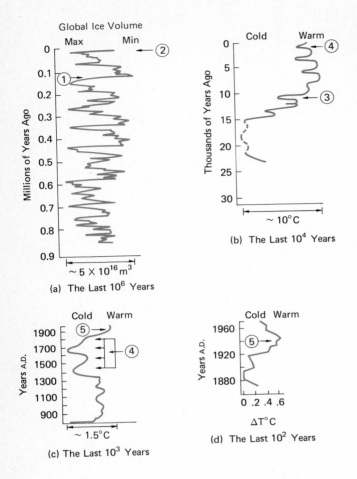

Global Ice Volume

Max   Min

(a) The Last 10⁶ Years

$\sim 5 \times 10^{16} \, m^3$

(b) The Last 10⁴ Years

$\sim 10°C$

Cold   Warm

(c) The Last 10³ Years

$\sim 1.5°C$

Cold   Warm

(d) The Last 10² Years

$\Delta T°C$

① Penultimate Interglacial
(Riss–Würm–Sangamon in North America)

② Present Interglacial (Holocene)

③ Younger Dryas Cold Interval

④ Little Ice Age

⑤ Thermal Maximum of 1940's

(above) The main trends of global climate during the past million years. Graph *a* is based on isotope analyses of deep sea floor sediments; *b* on alpine tree lines, fluctuations of alpine glacier snouts, and fossil pollen; *c* on historical records of alpine glacier snouts and wine and grain production; and *d* on measured temperature data. (Vantage Art, Inc. from *Understanding Climatic Change: A Program for Action*, National Academy of Sciences, 1975, Washington, D.C.)

systems make it difficult to predict the net effect from a particular change. The amount of carbon dioxide in the atmosphere is expected to increase by at least 20 to 30 percent in the last half of this century, primarily because of industrial activities. The direct effect of the increase in carbon dioxide will be a decrease in longwave radiation losses from the earth's surface, which will raise the average temperature. The higher temperatures will increase evaporation, perhaps leading to increased cloudiness. Because it is difficult to determine the degree to which each of these modifications will affect other environmental properties, the final temperature change caused by the increase of carbon dioxide cannot be predicted.

Any small, long-term change in the atmosphere is critical. The present state of the environment is a function of its past history; there is no fresh start each year because the environment accumulates the effects of minor changes. Orbit wiggles, volcanic dust—these and other small and complicated interactions influence climatic trends. Now human activities can be added to the list.

(opposite) Temperature and precipitation variability in southeastern Louisiana. The temperature deviation of each month from the averages of the 30-year period between 1931 and 1960 is shown in graph *a*, and precipitation deviations are shown in graph *c*. For example, January 1940 was 7°C (12°F) below normal, and the rainfall during October 1937 was 28 cm (11 in.) above normal. Six-month running averages of temperature deviations are used in graph *b* to illustrate a "smoothed" interpretation of short-term temperature variation, and graph *d* is a similar smoothed interpretation of precipitation variation. Note especially the colder temperatures beginning in late 1957 and the fluctuations between warmer and colder, and wetter and drier conditions. (Vantage Art, Inc. from R. A. Muller and J. E. Willis, "Climate Variability in the Lower Mississippi River Valley," *Geoscience and Man*, 1978)

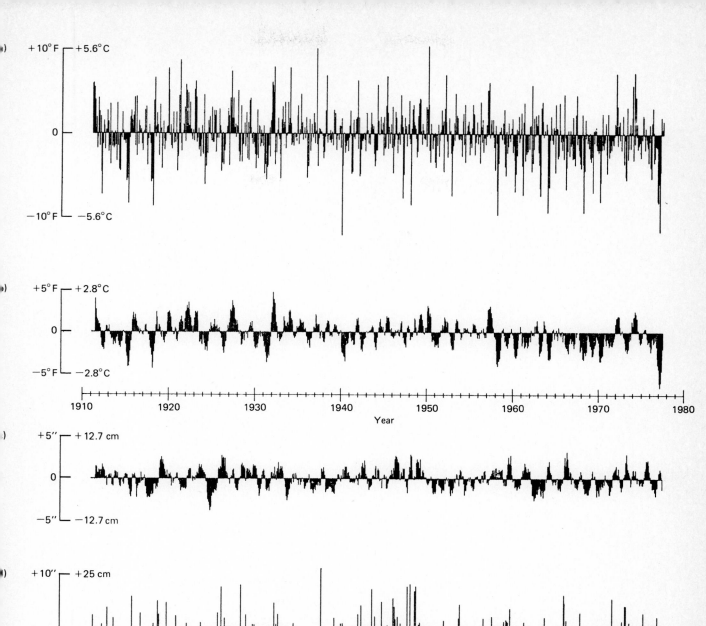

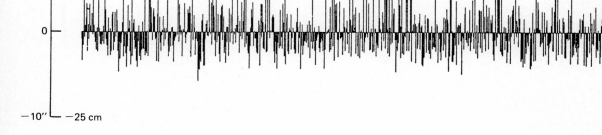

The Park by Gustav Klimt, 1910. (Collection, The Museum of Modern Art, New York; Gertrud A. Mellon Fund.)

Where both moisture and solar energy are abundant the earth's land surfaces are cloaked with a living blanket of green vegetation. This phenomenon, unique in our solar system, varies remarkably in form and behavior in response to our planet's diverse climates and surface characteristics. The appearance of plant life on the continents made possible the emergence of all other terrestrial life forms, whose food chains are ultimately based on vegetal matter.

CHAPTER 9
Climate and Vegetation

The chief wonder of our planet is its life, and the most visible part of the living world is the green blanket of vegetation that covers most of the land. It is liquid water that makes possible the chemical reactions that permit living cells to form and multiply. Elsewhere in the solar system, temperatures do not remain within the narrow but critical range that permits water to be present in liquid form, rather than being boiled away or permanently frozen. The availability of liquid water largely determines the nature and abundance of all life on our planet, and the lack of liquid water on other planets in the solar system makes it very unlikely that any form of life will be found on them.

When land plants evolved on earth, some 400 million years ago, they soon spread into all parts of the planet except those regions covered by glacial ice. To do so, they had to evolve adaptations to every combination of energy, moisture, and soil conditions found on the earth. Although these adaptations have caused plant life to vary in form from lichens to giant forest trees, most plants have certain similarities. Most have a root system to gather chemical nutrients dissolved in soil moisture, green leaves to convert solar energy into chemical energy for plant growth, and stems or branches to support the leaves and to channel nutrients and chemical energy throughout the plant.

These basic features must vary greatly to operate over a wide range of environmental conditions. The depth and spread of roots reflect the supply of moisture and nutrients. Leaves may be large or small depending upon light levels and moisture stress. Leaf form differs in response to environmental possibilities and constraints. Leaves may be retained all year by *evergreen* plants or dropped seasonally by *deciduous* plants to prevent frost damage or moisture loss when resupply is impossible. Leaves may even be dispensed with altogether, with stems and branches taking over their function. Stems themselves are adapted to local conditions and can take many forms. In later pages we shall see how these adaptations assist plants in specific environments.

In terms of their relationship to moisture supplies, plants may be classified as *hydrophytes*, which grow in water, *xerophytes*, which are structurally adapted to survive in extremely dry soils, or *mesophytes*, which occur where the water supply is neither scanty nor excessive. Likewise, plants may be *perennial*, enduring seasonal climatic fluctuations, or *annual*, dying off during periods of temperature or moisture stress but leaving behind a crop of seeds to germinate during the next favorable period for growth.

By their many adaptations plants have developed the variety allowing them to exist from the edge of wind-whipped snowfields to blackwater swamps and sun-baked deserts. This chapter first summarizes some of the ecological factors that are responsible for the development of natural vegetation types seen in varying environments. Following this, the general characteristics of the principal global vegetation formations and their associated climates are described. It must be kept in mind that the natural vegetation that has evolved in equilibrium with undisturbed environments has been vastly altered by human activities. These activities have caused the virtual disappearance of some natural vegetation formations and threaten the extinction of others in the not-too-distant future. We shall return to this problem later.

ECOLOGY OF VEGETATION

Plant Communities

The associated plant species that form the natural vegetation of any one place are known as a *plant community*. In any midlatitude forest, for example, many kinds of trees, shrubs, ferns, grasses, and flowering herbs all live together in one plant community. Numerous genetically unrelated species of plants not only live in association with one another but also with bacteria, fungi, insects, and burrowing, seed-collecting, grass- and herb-eating (grazing), and twig- and leaf-nibbling (browsing) animals. Besides providing food and shelter for animals, a plant community affects its local environment by modifying soils and moisture storage conditions, by shading the ground, and by cooling the air through latent heat transfer in the process of transpiration. In such ways individual plants also affect one another so that the plant community behaves as a system composed of many interacting parts.

Each species in a plant community has its own particular way of utilizing energy and moisture. In a broadleaf forest, ferns are able to use the subdued light that filters through the high leafy canopy. They do not compete with trees for direct sunlight. Mosses, lichens, and fungi growing on rocks utilize a different moisture supply than ferns growing on the forest floor. On this same principle, farmers in the tropics often successfully grow several dissimilar crops—such as bananas, a tall open plant, and cassava, a low root crop—in the same field. As long as the crops do not compete for the same moisture and sunlight, the crop yield per acre can be greater than if the fields were planted in a single crop.

Plant Succession and Climax

Hikers walking through the New England woods sometimes come upon the foundation stones of farm buildings. These farms were abandoned more than a hundred years ago. Although the land was once cleared for agriculture, it has become forest again. But forest is not immediately reestablished when cleared land is abandoned. A *succession* of plant communities occurs. Each community alters the local microclimate and surface condition, making possible the appearance of a more demanding community. Eventually a stable community is attained, and there is no further change in its composition.

The particular order in which plant communities succeed one another depends on whether the succession begins in cleared land, filled-in marsh, burned-over forest, or some other condition. It also depends on the climate and the kind of stable plant community that is characteristic of the region as a whole. In a succession that converts cleared land to forest, the early communities are dominated by grasses and low shrubs, as shown in Figure 9.1. In time, the shrubs increase in number and small-statured trees appear. These, in turn, are replaced by larger trees. The usual trend in plant succession is toward taller, more diverse, more stable vegetation. As more advanced communities become established, the pace of succession slows because the larger individual plants mature later and have longer lives. The larger trees of deciduous

forests may live several hundred years and not produce seeds for their first two decades.

The stable plant community that results at the end of a long undisturbed succession is called the *climax vegetation*. It may take hundreds or even thousands of years to establish a climax vegetation. Environmental change may disturb the climax once it has developed, initiating a renewed successional sequence. Such disturbances include fires, floods, severe droughts, hard freezes, insect or parasite infestations, and freezing rain that breaks down ice-laden trees. Pleistocene glaciation and associated climatic changes produced world-wide disturbance of plant communities. Stable communities have not yet been reestablished over large areas in the middle latitudes. More recently, human ac-

Figure 9.1
This diagram illustrates a typical plant succession from open land to forest in the middle latitudes. In the initial stages of the succession, low, fast-growing grasses and shrubs dominate. Each stage alters the soil and microclimate of the land, enabling other species to establish themselves and become dominant. The final stage consists of high trees that overshade and force out some of the low shrubs of earlier stages. (John Dawson after E. P. Odum, *Fundamentals of Ecology*, 3rd ed., © 1971, W. B. Saunders Co.)

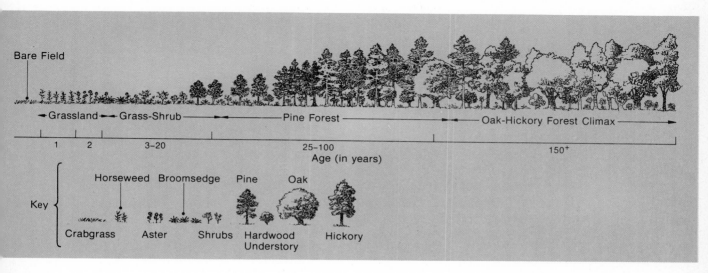

tivities, especially logging, agriculture, and live-stock grazing, have resulted in interruption of the plant succession or complete replacement of the climax vegetation.

PRINCIPAL VEGETATION FORMATIONS AND THEIR CLIMATES

Vegetation that has evolved in response to the average climatic and soil conditions of a region, as well as its occasional temperature and moisture extremes, is called *natural vegetation.* Figure 9.2 shows the close interrelationship between global climate and natural vegetation patterns. As Figure 9.2 demonstrates, energy availability is crucial in determining natural vegetation at higher latitudes, whereas moisture supply is more important at lower latitudes. In subhumid climates, where the moisture index is near zero, grasslands are predominant (Figure 9.3). Where the moisture index is somewhat greater than zero, the natural vegetation is usually forest. In polar and subarctic climates, where solar energy input is more critical than moisture, the natural vegetation is tundra or subarctic coniferous forests known as *taiga.*

Over long periods of earth history, vegetation in widely separated regions with similar climates has evolved similar characteristics. Many species of the New World cactus family and the Old World euphorbias have evolved identical characteristics, yet these species result from completely separate evolutionary streams. Their common characteristics allow species of both families to thrive in desert climates that exclude most other types of plants. Only an expert can tell them apart. This type of *convergent evolution* of different plant families toward similar forms is especially noticeable in climates having temperature or moisture extremes that require special plant adaptations for survival.

A global map of natural vegetation types is shown in Figure 9.4. This simplified map does not include natural and artificial disturbances

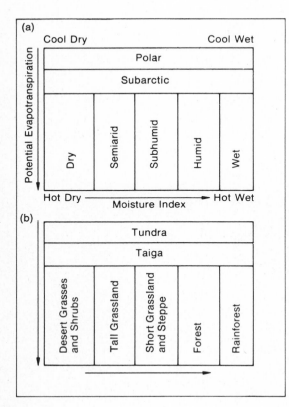

Figure 9.2
(a) The schematic diagram shows the principal climatic regions according to the Thornthwaite system of classification. (After Blumenstock and Thornthwaite, 1941)

(b) The principal vegetation regions are diagrammed according to the same *PE* and moisture index scales used to show Thornthwaite's climates. At higher latitudes, energy availability determines the presence of tundra or forest; at middle and lower latitudes, the vegetation is determined primarily by the availability of moisture. (Doug Armstrong adapted from Blumenstock and Thornthwaite, "Climate and the World Pattern," *Yearbook of Agriculture*, 1941, Government Printing Office)

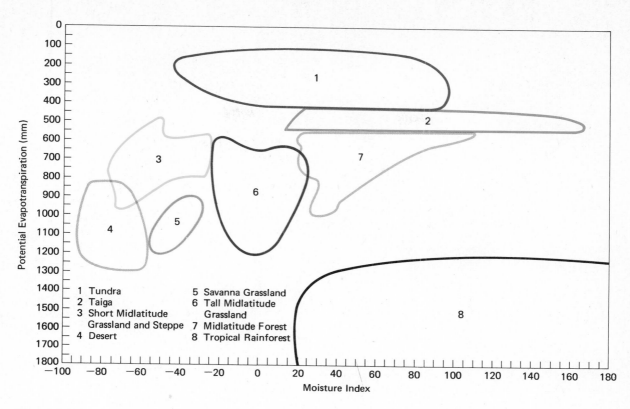

Figure 9.3
This figure shows the actual relationships between *PE*, the moisture index, and selected natural vegetation in North America and the tropics. Only some of the vegetation types discussed later in the chapter are included. The figure shows that the distribution of natural vegetation is quite similar to the simple theoretical scheme in Figure 9.2b and that there is some overlap among the types. (Vantage Art, Inc. after John R. Mather and Gary A. Yoshioka, "The Role of Climate in the Distribution of Vegetation," *Annals of the Association of American Geographers*, vol. 58, 1968)

and variations in terrain and soil conditions. Here the very complex distribution of plant communities has been simplified into nine general types, sometimes called *biomes*, a term that includes not only vegetation associations but related animals as well. Regional boundaries should be regarded as zones of intermixture between neighboring vegetation formations.

Many of the major vegetation regions portrayed in Figure 9.4 (see p. 214) and discussed below correspond to climatic regions discussed in Chapter 8. In the following pages we shall outline the nature of the relationships between climate and the vegetation formations on the earth, beginning at the equator and progressing poleward. Each vegetation formation is described and discussed in terms of its energy and moisture regime.

Tropical Rainforests

The tropical rainforest is the vegetative response to a climate that exerts virtually no limits on growth—a climate that provides abun-

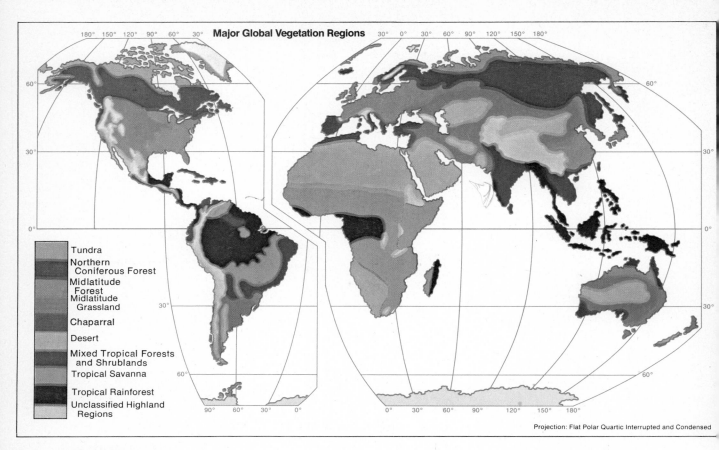

Major Global Vegetation Regions

Tundra
Northern Coniferous Forest
Midlatitude Forest
Midlatitude Grassland
Chaparral
Desert
Mixed Tropical Forests and Shrublands
Tropical Savanna
Tropical Rainforest
Unclassified Highland Regions

Projection: Flat Polar Quartic Interrupted and Condensed

Figure 9.4
This generalized global map of the distribution of major vegetation types should be compared with the global map of climates (Figure 8.3, pp. 186–187). Climate, soils, and vegetation are strongly interacting systems, and their global distributions bear marked similarities. (John Odam and Andy Lucus after Vernon C. Finch and Glenn T. Trewertha, *Physical Elements of Geography*, © 1949, McGraw-Hill Book Co.)

canopy of leaves, as well as high rates of evapotranspiration.

Perhaps because of the absence of climatic constraints, the number of plant and animal species per unit of area reaches its maximum in the tropical rainforest. Here competition among species and predation of one species upon another are the chief problems for all organisms.

dant energy and moisture in every month of the year. The resulting vegetation consists mainly of trees of varying heights that manufacture and shed leaves throughout the year. The large leaves of some species permit photosynthesis at the reduced light levels below the uppermost

Figure 9.5 (opposite)
The dense vegetation in this rainforest in Ecuador often reaches great heights and may have specially adapted forms such as climbing vines. (Sergio Larrain/Magnum Photos)

With no winter freezes, insects thrive and multiply the year around, consuming enormous amounts of plant material.

Plant species are adapted to differing light intensities and reach varying heights. The tall tree species form a comparatively dense canopy more than 30 m (100 ft) above the ground. A few still higher light-seeking trees protrude above the canopy here and there. A dense jungle of smaller trees and vines commonly makes a wall along sunlit riverbanks and around forest clearings, but the true rainforest is fairly open, having little underbrush because of the low light levels at the forest floor. Exceptions occur where large and small ferns utilize the dim light and abundant moisture near ground level (see Figure 9.5).

Tropical rainforests usually have an abundance of vines, some rooted in the ground and twining upward around the trunks of large and small trees in their quest for sunlight, and some, with no roots at all, hanging down from the branches of the forest giants. These latter types extract moisture from the air and nutrients from the litter of plant debris that lodges in the branches of the trees. Such arboreal plants, known as *epiphytes*, depend on rapid decay of organic material and are widespread only in the humid tropics.

The major limitation in the environment of the wet tropics is the poverty of the soils, which usually lack mineral nutrients. In Chapter 10 we shall see that the heavy rainfall of the wet tropics dissolves soluble mineral matter in the soil and flushes (leaches) it away, leaving behind a very infertile residue in which plants must grow. The most fertile part of the soil is the surface layer, which receives nutrients from the plant litter (leaves, branches, fruit, etc.). The roots of rainforest trees are generally shallow because of this nutrient distribution and because roots will not penetrate perpetually wet soils. Many tall tree species in rainforests have buttressed trunks that flare at the base to support a shallow-rooted, massive structure reaching high toward the sunlight.

It may seem hard to believe that the survival of the great rainforests of the wet tropics is not at all certain, but such is the case. Human activities such as logging and clearing for agriculture and cattle pasturing are destroying the rainforest at an alarming pace. Reestablishment of a cleared rainforest may be next to impossible since the soil deteriorates rapidly as soon as the forest cover is removed (Chapter 10).

In tropical rainforest areas, most of the precipitation falls as heavy showers and thunderstorms between late morning and early evening, when solar heating at the surface makes the humid tropical air most unstable. Clusters of wetter days are sometimes followed by days with only scattered showers. The wetter periods are associated with the westward passage of weak low-pressure disturbances within the trade-wind systems and intertropical convergence zone.

Air temperatures vary little either daily or seasonally. For example, the average monthly temperature at Singapore, located at latitude 1°N, shows little variation from season to sea-

Figure 9.6
Singapore is a station in a tropical rainforest region (Köppen *Af* climate). As the graph shows, the average monthly temperature and precipitation are high and nearly constant through the year. (Doug Armstrong after H. Nelson, *Climatic Data for Representative Stations of the World*, © 1968, by permission of University of Nebraska Press)

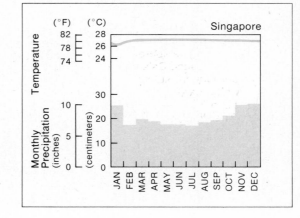

son, and the rainfall exceeds 15 cm (6 in.) in every month (Figure 9.6). Such a climate can be described by a *thermoisopleth* diagram (Figure 9.7), which shows the variation in air temperature both through the day and through the year. The pattern of the temperature contour lines, or *isotherms*, on a thermoisopleth diagram reflects the climatic characteristics. When the isotherms on the graph are primarily horizontal, temperature variations during a day are greater than variations from season to season. The thermoisopleth diagram for Belém, Brazil (Figure 9.7), located in a tropical rainforest near the mouth of the Amazon River, shows that the temperature changes by 5°C (9°F) or more during each day but varies by no more than 2°C (3.6°F) at any given hour through the year.

The local water budget can be used to pro-

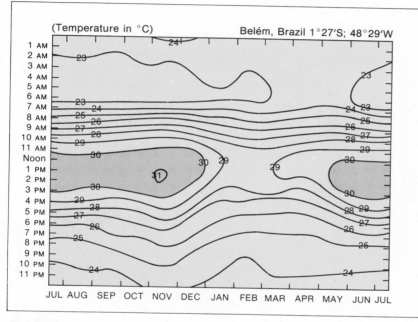

Figure 9.7
Belém, near the mouth of the Amazon River, is located in a tropical rainforest (Köppen *Af* climate). This thermoisopleth for Belém gives the average hourly temperature through a year. The scale of months begins with July because Belém is in the southern hemisphere, so that January and February are summer months. The noon temperature through the year, which can be read by tracing across the diagram from left to right, is between 30° and 31°C (86° and 88°F) until December, when it falls 1° or 2°. The diagram can also be used to trace the temperature through a given day. On January 1, for example, the temperature after midnight remains at about 23°C (73°F) until the early morning, then rises to nearly 30°C (86°F) at midday. The temperature falls again in the evening. In a tropical rainforest, the temperature range through a day is greater than the range of average monthly temperature through the year. (Doug Armstrong after C. Troll, *Oriental Geographer*, vol. 2, 1958, by permission)

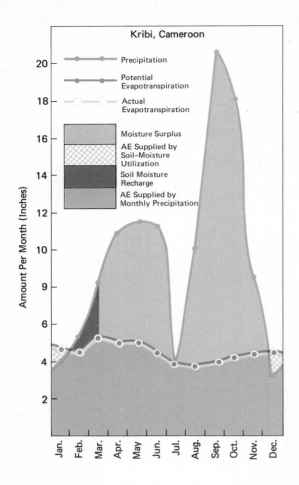

Kribi, Cameroon

Amount Per Month (Inches)

Legend:
— Precipitation
— Potential Evapotranspiration
— Actual Evapotranspiration
Moisture Surplus
AE Supplied by Soil–Moisture Utilization
Soil Moisture Recharge
AE Supplied by Monthly Precipitation

Figure 9.8
Kribi, Cameroon, is located at 3°N latitude in the tropical rainforest of Africa (Köppen *Af* climate). Monthly *PE* is high throughout the year, but during most months precipitation is even greater and large surpluses are generated. The ITC and associated showery precipitation tend to follow the seasonal migration of the sun; thus the two very rainy seasons at Kribi occur when the sun is overhead at noon near the equator. (Doug Armstrong after D. Carter and J. Mather, *Publications in Climatology*, vol. 19, 1966)

Mixed Tropical Forest and Shrubland

Despite the overall wetness of the climate, large portions of the tropics experience a short dry season. This is largely the result of seasonal migration of the intertropical convergence zone. The vegetation of such regions must adapt to moisture deficiency during this brief dry season.

vide another perspective on climate. The water budgets presented in this chapter are based on a realistic model of soil moisture availability in which the rate of removal of soil moisture by evapotranspiration diminishes as the amount of soil moisture decreases. Thus both a deficit and a utilization of soil moisture may occur in a given month. Figure 9.8 gives the water budget for Kribi, Cameroon, located in the western section of the tropical rainforest region of Africa. This water budget shows no deficits, on the average, and large surpluses nearly every month. The large moisture surpluses are clearly reflected in the richness of life in tropical rainforest regions.

Many of the trees are deciduous, shedding their leaves for a month or two when soil moisture is unavailable for evapotranspiration (see Figure 9.9). The canopy of this forest is lower and more open than that of the rainforest, so that more light penetrates to the ground. As a result there may be dense thickets of low-growing shrubby vegetation. Where the dry season is accentuated, the forest gives way to low thorny trees and shrubs with small hard-surfaced leaves that resist water loss. Figures 9.4 (p. 214) and 8.6 (pp. 192–193) show that this vegetation type corresponds to Köppen's tropical monsoon (*Am*) and portions of the tropical wet and dry (*Aw*) climates. Mixed tropical forest and shrubland vegetation is widespread in the monsoon regions of Southeast Asia and India, with smaller areas in Central and South America.

Figure 9.9
These photographs of an ancient Maya site in Honduras show a mixed tropical forest during the wet season (left) and the dry season (right). (James S. Packer)

Tropical Savannas

A *savanna* is a tropical grassland, generally with scattered trees (Figure 9.10, see p. 220). Although not all savannas correspond to a single climatic region, most are located within the tropical wet and dry climatic type, equivalent to Köppen's *Aw* climate. Thus the savannas lie between the tropical rainforests and the deserts centered near latitudes 30°N and S. In the savannas summer means the wet season and winter means the dry season. The rainy summer season is associated with the intertropical convergence zone (ITC), which invades the region during the high-sun period. The dry season results from subsidence of air in the subtropical highs that cover the region during the period when the sun's rays are most oblique. Extensive savannas are located north and south of the rainforests of both the Zaire (Congo) Basin in Africa and the Amazon Basin in South America.

Savannas vary from open woodland with a ground cover of grass to open grassland with isolated trees. Most of the trees shed their leaves during the dry season. Savanna trees commonly have thick fire-resistant bark and small drought-resistant leaves. During the long winter drought, the grass dies off above the surface, but the root systems survive and send up new shoots when moisture conditions improve.

Most savanna vegetation occupies flat plains. Often these plains are ancient land surfaces whose soils are crusted with iron and aluminum compounds (see Chapter 10) and are relatively infertile because the heavy rains of the high-sun period wash away soluble nutrients. Some savannas, then, may be *edaphic* features—related to soil conditions that exclude forests rather than to climate alone. The absence of forest trees also results from widespread grass fires that sweep over the plains, destroying young trees. Human use of savanna land for livestock pasturing also tends to prevent forests from becoming established.

The African savannas are famed for their enormous herds of hoofed animals, including ze-

Figure 9.10
The grass savanna in Amboseli Park in Kenya, Africa, is an open grassland dotted with acacia trees. (John Lewis Stage/Photo Researchers, Inc.)

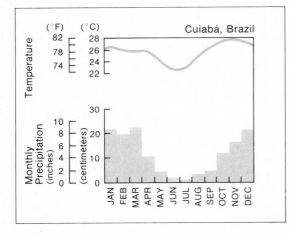

Figure 9.11
Cuiabá, in the tropical savanna of western Brazil, has a tropical wet and dry climate (Köppen *Aw* type). The graphs show a dip in average monthly temperature and precipitation during the winter months of May through September. Precipitation is nearly zero during June and July. (Doug Armstrong after H. Nelson, *Climatic Data for Representative Stations of the World*, © 1968, by permission of University of Nebraska Press)

bras, giraffes, buffalo, and many varieties of gazelle—always followed by predators such as lions, cheetahs, and, of course, men. This host of short and tall creatures is collectively equipped to chew at everything from the lowest herbs to the leaves and twigs of the spreading tree tops. Strangely, no such collection of herbivores is present in the extensive savanna regions of South America. Due to hunting and rapid agricultural encroachment, the great African herds of game animals seem doomed and may shortly be restricted to a few protected reservations.

Savanna regions may have an annual precipitation as high as 150 cm (60 in.) and as low as 50 cm (20 in.). Figure 9.11 shows monthly temperature and precipitation data at Cuiabá, Brazil, located in a tropical savanna region at latitude 16°S. As in tropical rainforests, the temperature at Cuiabá exhibits little seasonal variation. However, the rainfall greatly diminishes during the low-sun season from May through September. Figure 9.12 shows the water budget for Caracas, Venezuela, located at latitude 10°N in the

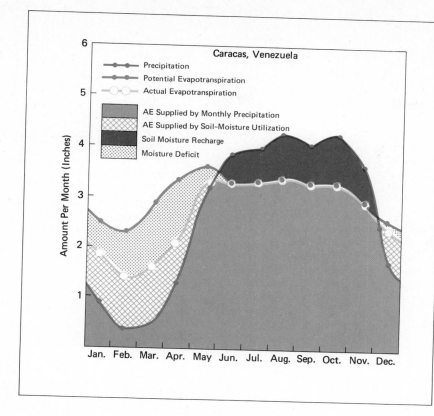

Figure 9.12
Caracas, Venezuela, has a tropical wet and dry climate (Köppen *Aw* type). The average annual precipitation is somewhat less than the annual potential evapotranspiration, but as the water budget shows, the seasonality of precipitation causes moisture deficits to occur during the months of December through May. Soil moisture stores are recharged during the remaining months. (Doug Armstrong after D. Carter and J. Mather, *Publications in Climatology*, vol. 19, 1966)

savanna of Venezuela. The annual precipitation at Caracas is 80 cm (32 in.). The water budget shows some deficiency of moisture during the winter months, but soil moisture is never completely exhausted.

Deserts

In Chapter 8 we saw that dry climates occur in several specific locations. In the subtropical latitudes, deserts are produced by subsidence of air on the eastern sides of the subtropical highs. In the middle latitudes, deserts occur in continental interior locations that are distant from oceanic moisture sources. Deserts also occur along coasts bathed by cold upwelling waters and in rainshadow areas on the lee sides of mountain systems where air is constantly descending and being warmed adiabatically.

The deserts of the subtropical dry realms are the most extensive on earth. The Sahara (Arabic for "the desert") stretches 5,500 km (3,500 miles) across Africa from the Atlantic coast to the Red Sea, its eastern sections sometimes being known as the Libyan and Egyptian deserts. The same dry realm continues across the Arabian Peninsula, Iran, and Afghanistan, extending still farther east into the Thar Desert of Pakistan. The desert zone stops there because of the effect of the rain-bearing summer monsoon of southern Asia. Subtropical deserts are also located in the southwestern United States (California, Arizona, New Mexico) and northern Mexico, and in the southern hemisphere in Australia, South America (Chile and Peru), and southern Africa.

The midlatitude deserts are concentrated in the northern hemisphere where continents are most massive. The Gobi Desert in Mongolia comes quickly to mind, along with the Central Asian deserts of Chinese Sinkiang and Soviet Turkestan. But there are also vast dry regions in the western United States in the basins and plateaus between the coastal mountain ranges and the Rocky Mountains. Although these mid-latitude deserts extending across Nevada and Utah experience cold winter temperatures, their summers are hot.

Desert regions also occur along coasts in both the tropics and subtropics where the oceanic circulation produces upwelling of cold subsurface waters. Such coastal deserts are present south of the equator in Peru and northern Chile, and in Namibia and Angola in southwest Africa; and north of the equator in Baja California, Morocco, and the Somali Republic in eastern Africa. All but the last tend to be cloudy, foggy, cool, and damp. However, precipitation is uncommon because of temperature inversions and atmospheric stability associated with the subtropical highs and cold coastal waters.

The vegetation cover developed in dry climatic areas consists of some plants that are drought-resistant and some that are drought-evading. Drought-resistant species have specialized adaptations that allow them to conserve moisture during long rainless periods. Drought-evading plants either germinate and complete their life cycles quickly during those brief periods when rain dampens desert soils, or they drop their leaves and become dormant until water is again available.

The drought-resisting plants are mainly shrubs. These are widely spaced (Figure 9.13), with extensive root systems to gather moisture, small and often waxy-surfaced leaves to conserve moisture, and in some cases tissues that can store moisture. In a few instances there are no true leaves at all, but only an enlarged green stem that takes over the function of leaves in the photosynthesis process. The giant saguaro cactus of southern Arizona (Figure 9.13) is the prime example. Since the moist, fleshy tissue of cacti is attractive to animals, most cacti are

Figure 9.13
(above) Little or no plant life can survive in this portion of Death Valley, California, because it receives so little moisture. The shrubs here are bursage (*Franseria dumosa*).

(below) These 5- to 8-meter-tall (15 to 25 ft) saguaro cactus near Phoenix, Arizona, have no leaves, a waxy skin, and are able to store water within their tissues, enabling them to survive in the desert. (T. M. Oberlander)

armed with sharp spines. Some desert shrubs, such as the creosote bush and sagebrush, send their roots many meters downward in search of moisture. Others, such as the saguaro cactus and Joshua tree, have shallow wide-spreading roots that take advantage of the surface moisture produced by light rain. Some remarkable shrubs can undergo almost complete dehydration without injury, becoming "resurrected" from a dormant state only when rains fall.

Much of the drought-evading vegetation consists of ephemeral grasses and flowering herbaceous plants. These vanish entirely during dry periods, and their seeds resist germination until guaranteed a moisture supply sufficient to take them through their life cycle. This is often accomplished by chemical germination inhibitors that must be leached away by water before germination can proceed. Following damp periods, the desert surface may be covered by flowers as the annuals spring into activity. Once their seeds are produced and distributed, the annuals wither away, leaving the ground bare until the next wet period causes the new crop of seeds to germinate. While the perennial shrubs remain, the cover of ephemeral annuals varies enormously from season to season and from year to year. This makes desert life difficult for animals (rodents, rabbits, antelope, Bighorn sheep) dependent upon vegetation or seeds for food, as well as for predators dependent upon the population of smaller animals, and for desert nomads whose sheep, goats, and camels are at the mercy of the unreliable rains.

The most common characteristic of each region of the dry realm is that precipitation is so much less than potential evapotranspiration. Most precipitation goes briefly into soil moisture storage and then is quickly returned to the atmosphere by evapotranspiration. During heavy rainshowers there may be some surface runoff. Most of this water drains into local basins and is evaporated, but some percolates down to the subsurface water table.

Figure 8.12 (p. 199) shows that at Phoenix, Arizona, moisture surpluses are rare. Where soil conditions are favorable, irrigation agriculture can be very productive, but irrigation water must be obtained from non-desert upland areas where orographic precipitation is much greater than potential evapotranspiration. Because there is little water available for evapotranspiration, most of the net radiation gain in desert areas goes into heating of the land surface and lower troposphere. At Phoenix, afternoon summer temperatures typically exceed 40°C (104°F), and soil temperatures in places like California's Death Valley can reach 90°C (194°F).

Chaparral

An almost unique vegetation assemblage has evolved where winters are mild and rainy and summers are hot and dry. This dry summer subtropical climatic region appears along the western coasts of continents between about 30° and 45° latitude on both sides of the equator. Both the climate and associated vegetation are often described as Mediterranean, because the Mediterranean coasts exhibit the largest area of the dry summer subtropical realm. This climate and its associated vegetation are also found in smaller coastal areas of California, Chile, South Africa, and Australia.

The distinctive vegetation of the dry summer subtropical climate is known in North America as *chaparral*. Chaparral consists of an almost impenetrable mat of brush ranging from knee-high to twice the height of a man. The shrubs are small-leaved and deep-rooted to survive the summer drought period and are generally evergreen (Figure 9.14, see p. 224). Since summer drought increases the danger of fire, most chaparral species have evolved the capability of resprouting from subsurface roots after being burned off above ground. Damp north-facing slopes may be covered by oak woodlands, and flat areas may be virtual savannas of large oaks rising from grass or scrub. As in the case of the deserts, chaparral vegetation on different continents and in different hemispheres is remark-

Figure 9.14
This chaparral vegetation is located in the Rif Mountains of northern Morocco, with the Straits of Gibraltar in the distance. The plants in such regions are adapted to Mediterranean climates and are protected from prolonged drought by small, leathery leaves. Nevertheless, climate is not the sole factor involved in the establishment of chaparral. Many chaparral areas were once forested, and it is believed that land clearing, overgrazing, and fire altered the ecological balance and made it difficult for trees to survive. (T. M. Oberlander)

ably similar in appearance, even though the plants included may be unrelated. Convergent evolution of plants is especially clear in this distinctive climatic realm.

The chaparral vegetation region in California often receives nationwide attention when late summer fires sweep over mountain slopes. Dry winds from the interior deserts can make firefighting in chaparral extremely difficult and dangerous. If a heavy winter rain occurs before the burned chaparral has resprouted, rapid erosion sends torrents of mud and boulders rushing into valleys. This has wreaked havoc in several communities along the California coast.

The monthly temperature and precipitation regime for Palermo, Italy, shown in Figure 9.15, is representative of dry summer subtropical climatic regions. Winter rainfall is associated with midlatitude cyclones, and the mild tempera-

tures at this location are due mostly to the protection from cold polar continental air offered by neighboring mountain barriers. Because of atmospheric subsidence associated with the subtropical highs, summers are hot and dry. In other locations, such as San Francisco, upwelling ocean water along the coast keeps summer temperatures cool.

The average water budget for San Francisco (Figure 9.16) shows a large moisture deficit during summer and fall and a smaller surplus during spring after soil moisture storage has been recharged. In most dry summer subtropical climatic regions, local water surpluses are not great enough to sustain irrigation agriculture during the summer. Winter surpluses from mountain regions must be stored in reservoirs to be delivered to irrigated cropland in the summer.

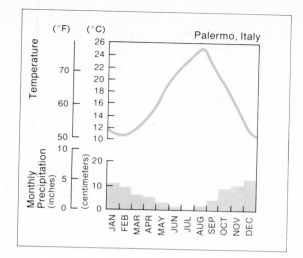

Figure 9.15
Palermo, Italy, has a Mediterranean climate characterized by moderate temperatures and a dry summer (Köppen *Cs* climate). As the graphs show, winter in Palermo is cooler and more moist than summer. The average annual precipitation is nearly 80 cm (31 in.), but June, July, and August combined receive only 4 cm (1.5 in.). (Doug Armstrong after H. Nelson, *Climatic Data for Representative Stations of the World*, © 1968, by permission of University of Nebraska Press)

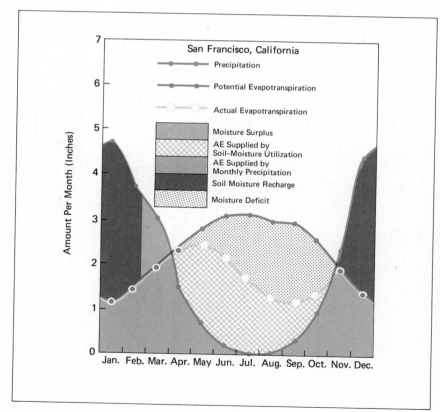

Figure 9.16
The local water budget for San Francisco, California, is characteristic of stations with a Mediterranean climate (Köppen *Cs* climate). A deficiency of moisture persists through the dry summer months, and plants that live through the summer must draw upon fog drip and stored soil moisture. From October through March, when most of the precipitation is received, soil moisture stores are replenished, and a surplus is generated. (Doug Armstrong after D. Carter and J. Mather, *Publications in Climatology*, vol. 19, 1966)

Midlatitude Forests

In the pre-agricultural period hardwood forests were dominant across the eastern United States, western Europe, Japan, Korea, and eastern China. Similar forests occupied much smaller areas of South America, South Africa, Australia, and New Zealand. These forests tend to be located within humid continental climatic regions, but they also occur in most of the humid subtropical and marine west coast climatic regions. Most of these regions have now been cleared for cropland and pasture and their upland forests have been cut over for lumber and firewood, so that very little of the original forest remains.

The midlatitude forests on the different continents include different combinations of species. In eastern North America, oak, hickory, maple, and beech each tend to be dominant in various areas. The multi-storied canopy of trees usually rises 30 m (100 ft) or more above the ground. Some shrubs and shade-tolerant annuals occupy the ground surface, but the forest tends to be relatively open below the canopy. Most of the trees drop their leaves before the onset of winter, so the appearance of the forest changes dramatically through the seasons (see Figure 9.17). In a portion of the region, soil water in the root zone of the forests is frozen in the winter and is therefore unavailable to the trees. With no uptake of moisture from the soil, continued transpiration of moisture by plant leaves would be fatal to the plants. But even before this occurred, the moisture-laden cells composing plant leaves would be ruptured and destroyed by winter freezes, causing the death of the plant. Therefore broadleaf plants must drop their leaves and remain dormant until the threat of frost is past. Thus the deciduous habit can result from either seasonal drought, as in the mixed tropical forests, or seasonal cold.

In the southern hemisphere, where winters are mild because of the strong maritime influ-

Figure 9.17
(left) The mixed broadleaf deciduous forests are alive with color for a few weeks in autumn; this example is from southern Wisconsin. Each species progresses through a sequence of color changes and leaf fall. Variable temperature and moisture conditions cause the date of maximum color to vary by as much as four weeks or more. (R. A. Muller)

(opposite) This hardwood forest in southern Kentucky is seen in the spring before the deciduous trees produce a new set of leaves. Note the amount of sunlight reaching the forest floor. Such bright conditions are never encountered within the world's evergreen forest types. (T. M. Oberlander)

CHAPTER 9

ence, the midlatitude forests are dominated by broadleaf evergreens such as Eucalyptus and Nothafagus (evergreen beech).

The broadleaf deciduous forests in the eastern United States and western Europe become mixed with coniferous (cone-bearing) evergreens on their northern margins and with broadleaf evergreens on their southern margins. Pines tend to be dominant on the higher portions of the coastal plain of the southeastern United States, where sandy soils lack nutrients and store only limited amounts of moisture. The southern pines are an important timber resource, but they are also more susceptible than deciduous trees to fires that sweep areas of the coastal plain from time to time. Some scientists believe that the southern pines are only an intermediate successional stage toward the broadleaf deciduous forest climax.

Most midlatitude forest regions experience large temperature ranges from winter to summer and a relatively even distribution of precipitation throughout the year. Pittsburgh, Pennsylvania, is near the climatic boundary between Köppen's humid subtropical (*Cfa*) and humid continental (*Dfa*) climates; its monthly temperature and precipitation regimes are displayed in Figure 9.18. In general, summers are hot, but temperatures well below freezing are common during winter months. Monthly precipitation varies little. Klagenfurt, Austria, is representative of midlatitude forests with humid continental climates; the large seasonal temperature range is shown in the thermoisopleth diagram in Figure 9.19 (see p. 228). This diagram should be compared with that in Figure 9.7 (p. 217), which illustrates the temperature regime in a tropical rainforest. The average water budgets of most locations with midlatitude forests show small

Figure 9.18
Pittsburgh, Pennsylvania, is near the boundary between the humid subtropical and humid continental climates (Köppen *Cfa* and *Dfa* climates). The graph shows the great range of the average monthly temperature between summer and winter. The monthly precipitation is nearly constant through the year, however. (Doug Armstrong after H. Nelson, *Climatic Data for Representative Stations of the World,* © 1968, by permission of University of Nebraska Press)

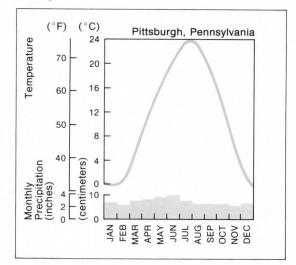

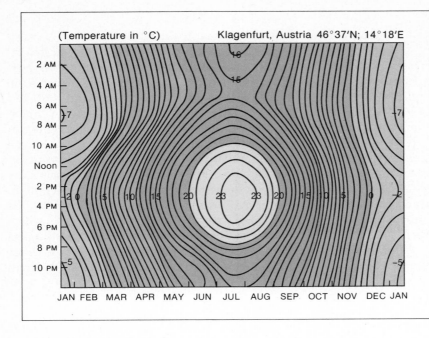

Figure 9.19
Klagenfurt, Austria, has a humid continental climate (Köppen *Dfa* climate). As the thermoisopleth shows, the range of temperature through the year exceeds the range through the average day. In January the noon temperature is approximately −4°C (25°F), whereas in July the noon temperature is 22°C (72°F). Note that in November the temperature through the day varies by only a few degrees, partly because of the moderating effect of cloudy weather. (Doug Armstrong after C. Troll, *World Maps of Climatology*, © 1965, Springer-Verlag Publishing)

Figure 9.20
These local water budgets for Middlesex County, in central New Jersey, illustrate variability in water budgets in a midlatitude forest at the equatorward margin of the humid continental climatic region. During the drought years of 1962 through early 1966, deficits were large, and winter-spring surpluses were not great enough to meet the water-resource needs of the region. Water restrictions were common, and there was some loss of shrubs and trees in suburban areas. (Doug Armstrong after Muller, 1969)

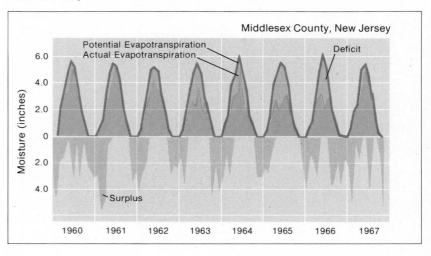

summer deficits and relatively large winter and spring surpluses. These vary considerably from year to year, as shown in Figure 9.20.

Midlatitude Grasslands

Grasses are the dominant vegetation where precipitation does not meet the needs of trees and shrubs or where repeated fires prevent tree regeneration from seedlings. Vast areas of continuous grasslands once extended from Texas to Alberta and Saskatchewan in central North America and across Eurasia from the Soviet Ukraine to Manchuria. These grasslands included *tall-* and *short-grass prairies* and *steppes*. Similar extensive grasslands were also present in South America, particularly in Argentina and Uruguay. Intense agricultural exploitation has left very few areas of natural grassland in any of these regions.

The midlatitude grasslands contained a large number of plant species that are different from the grasses of the tropical savannas. The dominant midlatitude grasses are usually perennials that lie dormant during the winter and continue their growth in the next growing season. Near the boundary between forest and grassland, where moisture is comparatively abundant, the natural grassland vegetation is usually prairie grass 1 to 2 meters in height. Most tall-grass prairie areas throughout the world are now utilized for intensive agriculture or wheat farming. Where there is less moisture, the dominant vegetation is short prairie grass, generally less than a meter high. In the driest grassland the grass grows in bunches or tufts, with bare ground often visible between the clumps. This is the common Eurasian type of grassland, known as *steppe* grassland. Patches of trees may occur here and there in all grassland types, especially along streams.

Like the tropical savannas, the temperate grasslands of Eurasia and North America were once immense pastures, supporting vast numbers of grazing animals. Upwards of 40 million bison and 50 to 100 million antelope pastured on the American grasslands, with single herds of 100,000 to 2 million animals having been seen by explorers and fur traders in the early 1800s. All modern breeds of horses appear to have descended from the vast herds of horses that once dominated the Eurasian grasslands.

The tall-grass prairies in Illinois and Iowa are utilized now for corn, small grain, and hog farming. Many ecologists believe that the climate of this region would have supported forests, if it were not for the fires that repeatedly swept over the open plains (see also Figure 9.21, p. 230). Farther west in Saskatchewan, the Dakotas, Nebraska, Kansas, and Oklahoma, wheat crops have replaced medium-height prairie grasses.

Beyond the wheat belt, on the high plains of Alberta, Montana, Wyoming, Colorado, and Texas, short-grass prairies are still present. Here tree growth is precluded (except along stream courses) by complete drying out of the subsoil during periodic droughts. This phenomenon is very clear in the morphology of the soils developed in this area, as will be seen in Chapter 10. The area is too dry for unirrigated agriculture, and its predominant use is as unfenced pasture for beef cattle. The medium-height prairie grass grades into short grass through a broad transition zone centered just east of the 100th meridian from central Texas to the Dakotas, with a westward swing into Saskatchewan and Alberta. This transition zone is tempting but dangerous for agriculture. Cycles of wet years repeatedly lure wheat farmers onto the short-grass prairies only to meet disaster when the inevitable dry years follow, transforming the ploughed land into a "dust bowl."

Most of the world's grassland areas are located where average precipitation barely equals potential evapotranspiration. In the grasslands of continental interiors in the northern hemisphere, summers tend to be hot, with periodic thunderstorms. Winters are very cold and dry.

The average water budget for Huron, South Dakota, located in the transition zone between the tall and short grasslands, is shown in Figure 9.22 (see p. 230). Despite a summer rainfall

Figure 9.21
This tussocked grassland is in South Island, New Zealand. Radiocarbon dating of wood fragments indicates that extensive forests occupied some of this region 500 years ago. The conversion to grassland may have occurred as a result of human use of fire. (G. R. Roberts, Nelson, N. Z.)

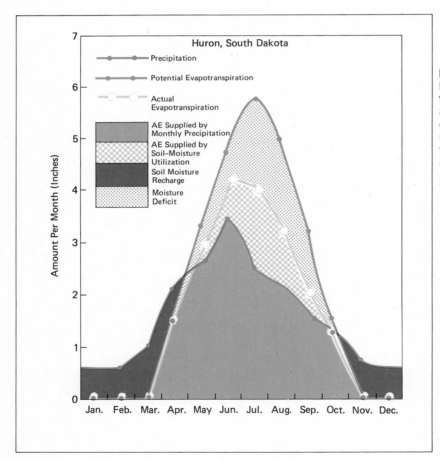

Figure 9.22
Huron, which is located in the prairie grasslands of South Dakota, has a humid continental climate with moderate moisture and cold winters (Köppen *Df* region) and a subhumid climate, according to Thornthwaite. The local water budget for Huron shows that precipitation occurs in all months, and is most plentiful during early summer. However, high summer temperatures cause potential evapotranspiration to be high then as well, and moisture deficits normally occur. The soil moisture is recharged during the cooler months, when the vegetation's demand for moisture is small, but winter precipitation is not large enough, on the average, to generate surpluses for runoff. (Doug Armstrong after D. Carter and J. Mather, *Publications in Climatology*, vol. 19, 1966)

maximum there is still a relatively large summer moisture deficit, and precipitation is not great enough during winter to generate a significant surplus. In addition, summer rainfall from convective thunderstorms is highly variable from year to year. Some climatologists and ecologists believe that clusters of drier-than-normal years prevent the establishment of forests within the wetter margins of the subhumid climatic regions.

Northern Coniferous Forests

Coniferous (cone-bearing) forests dominated by spruce, fir, and pine extend in a broad band across North America, Europe, and Asia between about 50° and 65°N latitude in the subarctic climatic region. These forests, also known as *taiga* or northern boreal forests, endure the largest annual temperature ranges encountered on earth. Winters are bitterly cold and very dry, with only light snowfalls. These areas are dominated by the Canadian and Siberian highs and are the source regions for continental polar air masses. The summer is brief, but there are many hours of daylight and temperatures are mild, even warm, occasionally exceeding 25°C (77°F).

The conifers of the taiga are tall, slim, and tapered, as shown in Figure 9.23. Most conifers are evergreen and do not lose their leaves during winter. Their small needle-shaped leaves and thick bark resist moisture losses during the long cold winters. Since they are not required to manufacture new leaves before they can begin photosynthesis and growth in the spring, they

Figure 9.23
Coniferous evergreens, commonly spruce, fir, and pine, occupy large regions of the northern middle latitudes and the subarctic. The uniform appearance of this coniferous forest in Washington's Cascade Range shows how only a few species of conifers form the dominant vegetation. Mt. Rainier is in the background. (T. M. Oberlander)

CLIMATE AND VEGETATION

conserve energy and are able to commence growth as soon as temperatures rise sufficiently. This is a clear advantage where the growing season is very short.

Coniferous forests contain a comparatively small number of plant species. Low sun angles and the dense foliage allow little light to reach the ground, and the cool temperatures also limit plant growth. While spruce is common in the coniferous forests of North America, larch, which drops its needle-like leaves in the winter, is dominant in eastern Siberia, where winter conditions are too severe even for needleleaf evergreens. Here average January temperatures plummet to −40 to −50°C (−40 to −60°F). On the southern margins of the coniferous forests, the trees are tall and densely packed. Farther poleward, the trees are smaller and the forest is more open.

Figure 9.24 shows the monthly regimes of temperature and precipitation at Moose Factory in central Canada. Monthly mean temperatures range from −20°C (−35°F) in winter to 16°C (60°F) in summer. Much of the annual precipitation falls as rain during summer, when potential evapotranspiration is nearly as high as in midlatitude regions. Hence, most summer rainfall is utilized for evapotranspiration. Local water budgets show that the only moisture surplus is in the late spring at the time of snowmelt. Nearly rainless periods occasionally result in small deficits. Some summers are dry enough for the danger of forest fires to be serious.

In North America the coniferous forests of the subarctic climatic region spread southeastward from the Mackenzie River valley in northwestern Canada across the Canadian border into sections of Michigan, New York, and New England. Along the northwestern coast of North America, the coniferous forest extends southward into Washington, Oregon, and northern California. Here winters are wet and summers are very dry. The virgin forests of the Pacific coast are often composed of giant trees, and the species differ from those of the taiga. Redwoods (*Sequoia*) and Douglas fir are dominant

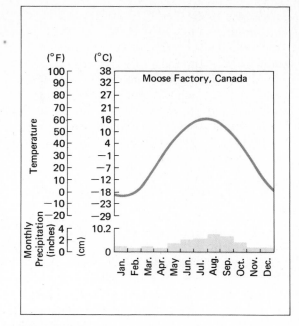

Figure 9.24
Mean monthly temperature and precipitation regimes at Moose Factory, in central Canada. The range between winter and summer temperatures in the subarctic climate is the largest of all global environments. The northern coniferous forest is well adapted to short but warm summers, when much of the annual precipitation falls. (Vantage Art, Inc. after Glenn Trewartha, *Introduction to Climate*, 4th ed., 1968, McGraw-Hill Book Co.)

in the California coast ranges, with fir, spruce, cedar, and hemlock prevailing in the forests farther northward. Moisture stress during the growing season has favored the survival of an evergreen coniferous forest in this midlatitude region of summer drought.

Tundra

Trees cannot survive unless the average temperature of the growing season exceeds 10°C (50°F) for a period of two to three months. Near the Arctic Ocean, the trees of the northern coniferous forest give way to low shrubs, grasses,

Figure 9.25
Low shrubs, grasses, and flowering herbs are the dominant vegetation types in tundra regions, as shown here in Norway. The tundra is wet during the period of thaw because water cannot drain through the permanently frozen ground below the surface. (Brian Hawkes/Carl Ostman Agency)

and flowering herbs, with mosses and lichens on rock surfaces (Figure 9.25). This vegetation formation is known as *tundra*.

Although winter temperatures in the tundra regions are not as low as those in the more continental taiga of eastern Siberia, they are nevertheless extremely severe and are commonly accompanied by gale-like winds from which there is no shelter. At −18°C (0°F) a 30 km per hr (20

mph) wind produces the equivalent temperature of −40°C (−40°F). This change in effective temperature is the *windchill* factor. At −30°C (−24°F) the same wind lowers the equivalent temperature to −55°C (−68°F). This combination of wind and temperature produces extreme moisture stress and danger to water-bearing tissues because evaporation induced by the wind cannot be offset by water intake from the frozen ground. As a consequence, tundra plants are small-leaved, like desert plants, and low-growing so that they will be covered by snow when icy winter winds sweep over the land surface.

Figure 9.26 is a thermoisopleth diagram for

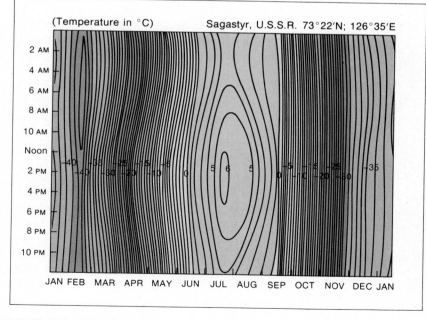

Figure 9.26
Sagastyr, U.S.S.R., is a station in the tundra region of Siberia north of the Arctic Circle (Köppen *ET* climate). The temperature during the year varies through an extreme range because of the lack of sun near the time of winter solstice and the continual daylight at summer solstice. The nearly vertical temperature contour lines imply that the temperature during any given day is essentially constant. (Doug Armstrong after C. Troll, *World Maps of Climatology*, © 1965, Springer-Verlag Publishing)

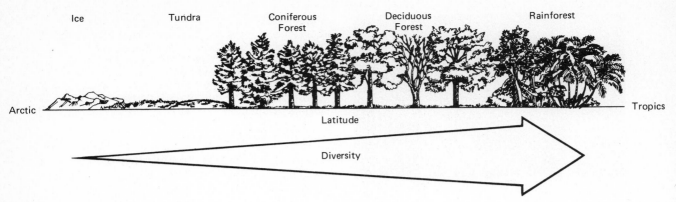

Ice   Tundra   Coniferous Forest   Deciduous Forest   Rainforest

Arctic   Tropics

Latitude

Diversity

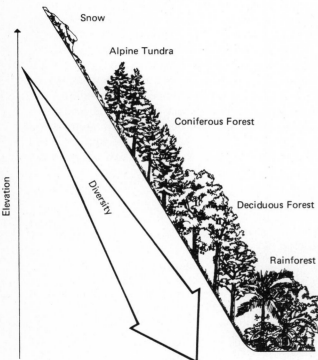

Snow

Alpine Tundra

Coniferous Forest

Diversity

Deciduous Forest

Rainforest

Elevation

Figure 9.27
Relationship of latitude and elevation to vegetation type and community diversity. Increasing latitude or elevation results in a shift to cooler vegetation types with a decrease in community diversity. Climbing 4,500 m (15,000 ft) up a tropical mountain will reveal changes in communities analogous to those observed on a trip from the tropics to the pole. (Barbara Hoopes)

In tundra regions precipitation tends to be low throughout the year. Winters are long, and the tundra is thinly covered by snow for six to eight months. Most of the sparse precipitation falls as rain during the brief summer. Despite the relatively low summer precipitation, the tundra is usually moist and waterlogged during the warm months, when the surface layer of soil thaws but cannot drain downward because the subsoil is permanently frozen.

Highland Vegetation

Under average conditions, temperatures decrease with increasing elevation. Thus one can progress through different climatic and vegetation zones while climbing upward in highland regions.

Figure 9.27 shows that in humid climatic regions the sequence of vegetation types encountered with increasing elevation is similar to the sequence met in traveling from the equator to

Sagastyr, at latitude 73°N, in the tundra of northern Siberia. The nearly vertical pattern of isotherms shows that the daily variation of temperature is only a few degrees because of the long daylight periods in summer and the absence of sunlight during midwinter. The average midday temperatures vary from 6°C (43°F) in July to −40°C (−40°F) in February.

the poles. In dry climates the increase in precipitation with elevation produces spectacular vegetation changes within short horizontal distances. In the San Francisco Peaks region of northern Arizona, where the vertical sequence of North American life-zones was first studied, the upward progression begins with desert shrubs at an elevation of 1,200 m (4,000 ft) and ends in a spruce-fir forest at 3,000 m (10,000 ft) that is not unlike the spruce-fir forests of subarctic Canada. Desert shrubs and spruce-fir forest, which normally grow thousands of miles apart, here exist within a distance of a few miles. In tropical uplands the diversity of contrasting vegetation is even greater. In the Peruvian Andes, which rise to glacier-clad summits from rainforests on one side and deserts on the other, nearly all the world's latitudinal temperature and vegetation zones can be recognized in vertical succession.

Local climates and vegetation in highland areas are very complex, however. Adjacent slopes with different orientations differ dramatically in terms of solar radiation, temperature, and precipitation. North-facing slopes of deep valleys in the middle latitudes receive little direct sunlight over the year, while nearby south-facing slopes may bake in many hours of sunshine each day. Leeward slopes may be rather dry, while precipitation on windward slopes tends to increase sharply with elevation.

For at least the first 1 or 2 kilometers of elevation, upland areas produce much larger water surpluses than surrounding lowlands. A flourishing vegetative cover in upland regions retards water runoff, reducing soil erosion on steep slopes and decreasing flood hazards in adjacent lowlands. At the same time, forested uplands release water slowly to streams, sustaining their flow through rainless periods. Thus the management of upland vegetation, the control of soil erosion, and the stabilization of streamflow all go hand-in-hand. This is a matter of critical concern to people who live within the highland regions and far beyond them on the great river floodplains of the earth.

SUMMARY

Solar energy and moisture determine the general type of vegetation that can grow in a particular location. Normally the vegetation of any place is a mosaic of different plant species that form a plant community. When disturbance of natural vegetation occurs, or when agricultural land is abandoned, the area becomes occupied by a largely predictable succession of plant communities. Eventually a mature, or climax, vegetation is attained. The climax vegetation is assumed to be in equilibrium with the environment and does not change further. The climax vegetation is determined primarily by soil conditions and the availability of moisture and energy, but climatic variability and various disturbance factors also have a significant impact.

In this chapter the natural vegetation formations have been classified into nine global types. Forests are restricted to the humid climatic realm, but various types have evolved differently in response to seasonal energy and moisture regimes. Thus there are tropical rainforests, tropical mixed forests, midlatitude forests, and northern coniferous forests. Grasslands and specialized xerophytic vegetation dominate the dry climatic realm. Savannas occur where there is winter drought and summer rain, and chaparral vegetation is present where there is winter rain and summer drought. Energy availability is so restricted toward the polar margins of the humid climatic realm that forests give way to low-growing forms of vegetation collectively called tundra. In highland areas the distributions of local climates and vegetation are complex, but the vertical sequence of plant communities shows similarities to the poleward gradations of climate and vegetation.

APPLICATIONS

1. In the vicinity of your campus there is probably land, once disturbed or cleared,

that fairly recently has been allowed to revert to a more natural or uncontrolled vegetation cover. Locate two such sites and compare the composition of their vegetation. Is there any good explanation for the differences you can see?

2. Analyze the map of global vegetation (Figure 9.4) in comparison with the map of global climate (Figure 8.3). Make a list of the climates of the natural vegetation types. Now make a final list of global regions where the general associations of climate and vegetation do not seem to fit, speculating about possible reasons.

3. Are there any preserved remnants of the prehistoric natural vegetation of your region? Was the prehistoric vegetation actually regarded as a climax type, or was it part of a plant succession after some natural disturbance? In what areas of North America would you expect the prehistoric vegetation to have been other than a climax form?

4. How does the size of plants appear to be related to the availability of energy and moisture?

5. There are similarities in the appearance of the vegetation in desert and tundra regions. What are the resemblances? Is this a result of the same or different climatic stresses?

6. One peculiar aspect of the climatic regime of tundra regions is also seen in the tropical rainforests. This climatic characteristic is especially evident in a comparison of thermoisograms from different vegetation regions. What is the similarity? Are its causes the same in the two regions?

FURTHER READING

Bennett, Charles F., Jr. *Man and Earth's Ecosystems: An Introduction to the Geography of Human Modification of the Earth.* New York: Wiley (1975), 331 pp. This unique text is organized by world regions in order to focus on the ecological impacts of human use of the earth. The book specifically analyzes the geographical and historical background of the environmental crisis.

Billings, W. D. *Plants, Man, and the Ecosystem.* 2nd ed. Belmont, Calif.: Wadsworth (1970), 160 pp. This brief paperback, part of the Fundamentals of Botany series, includes many succinct sections that supplement this chapter.

Eyre, S. R. *Vegetation and Soils: A World Picture.* 2nd ed. Chicago: Aldine (1968), 328 pp. This book, written from the perspective of the British Isles, focuses on relationships between vegetation and soils. It is organized by global vegetation types and is especially useful for Chapters 9 and 10 of this text.

Shelford, Victor E. *The Ecology of North America.* Urbana: University of Illinois Press (1963), 610 pp. Shelford's book is a detailed account of North American biomes. It is packed with information on plant and animal interactions, and is a valuable reference book, now available in a paperback edition.

Tivy, Joy. *Biogeography.* New York: Longman (1971), 394 pp. Plant distribution and vegetation formations are explained in terms of environmental factors.

Vankat, John L. *The Natural Vegetation of North America: An Introduction.* New York: Wiley (1979), 261 pp. This compact paperback presents the ecological basis of the major vegetation formations of North America from tundra to tropical rainforest. Very informative for its size.

Walter, Heinrich. *Vegetation of the Earth in Relation to Climate and the Eco-Physiological Conditions.* Translation of 2nd German ed. New York: Springer-Verlag (1973), 237 pp. This work stresses relationships between climate and vegetation and also includes short descriptions of the vegetation formations of each natural vegetation type.

Whittaker, Robert H. *Communities and Ecosystems.* 2nd ed. New York: Macmillan (1975), 385 pp. This is another outstanding paperback on ecology, especially pertinent to this chapter.

CASE STUDY: Slash-and-Burn Agriculture: An Efficient or Destructive Land Use?

In vast areas of the tropics, a method of agriculture is practiced that is largely unknown to midlatitude populations. Slash-and-burn agriculture—also known as shifting cultivation, *swidden* cultivation, *milpa* cultivation (Latin America), and *ladang* (Southeast Asia)—constitutes the principal method of farming practiced on one-third of the total land area currently used for agriculture in Southeast Asia. It has been estimated that in some countries, among them the Philippines, up to 10 percent of the population depends on shifting cultivation for its food.

In Latin America, typical crops produced by shifting cultivation are maize, manioc, and squash. In Southeast Asia, rice, cucumbers, and maize are often planted the first year, followed by second-year crops of cassava, sugarcane, and squash.

The actual technique of slash-and-burn agriculture is essentially what the name implies. Several months before the region's rainy season begins, the trees and undergrowth on a selected plot of forested land are cut. The remaining vegetation is left on the ground to dry and is later cleared away by burning. At the beginning of the rainy season, crops are planted, usually by using sticks or hoes to dig holes for the seeds.

The same plot is replanted for several years, until the decrease in soil fertility or the growth of weeds and grasses reaches a point at which cultivation is no longer worthwhile. Cultivators then abandon the field, leaving it to fallow for several years, and move on to another plot to repeat the process of cutting, burning, planting, and harvesting.

Shifting cultivation is not feasible for a high-density population because crop yields are low and the long fallow period required for reestablishment of the cleared vegetation means only a fraction of the arable land can be productive at any given time.

The overall effect of slash-and-burn agriculture on the ecology of a region is difficult to evaluate. Many experts argue that shifting cultivation is an efficient and, if properly practiced, ecologically sound method of utilizing tropical land that would be difficult to cultivate by most other methods. They are also quick to point out that in underdeveloped countries, slash-and-burn agriculture may be necessary to support the people who eke out a subsistence living by practicing it.

Other experts, however, argue that this method of agriculture is a reflection of cultural level rather than of any unavoidable or unalterable physical limitations of the environment, and that on the whole the method is wasteful and inefficient. They point out that the clearing of vegetation exposes the already thin, infertile topsoil of the slopes to the powerful erosive force of tropical rainfall. The heavy erosion results in an increased silt load in stream channels, which decreases their depth and causes flooding. In some regions, slash-and-burn agriculture is

(top) The natural vegetation on this Honduran slope has been cut, dried, and burned, and replaced by the *milpa* crops shown here. The farmers who cultivate this plot will abandon it for a new area after only a few harvests. (James S. Packer)

(bottom) Most of this formerly forested area near Oaxaca, Mexico, has been transformed into a wasteland as a result of *milpa* agricultural practices. (James S. Packer)

also responsible for the depletion of various species of timber.

Most scientists would agree, however, that any damage wrought by slash-and-burn agriculture is minimal—perhaps even negligible—under certain conditions and if properly practiced. The amount of available land must be sufficient to allow the abandoned plots adequately long fallow periods, and there must be sufficient seasonal variation in climate and rainfall so that cut vegetation will dry and newly planted crops will be watered. The method can be used only to support a low-density population with a subsistence economy that does not require a large surplus of food for trade, and there must be minimal influence or pressure from external economic groups, such as loggers, whose practices conflict with the system of shifting cultivation.

In addition, certain practices must be followed: a safety path must be made around the clearing to prevent the spread of fire during the burning process; to avoid depletion of certain soil nutrients the same plot must not be used over and over again for the same crop; and secondary rather than primary forests must be used as cultivation sites. If these practices are not followed, and if the proper conditions are not present, slash-and-burn agriculture can indeed do great damage to soil, biotic, and water resources, reducing vast areas to economic uselessness in a short period of time under the climatic extremes of the tropics.

A thin layer of soil cloaks the earth's surface,
softening its contours and supporting its
vegetation. The soils we see today are products of
the interactions of living and nonliving materials
with energy and moisture over thousands of years.

Lime Banks by Andrew Wyeth, 1962. (Collection, Mr. Smith W. Bagley; copyright © by Andrew Wyeth)

The Soil System

A vital factor in the productivity of any region is the nature of its soils. Soils are the medium in which land plants grow, and plants support the food chains that sustain the web of life on the earth. In fact, soil is a much more complex phenomenon than is generally realized. It is not just loose "dirt" that one can dig into with a shovel or push around with a bulldozer. A true soil is the product of a living environment. Where there is no life, as on the moon, there is no true soil. Since the earth teems with life, soils of some type are present almost everywhere.

For soil to develop two things must happen. First, water percolating down through loose rock material must cause physical and chemical modifications in the original material. Second, the action of living organisms must cause further changes. The less water and organic activity present, the weaker the development of the soil. The more strongly developed the soil, the more apparent is the fact that percolating water and the activity of organisms have caused the original material to develop visible layers of varying physical and chemical characteristics. The presence of these differentiated layers is what identifies a true soil.

Figure 10.1
(top) Granite rock usually breaks into blocklike masses along joints when it is exposed at the earth's surface. The joints in this granite at Pike's Peak, Colorado, have been enlarged by physical and chemical weathering processes that slowly transform such rock into sand and clay. (Ward's Natural Science Establishment)

(center) This photo illustrates the effect of frost weathering in cold climates in high latitudes and at high altitudes. The angularity of the forms produced here in the Teton Range of Wyoming is characteristic of frost weathering. (T. M. Oberlander)

(bottom) The reduction of confining pressure due to erosion into masses of granitic rock causes the rock to expand by breaking into parallel sheets. These are often curved and resemble an onion structure. This process, called *exfoliation*, only affects unjointed rock, as here in California's Yosemite region. (T. M. Oberlander)

SOIL-FORMING PROCESSES

Soils develop on rock material that has already been reduced to fine fragments. This material may be either a mass of decomposed rock, or sediment that has been transported and deposited by an agent of erosion such as running water, wind, or glacial ice. Thus the story of soil development begins with the initial fragmentation of the solid rock of the earth's crust, a process known as *weathering*.

Weathering

Both mechanical and chemical processes can fragment solid rock. All rocks are brittle and break into networks of cracks, called *joints*, as a result of slow stretching or twisting motions of the earth's crust. Some massive rock, such as granite, expands when erosion of the land surface above it reduces the confining pressure on the rock (Figure 10.1). Massive unjointed rock expands by breaking into sheets parallel to the land surface in a process called *exfoliation*.

Water enters rock masses by way of joint systems, as well as through tiny pores between the mineral grains composing most rocks (see Chapter 11). When water freezes in joints and pore spaces, it splits rocks into smaller fragments, just as it bursts pipes when heating systems fail in winter. This is because the phase change from liquid water to ice produces nearly a 10 percent increase in volume. Wherever winters are severe, this so-called *frost weathering* is an important agent of rock disintegration (Figure 10.1). The prying action of roots and the burrowing activities of animals also help to widen the joints in rocks.

In dry climates the evaporation of water carrying dissolved salts causes the growth of saline crystals on rock surfaces. This crystal growth wedges apart the mineral grains composing the rock. The subsequent expansion and contraction of the saline crystals due to daily temperature changes help to loosen adjacent particles.

The product of all these forms of mechanical weathering is chemically unchanged rock debris of all sizes: boulders, gravel, sand, and silt.

Except in very cold climates, chemical weathering is generally more important than mechanical fragmentation of rock. Chemical weathering is assisted by plants, which give off CO_2 and organic substances that transform downward percolating water into a weak acid. This makes the water more reactive with most rocks. Some of the products of these chemical reactions are very common, such as the orange rust that forms on iron, or the green deposit that appears on copper roofs and leaky copper plumbing. Both of these are products of the chemical reaction known as *oxidation*. Most soils and decayed rock are shades of brown due to the oxidation of iron-bearing minerals. In general, chemical reactions cause rock disintegration by swelling and softening some minerals and causing others to alter to new types.

Chemical weathering is most effective where moisture is plentiful and high temperatures speed up chemical reactions. The result is a *weathered mantle* composed of fine particles of chemically altered material overlying the solid bedrock (Figure 10.2, see p. 244). This mantle, which may be several meters deep, often has a high clay content. Although only certain minerals weather to clay, these minerals are abundant in many rock types. The proportion and type of clay formed indicate the kind of environment in which the weathering occurred.

Translocation

The mantle of fine particles produced by weathering is not yet a true soil. Additional changes are required. Some of these result from the process of *translocation*, in which both solid and dissolved materials are moved downward by water sinking into the ground. Translocation involves both the loss of material from the upper part of the soil and the arrival of translocated substances in the lower part of the soil. The process of *eluviation* is the downward flushing of both solid and dissolved (or *leached*) matter. Ev-

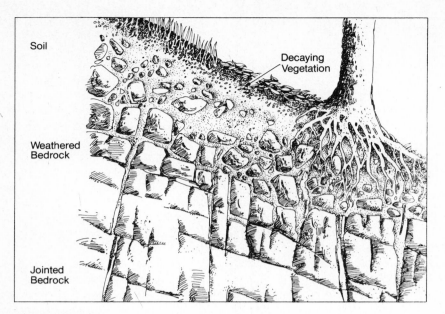

Figure 10.2
Chemical and physical weathering processes break down massive rock into the small particles that form the inorganic component of soil. In warm, moist climates, weathering proceeds actively along the exposed surfaces of jointed rock. Some of the mineral nutrients supplied to the soil by the parent rock are incorporated into growing vegetation. The nutrients are returned to the surface of the soil when plant litter decays, and rainwater subsequently washes them downward into the soil. (John Dawson after Arthur N. Strahler, *Physical Geography*, 3rd ed., © 1960 by John Wiley & Sons, by permission)

ery soil has a somewhat porous *eluvial layer* (or zone) from which translocated material has been lost.

The deposition of translocated material at a lower level in the soil is known as *illuviation*. The fine particles arriving in the *illuvial layer* (or zone) of the soil make this layer denser than the eluvial zone above it. The deposition in the illuvial zone of iron and aluminum oxides leached from the eluvial layer by acidic waters sometimes imparts a yellow or orange color to the illuvial zone. In dry regions calcium carbonate is translocated into the illuvial zone, producing whitish flecks or veins of lime.

It is the translocation of material that causes well-developed soils to become differentiated into layers of varying density, color, texture, and chemical composition. These layers are known as *soil horizons*.

Organic Activity

Soil horizons also reflect the influence of plant and animal life. Although the translocation process began with the earth's first rains, no real soils existed until life appeared on the land. Organic activity is crucial to soil development, and organic matter is a vital component of soils. Several types of organisms play a role in soil formation: plants contribute vegetative matter; fungi and bacteria reduce this vegetative matter to humus; and soil animals, such as ants, termites, and earthworms, mix the humus into the mineral matter of the soil.

Humus is the dark brown to black organic substance that makes the upper portions of the best soils so much darker than the deeper sub-soils. Humus becomes so mixed with the mineral matter of the soil that it is almost impossible to isolate. It has many beneficial properties. It increases the soil's ability to retain moisture and soluble plant nutrients. It is an important source of the carbon and nitrogen required by plants. And it maintains a soil structure that is neither too compact nor too porous for plant growth.

Aside from producing humus, soil bacteria are vital in *nitrogen fixation*—the conversion of gaseous nitrogen to forms that can be utilized by plant life. Nitrogen, which is absolutely essential to plant growth, is present in the air occupying pore spaces in the soil but cannot be taken up by plants in this form. Usable nitrogen must be liberated by the action of bacteria that are parasitic on the roots of the large family of plants known as *legumes*, which includes peas, soybeans, clover, alfalfa, and many other useful plants. Progressive farmers rotate their other crops with legumes to add nitrogen to the soil. It is estimated that in the eastern United States organic activity fixes more than 135 kilograms of nitrogen per hectare (or 120 lbs per acre) each year.

The abundance of organisms in soils is far greater than one would imagine. In addition to parasitic bacteria on plant roots there may be a million earthworms and 25 million insects in 1 hectare (2.47 acres) of pasture land. As many as a million bacteria may inhabit one cubic centimeter of soil. All play essential roles in the soil system.

PROPERTIES OF SOILS

It is possible to compare soils of various types by focusing on several properties, including tex-ture, structure, chemical characteristics, color, and profile development. These properties result from the environment of the soil and the parent material from which it was derived. They also reflect a number of soil-forming factors to be discussed subsequently.

Soil Texture

Soil *texture* refers to the size distribution of the mineral particles composing soils. Soil particle sizes are classified into four general categories: gravel, sand, silt, and clay—with subdivisions within these (Figure 10.3). Table 10.1 indicates the size ranges included in the categories and their subdivisions.

The United States Soil Conservation Service has established a classification of soil textures that describes the various mixtures of different particle sizes. Figure 10.4 shows the standard

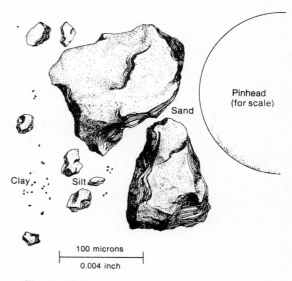

Figure 10.3
The particles that constitute the inorganic component of soil are classified according to size. Particles smaller than 2 microns in diameter are called clay, particles from 2 to 62 microns in diameter are called silt, and larger particles are considered sand or gravel. (John Dawson)

Table 10.1
Particle-Size Classification

CLASS	MILLIMETERS	MICRONS
Gravel	greater than 2.0	greater than 2,000
Very Coarse Sand	2.0–1.0	2,000–1,000
Coarse Sand	1.0–0.5	1,000–500
Medium Sand	0.5–0.25	500–250
Fine Sand	0.25–0.125	250–125
Very Fine Sand	0.125–0.062	125–62
Very Coarse Silt	0.062–0.031	62–31
Coarse Silt	0.031–0.016	31–16
Medium Silt	0.016–0.008	16–8
Fine Silt	0.008–0.004	8–4
Very Fine Silt	0.004–0.002	4–2
Clay	less than 0.002	less than 2

Source: Ruhe, Robert V. 1969. *Quaternary Landscapes in Iowa.* © Iowa State University Press, Ames.

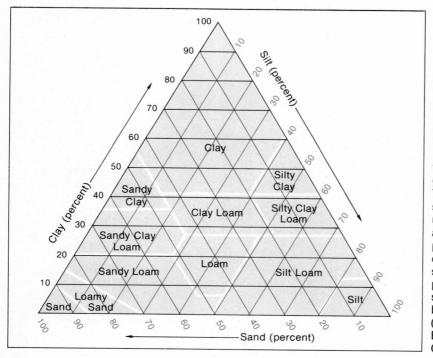

Figure 10.4
The texture of a soil is determined by measuring the proportions of clay, silt, and sand in the inorganic part of the soil. Texture is measured by sifting the soil sample through a series of screens graded from coarse to fine. The soil texture triangle shown in the figure can be used to classify the texture of a soil sample once the percentages of the components are known. If a soil sample contains 30 percent clay and 40 percent sand, for example, it would be classified as a clay loam. (Doug Armstrong after E. M. Bridges, *World Soils*, © 1970, Cambridge University Press)

classification scheme. The term *loam* indicates a mixture of sand, silt, and clay that has a more favorable structure for agriculture than does a soil composed of a more uniform particle size, such as sand or clay. Coarse-grained sandy soils are permeable and absorb water easily, but dry out rapidly. Their permeability also permits rapid leaching of soluble nutrients. Fine-grained clay soils are very dense, making them hard to work; they accept moisture very slowly and then hold it tenaciously.

Soil Structure

The mineral particles composing most soils cling together in masses called *peds*. The size and form of the peds produce the *structure* of the soil. As shown in Figure 10.5, soil peds vary considerably in form. Most plants grow best in soils with a granular or crumb structure having peds measuring 1 to 5 mm (0.04 to 0.2 in.) in diameter. Soil structure determines the rate of absorption of water and the ease of root penetration.

Another aspect of soil structure is soil *bulk density* (mass per unit of volume, including pore space). Bulk density increases with clay content and is a measure of soil compactness. High bulk density is a problem since it reduces the rate of acceptance of water by the soil. Unfortunately, bulk density is often increased unintentionally by the use of heavy agricultural machinery on

a

b

c

d

Figure 10.5
Four common soil ped structures are shown here. Scales are in inches. (a) Platy. (b) Prismatic. (c) Blocky. (d) Granular. (Roy W. Simonson, Courtesy USDA)

clay-rich soils, resulting in soil compaction. This can lead to increases in surface runoff that accelerate soil erosion and gully development on slopes.

Soil Chemistry

The chemical behavior of soil, which we think of as soil fertility, is closely related to the clay-humus complex in the soil. The *clay-humus complex* consists of microscopic particles of humus and clay, bound together so that they behave like large molecules. The extremely small (less than 1 micron) clay and humus particles carry negative electrical charges and remain in suspension in soil moisture. As a result, they attract positively charged ions, such as those of the chemical bases calcium, magnesium, potassium, and sodium. These essential plant nutrients are easily dissolved and would be washed from the soil were it not for their retention by the clay-humus complex.

In a fertile soil the clay-humus complex maintains a delicate balance. It must hold nutrients strongly enough to keep them from being leached away, but not so strongly that plants cannot extract them from the soil. Many different types of positively charged ions, or *cations*, are attracted and held in the clay-humus complex with various degrees of strength. Weakly bound cations may be replaced by others that are more strongly attracted. Basic cations, such as sodium, can be replaced by metallic cations, such as iron or aluminum, or by hydrogen ions from water. The ability of a soil to acquire and retain cations is known as the *cation exchange capacity*, which is an indicator of soil fertility (Figure 10.6).

Not all soils have an active clay-humus complex. Desert soils are rich in soluble nutrients because so little water moves through them. But the soils of desert regions contain almost no humus and often very little clay. When such soils are watered artificially, their nutrients are rap-

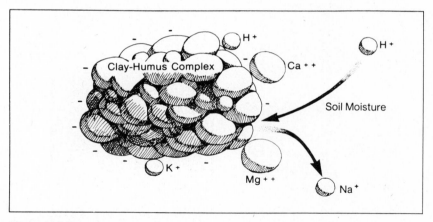

Figure 10.6
The finest inorganic particles and humus in a soil bind together to form the clay-humus complex. On a submicroscopic scale, the particles of the clay-humus complex act like giant molecules with the power to attract cations electrically. The complex performs an important function in a soil by preventing chemical nutrients needed by plants from washing out of the soil. Acidic soil moisture contains numerous hydrogen ions that can replace basic cations on the surface of the complex, as the figure shows. Hence acidic soil moisture removes basic inorganic nutrients from the soil. (John Dawson)

idly flushed away. This is a major problem of desert irrigation agriculture.

The concentration of hydrogen ions in a soil may be a problem, since they may crowd out plant nutrients. Hydrogen ion concentration, known as soil *acidity*, is measured in terms of a pH scale, ranging from about 3 to 10. Pure water has a pH of 7. This means it contains 10^{-7} grams of hydrogen per 1,000 cc. The lower the pH, the more acidic the soil.

The best agricultural yields are on soils with pH values between 5 and 7 (Figure 10.7). Acidic soils with low pH values occur in wet areas where there is abundant partially decayed vegetation on the soil surface. It is possible to raise the pH of acidic soils by adding lime ($CaCO_3$) to them. Soils with pH values greater than 7 are known as *alkaline* soils. They are a problem because they may contain toxic amounts of sodium, which also produces high bulk density and very poor soil structure. Alkaline soils, which commonly develop in semiarid regions, can be remedied by adding a source of hydrogen, such as ammonium sulfate ($[NH_4]_2SO_4$).

Soil Color

An obvious characteristic of any soil is its color. Soil color often changes as one digs downward. It also varies from place to place. Soils of the humid tropics commonly are orange or red. In the temperate grasslands, soils are dark brown to black. Soils under coniferous forests tend to be gray at the top and orange or yellow deeper down.

Soil color results almost entirely from the amount of organic matter and iron in the soil, and from the chemical state of the iron. Organic matter colors the soil dark brown to black. Iron that has been oxidized (combined with oxygen) produces reds, yellows, and browns. Where oxygen has been excluded, iron produces greenish and gray-blue hues. This most often occurs where waterlogging has kept air from moving through the soil. Other coloring matter is sometimes present, especially the white of calcium carbonate (lime). Soil color identifications are made by comparisons with color charts designed especially for the purpose of soil description.

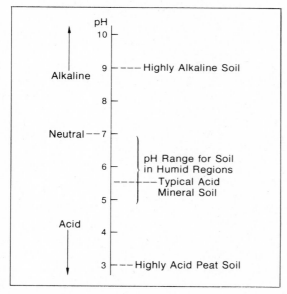

Figure 10.7
The pH value of a soil, or its hydrogen ion concentration, is one of the measures that can be used to estimate a soil's suitability for agriculture. A low pH indicates that a soil is acidic and may have lost many of its nutrients by exchange with hydrogen ions. A high pH indicates that a soil contains strong alkalis, which may be damaging to plant root tissues. (Doug Armstrong after Lyon and Buckman, *The Nature and Property of Soils*, 4th ed., © Macmillan Co.)

Soil Profiles

Every soil has its own distinctive sequence of layers, or horizons, resulting from the processes of translocation and organic activity. These horizons make up the *soil profile*. Five separate layers can usually be distinguished, termed the *O, A, B, C,* and *R* horizons. Each is further subdivided, as shown in Figure 10.8 (see p. 250).

The *O* horizon consists of undecomposed plant litter or raw humus at the soil surface. Be-

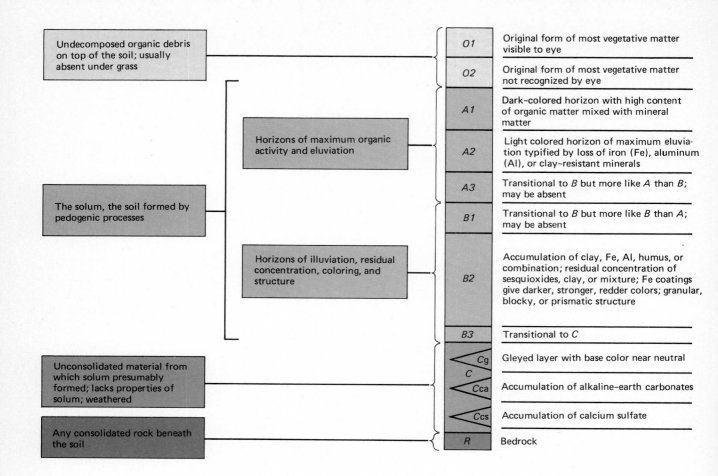

low it is the *A* horizon, at the top of which humus is mixed with mineral particles. The presence of organic matter makes the *A* horizon dark in color at the top; farther down the loss of fine mineral and soluble substances by eluviation causes it to be lighter in color and relatively porous and light in texture (sandy or silty). The *B* horizon receives material translocated from the *A* horizon, giving it a higher clay content and a denser texture than the *A* horizon. It is in the *B* horizon that we encounter blocky and prismatic soil structures and sometimes dense clay pans as much as a meter thick. The *B* horizon may also be vividly colored by iron oxide leached from the *A* horizon. The *C* horizon is composed of weathered material or loose deposits that have not yet been affected by soil-forming processes. In dry regions some whitish calcium carbonate (lime) may be deposited in the *C* horizon. The *R* horizon, where present, consists of unweathered bedrock.

From one soil profile to another, the characteristic horizons vary greatly in thickness, depth below the surface, and strength of development. Laboratory procedures are necessary to determine such important aspects of soil profiles as bulk density and the content of organic matter, carbonate, clay, oxides, and other constituents at different levels. Since many soil profiles reveal changes in the soil-forming processes over time, the profiles are indicators of past as well as present environmental conditions.

Figure 10.8 (opposite)
Standard horizons in soil profiles. No single profile contains all of the horizons shown. Additional subhorizons similar to those indicated for the *C* horizon include: *B2t*—illuvial clay; *B2ir*—illuvial iron; *B2h*—illuvial humus; *B2m*—strong cementation; *Csi*—cementation by silica; *sa*—enriched by salts; *f*—permanently frozen; *x*—hardpan composed of sand and/or silt. (Vantage Art, Inc. modified from Robert Ruhe, *Geomorphology*, 1975, Houghton Mifflin, and Soil Survey Staff, 1951, 1962)

FACTORS AFFECTING SOIL DEVELOPMENT

Our present understanding of soils has developed largely out of work begun by Russian soil scientists more than a hundred years ago. Over the vast area of Russia, soil characteristics seem closely associated with large-scale patterns of climate and vegetation. The studies of the early Russian soil scientists and their more recent successors around the world have indicated that soil profiles show the influence of five separate factors: (1) parent material, (2) climate, (3) site, (4) organisms, and (5) time. Soil scientists today call these the *factors of soil formation*.

Parent Material

The inorganic material on which a soil develops is called the soil's *parent material*. This can be rock that has decomposed in place or loose material that has been deposited by streams, glaciers, rockfalls, landslides, or the wind. The parent material determines what chemical elements are initially present in the soil. Some parent materials have an abundant supply of the nutrients most needed by plants; others do not.

If the parent material is rich in soluble bases—calcium, magnesium, potassium, and sodium—which are easily dissolved by water, it can continually replace these nutrients despite constant leaching from the soil. Rocks like limestone and basaltic lava are composed largely of soluble bases and in humid areas produce the

most fertile soils. In the eastern United States outstanding examples are seen in southeastern Pennsylvania near Lancaster and Harrisburg and in Alabama's Black Prairie, which are areas of limestone bedrock; in regions of basaltic rock in Virginia, Maryland, Delaware, New Jersey, and Pennsylvania; and on the calcareous (lime-

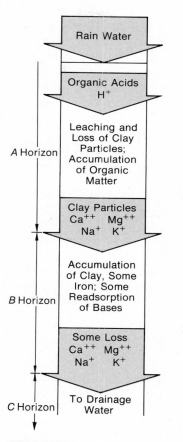

Figure 10.9
In the eluviation process shown here schematically, rainwater, which has become acidic from dissolved carbon dioxide or from organic acids in the humus on the ground, infiltrates the soil. The hydrogen ions in the acidic water displace basic cations from the clay-humus complex, causing a downward movement of soluble nutrients. (Doug Armstrong after E. M. Bridges, *World Soils*, © 1970, Cambridge University Press)

rich) glacial deposits of the "corn belt," from Iowa to Ohio.

If soluble nutrients are not abundant in the parent material, water moving through the soil removes bases and replaces them with hydrogen ions (Figure 10.9). The soil thus becomes increasingly acidic through leaching and less suitable for agriculture. Sandstone often produces infertile soils, since it is poor in nutrients as well as coarse in texture, which facilitates leaching. This can be seen in the fact that the natural vegetation of the sandy coastal plain from Delaware to New Jersey supports only pine forest, which has very low nutrient demands. The soils of this area are too poor for any type of agriculture. Similar "pine barrens" occur on sands in the Carolinas, Georgia, and Florida.

In addition to soluble bases, crop plants require certain amounts of iron, phosphorus, and sulfur, and trace amounts of such elements as boron and copper. If these are not available in the soil's parent material, farmers must supply them artificially.

Abrupt changes in natural vegetation commonly reflect a change in soil due to variations in parent material. Even so, soils from dissimilar parent materials may become quite similar with the passage of time if other soil-forming factors are the same.

Climate

On a global scale, major soil types show a close relationship to climatic zones. The energy and moisture delivered by the atmosphere influence many aspects of soil formation. These include translocation, the rates of chemical reactions, and organic activity in the soil. Abundant rainfall aids translocation, warm and wet conditions favor chemical reactions, and moderately warm and moist climates encourage organic activity.

Both vegetative production and the activity of soil bacteria and larger organisms are low in desert areas and in tundra regions at high latitudes or lofty elevations. In the dry desert environment plant litter oxidizes to dust, and in the tundra it decays slowly, often forming acid peat rather than a clay-humus complex. Plant production is at a maximum in the warm and wet tropics, but here the destruction of litter by organisms is so rapid and thorough that humus cannot form, and the soil is actually poor in organic matter.

Climate affects the chemistry of soil moisture, which, in turn, affects the solubility of various substances in the soil. For example, iron can be removed only by acidic water. Soil water tends to be acidic in cool, wet areas. Therefore iron tends to be leached from the eluvial layer in such areas, which are normally covered by coniferous forest. In dry regions lime leached from the upper portion of the soil is redeposited at a lower level, where the moisture evaporates, rather than continuing down to the water table.

Many soils contain features left over from past environmental conditions. Such "relict" features are important clues to environmental changes through time, especially shifts in the earth's climatic zones. Soil features have given evidence of past changes in the forest-tundra boundary in high latitudes and in the forest-grassland boundary in the midcontinent regions. Soils also reflect shifts in the world's deserts, as well as less severe climatic fluctuations in nearly all parts of the world. The interactions between climate and soil formation will be made clearer in the discussion of major pedogenic regimes later in this chapter.

Site

The specific location of a soil helps determine the soil type. Since water drains downward, soil at the foot of a slope will evolve in a wetter environment than soil on the hillcrest or on the slope itself. The material at the slope foot is finer than that upslope. There are two reasons for this. Fine material is washed downslope on the surface, and the greater dampness at the slope foot causes clay formation by the chemical weathering process.

Generally, soils on slopes are thinner, stonier, lower in organic matter, and less well developed than those on level or low-lying land. On level surfaces the effects of translocation are much stronger, producing an eluviated *A* horizon resting on a more massive clay-rich *B* horizon.

Organisms

The intensity and type of organic activity vary geographically. Since much of this variation is due to climate or microclimate, the climatic and organic factors in soil development are sometimes hard to separate.

We have already noted the general influences of organisms in the discussion of soil-forming

Figure 10.10
The nature of the nutrient cycle helps determine soil fertility.

(left) Plant species with high nutrient demands prevent soluble compounds from being leached from the soil. The plant extracts nutrients and then returns them to the soil in the form of plant litter. The constant two-way exchange between the plant and the soil maintains a high saturation of nutrients in the soil.

(right) Plant species with low nutrient demands permit unused soil chemicals to be leached away and return few nutrients to the soil. Thus the soil's fertility declines. (Vantage Art, Inc.)

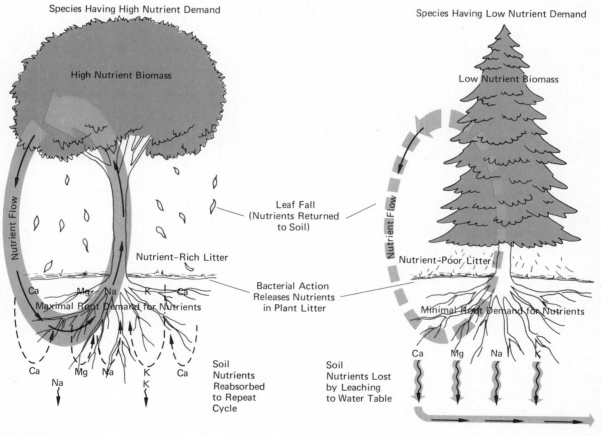

Species Having High Nutrient Demand

High Nutrient Biomass

Nutrient Flow

Leaf Fall (Nutrients Returned to Soil)

Nutrient-Rich Litter

Ca   Mg   Na   K   Ca
Maximal Root Demand for Nutrients

Bacterial Action Releases Nutrients in Plant Litter

Ca   Mg   Na   K   Ca
Na

Soil Nutrients Reabsorbed to Repeat Cycle

Species Having Low Nutrient Demand

Low Nutrient Biomass

Nutrient Flow

Nutrient-Poor Litter

Minimal Root Demand for Nutrients

Ca   Mg   Na   K

Soil Nutrients Lost by Leaching to Water Table

processes, but there remains one aspect of special importance—the *nutrient cycle*.

Wherever plants and animals exist, there is a constant cycling of material and energy between life forms and their environment. Organisms need the nutrients in soils to carry out their life-sustaining processes. They return these same nutrients to the environment as waste products, or as litter, or in the form of their own bodies when they die. This establishes a nutrient cycle. Without this constant uptake and return of nutrients, soluble compounds would soon be leached out of the soils of humid regions. This would cause a steady decrease in the soils' capacity to support life.

Nutrients not used by organisms are indeed gradually flushed from the soil (Figure 10.10). These vary depending upon the climate and life forms present. For example, we have noted previously that pines do not require nutrient-rich soils. Therefore the soils under pine forests gradually lose their soluble nutrients and become acidic in nature. On the other hand, tropical forests have high nutrient demands. Despite the heavy rainfalls of the humid tropics, the forest soils are stabilized by rapid nutrient cycling between the vegetation and the soil. Removal of such forests, where the bulk of the nutrients are stored, results in rapid soil deterioration.

Time

Time is required for translocation and organic activity to produce strong horizon development in soils. The only question is: do soil horizons continue to strengthen through time, or does the soil eventually reach an equilibrium with its environment and cease to be altered further? Because the environment and soil-forming factors frequently change, this question is hard to answer.

It has been possible to study the rate of soil formation on parent material deposited in historic time. When the volcano Krakatau, between Java and Sumatra, exploded in 1883, large amounts of volcanic ash fell on the sur-

rounding land. Under the moist tropical climate prevailing there, soil development on the ash was rapid—the soil thickening at a rate of about 1 cm (0.4 in.) per year. In drier areas in Central America, it has taken a thousand years to produce a soil about 30 cm (1 ft) thick on volcanic ash.

Many of the world's most valuable soils are

Figure 10.11 (opposite)
A calcium carbonate duricrust, called calcrete, or *caliche* in the southwestern United States, forms a resistant cap rock and produces a bold ledge along the valley of the Virgin River in southwestern Nevada: **(top)** general view; **(bottom)** detail of the calcrete cap rock. In tropical savanna regions, ferricrete (laterite) crusts produce similar landforms. Such crusts form within the soil and are exposed by later erosion of the upper part of the soil profile. (Dr. John S. Shelton)

fertile because they are young. They have formed too recently to be strongly affected by chemical and mechanical eluviation. The soils of river floodplains are usually productive. These soils have developed upon *alluvium* composed of sand, silt, and clay deposited by streams during floods. Soils on recent volcanic ash and glacial deposits are also usually very productive because they have not yet been strongly modified by translocation.

By contrast, soils on flattish land surfaces of great age tend to be infertile. Porous impoverished *A* horizons overlie impermeable *B* horizons composed of dense clays or cemented into a rock-like mass by translocated lime, iron oxide, or silica. In some areas the land surface has been eroded down to these "hardpans," which produce a resistant crust over the weathered parent material. These so-called *duricrusts* are best developed in areas having well-defined dry seasons. Figure 10.11 shows a duricrust layer cemented by calcium carbonate.

MAJOR PEDOGENIC REGIMES

Most of the earth's soils have been created by one of a small number of distinctive *pedogenic* (soil-forming) *regimes*. These are related to climate both directly and indirectly through the influence of vegetation. Each of the major pedogenic regimes produces a soil of a distinctive general type. Each of these soil types reflects major geographic variations in energy and moisture budgets.

Laterization

Red-colored soils are dominant in the portions of the tropics that are wet either seasonally or year-round. These soils result from the process of *laterization* (Figure 10.12, p. 256). The unique feature of these red soils is their high content of iron in relation to silica. Silica is normally the most abundant mineral in both solid and decomposed rock because it is highly resistant to chemical alteration and solution. To remove silica from soils requires a very aggressive regime of weathering and soil eluviation.

In the humid tropics decomposition and insect consumption of plant litter are too rapid to permit the formation of humus. Lack of humus decreases the soil's ability to retain the soluble nutrients not immediately taken up by plants. When forests are cleared, the laterization process is carried to its extreme as eluviation removes nearly all cations in solution and even the silica from decomposing clays. The residue of the process is quartz sand and oxides and hydroxides of iron and aluminum that often form a brick-like duricrust known as *laterite*. Occasionally the residue is *bauxite*, the principal commercial ore of aluminum. Laterite crusts cannot form under tropical forests due to nutrient cycling, but may develop in as little as thirty years when forests are removed. They are most common in the savanna regions fringing the Amazon and Zaire (Congo) basins and in Southeast Asia, sometimes attaining thicknesses of as much as 6 m (20 ft).

The process of laterization in a less extreme form has created the red soils on old land surfaces in the southeastern United States. These are the soils that were rapidly exhausted by intensive cotton farming in the 1800s. The laterized soils extending from Alabama through the Carolinas are usable agriculturally with proper fertilization and erosion control measures.

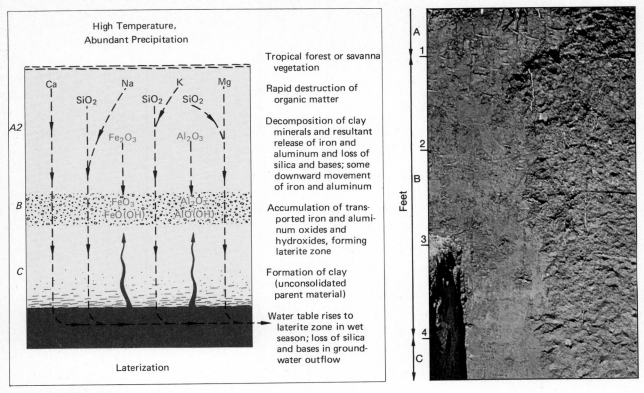

High Temperature, Abundant Precipitation

Tropical forest or savanna vegetation

Rapid destruction of organic matter

Decomposition of clay minerals and resultant release of iron and aluminum and loss of silica and bases; some downward movement of iron and aluminum

Accumulation of transported iron and aluminum oxides and hydroxides, forming laterite zone

Formation of clay (unconsolidated parent material)

Water table rises to laterite zone in wet season; loss of silica and bases in groundwater outflow

Laterization

Figure 10.12
(left) Diagrammatic illustration of the process of laterization, showing movements of chemical substances. Substances in red in this figure are accumulating in place as residuum or by illuviation. (Vantage Art, Inc.)

(right) This laterized soil has developed on deeply weathered rock in central Puerto Rico. The soil is low in nutrients but is permeable to water and easily worked in agriculture. (Reproduced from Soil Science Society of America, C. F. Marbut Memorial Slide Collection)

Podzolization

The vast coniferous forests of the higher middle latitudes in North America and Eurasia do not have high nutrient demands. They produce acidic plant litter and permit soluble bases to be leached from the soil. Since soil organisms do not thrive in the acidic environment of the soil (which is frozen during the winter), humus production is discouraged. As in the moist tropics, the lack of an active clay-humus complex is a factor in the leaching of soluble nutrients. The acidic soil moisture is able to decompose clay and remove iron in solution, leaving a residue of silica. The process of iron removal and silica concentration produces a light gray ashy-looking *A* horizon. Thus the general soil-forming process is called *podzolization* after the Russian word for wood ash. The iron and clay eluviated from the *A* horizon are deposited in the *B* horizon, coloring it orange or yellow in sharp contrast to the bleached *A* horizon (Figure 10.13).

Podzolization can also occur where parent materials are poor in nutrients, resulting in acidic soil moisture. This is often the case on the sands of coastal plains. The soils of much of eastern and northern North America reflect

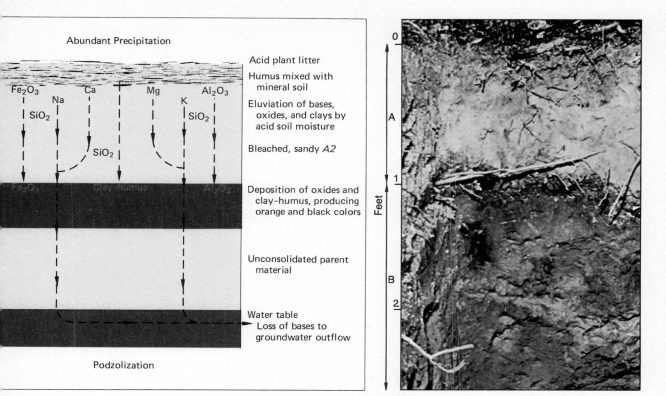

Podzolization

Figure 10.13
(left) Diagrammatic illustration of the process of podzolization, showing movements of chemical substances. Substances in red in this figure are accumulating in place as residuum or by illuviation. (Vantage Art, Inc.)

  (right) This podzolized soil in New York State has developed on sandy glacial outwash. The bleached eluvial *A2* horizon and the iron-enriched illuvial *B* horizon are well shown. (Reproduced from Soil Science Society of America, C. F. Marbut Memorial Slide Collection)

some degree of podzolization, being acidic and leached to varying degrees. Moderately podzolized soils developed under deciduous or mixed forest vegetation can be highly productive with proper management, but strongly podzolized soils developed under coniferous forests are very acidic and low in productivity.

Calcification

In areas of little rainfall there is too little moisture moving through the soil to remove soluble bases completely. In dry regions there may be some translocation from the *A* horizon, but water evaporates before dissolved material can be carried deeper than the *B* and *C* horizons. The most highly visible mineral is calcium carbonate, which is deposited in various forms in the *B* and *C* horizons, forming a subsidiary

calcic horizon. This subsoil enrichment process is called *calcification* (Figure 10.14, p. 258).

  Soils produced by the calcification process show varying degrees of horizon development. In midlatitude grasslands dark brown to black humus-rich *A* horizons normally occur above light-colored *B* horizons that contain only flecks or nodules of calcium carbonate. In semidesert

THE SOIL SYSTEM

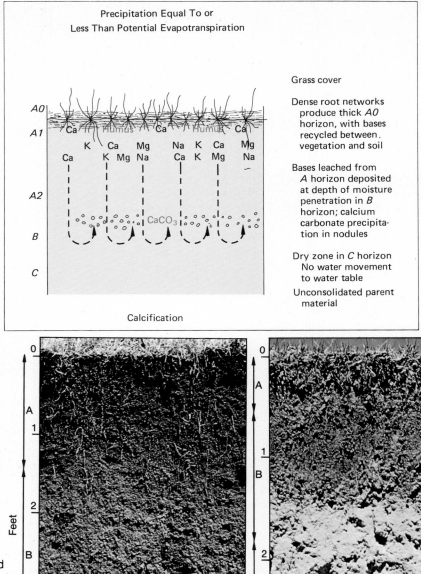

Precipitation Equal To or
Less Than Potential Evapotranspiration

Grass cover

Dense root networks
produce thick A0
horizon, with bases
recycled between
vegetation and soil

Bases leached from
A horizon deposited
at depth of moisture
penetration in B
horizon; calcium
carbonate precipita-
tion in nodules

Dry zone in C horizon
No water movement
to water table

Unconsolidated parent
material

Calcification

Figure 10.14
(top) Diagrammatic
illustration of the process of
calcification, showing
movements of chemical
substances. Substances in
red in this figure are
accumulating in place as
residuum (humus) or by
illuviation (CaCO₃). (Vantage
Art, Inc.)

(bottom left) This calcified
soil has developed under
tallgrass prairie in central
Iowa and is used for growing
crops such as corn and
soybeans. The parent
material is loess, which is
fine dust transported long
distances by wind.

(bottom right) This calcified
soil has developed under
shortgrass prairie in eastern
Colorado. The organic
content and the soil depth
are much lower than those of
the soil in Iowa. This soil is
used primarily to grow small
grains. (Reproduced from
Soil Science Society of
America, C. F. Marbut
Memorial Slide Collection)

areas there may be almost no organic horizons, but strongly developed *B* horizons containing conspicuous veins of calcium carbonate. The depth of the calcic horizon reflects the depth to which moisture penetrates in the soil before evaporating.

The calcification process produces soils that are rich in nutrients and are especially productive where there is a thick organic horizon. Such soils were developed under the grass prairies and the transition to steppes in the middle latitudes. The outstanding examples of this are the Russian Ukraine, the North American Great Plains east of the 100th meridian, and the Argentinian Pampas. The soils of these regions are so valuable that they are almost all in cropland. Fertile soils produced by calcification continue into the short-grass prairie west of the 100th meridian in the United States, but here there is insufficient rainfall for reliable crops. This is Dust Bowl country—tempting to the farmer but the scene of repeated droughts resulting in recurring economic and human disasters.

Salinization

In dry regions water that runs off hillslopes is often trapped in depressions with no drainage outlet. This may produce a temporary lake that soon evaporates. In these depressions the water table may rise close to the land surface. Above the water table there is always a *capillary fringe*, in which water rises upward through pore spaces in the soil. If the water table is close to the land surface, capillary action causes water to rise all the way to the surface, where it evaporates. Salts contained in the water are left behind as a white powder covering the dry soil. This process is known as *salinization*. The deposits consist of various chlorides, sulfates, and carbonates of calcium, magnesium, and sodium, including common salt ($NaCl$) and gypsum ($CaSO_4$). Because high concentrations of salts are toxic to most plants and soil organisms, salinization causes land to become barren and useless.

Unfortunately, salinization is often caused by human use of the land. When desert lowlands are irrigated for agriculture, some of the water sinks through the soil to the water table. This raises the water table, causing waterlogging and artificial salinization of the soil. This has been a problem in irrigated areas since ancient times; we first read of it in 4,000-year-old cuneiform records from Mesopotamia. Russian scientists have estimated that every year artificial salinization ruins as much as 300,000 hectares (741,000 acres) of cropland. The Mesopotamian Plain in Iraq, Egypt's Nile Valley, the Indus Valley in Pakistan, and the San Joaquin and Imperial valleys in California are all seriously affected by artificially induced salinization.

Gleization

In regions of high rainfall, low-lying areas are often waterlogged naturally. Here the water is not saline, as salts are leached away in the constantly wet environment, and vegetation can grow thickly. The litter from this vegetation cannot be decomposed normally due to the low level of bacterial activity in the oxygen-poor muck. As a result, the upper soil takes the form of dark *peat*, composed of partially rotted plant matter. Decay of plant parts in the damp environment releases organic acids that react with iron in the mineral soil. The iron is reduced, rather than oxidized, giving wetland soils a black to bluish-gray color. Acidic wetland soils of such colors are called *gley* soils, and the formative processes involved are called *gleization*. Gley soils occur in limited areas throughout the world. In North America they are most often related to depressions produced by irregular deposition of debris by continental ice sheets. Therefore, they are scattered throughout the Midwest north of the Ohio and Missouri rivers, as well as occurring in marshy areas on river floodplains, especially in the Mississippi River valley.

To make gley soils productive, it is necessary to drain them and raise their pH values by adding lime. As we saw earlier, this creates the proper environment for humus-producing microorganisms.

CLASSIFICATION OF SOILS

The classification of soils is even more difficult than the classification of climates. For one thing, there are many more variables to take into account. Whereas a climate may be characterized by its temperature and moisture regimes, soils may vary in texture, structure, organic content, cation exchange capacity, and the details of horizon development. Their origins vary due to differences in parent material, climate, site, organic activity, and relative age. Furthermore, soils, unlike climates, can differ markedly from point to point within an area the size of a city block. Finally, human activities can completely alter soils both physically and chemically, either by design or unintentionally.

In the United States, soil classification begins with detailed field work that involves the description and mapping of individual soil types that form a *soil series*. This is named after the geographic location in which the soil is found. Soil series having features in common are subsequently grouped into *soil families*. Soil families are regarded as subdivisions of *great soil groups* that have world-wide distributions. The names of the great soil groups were established in 1938 by the U.S. Department of Agriculture and for many years constituted the terminology used in referring to soils on a global scale.

The 1938 great soil groups in turn constituted three general *soil orders*: zonal, azonal, and intrazonal soils. Soils with good profile development that seemed to show the direct influence of the principal climates and vegetation types around the world fell into the category of *zonal soils*. Soils that had not had time to develop a profile in response to the soil-forming factors were *azonal soils*. Soils dominated by special non-climatic influences, such as drainage conditions or parent material, were called *intrazonal soils*.

The explanatory-descriptive classification of 1938 neither explained nor described soils to the satisfaction of soil scientists. It was superseded in 1960 by a new system based on terms that describe a soil's properties, not its environment or origin. But this system, while in broad use today, is much more difficult to learn and has not been accepted by all soil scientists. As yet there is no agreement on a comprehensive world-wide soil classification system or terminology, and each country has its own official classification scheme. For our purposes, it is sufficient to understand the soil-forming factors and regimes and the ways in which they produce soils of different types and different productive capacities. Figures 10.15 and 10.16 summarize the most important aspects of soils in relation to the environments in which they form.

SOIL MANAGEMENT

The soils in which plants grow are one of our planet's most priceless resources, surpassed in importance only by air and water. Human replacement of natural vegetation with crop plants has seriously affected the earth's soils. As we have seen, the soil is a dynamic system that is in constant interaction with the things living and growing in it. In their natural state many soils seem to reach an equilibrium condition.

Figure 10.15 (opposite)
Latitudinal distribution of influences on soils. This diagram summarizes how precipitation, energy, evaporation, and vegetation affect rock weathering and soil formation. (Vantage Art, Inc. modified from N. M. Strakhov, *Principles of Lithogenesis*, vol. 1, 1967, Plenum Publishing Corp., and P. W. Birkeland, *Pedology, Weathering, and Geomorphological Research*, 1974, Oxford University Press)

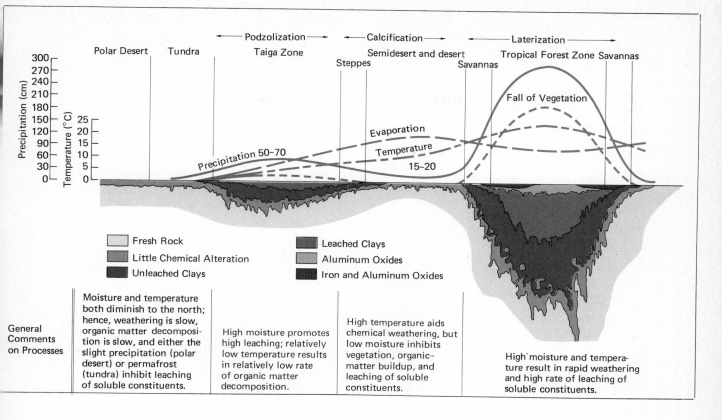

Polar Desert Tundra ←— Podzolization —→ ←— Calcification —→ ←————— Laterization —————→
                        Taiga Zone   Semidesert and desert Tropical Forest Zone Savannas
                                 Steppes   Savannas

Precipitation (cm): 300, 270, 240, 210, 180, 150, 120, 90, 60, 30, 0
Temperature (°C): 25, 20, 15, 10, 5, 0

Fall of Vegetation
Evaporation
Precipitation 50–70
Temperature 15–20

Fresh Rock
Little Chemical Alteration
Unleached Clays
Leached Clays
Aluminum Oxides
Iron and Aluminum Oxides

General Comments on Processes

Moisture and temperature both diminish to the north; hence, weathering is slow, organic matter decomposition is slow, and either the slight precipitation (polar desert) or permafrost (tundra) inhibit leaching of soluble constituents.

High moisture promotes high leaching; relatively low temperature results in relatively low rate of organic matter decomposition.

High temperature aids chemical weathering, but low moisture inhibits vegetation, organic-matter buildup, and leaching of soluble constituents.

High moisture and temperature result in rapid weathering and high rate of leaching of soluble constituents.

Over the space of a year the nutrients taken from the soil by its vegetative cover are returned to the soil in the process of nutrient cycling.

Clearing the land for agriculture interrupts the cycling of nutrients. We keep most of the products of our croplands, so only a fraction of the nutrients that plants remove from the soil is recycled into it in the form of organic litter. Thus agriculture inevitably produces a loss in soil fertility. At the same time, soil erosion is increased by exposing bare ground to the impact of rain and the force of wind and running water, and by disturbing the soil through plowing and cultivating. Erosion by water and wind removes the vitally important A horizon that contains the humus needed for moisture retention and soil fertility. Loss of the A horizon reduces the soil's water-accepting and water-holding capacities. This increases the proportion of water that runs off on the surface, which stimulates still more erosion.

In eroded soils plant roots must seek nutrients in dense clay subsoils instead of the loose, humus-rich A horizons to which they are adapted. As soils become thinner and less fertile, plants become increasingly sensitive to periodic variations in moisture. Thus many "droughts" are actually normal events that have a disastrous impact because the soil system has been weakened by human activities.

Loss of soil by erosion creates other problems. The erosion of deforested hillslopes has choked streams with sediments, causing them to become shallower. This increases the hazard of flooding on the plains beyond the mountains.

THE SOIL SYSTEM

Figure 10.16
Characteristic types of soil are associated with certain climatic, vegetation, and drainage conditions, as these illustrations suggest.

(a) *Tundra Region*. Tundra vegetation consists of mosses, lichens, low shrubs, grasses, and flowering herbaceous plants. Tundra soils are developed over ground that is permanently frozen and thus impermeable to water. Repeated disturbances of the soil by freeze and thaw, as well as seasonal waterlogging, are characteristic. Networks of ice wedges are also common in low-lying areas. The soils have poor horizonation.

(b) *Coniferous Forest Region*. Coniferous forests produce acidic litter and do not recycle soluble bases in large amounts. Consequently, the soils beneath such forests are acidic, with clay, humus, bases, and even the oxides of iron and aluminum eluviated by acidic soil moisture. The result is the distinctive podzolized soil type, in which a bleached sandy *A2* horizon overlies an oxide-colored illuvial *B* horizon. Such soils are infertile unless neutralized by the addition of lime.

(c) *Short-Grass Prairie Region*. Grassland soils are generally high in nutrients due to low precipitation, rapid nutrient cycling, and a rich clay-humus complex. The organic horizon is well developed and dark in color due to its humus. Short-grass prairies have shallower soils than tall-grass prairies because soil moisture does not penetrate as deeply. The soils formed in the transition between short- and tall-grass prairies are exceptionally fertile. Short-grass prairies have a visible zone of lime accumulation in the *B* or *C* horizon due to shallow penetration of moisture.

(d) *Tall-Grass Prairie Region*. Tallgrass prairies receive more precipitation and develop soils that are deeper and even richer in organic matter than the soils of short-grass prairies. They have high

(a) Tundra Region

(b) Coniferous Forest Region

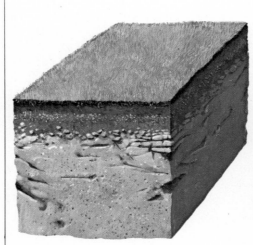

(c) Short-grass Prairie Region

(d) Tall-grass Prairie Region

Midlatitude Forest Region

(f) Tropical Forest Region

Desert Region

(h) Bogs and Meadows

base saturation, however, and are somewhat more leached than the short-grass prairie soils. This fact suggests that at times these soils have been occupied by forests, which have affected the soil's development. They are the most fertile soils located within a zone of reliable rainfall.

(e) *Midlatitude Forest Region.* Beneath the broadleaf deciduous forests of the midlatitudes, soils tend to be well supplied with mineral nutrients due to nutrient cycling where deep roots penetrate the parent material and nutrients stored in the vegetation are returned seasonally to the soil in leaf fall. In midlatitudes, bacterial action supplies abundant humus.

(f) *Tropical Forest Region.* Plant litter decomposes so rapidly under humid tropical conditions that humus is not formed. Nevertheless, the steady decay of plant litter on the forest floor returns nutrients to the soil, where they are immediately taken up by vegetation. Thus the soil nutrients are largely locked up in the vegetation itself, with the subsoil being eluviated of clays, bases, and even silica. Oxides and hydroxides of iron and aluminum color the soil orange or red.

(g) *Desert Region.* Desert soils are low in clay due to the lack of water for chemical weathering and low in organic matter as a consequence of the sparse plant cover. Thus they lack a clay-humus complex. They are high in nutrients since there is little eluviation, but are easily leached when irrigated. Desert soils commonly include calcrete (caliche) layers.

(h) *Bogs and Meadows.* Decaying plant material makes most waterlogged soils strongly acidic, and the upper horizon may be dark peat. Changes in water level produce color changes as iron is oxidized by contact with air or reduced by the exclusion of air. (John Dawson, adapted from S. R. Eyre, *Vegetation and Soils,* Edward Arnold Publishers)

Figure 10.17
(top) Gullying of this cornfield in Missouri resulted from the improper practice of plowing and harrowing up and down the slope. The damage was done in only a few days by several heavy rains. At the time of the photo, this land, which had a natural cover of grass, had been used agriculturally for only three years.

(bottom) This scene in southern Wisconsin illustrates contour strip cropping, which retards runoff and greatly reduces soil erosion. According to the Soil Conservation Service, agricultural yields have doubled where contour strip cropping has been introduced in this area. (U.S. Soil Conservation Service)

Deforestation and soil erosion in upland areas have become an acute problem in the foothills of the Andes and Himalaya mountains and in the East African highlands because rapid population growth has pushed agriculturalists onto ever-steeper land. This is clearly increasing flooding in the nearby heavily populated lowlands. Soil erosion also shortens the life of reservoirs, many of which are filling with silt two or three times more rapidly than predicted. Satellite images show that a major new reservoir in northern China has been entirely filled with sediment in little more than five years. As we saw earlier, land deterioration due to salinization is also a very serious problem in dry regions all around the world, possibly affecting as much as 75 percent of all irrigated land.

Of course, measures are being taken to preserve our soil resource. Fertilization and crop rotation can stave off soil exhaustion; the cost of fertilizers of various types is a major expense on any modern farm. To reduce wind and water erosion, crop stubble and litter may be left on the fields or cover crops planted after harvesting. In rolling country, plowing along the contour rather than up and down the slope will hold back runoff and reduce soil erosion by as much as two-thirds. Contour plowing can be combined with stip cropping, in which different crops are alternated in bands along the contour (Figure 10.17).

In deforested uplands massive tree plantings can reduce erosion and sediment input into streams. In irrigated areas deep drains will prevent the artificial rises in water tables that cause salinization of dryland soils.

All these procedures are being carried out in many parts of the world, but often only after centuries of neglect. Unfortunately, in some regions it is already too late, causing an exodus of people from rural areas. Ironically, the deterioration of soils around the world has become evident at the very time that the world's population and food demands are increasing explosively. While crop yields have increased markedly in many areas, the cost of producing such yields has risen even faster, and crop failures have become more frequent. The future of humankind probably depends less on world political events than on how we handle the fragile soil resource that sustains life on our planet.

SUMMARY

Soil is a mixture of mineral and organic matter and is the product of a living environment. The mineral matter is produced by rock decomposition, and the organic matter results from bacterial decomposition of plant material. Soil development begins with the fragmentation of rock by mechanical and chemical weathering processes. The nature of weathering, which produces the raw material for soil development, varies with changing energy and moisture conditions. Translocation and organic activity convert the decomposed rock to soil.

Soil differs from weathered rock by having a profile consisting of separate soil horizons. These vary in texture, structure, bulk density, color, and chemical composition. Soil pH is a measure of acidity, or hydrogen ion concentration, and cation exchange capacity. The clay-humus complex of a soil is of special importance in the soil's ability to retain basic cations that are essential plant nutrients. The nutrient cycle between plants and their soils is another factor in the maintenance of soil fertility.

The five principal influences on the formation of soil are parent material, climate, site, organic activity, and time. Climate, together with organic activity, produces three different soil-forming regimes: laterization, podzolization, and calcification. Drainage conditions combine with climate to produce two additional soil-forming regimes: salinization and gleization. Because of the great number of variables in soil formation and soil character, soil classification is extremely difficult, and several classification schemes are in use.

Soil management is an important task. Any

interference with the natural soil system disturbs soil equilibrium and increases erosion and nutrient losses. Agricultural use of soils demands careful attention to these problems. Several highly effective procedures have been developed to combat soil erosion and soil exhaustion. However, in many areas in all parts of the world, remedial action has been too late in coming, and the most productive portion of the soil has already been lost.

APPLICATIONS

1. Cemeteries are good places to observe the effects of weathering on various types of fresh rock. Go to an old cemetery in your area; staying on the cemetery paths, look at the weathering phenomena on the grave markers. What weathering effects are visible? What types of stone are most resistant to weathering? What types are least resistant? Can you explain the varying responses to weathering of the different types of stone?
2. What soil series occur in the agricultural areas nearest to your campus or your home? How would the soils be classified according to the U.S. Comprehensive System (7th Approximation System)? Do these soils contain the standard O, A, B, and C horizons? Do they contain any special features like Cca horizons?
3. Are the soils in your area derived from varying parent materials? If so, examine the soil texture and pH at varying depths on the different parent materials. In what ways do the soils reflect their parent materials?
4. What are the principal soil management problems in your area? Are there any unusual problems related to the physical or chemical properties of local soils?

5. How can the water budget concept be applied to the various degrees and types of soil development?
6. What changes in soils would occur where a midlatitude forest advances into a former grassland area? Where a grassland expands into a former forest area?
7. How is it possible for a calcrete layer 1 m thick to be found 1 m below the surface in a desert region where the heaviest rainfalls only wet the soil to a depth of 0.5 m?
8. How could natural changes in the environment duplicate human effects on soil systems in varying settings?

FURTHER READING

Bennett, Hugh H. *Soil Conservation.* New York: McGraw-Hill (1939), 933 pp. This abundantly illustrated classic, dealing with the processes and results of soil erosion in the 1930s, was written by the first Chief of the U.S. Soil Conservation Service. It is useful for its regional treatment for soil erosion problems in the United States and foreign areas and for its wealth of detail concerning specific localities.

Birkeland, Peter W. *Pedology, Weathering, and Geomorphology.* New York: Oxford University Press (1974), 285 pp. This work provides a thorough treatment of weathering, the factors of soil formation, and the study of soils as geological deposits that convey considerable information on recent environmental history. It is not concerned with soil-plant relations or agricultural applications.

Bridges, E. M. *World Soils.* 2nd ed. New York: Cambridge University Press (1978), 128 pp. This brief work is a well-illustrated (with color) outline of soil development and soil types in major natural regions of the world.

Buol, S. W., F. D. Hole, and **R. J. M. McCracken.** *Soil Genesis and Classification.* Ames: Iowa University Press (1973), 360 pp. A thorough treatment of soil-forming factors, processes, and resulting soil types classified according to the U.S. Comprehensive System.

CHAPTER 10

Carter, V. G., and **T. Dale.** *Topsoil and Civilization.* 2nd ed. Norman: University of Oklahoma Press (1974), 292 pp. The authors interpret the rise and fall of great world civilizations as due to human-induced deterioration of their resource bases—particularly the soils that supported their agricultural systems.

Clarke, G. R. *The Study of Soil in the Field.* 5th ed. Oxford: Clarendon Press (1971), 145 pp. This is an excellent manual outlining in very readable fashion the exact techniques used in analyzing soils in the field, with information on soil mapping and the use of aerial photographs.

Eckholm, Erik P. *Losing Ground: Environmental Stress and World Food Prospects.* New York: W. W. Norton (1976), 223 pp. The author presents an ominous view of the problem of current human-induced soil deterioration in varying environments, with special stress on future problems of food production, especially in the Third World. This book should be on every geographer's reading list.

Soil Survey Staff, U.S. Dept. of Agriculture. *Soil Taxonomy: A Basic System of Soil Classification for Making and Interpreting Soil Surveys.* U.S.D.A. Handbook 436. Washington, D.C.: U.S. Government Printing Office (1975), 754 pp. This is the updated official presentation of the U.S. Comprehensive Soil Classification System (7th Approximation), providing the system's rationale and detailed application.

Steila, Donald. *The Geography of Soils.* Englewood Cliffs, N.J.: Prentice-Hall (1976), 222 pp. This valuable paperback is a good introduction to the U.S. Comprehensive Soil Classification System (7th Approximation), including material on the uses and management of soils of the different orders in the classification scheme.

United States Department of Agriculture. *Soils and Men.* Yearbook of Agriculture, 1938. Washington, D.C.: Dept. of Agriculture (1938), 1,232 pp. Although dated, this work contains a wealth of information on soils, their classification, characteristics, uses, and maintenance.

CASE STUDY: **The Terrace Builders**

For any region to support sizable numbers of people, crops must be grown, requiring extensive areas of farmable soil. The value of arable land, which inhabitants of the benign environments of the middle latitudes accept unquestioningly, is best seen in the harsh lands that lack it. Skellig Michael, a snag of rock off the Irish coast, is a vivid example. Sometime before the ninth century, Skellig Michael's solitude and isolation caused it to be chosen by Irish monks as the site of a monastery. But no food could be produced there. The island was completely barren, exposing naked rock everywhere. Not to be denied, the monks labored to create productive patches by building retaining walls in rock niches and filling them with soil brought by boat from the mainland, 10 km (6 miles) distant. We know no more than this, for the island was abandoned more than 1,000 years ago, probably due to attacks by Viking raiders. But we can still see the rock niches and retaining walls, packed with foreign soil and now covered with wildflowers.

Vaster attempts to create productive cropland are seen in many parts of the world. Steep hillslopes have been reconstructed into staircase-like flights of horizontal terraces where only steep slopes and thin moisture-deficient soils existed naturally.

Agricultural terracing to create deep soils on steep slopes is widespread where population pressure is high. In some areas, such as the Philippine island of Luzon, the landscape has been completely transformed by terracing. Although requiring great labor to construct and maintain, terraces are found in a variety of climates, being conspicuous along the Mediterranean coastlands, especially in Spain, Italy, and Lebanon, on the slopes of the Pyrenees and Sierra Nevada of Spain, and on the southern slopes of the Atlas Mountains overlooking the Sahara Desert. They also appear in the Canary Islands, the Andes of South America, the highlands of Yemen, the Himalayan foothills, and the uplands of Sri Lanka (Ceylon), and on the volcanoes of Java and Sumatra. In some regions, terraces contour the hillsides; in others, they are built into the floors of narrow, steeply rising valleys and catch the runoff from adjacent rocky slopes. As at Skellig Michael, the terrace fill was sometimes carried from a great distance. The objective was not to create mere flat land, but deep, moisture-retentive soils—possibly to be irrigated by elaborate canal systems.

Geographers have long disputed whether the technology involved in agricultural terracing originated in a single location and diffused outward or was developed independently in many widely separated regions. Logic seems to favor the latter since the idea of terracing is a simple one, with natural examples provided by rock ledges and steplike landslide terrain, both of which are frequently seen in the very areas noted for artificial terraces. Terracing in the Andean area goes back at least 2,500 years, in Asia perhaps much further.

In many regions, terraces are irrigated, necessitating careful construction, leveling, and maintenance—along with the intricate canal and aqueduct systems that water them. Even in wet areas of the tropics, irrigation is practiced, for rice is grown in paddies on the terraces. These upland rice paddies are marvels of primitive technology, as were the irrigated terraces for maize built by the Incas in the Peruvian Andes. From the tropics to the middle latitudes, orographic rainfall may be adequate to nourish maize, potatoes, vegetables, tea, coffee, cocoa, sugarcane, vineyards, orchard crops, and cereals. Even unirrigated terraces are gardened rather than farmed, being carefully fertilized with animal manure, human waste, bird or bat droppings (guano), and even fish heads and volcanic ash. Constant attention is devoted to the state of terrace walls often made of well-fitted stone. Two or more crops may be grown in the same field—a tree crop and some lower-growing form.

The creation of agricultural terraces transforms steep mountain slopes into lush gardens, sometimes studded with villas surrounded by ornamental plants—one of the most attractive of all of the landscapes created by human activity. In areas of frequent strife, fortified villages may be perched on projecting mountain spurs between the green fields.

The effort expended on terracing is extreme, but provides an enduring relationship between the land and its occupiers. Where terracing is not practiced the results are far less satisfactory. Fields may be cleared on 45° slopes where just standing erect is difficult. The rate of erosion is catastrophically high, and the plots are soon abandoned with new ones being sought out every few years. This form of shifting cultivation produces a highly visible patchwork of active and abandoned fields in various stages of reversion to brush or forest. This pattern is conspicuous in parts of Central and South America, East Africa, South and Southeast Asia, and even the southern portions of Europe. Under such treatment, the entire soil resource can be lost in a few decades, with bare rock becoming exposed where forests once prevailed. With the rapid growth of village populations in such settings, forest clearing for fuel often precedes agricultural use of the land. This is

certainly the case in portions of the Andes, East African highlands, India, and Nepal. Here the use of the land seems purely exploitive, with little thought of a permanent relationship to a particular plot.

It is difficult to explain the variations in local traditions that lead one group to mine the land and another to tend it as the terrace builders do. Part of the explanation lies in the system of land tenure, with those having secure title finding it to their advantage to improve the land, while those with no enduring claim take what they can and move on. But this by no means accounts for all local variations seen. Everywhere there is a tendency to do things as they have traditionally been done, even if the traditions were established in ignorance of their effect or under circumstances that have changed profoundly.

This extremely steep land in Ifugau Province on the Philippine island of Luzon has been made suitable for agriculture by slope terracing. The stone-faced terraces are 2 to 6 meters high, and support irrigated rice crops. (Robert Reed)

The large scale features of the earth's crust—
continents, ocean basins, mountain ranges, and
plateaus on land and below the sea—have been
formed by the earth's internal energy, sometimes
manifested catastrophically by earthquakes and
volcanic eruptions.

Cotapaxi, by Frederick Edwin Church. (On loan to the Metropolitan Museum of Art from John Astor)

CHAPTER 11
The Lithosphere

To this point we have been discussing physical processes powered by radiant energy from the sun. But not all change on the earth is due to outside forces. A California winery building is slowly being torn apart by movements of the land under it. The Yellowstone region has risen .6 meters (2 ft) in the last 50 years. Earthquakes in Iran, Turkey, Central America, and China repeatedly shatter villages and cities alike. Volcanoes spring to life violently and then rest for centuries. Islands along the Norwegian coast are growing larger as the land slowly rises out of the sea. All these changes require a source of energy within the earth itself.

The *lithosphere*, the solid outer shell of the earth, including the crust and upper mantle, is affected by energy released by the radioactive decay of unstable elements within the earth's mantle. Without this internal energy and the motions it generates, the earth would be flat and featureless—all its relief worn away by rain, wind, and running water. Deep canyons, rough hills, and soaring mountain ranges all tell of an ongoing contest between two sets of forces. One of these, powered by gravity and radiant energy from the sun, tends to wear down and smooth out the earth; the other, powered by the earth's internal energy, heaves up the land, creating irregularities that we call "scenery."

Catastrophic earthquakes and volcanic eruptions have played their role in the formation of the earth's grandest scenery. But most relief features associated with the earth's internal forces are the result of small unspectacular movements repeated again and again over millions of years. It is worth remembering that most of the rocks seen at the earth's highest elevations were formed on the ocean floors or many kilometers below the surface of the land.

In this chapter we shall discuss the general types of relief features found on the earth and outline the origin and composition of the rocks that are the foundation of these features. Subsequent chapters will treat the motions of the earth's crust and the gradational processes in more detail.

ORDERS OF RELIEF

Irregularities in the earth's surface occur on several different scales. These can be called *orders of relief*. When we speak of relief, we mean differences in elevation between high and low places on a surface. However, relief features have horizontal as well as vertical dimensions. Relief features on the earth range in scale from raindrop imprints and worm trails to the continents and ocean basins, which are the largest-scale relief features on the earth, both vertically and horizontally. These major divisions of the earth's crust are regarded as the *first order of relief*.

First Order of Relief: Continents and Ocean Basins

The arrangement of the earth's continents and ocean basins is one of the most important facts of physical geography. The distribution of land and water masses is a major influence on the atmospheric circulation and the distribution of plant and animal species around the world. But the arrangement of the continents and oceans is not permanent. These features have changed in size, form, and position, and are continuing to change today (Figure 11.1).

Figure 11.1
This photograph, taken from a spacecraft orbiting more than 480 km (300 miles) above the earth's surface, shows a portion of the east coast of Africa looking toward Saudi Arabia, with the Red Sea at the left and the Gulf of Aden at the right. The opening of the Red Sea is believed to have begun 5 to 10 million years ago when the lithospheric plates carrying Africa (bottom) and Arabia (top) began to drift apart. (NASA)

The continents and ocean basins are features of the thin outer layer of the earth known as the *crust* (Chapter 1). The crust is not an unbroken shell. In fact, it is divided into a dozen or more adjoining plates. These plates, which are 70 to 100 km (40 to 60 miles) thick, include the upper portion of the earth's mantle. The continents are embedded in the somewhat denser crustal material that forms the ocean basins. The general name for the rocks of the continents is *sial*—meaning that their chemical make-up is dominated by silica (SiO_2) and aluminum. The rocks of the ocean floors, and the mantle beneath, are termed *sima*—silica-magnesium rocks. Sial is about four-fifths as dense as sima; therefore the separate continents "float" about

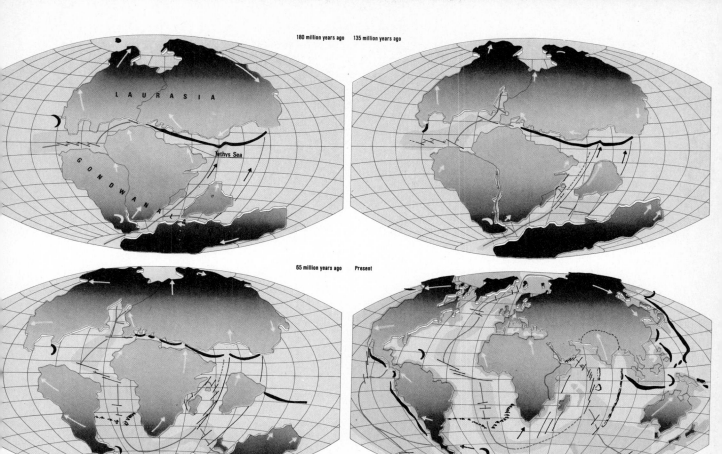

180 million years ago   135 million years ago

65 million years ago   Present

Figure 11.2
Alfred Wegener suggested in 1912 that about 200 million years ago all of the landmasses of the earth were grouped together, forming a supercontinent, Pangaea. Pangaea subsequently separated into two continents, Laurasia in the northern hemisphere and Gondwanaland in the southern hemisphere. The rifting continued, with crustal plates and associated continents moving in the directions indicated by the arrows. The plate that carries India, for example, separated from Antarctica and drifted northward until it collided with the Asian plate. The drift of the continents is now thought of in terms of plate tectonics, sea floor spreading, and subduction. In this figure, subduction zones are shown in black and spreading centers in red. (*The Atlas of the Earth*, p. 36, © 1971, Mitchell-Beazley, Ltd.)

four-fifths submerged in the continuous layer of sima.

When the first accurate world maps were constructed some 400 years ago, it was immediately noticed that the facing coasts of the continents of Africa and South America fit together, as in a jigsaw puzzle. This suggested that these continents had broken apart and somehow moved to their present positions. The hypothesis of "continental drift" was first set forward in a formal way in 1912 by the German meteorologist Alfred Wegener (Figure 11.2). Wegener pointed out that coal beds in temperate regions were composed of tropical vegetation types, suggest-

THE LITHOSPHERE

Figure 11.3
This map indicates the probable locations of lithospheric plate boundaries in the earth's crust and the relative motion of the plates. The red lines represent spreading centers where new crust is being formed, and the blue lines are regions where plates are descending into the mantle in the subduction process. The arrows indicate the general relative directions of plate motion. The seven major plates are identified in bold type and several of the smaller plates in lighter type. Note that the Pacific basin is largely rimmed by subduction zones.

Many features of the earth's crust can be understood in terms of the motion and interaction of plate boundaries. Mountain uplift is active at the plate boundaries along the west coasts of North and South America, where the mountains are young. Along the east coasts, however, there are no plate boundaries and no mountain-building activity. Earthquake and volcanic activity is associated with converging plates such as those around the rim of the Pacific Ocean basin. (Calvin Woo from John F. Dewey, "Plate Tectonics," *Scientific American*, copyright © 1972 by Scientific American, Inc. All rights reserved)

ing that today's mid-latitude areas had once been located near the equator. Likewise, ancient glacial deposits are seen in regions that are now tropical; thus low-latitude land masses once had polar or sub-arctic locations. Furthermore, geological structures and rock types along the west coast of Africa and the east coast of Latin America seemed to match. Finally, the fossilized remains of ancient plants clearly indicated former connections among all the widely separated southern hemisphere continents.

For almost half a century most geologists scoffed at Wegener's ideas. The problem was that there seemed to be no conceivable force that could move an entire continent. The fact that the sial of the continents projects downward into the denser sima beneath made horizontal movement of the continents all the more unlikely.

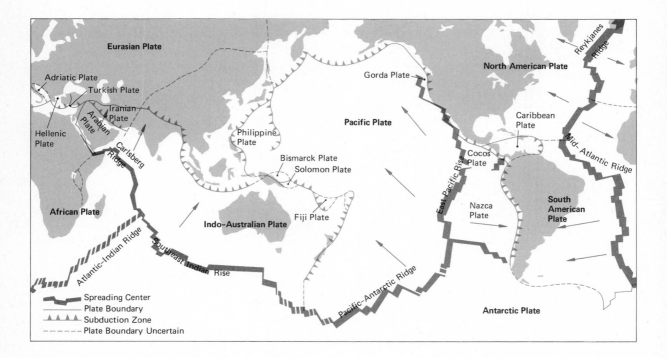

Figure 11.4

(a) A newly developed spreading center causes continental plates to rift and move apart. (b) The development of new crust at a spreading center causes a sizeable ocean basin to form over tens of millions of years. The sediment deposits are thickest near the continental plates, where the oceanic crust is oldest. This diagram represents the Atlantic Ocean basin. The magnetic stripe pattern indicates the record of the earth's magnetism frozen into new crustal rock as it cools. The pattern is the same on both sides of the spreading center; such patterns occur in all ocean basins.

(c) Deep ocean trenches, such as those in the western Pacific Ocean, occur where a plate of oceanic crust plunges downward into the mantle beneath either a continental plate or another oceanic plate (plate convergence). Crustal deformation at a convergent boundary leads to mountain uplift, formation of island arcs, and earthquakes and volcanic activity.

(d) A collision between two continents produces high mountain ranges such as the Himalayas. The continental crust is doubled in thickness and the surficial sediments are compressed to form contorted geologic structures. (Calvin Woo from Robert S. Dietz, ''Geosynclines, Mountains, and Continent Building,'' *Scientific American*, copyright © 1972 by Scientific American, Inc. All rights reserved)

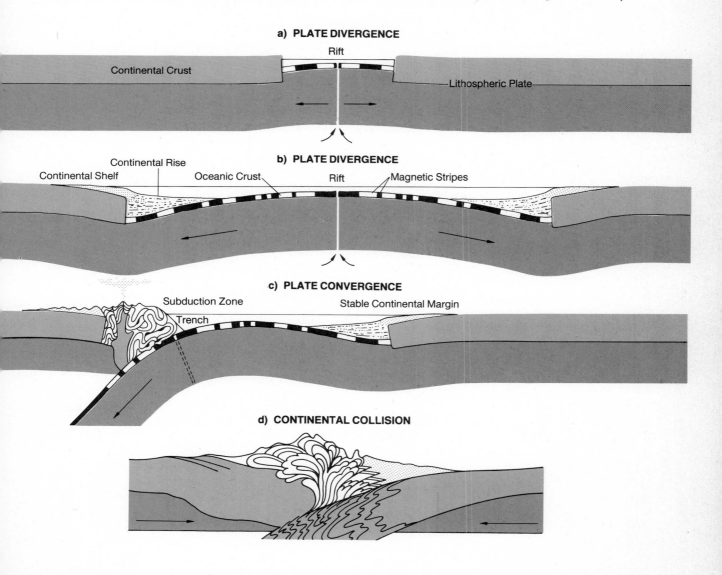

a) PLATE DIVERGENCE

Rift

Continental Crust

Lithospheric Plate

b) PLATE DIVERGENCE

Continental Rise

Continental Shelf

Oceanic Crust

Rift

Magnetic Stripes

c) PLATE CONVERGENCE

Subduction Zone

Stable Continental Margin

Trench

d) CONTINENTAL COLLISION

In 1944 the geologist Arthur Holmes produced the key to the puzzle. He suggested that it is the sea floors that are moving—merely dragging the continents along with them. Holmes proposed that the driving force was a process of convection ("boiling") in the upper part of the earth's mantle. This produced slow horizontal flows of molten material below the earth's crust that dragged the crust this way and that.

During World War II, it was discovered that all the oceans are divided by continuous undersea ridge systems. The best example is the Mid-Atlantic Ridge. In addition, trenches deep enough to hold the world's highest mountains almost completely ring the Pacific Ocean. Holmes suggested that volcanic eruptions along the oceanic ridge systems create new areas of ocean floor. Old areas of ocean floor eventually disappear by descending into the oceanic trenches, a process that later came to be known as *subduction* ("underflow"). The continents ride passively on lithospheric plates that move like a conveyor belt from the oceanic ridges toward the oceanic trenches in the process now known as *sea floor spreading*. The continents cannot themselves be subducted downward because of their low density relative to the material lower in the crust and mantle. The global pattern of crustal plates, spreading centers, and subduction zones was established by oceanographers in the 1960s, and is shown in Figure 11.3. The movement of crustal plates and the interactions between them are known as *plate tectonics*.

The sea floor spreading hypothesis has been supported by many findings since the 1960s. Scientists have drilled into the sea floors hundreds of times, principally from the research vessel *Glomar Challenger*, recovering cores of sediment and rock for analysis. These have shown that the sea floors are extremely young compared to the continents. Whereas the earth itself is almost 5 billion years old and 3.8 billion-year-old rocks have been found on the continents, the oldest portions of the sea floors have an age of only about 170 million years. Furthermore, the thickness and age of oceanic sediments increase with distance from the oceanic ridges. This supports the idea that the sea floors are being formed at the ridges and are moving away from them (Figure 11.4).

Some of the most convincing evidence for sea floor spreading comes from the record of ancient magnetism in the volcanic rocks of the ocean floors (Figure 11.4). When volcanic rocks solidify from a molten condition, tiny grains of the iron-bearing mineral magnetite align with the earth's magnetic field. Studies of volcanic rocks reveal that the magnetic field has changed in strength and has reversed in polarity (so that a compass needle would point south instead of north) at intervals of hundreds of thousands of years. The lavas on the sea floors show alternating strips of normal and reversed polarity. These strips are parallel to the oceanic ridges and are strikingly symmetrical on the opposite sides of the ridges. Changes in the direction of the strips indicate changes in the direction of sea floor spreading.

In general, the sea floors appear to be moving laterally at rates of 1 to 10 cm (.4 to 4 inches) yearly, dragging the continents with them. The Atlantic Ocean is widening by this process, whereas continents are being forced toward the Pacific Ocean from two sides. The floor of the Pacific Ocean is detached from the adjacent continents and is forced to descend under them in the submarine trench system that encircles the Pacific.

Second Order of Relief: Major Subdivisions of the Continents and Ocean Basins

The continents themselves contain large-scale relief features in the form of mountain systems like the Alps or Rockies, large depressions on the scale of the Mississippi Valley and the West Siberian Lowland, and great plateaus such as those of Tibet or South Africa. Features on this scale constitute the *second order of relief*.

The ocean basins also possess a second order of relief in the form of the oceanic ridge systems, the submarine trenches, and other large-scale irregularities (Figure 11.5, see pp. 278-279).

All second-order relief features have been produced by either vertical or lateral motions of the earth's crust. All can be explained in terms of plate tectonics. The great continental mountain systems were created by compression and volcanic activity where one crustal plate pushed against another. The rugged Andes Mountains bordering the west coast of South America are currently being forced up as the eastward-moving Nazca plate descends under the westward-moving South American plate (see Figure 11.3). A similar relationship in the past explains the rise of the mountain systems of the Pacific coast of North America. The Himalaya Mountains and Tibetan Plateau were raised when the crustal plate that carries India was forced northward against the Eurasian crustal plate.

Along the Atlantic coasts of the Americas, Europe, and Africa, the continents ride on crustal plates that have been moving away from the Mid-Atlantic Ridge for the past 100 to 200 million years. Since there are no plate boundaries near these coasts, they are low-lying for the most part. However, 200 to 300 million years ago these same coasts were colliding as an older Atlantic Ocean was squeezed closed. These flat coasts and their borderlands were then the sites of mountain systems like those of the Pacific coasts today.

Between the continents and ocean basins is a series of second-order relief features submerged by the seas that are actually portions of the continents. These are the continental shelves, slopes, and rises (Figure 11.6, see p. 278).

The *continental shelves* are the submerged rims of the continents. They descend from coastlines to an average depth of about 130 m (430 ft). Although they have an average slope of less than a tenth of one degree, they are far from smooth in many areas. They were exposed land as recently as 15,000 years ago. At that time, close to 3 percent of the water that is pres-ently stored in the sea was held on the land in the form of vast ice sheets. The average width of the shelves is 75 km (46 miles), but some areas have shelves far wider than this, while others lack any shelf at all.

The continental shelves have long had great economic importance. Because the shallow water over the shelves allows sunlight to penetrate to the bottom of the inner portions, and because nutrients are constantly supplied by erosion on the land, the shelves are able to support an abundant marine food chain. This has generated controversies among fishing nations over who owns the continental shelves. More recently, large deposits of oil have been found in the sediments blanketing the shelves in the Gulf of Mexico, the North Sea, the Yellow Sea, and other places. This has intensified ownership controversies among maritime nations.

The continental shelves are succeeded by the slightly steeper (4 to 5 degrees) and more irregular *continental slopes*, which descend to some 3,000 to 3,600 m (10,000 to 12,000 ft). The less-steeply sloping *continental rises* below the continental slopes seem to be embankments of sediment washed from the land and resting against margins of the sialic continental slabs. It is the sediment blankets of the continental shelves, slopes, and rises that are crumpled and upheaved by compression during the collisions of crustal plates, thereby generating the earth's great mountain chains.

Stretching out from the bases of the continental rises at depths of 5,000 m (16,000 ft) or more are the deep ocean floors. Some portions of these, the so-called *abyssal plains*, are flat and featureless blankets of marine sediments that cover the older volcanic rocks of the sea floors. At one time it was assumed that the sea floors were composed almost entirely of abyssal plains. But this is far from true: there is a highly diversified relief beneath the waters of the seas, including the ocean trenches and ridge systems, plateaus, great fault scarps, clusters of undersea mountains, and mysterious erosional canyons (see Chapter 16).

Figure 11.5
The ocean basins, which are first-order relief features, contain second-order relief features, such as deep trenches and extensive undersea mountain ranges. This physiographic diagram of the North Pacific Ocean basin shows the plains, trenches, and seamounts characteristic of the ocean floor. Heights above and below sea level are given in feet. Note in particular the island arcs along the continental side of the great trenches. Such arcs are located at the boundaries where oceanic crustal plates are plunging downward into the mantle under continental plates, and they are sites of volcanic and earthquake activity. (Courtesy of the National Geographic Society)

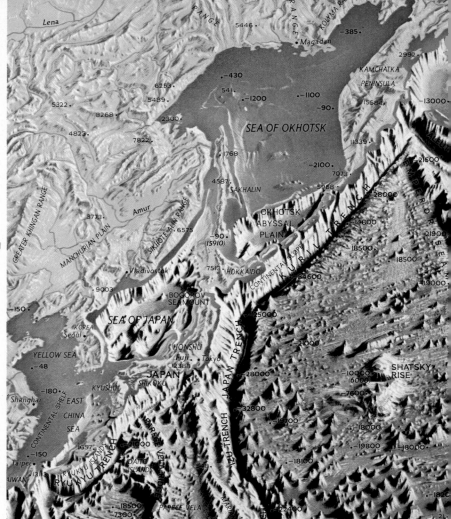

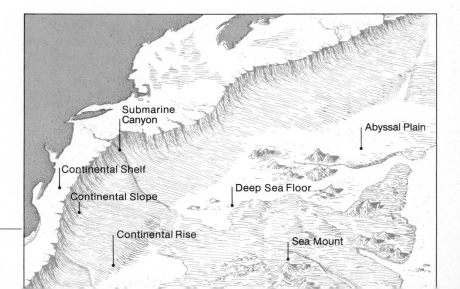

Submarine Canyon

Continental Shelf

Continental Slope

Continental Rise

Deep Sea Floor

Sea Mount

Abyssal Plain

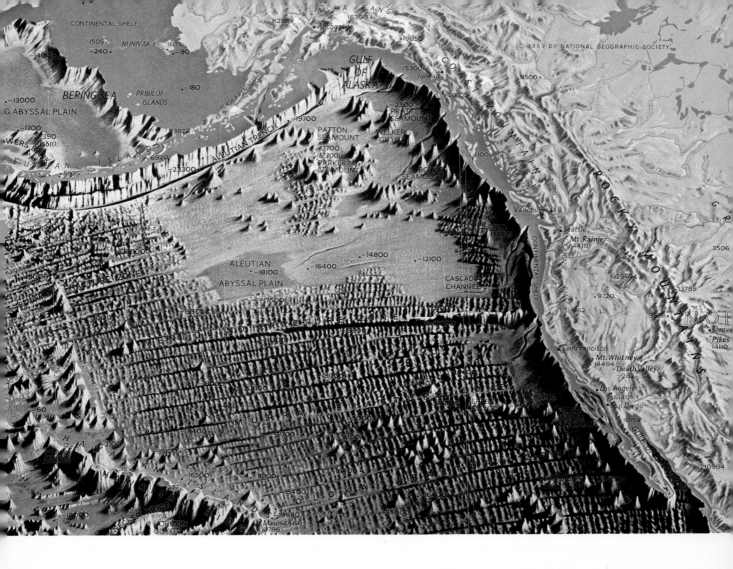

Figure 11.6 (opposite)
The topography of the ocean bottom off the coast of the northeastern
United States exhibits several major divisions. The vertical scale is
exaggerated approximately 20 times for clarity. Near the margin of the
continent, the water is shallow and the ocean floor descends very
gradually as the *continental shelf*. At the edge of the shelf, the bottom
slopes steeply down the *continental slope* to the deep sea floor. The
continental rise, where the continental slope meets the sea floor, is a
region where sediments accumulate. Note also the smaller features of
relief such as the deeply trenched *submarine canyon* associated with the
Hudson River and the relief forms on the sea floor. (John Dawson after
Bruce C. Heezen and Marie Tharp, Lamont Geological Observatory)

THE LITHOSPHERE   **279**

Third Order of Relief: Landforms

The individual hills, valleys, cliffs, and plains of landscapes constitute the *third order of relief.* These sculptural details of the second-order relief features are called *landforms.* Landforms are features that can be seen in their entirety in a single view. The largest third-order relief features would be plains created by erosion or the deposition of sediments, individual mountains within a larger mountain range, or particular valleys within a valley system. There is no lower limit to the size of a landform.

Most landforms have been produced by erosion or deposition of material rather than by motions of the earth's crust. Chapters 12 through 16 focus on the origin of landforms, which set the stage on which the various physical systems interact, and which influence these interactions in many ways. Their slopes provide gravitational energy; their heights are sources of water and sediment; they channel or block movement; their form and composition provide problems or possibilities for human activity. But before we can look more closely at landforms, it is necessary to examine the raw material for landform development—the materials composing the earth's crust.

THE STRUCTURAL FOUNDATION

We inhabit the surface of the *lithosphere,* or "sphere of stone"—the outer shell of the earth that is composed of either solid or fragmented rock material. Rocks differ greatly in origin and composition. Because of this, they vary in physical strength, chemical stability, and permeability to water. The weathering of different rock types creates different end-products. Finally, rock masses can be arranged in many different ways: in layers that are horizontal, wrinkled, or broken and offset, or in seemingly uniform masses. All these variations are influences on

landform development. To understand the appearance of the earth's surface it is necessary to know something about rocks.

Rocks and Minerals

When one looks at ordinary rocks under a microscope, it becomes apparent that they are composed of small particles. These either interlock or are cemented together by some other substance. Each of these particles is one of about 2,000 naturally occurring inorganic substances called *minerals.*

Minerals are natural combinations of chemical elements in solid form. The atoms of each mineral have a characteristic three-dimensional arrangement that gives its crystals a certain geometric form, as well as a particular luster, color, hardness, and specific gravity (density relative to an equal volume of water). Many minerals can be extremely beautiful, and perfect specimens are sought by collectors. Certain minerals are very familiar, such as *halite* (common salt, NaCl), *ice* (solid H_2O), and *geothite* (iron rust, FeO(OH)). The high-density metals, such as iron, aluminum, copper, manganese, lead,

Table 11.1

Major Elements of the Earth's Crust

ELEMENT	WEIGHT (PERCENT)	VOLUME* (PERCENT)
Oxygen (O)	46.60	93.77
Silicon (Si)	27.72	0.86
Aluminum (Al)	8.13	0.47
Iron (Fe)	5.00	0.43
Calcium (Ca)	3.63	1.03
Sodium (Na)	2.83	1.32
Potassium (K)	2.59	1.83
Magnesium (Mg)	2.09	0.29
Totals	98.59	100.00

*Computed as 100 percent, hence approximate. After Brian Mason, *Principles of Geochemistry,* John Wiley & Sons, Inc., 3rd ed., 1966.

zinc, nickel, tin, and silver, rarely form separate minerals. They are usually combined with other elements to form minerals that must be artificially decomposed, or "refined," to extract the metal. Such minerals are known as the *ores* of the metals. Gold and platinum are unusual elements in that they occur only in pure form.

The elements combined in each mineral are held together by electrical bonds. Thus all minerals are composed of a combination of positive and negative ions. The most abundant ions are those of silicon (Si), which has a positive charge, and oxygen (O), which carries a negative charge. Silicon and oxygen dominate in mineral structure (see Table 11.1), producing two large families of minerals: the silicates and the oxides. In the silicates, ions of basic and metallic elements combine with silica (SiO_2). In the oxides, basic and metallic elements are combined with oxygen. Most rock-forming minerals are silicates,

and about 92 percent of the earth's crust is composed of silicate minerals (Table 11.2). Oxides are created by the chemical decay of silicate minerals, which releases ions of metals and bases that combine with oxygen ions carried in water. Since oxygen is even more abundant than silicon in the silicate minerals, almost 94 percent of the volume of the earth's crust consists of oxygen ions.

Of the vast number of different minerals known to occur on the earth, only those few shown in Table 11.2 are common constituents of rocks. Rocks are themselves broadly classified as igneous, sedimentary, or metamorphic, referring to their mode of origin.

Igneous Rocks

Much of the rock of both the continents and ocean basins has solidified from a molten condi-

Table 11.2
Major Rock-forming Minerals

MINERAL GROUP	MINERAL	GENERALIZED CHEMICAL COMPOSITION	
		POSITIVE IONS	NEGATIVE GROUP
Silicates	Olivine	Mg, Fe	(SiO_4)
	Garnets	Mg, Al, Ca, Fe	
	Pyroxenes	Na, Mg, Al, Ca, Fe	(SiO_3)
	Amphiboles	Na, Mg, Al, Ca, Fe	(Si_4O_{11}), (OH)
	Micas	Mg, Al, K, Fe	(Si_2O_5), (OH)
	Clay Minerals	Al, K	
	Plagioclase Feldspar	Na, Al, Ca	(SiO_2)
	Orthoclase Feldspar	Al, K	
	Quartz	Si	O
Carbonates	Calcite	Ca	(CO_3)
	Dolomite	Ca, Mg	

After A. Lee McAlester, The Earth, New York: Prentice Hall, 1973.

Figure 11.7
Red-hot lava erupted at the earth's surface at temperatures of 900° to 1,200°C (1,600° to 2,200°F) rapidly congeals into extrusive igneous rock (foreground). "Fire fountains" such as this one seen in Hawaii commonly accompany extrusions of basaltic lava and may erupt for periods of several minutes at a time, rising and falling, and often being intermittently active for several days and occasionally even months. (Warren Hamilton/U. S. G. S.)

tion. Molten rock-forming material, or *magma*, is present everywhere on earth below a depth of about 70 km (40 miles). Occasionally this molten material, with a temperature of 900° to 1,200°C (1,600° to 2,200°F), forces its way through the crust and spills out at the surface as red-hot lava (Figure 11.7). Much larger volumes of magma solidify below the surface, deep within the crust. In either case the product is *igneous rock* (from the Latin *ignis*, fire); those rocks solidifying below the surface are *intrusive* igneous rock, while those that solidify after flowing out at the surface are *extrusive*.

As fluid magma cools, different silicate miner-

als "precipitate" out in crystal form at successively lower temperatures (Figure 11.8). Magma that crystallizes far below the earth's surface cools more slowly than magma that pushes closer to the surface. The greater the time required to solidify, the larger the mineral crystals that result. Igneous rock formed at a depth of many kilometers is coarse-grained *plutonic rock* (after Pluto, the Latin god of the lower world). The mass of rock is itself called a *pluton*. Deep-seated plutonic rocks, such as granite, gradually come to be exposed at the surface by crustal upheaval and the erosion of covering rock masses.

When magma forces its way to the land surface the result is a volcanic eruption from a central vent or from linear cracks called fissures. Central vent eruptions build cone-shaped volcanoes, while eruptions from fissures simply create deep masses of volcanic rock as in the Columbia Plateau of Washington, the Snake River Plain of Idaho, and the Yellowstone Plateau in Wyoming (see Chapter 14). Volcanic eruptions produce two different types of extrusive igneous rocks: lava and tuff. *Lava* is fluid or semi-fluid magma that flows out at the surface. Lava solidifies into rock in a matter of hours or days rather than the years required for magma below the surface (probably many thousands of years in the case of large plutons). Lava's rapid cooling produces much smaller mineral crystals than those of plutonic rock. Most lavas contain a scattering of 1- to 2-mm-long crystals (phenocrysts) embedded in a fine-grained matrix (groundmass) of other crystals that are too small to be visible to the naked eye.

Volcanic eruptions also hurl rock particles and bits of lava into the air. These rain down to form loose deposits of *volcanic ash*. Extremely violent eruptions vent great volumes of fine glowing particles that are welded together by heat when they settle to the ground. This produces the igneous rock known as *tuff*.

Almost every type of intrusive igneous rock has an extrusive equivalent (see Table 11.3, p. 284). This table also shows that a small number of silicate minerals (including quartz, feldspars, micas, hornblende, pyroxene, and olivine) are

Figure 11.8
Igneous rocks are classified by the proportions of their constituent minerals. Here, the approximate mineralogical composition of the more common igneous rocks is shown in terms of the percentage of the total volume occupied by each mineral. Thus a granite should contain approximately 40 percent orthoclase, 30 percent quartz, 15 percent plagioclase, 10 percent biotite, and 5 percent hornblende. Intrusive rock types are labeled across the top, with their extrusive equivalents in parentheses. Peridotite and dunite are rare, probably being formed only in the earth's upper mantle. (Doug Armstrong)

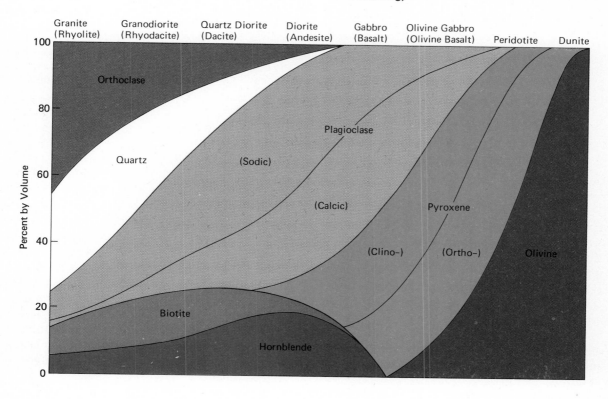

Table 11.3
Mineral Compositions of Igneous Rocks

INTRUSIVE ROCK	MINERAL COMPOSITION		EXTRUSIVE ROCK
	ABUNDANT	LESS ABUNDANT	
Granite	Quartz, K- and Na-feldspars	Biotite, hornblende, and muscovite	Rhyolite
Syenite	K-feldspar	Na-feldspar, biotite, hornblende, muscovite, and less than 5 percent quartz	Trachyte
Diorite	Na- and Ca-feldspars (plagioclase) and hornblende	Biotite and pyroxenes; quartz usually absent	Andesite
Gabbro	Ca-feldspar (plagioclase), pyroxenes, and olivine	Hornblende	Basalt
Peridotite	Olivine and pyroxenes	Oxides of iron; feldspars usually absent	No extrusive equivalent

dominant in igneous rocks. The most common igneous rocks are the extrusive *basalts* (Figure 11.9), which are dark lavas that compose the sea floors and are also widespread on the land; extrusive *andesites*, which are lighter colored lavas that form the bulk of the earth's large volcanic cones; and coarse-grained intrusive *granites* (Figure 11.10, p. 286) that form the cores of the continents and are often exposed where the crust has been upheaved in mountain systems. Widespread tuffs, such as those of the Yellowstone region, are generally composed of light-colored *rhyolite*, the extrusive equivalent of granite.

Although igneous rocks form about 80 percent of the earth's crust, they are commonly hidden by a veneer of sedimentary rock on both the continents and sea floors. As was pointed out in Chapter 1, there is a fundamental contrast between the rocks of the continents and those of the ocean basins. The igneous rocks of the continents are mainly low-density plutonic types, while those of the ocean floors are higher-density basaltic lavas erupted along the oceanic ridge system. We have seen previously that the basaltic lavas of the ocean basins are considerably younger than most of the plutonic rocks of the continents.

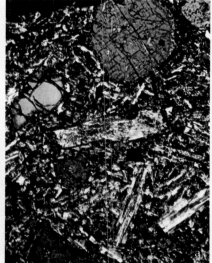

Figure 11.9

Basalt is an extrusive volcanic rock (or lava) and forms the floors of the ocean basins.

(a) The ten-times enlargement of a thin section shows basalt to consist largely of three different silicate minerals and volcanic glass. (M. E. Bickford, University of Kansas)

(b) This solidified volcanic lava flow on Kilauea crater, Hawaii, shows the rumpled surface of highly fluid basalt. (Warren Hamilton, U. S. G. S.)

(c) The Devil's Postpile in the Sierra Nevada of California consists of prismatic columns of andesite, an extrusive rock intermediate in chemical composition between granite and basalt. The columns were caused by thermal contraction as the rock solidified. (U. S. Forest Service)

THE LITHOSPHERE

Figure 11.10

Granite is the most common intrusive igneous rock on the continents.

(a) The thin section (enlarged about ten times) shows that granite is a coarse-grained rock consisting of many different minerals that have an interlocking structure. Silicate minerals are the main constituents of granite; when granite weathers, it first decomposes into sand. Further chemical alteration transforms some minerals to clay. (M. E. Bickford, University of Kansas)

(b) This dome of granite exhibits the characteristic way granite exposed at the earth's surface splits into sheets, or *exfoliates*, due to release of confining pressure. (G. K. Gilbert/U. S. G. S.)

(c) The Sierra Nevada range in California, shown here in the vicinity of Mount Whitney, consists largely of granite. The boulders in the foreground are separated from massive parent rock by chemical weathering along the joints of the granite, while the vertical slabs in the background are produced by frost weathering. (U. S. Forest Service)

Sedimentary Rocks

Rock weathering (Chapter 10) constantly produces fragmented rock material known as *clastic debris*. This is moved from higher to lower elevations by the energy of running water, wind, waves, and glacial ice, and the direct pull of gravity. All these agents of erosion eventually lose some of their energy, resulting in deposition of some or all of the debris they are carrying. Streams transport clastic debris to lakes, inland seas, and the oceans, where sediment accumulates in layers. Sediment is also deposited on the land—on river floodplains, in crustal depressions, at the margins of glaciers, at the bases of cliffs, and where sand-carrying winds lose velocity.

Sediments deposited on the continents consist mainly of clastic debris, including gravel, sand, and silt. Marine sediments formed close to shorelines include sand, silt, and clay; farther from the shore we find only clays, chemically precipitated calcium carbonate, and the limy remains of tiny marine animals. When both land and sea sediments become deeply buried under later deposits, compaction and the deposition of cementing substances cause them to become *lithified*, or converted into *sedimentary rock* of either clastic or chemical origin (see Table 11.4, p. 288).

Sedimentary rocks form in layers (Figure 11.11, p. 289). These may be millimeters to hundreds of meters thick. Each layer, called a *bed* or *stratum*, indicates a period of sediment deposition. The separations between strata are *bedding planes*, indicating a period of no deposition at that location. The strata above and below a bedding plane are often quite unlike, indicating a change in the conditions of sediment delivery. Individual beds may also change laterally as a result of variations in energy conditions and distance from the original source of the sediments. A *conglomerate* (cemented gravel) can change to *sandstone* (cemented sand), which passes into *siltstone* (cemented silt), which changes to *mudstone* (cemented silty clay), which becomes *shale* (cemented clay). The same bed may finally change from shale to *limestone*, formed either as a chemical deposit or as an accumulation of the skeletal remains of tiny marine animals.

Limestone, which is composed of calcium carbonate, has considerable economic value. From it is made the cement used to construct highways, buildings, sidewalks, patios, and swimming pools. *Dolomite* is a calcium-manganese carbonate rock that forms as a chemical precipitate or by later chemical alteration of limestone. Limestone is peculiar among common rock types in that it dissolves away completely where there is abundant moisture and vegetation. This creates very distinctive landscapes, as we shall see in Chapter 14. Dolomite is also soluble, but much less so than limestone.

Marine organic deposits that have an exceptionally high content of carbon form *hydrocarbons*—petroleum and natural gas. Their low density allows the petroleum and natural gas to migrate upward to fill openings in more porous rocks, especially sandstone. These are the "reservoir rocks" in oil and gas fields.

Economically, the most valuable of all sedimentary rocks is *coal*, which originates as luxuriant vegetation growing in freshwater lagoons and swamps. To be preserved, this organic material must accumulate in a stagnant-water environment and be acted upon by bacteria that can thrive without oxygen. To be transformed into coal, the resulting organic complex must be compressed by deep burial. The first stages of coal formation are probably occurring today in the swamps of the southeastern United States.

Two other sedimentary rock types of economic value are *rock salt* and *gypsum*. These are chemical deposits formed on the beds of evaporating lakes and inland seas in dry regions. The uses of salt are too numerous to mention. Gypsum likewise has many uses; among the most important is the manufacture of plaster of Paris, as well as gypsum board, the standard material used to sheath the walls of houses.

Table 11.4
Common Sedimentary Rocks

	UNCONSOLIDATED SEDIMENT	GRAIN SIZE	LITHIFIED ROCK
CLASTIC IN ORIGIN	Angular boulders, cobbles, pebbles	>2 mm	Breccia
	Rounded boulders, cobbles, pebbles	>2 mm	Conglomerate
	Sand	0.02–2.0 mm	Sandstone
	Silt	0.002–0.02 mm	Siltstone (Mudstone)
	Clay	<0.002 mm	Shale

	UNCONSOLIDATED SEDIMENT	MINERAL COMPOSITION	LITHIFIED ROCK
CHEMICAL IN ORIGIN	Calcareous parts of marine organisms and direct calcium carbonate precipitates	Calcite ($CaCo_3$)	Limestone
	Magnesium replacement of calcium and direct magnesium carbonate precipitates	Dolomite ($CaMg(CO_3)_2$)	Dolomite
	Amorphous silica	Chalcedony, Quartz (SiO_2)	Chert (Flint)
	Compacted plant remains	Carbon (C)	Bituminous Coal
	Salt left by evaporation of sea or saline lake water	Halite (NaCl)	Rock Salt
	Gypsum left by evaporation of sulfate-laden water	Gypsum ($CaSo_4–2H_2O$)	Gypsum

Figure 11.11

Sedimentary rocks consist of fragments of rock debris or organic material compacted and cemented together by various chemical substances, most commonly silica, calcite, and iron oxide.

(left) *Sandstone*, one of the most common sedimentary rocks, is usually composed of grains of silicate minerals cemented together by other minerals, as shown in this enlarged thin section. In general, sandstone is any rock composed of fragments that are in the size range of sand grains. When sandstone weathers, the cementing material disintegrates, releasing the individual grains. Sandstone is formed by the consolidation of beds of sand deposited by wind or water both on the land and in the sea. (M. E. Bickford, University of Kansas)

(below) This photograph illustrates the laminated nature of sedimentary rocks, which form *strata* that vary in thickness and physical and chemical characteristics. (Dr. Warren B. Hamilton/U. S. G. S.)

Metamorphic Rocks

Crustal motions related to plate tectonics sometimes cause rock masses to be forced deep down into the lowest portions of the earth's crust. Here pressures and temperatures are hundreds of times those at the earth's surface. This causes the rock material to deform and flow in a plastic manner or to be melted and recrystallized in different minerals. This process of *metamorphism* of rock material transforms limestone to *marble*, shale to *slate* or *schist*, sandstone to *quartzite*, granite to *gneiss*, and lava, like shale, to schist (Table 11.5, p. 290). Metamorphism most often occurs in or near subduction zones, where crustal plates converge.

In many metamorphic rocks the minerals are "smeared out," or oriented along visible planes of flow (Figure 11.12, p. 290). Where the original rock contained a mixture of minerals, as in granite, metamorphism may segregate them in wavy bands of contrasting color. The result is gneiss. Complete melting of the original rock generates new fluid magma, which may force its way up-

Submarine volcanic activity can saturate sea water with silica, resulting in the chemical precipitation of silica as thin beds of *chert*, often called "flint." Masses of chert are often found in limestone. Because of its hardness and ability to retain sharp edges, chert was used in the manufacture of stone tools by primitive peoples.

THE LITHOSPHERE

Table 11.5

Structure and Composition of Metamorphic Rocks

	METAMORPHIC ROCK	TEXTURE	MINERAL COMPOSITION	DERIVED FROM
FOLIATED	Slate	Fine-grained; smooth, slaty cleavage; separate grains not visible	Clay minerals, chlorite, and minor micas	Shale
	Schist	Medium-grained; separate grains visible	Various platy minerals, such as micas, graphite, and talc, plus quartz and sodium plagioclase feldspar	Shale, basalt
	Gneiss	Medium- to course-grained; alternating bands of light and dark minerals	Quartz, feldspars, garnet, micas, amphiboles, occasionally pyroxenes	Granite
NONFOLIATED	Quartzite	Medium-grained	Recrystallized quartz, feldspars, and occasionally minor muscovite	Sandstone
	Marble	Medium- to course-grained	Recrystallized quartz or dolomite plus minor calcium silicate minerals	Limestone or dolomite

Figure 11.12

Metamorphic rocks form by the transformation of preexisting rocks under conditions of heat and high pressure.

(a) This outcrop of *gneiss* exhibits the swirled patterns often seen in metamorphic rock. (M. E. Bickford, University of Kansas)

(b) *Schist* is a crystalline rock dominated by a layered arrangement of platy minerals as seen in this thin section enlarged ten times. Because weathering tends to be most effective between the sheets, schists tend to break into thin flakes. (Warren Hamilton/U. S. G. S.)

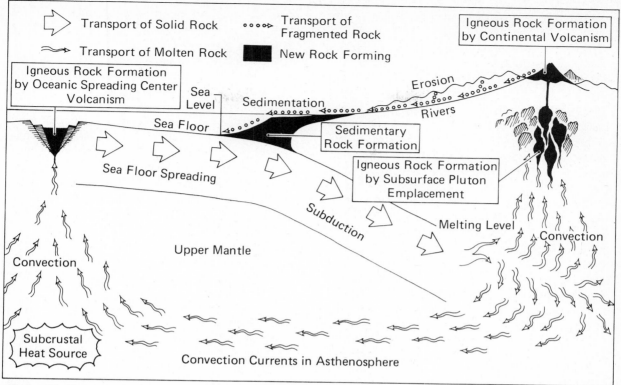

Figure 11.13
This diagram displays the flow of material in the rock cycle. The diagram represents a vertical plane cutting through the earth from its surface to the asthenosphere in the upper mantle. The circulation is driven by subcrustal heat sources and gravity. Arrows show the transport of rock material in different forms. Dark areas represent locations in which new rock masses are formed. Metamorphic rocks are produced by partial melting and recrystallization of older rocks in subduction zones and regions of pluton formation. Blank areas consist of older rock previously formed by one of the processes indicated in the dark symbol.

ward in the crust to become a pluton of igneous rock. The process of metamorphism, then, is transitional to complete melting.

Near the centers of all the continents are rigid areas of very ancient *crystalline* rocks (rocks clearly displaying mineral crystals: prin-

cipally, granite, gneiss, and schist). These ancient rocks may be exposed at the surface, as around Hudson Bay in Canada and in New York's Adirondack Mountains, or they may be covered by a thin blanket of younger sedimentary rocks. These oldest portions of the continents are known as *crystalline shields*. The rocks of the crystalline shields usually have been subjected to extreme metamorphism, and many of the plutonic rocks appear to have been created by the melting of even more ancient sedimentary rocks. The crystalline shields represent the deep roots of ancient mountain systems that were erased long ago by the processes of erosion. They contain the earth's oldest known rocks (about 3.8 billion years in age), all of which are metamorphic types exposed where erosion has removed thicknesses of tens of kilometers of overlying younger rocks.

The Rock Cycle

During the course of time, rock materials pass from one form to another in the *rock cycle* (Figure 11.13). Our understanding of the rock cycle has been greatly advanced by the discovery of sea floor spreading and the crustal subduction process. We have long known that most of the rocks exposed at the earth's surface are sedimentary types composed of the debris of older rocks of all types. We have seen that, despite the nearly 5-billion-year age of the earth, all the ocean floors are younger than 200 million years. The older oceanic crust is swallowed by subduction. This material descends deep into the lithosphere and becomes metamorphosed. At depths of 70 to 90 km (45 to 55 miles) it begins to be melted into new magma that rises into the continental crust in the form of igneous intrusions. Some of this igneous material erupts onto the surface as volcanic ash and lava. Erosion of the surface ash and lava and the metamorphic and plutonic rock below it provides new sediments. These sediments are gradually transformed into new sedimentary rock. In time this may be subducted, metamorphosed, melted, and recycled as new igneous rock. In this rock cycle the same atoms that have been part of the earth since its formation are used over and over to create generation after generation of rock material.

Differences in the rock types resulting from the rock cycle play a very important role in the appearance of landscapes. We shall explore this topic in Chapter 14. But first it is necessary to look at the landscape-making processes that are at work on these rocks. The next chapter is an introduction to the subject of landforms.

SUMMARY

Without disturbances by forces within the planet, the earth's surface would soon be worn smooth by its vigorous processes of erosion. Our planet's internal energy is most evident in the form of earthquakes and volcanic eruptions. However, the major differences in elevation on the earth are produced by very small crustal movements repeated over periods of millions of years.

The earth's relief features may be divided into three scales of magnitude. The largest-scale features, the first order of relief, are the continents and ocean basins. These are not permanent, but change in form and position over millions of years. The active elements are the sea floors, which are pushing outward from the oceanic ridge systems. The continents are embedded in the denser material that forms the ocean basins and are dragged along by the lateral motion of this material. This motion is thought to be produced by convection in the earth's mantle. The second order of relief consists of major continental and oceanic mountain systems and related large-scale relief features. The motions of the earth's crust related to sea floor spreading and the subduction of crustal plates produce these features. The third order of relief consists of individual landforms that are generally produced by processes of erosion and deposition.

Landforms are themselves composed of rock or rock debris. All rocks are composed of minerals, which are natural combinations of chemical elements in solid form. The most common rock-forming minerals are the silicates and the oxides, in which silicon and oxygen combine with other chemical elements.

Rock originates in several ways. Igneous rock crystallizes from a molten condition. It is generated as fluid magma far below the earth's surface and is either intruded into the crust or extruded at the earth's surface. Sedimentary rock is formed by the lithification of deposits of loose rock debris and organic matter, both on the land and on the sea floor. Metamorphic rock is a product of the recrystallization of older rocks that have been subjected to great heat and pressure deep within the earth's crust. The rock cycle is an endless sequence of material transformations, in which the same atoms are used over and over to form new rock material from older rocks.

APPLICATIONS

1. Poorly-informed persons periodically assert that California will someday "fall off" during an earthquake. Why is this impossible?
2. On an outline map of the world indicate the locations of offshore oil fields on the continental shelves.
3. Make a collection of the rocks present in the county in which your campus is located. What does each rock specimen tell you about the geological environment in which it was formed? Do any of the rock types create distincive landforms? If your campus is in an area of glacial deposition, see how many different rock types you can collect and determine the relative proportions of local and far-traveled rocks. Be aware that an occasional diamond has been found in glacial deposits in the midwestern United States!
4. What kinds of geological materials are quarried or mined in your county or state? In a state almanac or annual geological survey report, look up the annual value of your state's mined or quarried products. Are you surprised about the values of any of the various materials?
5. Geological studies show that in some areas of lithospheric plate convergence, volcanic island arcs (like Japan and the Philippines) have been "welded" onto a continental margin. Draw a series of cross sections through an island arc resulting from subduction, showing how an island arc distant from any continent could become welded onto a continent.

FURTHER READING

Glen, William. *Continental Drift and Plate Tectonics.* Columbus, Ohio: Merrill (1975), 188 pp. A concise and well-illustrated account of the essentials of plate tectonic theory.

Hamblin, W.K. *The Earth's Dynamic Systems.* Minneapolis: Burgess (1978), 470 pp. Extremely well-illustrated treatment of rocks, geological structures, and landforms. Notable for use of LANDSAT images.

Hurlbut, Cornelius S., Jr. *Minerals and Man.* New York: Random House (1970), 304 pp. Hurlbut's informative book on the occurrence, characteristics, and uses of minerals is renowned for its superb color photographs of mineral specimens.

Sorrell, Charles A. *Rocks and Minerals: A Guide to Field Interpretation.* New York: Golden Press (1973), 280 pp. An extremely handy guide to rock and mineral identification. The first seventy pages are an excellent introduction to minerals and rocks. Exceptional illustrations.

Sullivan, Walter. *Continents in Motion.* New York: McGraw-Hill (1974), 399 pp. The story of the development of the theory of plate tectonics and sea floor spreading, presented as the unraveling of a scientific mystery rather than in textbook fashion.

Wilson, J. Tuzo. *Continents Adrift and Continents Aground.* San Francisco: W.H. Freeman (1976), 230 pp. This is a collection of seventeen superbly illustrated articles on plate tectonics originally published in *Scientific American.*

Windley, Brian F. *The Evolving Continents.* New York: Wiley (1977), 385 pp. This advanced treatment outlines the geological development of all the continents in terms of plate tectonic theories. Very well illustrated with diagrams and cross sections that demonstrate the great variety of lithospheric plate interactions.

Wyllie, Peter. *The Way the Earth Works.* New York: Wiley (1976), 296 pp. Excellent thoughtful account of the earth's geological environment. Notable for its text rather than its illustrations.

CASE STUDY: **Earthquakes**

A little after 02:00 [2 A.M.] on December 16, the inhabitants of the region suddenly were awakened by the groaning, creaking, and cracking of the timbers of their houses and cabins, the sounds of furniture being thrown down, and the crashing of falling chimneys. In fear and trembling, they hurriedly groped their way from their houses to escape the falling debris. The repeated shocks during the night kept them from returning to their weakened and tottering dwellings until morning. Daylight brought little improvement to their situation, for early in the morning another shock, preceded by a low rumbling and fully as severe as the first, was experienced. The ground rose and fell as earth waves, like the long, low swell of the sea, passed across the surface, bending the trees until their branches interlocked and opening the soil in deep cracks. Landslides swept down the steeper bluffs and hillsides; considerable areas were uplifted; and still larger areas sank and became covered with water emerging from below through fissures or craterlets, or accumulating from the obstruction of the surface drainage. On the river, great waves were created which overwhelmed many boats and washed others high upon the shore, the returning current breaking off thousands of trees and carrying them into the river. High banks caved and were precipitated into the river; sandbars and points of islands gave way; and whole islands disappeared. (*Earthquake History of the United States*, 1973)

Obviously, this report must be from fault-shattered California or some other part of the tectonically-active mountainous west of North America. Or is it? In fact, the site of this description is New Madrid, Missouri—then a community of log cabins next to the Mississippi River—the date: December 1811. Further shocks occurred there in January and February of 1812. Reports of similar events are also available from Quebec (1638) and Charleston, South Carolina (1886), with only somewhat less catastrophic effects having occurred again in Missouri (1895), western Texas (1931), western Ohio (1937),

and southern Illinois (1968), to name only a few normally placid geologic settings that have experienced sudden seismic shocks.

Unlike volcanic eruptions, which have restricted geographic distributions, earthquakes can occur anywhere. Furthermore, unlike most other environmental hazards, they occur without warning. By their direct and indirect effects, including fires, flooding, landslides, and so-called "tidal waves," earthquakes have caused more destruction and taken more lives than any other single type of natural catastrophe.

What are earthquakes? The slow distortion of rock masses by tectonic stress causes a build-up of potential energy in the rock much as in a compressed or extended spring. When the deformation, or strain, surpasses the elastic limit of the rock, it ruptures suddenly or slips along preexisting faults, releasing the stored energy in the form of seismic shock waves that speed outward from the point of rupture. Several types of shock waves are involved, including rapidly moving body waves of varying character that pass through the mass of the earth and slower surface waves that follow the earth's outer skin. The surface waves create the ground motion we call an earthquake. The point below the earth's surface at which the energy is released is the earthquake *focus*, or *hypocenter*. The point on the earth's surface directly above the focus is the earthquake *epicenter*. It is at the epicenter that shock waves are most strongly felt.

The type of ground motion produced and the damage occurring during an earthquake vary according to the depth of the focus, distance from the epicenter, and nature of the local geologic material. In a small tremor, the sensation produced ranges from a gentle horizontal shaking to one or more sharp jolts. Near the epicenters of large earthquakes, one has the feeling of being in a rough sea in a small boat, with the ground visibly rising and falling in moving crests and troughs similar to ocean waves. Normally, the shaking lasts no more than a few minutes. However, smaller aftershocks, sometimes numbering in the hundreds, may continue over a period of days to months, indicating continuing subsurface rock displacements.

The violence of an earthquake is measured on two different scales that specify the magnitude and surface effect of the energy released. The magnitude scale devised in 1935 by Charles F. Richter of the California Institute of Technology is a logarithmic one in which each successively higher unit represents about 32 times the energy release of the preceding unit. Thus the difference in energy released by earthquakes of magnitudes 4 and 8 on the Richter scale is not 2 times, but more than a million times. The largest earthquakes felt since the invention of *seismographs* (instruments that record earthquake intensity) have attained magnitudes of about 8.9 on the Richter scale. Any earthquake of magnitude 8 or above is considered catastrophic.

The earthquake intensity scale, which measures the surface effects of earthquakes, was developed by the Italian geophysicist G. Mercalli in 1905. The Modified Mercalli scale of 1931, which is now in general use, rates these effects from I to XII (see table). There is no correspondence between the Richter and Mercalli scales because earthquake intensity at any moment can vary from point to point with no change in earthquake magnitude, due to the manner in which the local geologic material transmits seismic shock waves. For example, there is normally less ground motion on solid rock near the earthquake epicenter than on unconsolidated alluvium or artificial land fill some kilometers from the epicenter. Unconsolidated deposits

magnify the ground shaking and often collapse, behaving almost as a fluid in severe earthquakes. This phenomenon was conspicuous at Lisbon in 1755, San Francisco in 1906, Anchorage, Alaska, in 1964, and in the New Madrid earthquakes of 1811–1812, in which there was catastrophic shaking and rupturing of the ground on the Mississippi River floodplain but little effect in the adjacent bedrock hills. Artificial land fills are a ubiquitous part of our expanding urban scene, especially in hilly and coastal regions. Fortunately, North America has experienced few severe earthquakes since the explosive expansion of cities and their suburbs in the present century. Unfortunately, every day brings us closer to the first great earthquake that will strike one of our modern urban centers, as has recently befallen Managua, Nicaragua (1972), Beijing (Peking), Tangshan, and Tianjin (Tientsin), China (1976), and Guatemala City, Guatemala (1976).

Earthquakes can be anticipated in areas overlying subduction zones and lithospheric plate boundaries, and also along the oceanic ridge system. In all of these locations, great masses of rock are being moved past one another as part of the process of sea floor spreading. We may expect repeated news of earthquake disasters from Peru, Chile, Central America, California, Alaska, Japan, the Philippines, New Zealand, Indonesia, the Mediterranean, Turkey, Iran, the Himalayan foothills, Burma, and China. But some of history's most devastating earthquakes cannot be explained in terms of the gross motions of lithospheric plates—for example, those of the Mississippi Valley, Charleston, and Quebec, all apparently in the middle of the North American plate. And there is the greatest earthquake in European history, the Lisbon quake of 1755, which in 6 minutes took hundreds of thousands of lives and destroyed castles, fortresses, cathedrals, and mosques all over Portugal, southern Spain, Morocco, and Algeria. The epicenter of this earthquake was apparently offshore in the Atlantic, in a region where there is no major tectonic feature to explain such a geologic cataclysm. In time, geologists will uncover the reasons for earthquakes in such areas. For example, recent evidence suggests that the Mississippi Valley, a midcontinental area of frequent seismic activity, may be an ancient continental rift similar to those presently containing the Red Sea and the East African lake system. Several types of evidence suggest that this billion-year-old rift, which is now buried by later sediment accumulations, has been reactivated by the intrusion of mantlematerial at depth, followed by isostatic adjustments that may be triggering earthquakes.

While ground motion during earthquakes causes enormous damage, subsidiary effects may be equally dangerous. The enormous fires that are triggered by earthquakes—as in San Francisco in 1906 and Tokyo in 1923—may be more destructive than the ground shaking. The New Madrid earthquakes caused changes in elevation of 2 to 6 m (6 to 20 ft) over an area of 100,000 sq km (40,000 sq miles), creating temporary waterfalls in the Mississippi River and resulting in permanent submergence of 600 sq km (250 sq miles) of virgin forest by groundwater outflow and surface stream flooding. In coastal and undersea regions, earthquakes resulting from sudden movement along faults produce seismic sea waves, or tsunamis (see Chapter 16), popularly called tidal waves, which have swept over populous coastal regions, causing enormous loss of life. Earthquake tremors may set up resonance effects in large and small water bodies, causing them to slosh back and forth rhythmically, often with destructive effects on docks and ships at anchor. During the Lisbon earthquake of 1755, these resonance effects, termed *seiches*, were observed as far away as Switzerland, the British Isles, and Scandinavia.

Modified Mercalli (MM) Intensity Scale of 1931

I   Not felt except by a very few under especially favorable circumstances.

II   Felt only by a few persons at rest. Delicately suspended objects may swing.

III   Felt quite noticeably indoors. Standing motor cars may rock slightly. Vibration like passing of truck.

IV   Felt indoors by many, outdoors by few. Dishes, windows, doors disturbed; walls make cracking sound. Sensation like heavy truck striking building.

V   Felt by nearly everyone. Some dishes, windows, etc., broken; a few instances of cracked plaster; unstable objects overturned. Disturbances of trees, poles, and other tall objects sometimes noticed.

VI   Felt by all, many frightened and run outdoors. Some heavy furniture moved; a few instances of fallen plaster or damaged chimneys.

VII   Everybody runs outdoors. Damage slight to moderate in well-built ordinary structures; considerable in poorly built or designed structures; some chimneys broken.

VIII   Damage considerable in ordinary substantial buildings, with partial collapse; great in poorly built structures. Fall of chimneys, factory stacks, columns, monuments, walls. Heavy furniture overturned. Sand and mud ejected in small amounts.

IX   Damage great in substantial buildings, with partial collapse. Buildings shifted off foundations. Ground cracked conspicuously. Underground pipes broken.

X   Some well-built wooden structures destroyed; most masonry and frame structures destroyed; ground badly cracked. Rails bent. Landslides considerable from river banks and steep slopes. Shifted sand and mud. Water splashed (slopped) over banks.

XI   Few, if any, masonry structures remain standing. Bridges destroyed. Broad fissures in ground. Underground pipelines completely out of service. Earth slumps and land slips in soft ground. Rails bent greatly.

XII   Damage total. Practically all construction damaged greatly or destroyed. Waves seen on ground surface. Objects thrown upward into the air.

Source: Bolt, Bruce, et al. 1975. *Geological Hazards*. New York: Springer-Verlag, p. 9.

Almost all earthquakes trigger landslides, ranging from gigantic rockslides in mountain regions to stream bank cavings along flat floodplains. While the latter may create destructive waves in rivers, the former have often obliterated entire communities. The most tragic recent example occurred in Peru in May 1970, when an earthquake in the Peru-Chile oceanic trench produced an ice and rock avalanche from glacier-clad Mt. Huascaran in the Peruvian Andes that completely entombed one moderate-size town and a large portion of another, costing the lives of 18,000 people.

Landslides, devastating in themselves, often create still further dangers. In 1959 an earthquake sent a great rockslide crashing into the Madison River Canyon in Montana, just outside Yellowstone National Park. While this slide snuffed out the lives of 26 campers, an even greater hazard was created, as the slide blocked the Madison River, causing it to back up in a rapidly rising lake. Had the lake risen to the top of the landslide dam and spilled over, it would have washed away the dam, producing a flood of cataclysmic proportions in the valley beyond. Men and heavy earth-moving equipment were rushed to the scene to labor around the clock to excavate a controlled spillway for rising Earthquake Lake. Fortunately, the work was finished in the nick of time, and the second disaster was averted. The highest flood crest ever recorded resulted from the washout in 1895 of a landslide dam 274 m (900 ft) high in the Himalayan foothills in India; the resulting flood flow was 73 m (240 ft) deep.

It would be no comfort to victims of earthquakes, volcanic eruptions, hurricanes, and tornadoes to be told that each of these phenomena is the result of processes that help to maintain equilibrium in the earth's intertwining physical systems, but such is indeed the case. Some disequilibrium is building even as you read this, and before many months have passed, it will be compensated by a seemingly violent event that will take lives, destroy property, and be recorded in newspaper headlines as another natural catastrophe.

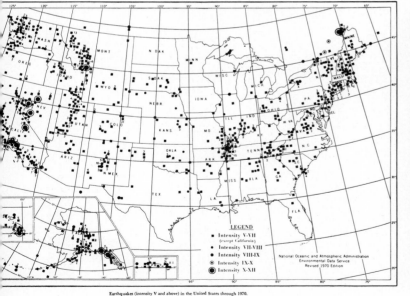

Earthquakes (intensity V and above) in the United States through 1970.

Historic earthquake epicenters in the United States are recorded on this map according to their intensities on the Modified Mercalli scale. It can be seen that no region of the United States is safe from earthquake damage. When looking at such a map, one must imagine a halo of destruction around each epicenter, often covering thousands of square kilometers. The Mississippi Valley earthquakes of 1811–1812 were felt over some 2,600,000 sq km (1,000,000 sq miles) to distances as remote as Washington, D.C., with chimneys being toppled as far away as Pennsylvania, Ohio, Georgia, and South Carolina. (Vantage Art, Inc., from *Earthquake History of the United States*, 1973, U.S. Dept. of Commerce, National Oceanic and Atmospheric Administration)

Winter Landscape from the Ch'ing Dynasty. (Courtesy of the Smithsonian Institution, Freer Gallery of Art, Washington, D.C.)

Towering pillars of limestone in southern China are an example of a unique landform produced by a particular combination of weathering processes and rock structures. Many different factors give landforms their characteristic shapes.

CHAPTER 12
Landforms

In addition to having oceans, unique climates, and a living biosphere, the earth is unlike all other planets in our solar system with respect to surface form. No known planet has scenery like the earth because no other planet in our solar system has a similar atmosphere, or hydrosphere, or such a constantly active interior. No other planet has been intricately sculptured by erosional processes, or has the variety of rock types, or geological structures, or surface forms seen on the earth. No other planet has had its early history so thoroughly effaced by continuing dynamic processes that constantly change the planetary surface.

The form of the land surface is the most visible and certainly one of the most influential of all aspects of the human environment. Thus it is an important area of study within physical geography. The variable form of the earth's surface, like other aspects of our environment, can be explained in terms of energy and materials. As we have seen in earlier chapters, the materials are rocks and minerals of various types and water in all its forms. The energy sources that affect these materials are solar radiation, gravitational force, and the heat produced by radioactive decay of unstable elements deep within the earth.

Solar radiation drives the atmospheric circulation, which produces rain, snow, and wind. Surface water and ice, moving downslope under the influence of gravity, pick up and transport rock fragments produced by weathering. The result is a constant shifting of fragmented rock material from high to low places on the earth's

Figure 12.1
Landforms of the type shown here are third order relief features. Such features are produced by a variety of erosional and depositional processes. The valley occupied by the stream was created by river erosion and enlarged by glacier scouring.

In the right background is the summit of Mt. Rainier, a volcano in the Cascade Range in Washington—a constructional landform that is composed of lava and volcanic ash. The peak to the left is on a rim created by the collapse of an earlier summit of the volcano. In the middle distance are moraines, ridgelike deposits left by the Emmons glacier when it was much larger than at present. In the foreground are cobbles washed out of the glacier by meltwater. The larger boulders and some of the cobbles were brought here by a rockslide in 1963. The rockslide covers the surface of the Emmons glacier in the center of the photo. (T. M. Oberlander)

surface. This tendency for high places to be worn down and low places to be filled in is known as *gradation*. If our planet had no internal energy, gradation would eventually produce a smooth earth. But energy within the earth constantly disturbs the earth's surface—creating new highs and lows and new work for the gradational processes.

There is unmistakable evidence that the form of the earth's surface changes constantly. Sometimes the changes are easily seen, as when a severe flood alters a river channel or a landslide tears away a hillside. Other changes are too slow to be visible over a human lifetime. Valleys deepen and expand. Ocean waves gradually gnaw away the edges of the land. Our hills seem everlasting, but after each spring thaw or heavy

Figure 12.2
The layers of sedimentary rock visible in this photograph of the Canadian Rocky Mountains have been subjected to intense deformation by movements of the earth's crust. The once-horizontal sedimentary rock strata have been compressed and folded by stresses caused by the motion of crustal plates. (Dr. John S. Shelton)

rain, streams are brown, yellow, or red with sediment eroded from the land. The hills, large and small, are being washed to the sea—slowly, grain by grain.

The remainder of this book focuses on the form of the land surface and on the processes that create individual relief features of the third order, which we call *landforms* (Figure 12.1). The study of landforms, known as *geomorphology*, draws on many of the facts presented in the earlier discussions of energy, the hydrologic cycle, climate, vegetation, and soils—all of which influence landform development. In turn,

terrain features affect other physical systems. They determine surface elevation, causing orographic rainfall and producing rainshadows. They influence the nature and amount of surface and subsurface water supplies. They affect

LANDFORMS

Figure 12.3
Water acting as an agent of erosion and deposition affecting the varied materials and structures of the earth's crust has generated widely different landforms, as these examples show.

(a) Motueka River valley in New Zealand was carved by a river, then filled with sediments washed out of a melting glacier. The present sinuous river slowly changes its course by lateral migration, alternately eroding and depositing a veneer of sediment and producing a wide floodplain between the valley walls. (G. R. Roberts, Nelson, N. Z.)

(b) Water acting on these sedimentary rocks in Red Rock Canyon, California, has cut a variety of shapes and patterns in layers of varying resistance to erosion. (Florence Fujimoto)

(c) Surface runoff and streamflow are carving a dendritic, or branched, drainage pattern into this landscape north of Christchurch, New Zealand. The relatively uniform resistance of the underlying rock favors the development of a treelike drainage pattern as each of the side tributary branches extends by headward erosion away from the main stream channels. (G. R. Roberts, Nelson, N. Z.)

(d) Half Dome, a famous natural feature in Yosemite National Park, California, was originally a complete granite dome, probably shaped by *exfoliation*. When granite is exposed at the surface by downward erosion, the confining pressure on it is released, and concentric sheets split off, or exfoliate, from the main mass. A later rockslide removed half of the dome, with the debris being carried away by a glacier. (David Miller)

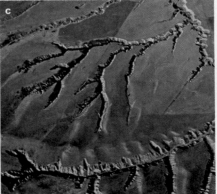

302

(e) This natural arch of sandstone is an erosional remnant formed by collapse and by the action of water, even though it is located in the arid southeast of Utah where the annual precipitation is only 20 cm (8 in.). The lower layers of rock weather more rapidly than the upper layers due to greater dampness at the base of the rock exposure. (David Miller)

(f) The attack of waves at the base of these limestone hills near Dorset, England, has removed the support for the overlying rock. The entire face has become a nearly vertical cliff because of removal of rock from the foot. (G. R. Roberts, Nelson, N. Z.)

(g) The rock spires in Bryce Canyon National Park, Utah, are products of water erosion of weakly consolidated sedimentary rocks. They exhibit a step-like structure because the rock layers have different resistances to erosion. (David Miller)

(h) Pillars have been cut into the soft pumice slopes of an ancient collapsed volcano at Crater Lake, Oregon. Such pillars are formed by rainfall on erodible material containing boulders that prevent portions of the surface from being eroded downward as rapidly as unprotected portions. (Florence Fujimoto)

LANDFORMS

the local character of vegetation and soils. Thus landforms are an active part of the physical environment.

This chapter considers the factors that cause landforms to differ, discusses the processes affecting slopes, and concludes with some general theories of landform development. Subsequent chapters will analyze the landforms created by specific geomorphic processes and geologic structures.

THE DIFFERENTIATION OF LANDFORMS

The limestone hills accurately portrayed in Chinese paintings, such as the one at the beginning of this chapter, certainly have a different appearance from hills made of the same rock in Kentucky or Missouri. Why is this? And why are landforms in widely separated parts of the world similar to one another but different from features only kilometers away? Four factors are involved: (1) the *geologic structure*, which includes the type and arrangement of the material composing the landform; (2) *tectonic activity*, the presence or absence of local movements of the earth's crust; (3) the *gradational processes* that are active; and (4) the amount of *time* the tectonic and gradational processes have operated in a particular way. These factors are considered in detail in subsequent chapters. We shall touch on them only briefly at this point.

Geologic Structure

Geologic structure refers to the physical nature and geometric arrangement of the rock materials from which landforms are built up, or in which they are sculptured by erosion. Geologic materials are extremely diverse, ranging from blankets of dust that have settled from the atmosphere to seemingly bottomless masses of uniform hard rock that have crystallized from subsurface magma. Some materials are chemically and physically simple; others are extremely complex. They may be arranged in layers or in uniform masses. Layered materials may be horizontal, tilted, or wrinkled (Figure 12.2); massive rocks may be shattered by fractures. The wide range of geologic structures affecting landforms is important enough to merit separate treatment in Chapter 14.

Tectonic Activity

The crustal motions that constantly retard gradation are known as *tectonic activity*, or simply *tectonism* (from the Greek word *tekton*, meaning "builder"). Tectonic activity is a reflection of the earth's restless interior, which seems to include slow "boiling" motions related to uneven heat production by radioactive decay processes. Tectonism influences scenery in two ways: by broadly elevating and depressing the earth's crust, and by creating the details of geologic structures that affect landscape development (Chapter 14). Because the geologic structures created by tectonic activity are attacked by erosion even as they are being formed, these structures are rarely seen intact. But it is tectonic uplift that provides the potential energy that stimulates this erosion. Most areas that are rough or mountainous are experiencing tectonic uplift at the present time. Areas that have little relief are either sinking and filling with sediments or have been free of vertical motion for many millions of years.

Gradational Processes

Whereas tectonic activity tends to cause the earth's surface to be uneven, gravity tends to smooth the surface. It does this by providing the energy for flows of water, ice, and loose rock material in the processes of gradation. These processes include the fragmentation of rocks by weathering, the detachment and removal of rock fragments by the agents of erosion, and the deposition of rock debris at lower elevations.

The chief physical agent of gradation is water

(Figure 12.3). It is the main factor in the break-up of rock by chemical and mechanical weathering, and the principal mechanism by which rock debris is moved from high to low elevations. While channeled flows of running water are the most widespread single agent of erosion and deposition, other important gradational agents are the direct force of gravity, wind, glaciers, and water waves and currents in lakes and seas. Any one process may be dominant over vast areas, producing a distinctive imprint upon the landscape. Different processes will often work jointly, or alternate in dominance through time.

Usually climate determines which of the gradational processes is most active in an area. Changes in the type or intensity of gradational processes can generally be traced to a climatic change. An example is the periodic cutting and filling of gullies in the arid southwestern United States, widely thought to reflect fluctuations in rainfall intensities over periods of years. A gully-cutting episode after about 1200 A.D. is thought to have destroyed the agricultural system of the Pueblo Indians who inhabited the impressive cliff dwellings of Arizona, New Mexico, and southwestern Colorado, causing the Indians to abandon their homes permanently. This particular episode is but one of a continuing sequence of gully cutting and filling in the region,

whose erosional systems appear to be very sensitive to climatic fluctuations.

Climate also affects gradation through the influence it exerts on the vegetative cover (Figure 12.4). Vegetation is the dominant factor controlling the rate of erosion by water and wind. A dense vegetative cover holds soil in place, breaks the impact of wind and rain, and maintains a surface that absorbs rainfall rather than allowing it to run off and erode the land.

On hillslopes not protected by vegetation,

Figure 12.4
These views illustrate variation in slope form in physically similar interbedded shales and sandstones in two different climatic regions.

(left) In California's Coast Ranges, the grass cover developed in response to an annual rainfall of about 50 cm (20 in.) protects a thin soil, increases moisture infiltration into the ground, and reduces erosive surface runoff.

(right) In southern Utah, a rainfall of about 25 cm (10 in.) is insufficient to produce vegetative protection of slopes, resulting in intricate gully erosion and exposure of bare rock ledges that shed coarse waste onto the slopes. The nature of the rainfall regimes in the two areas is also a factor. In California, most of the precipitation falls in low-intensity drizzles associated with the passage of winter frontal systems. In southern Utah, summer thunderstorm activity produces brief episodes of high-intensity precipitation on bare ground. (T. M. Oberlander)

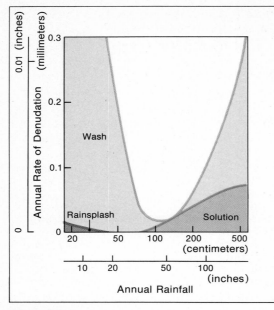

Figure 12.5
This diagram illustrates how the average rate of surface lowering by erosion, known as the rate of denudation, is affected by precipitation and vegetation. The example considered is a 10° slope in a warm region where the average annual temperature is 25°C (77°F). Soil wash by surface runoff is the most important process of denudation. The rate of lowering is least where the annual precipitation is about 100 cm (40 in.) because in such regions vegetative protection of the soil surface reduces the erosive impact of rainfall and the increase in soil permeability due to organic activity allows little surface runoff to be generated. The rate of denudation is greater in drier regions because the vegetation is sparse and surface runoff readily erodes the exposed bare ground. Rainsplash on bare ground also plays a role. In very moist regions, more moisture is available than the soil can absorb, and large moisture surpluses are generated. The abundant runoff of such regions carries away material in solution as well as by soil wash. (Doug Armstrong after M. G. Wolman)

loose material produced by weathering is periodically flushed away by water runoff. This prohibits soil development and generates a gullied, rocky landscape. The erosional loss from bare

soil is 50 to 100 times greater than that from a grass-covered area. Similarly, hillslopes covered by grass are eroded many times faster than hillslopes covered by temperate forest vegetation. But in the humid tropics the intensity of rainfall overcomes the protective effect of vegetation, and the rate of erosion rises despite a dense vegetative cover (Figure 12.5). Human activity has increased erosion rates in all environments, mainly as a result of direct or indirect modification of vegetation (Figure 12.6).

Time

Where vast areas of hard rocks have been eroded to very low relief, as in much of Africa and Australia, it is clear that the passage of time is reflected in the landscape. In the absence of tectonic activity, the effect of time is to permit gradation to reduce landscapes to monotonous plains.

We may also think of the effect of time in terms of response times and relaxation times, as described in Chapter 2. Some types of landforms respond very quickly to changes in geomorphic processes. River channels and beaches are examples. River channels enlarge rapidly during floods and then return to their pre-flood form when floodwaters recede. Beaches become steeper and diminish in size when storms send large waves against the shore and rebuild quickly after the storm subsides. Such adjustments, or responses, tend to reestablish equilibrium between the landform and the changed geomorphic process. The time required for restoration of equilibrium is the relaxation time. In many quick-response landform systems, the inputs of energy and materials fluctuate continuously. As a result, form changes are more or less continuous, with equilibrium being repeatedly approached but never actually attained.

In cases where the response and relaxation times are very long, landforms may preserve "leftover" features resulting from processes that are no longer active. These are called *relict* forms. The clearest examples are the landforms

Figure 12.6
This view illustrates landscape transformation in the humid climate of
Tennessee due to removal of the original vegetation cover. The area
formerly supported a mixed hardwood and pine forest. Deforestation
originally occurred to provide fuel for a copper smelter built in 1854.
Toxic fumes from this and subsequent smelters prevented vegetative
regeneration, keeping the land in a nearly barren condition for more than
100 years despite the region's abundant rainfall. An artificially established
grass cover offers some protection, but it has been largely removed by
gully expansion, as this view indicates. Although the locality receives
about 140 cm (55 in.) of rainfall annually, the landforms resemble those
seen in deserts. (T. M. Oberlander)

produced in alpine mountains throughout the
world by glacial erosion and deposition that
ended more than 10,000 years ago (Chapter 15).
To erase the evidence of glacial modification of
landscapes will require perhaps a million years
of gradation by other processes.

SLOPES: THE BASIC ELEMENT OF LANDFORMS

The chief problem in landform analysis is to
explain the conditions producing the different
types of slopes and flat surfaces that make up
the earth's landscapes. The views in Figure 12.7

(see pp. 308–309) provide a sample of the diversity of slope types found in different regions.

Most of the slopes we see around us were initiated by downward erosion by channeled flows
of running water. The same process that cuts
gullies in a sloping cornfield like that in Figure
10.17 (p. 264) has excavated the Grand Canyon
of the Colorado River. Stream incision provides
new vertical surfaces that are quickly transformed into slopes by other erosional processes.
These processes loosen material and move it
down the slopes and into streams, ever widening
the excavation made by vertical stream erosion.
Weathering of rock material, mass wasting, and
water-assisted erosion all play a part in the
shaping of slopes.

Figure 12.7
These photographs show the variety of slope types that result from varying geologic materials and climatically controlled slope-forming processes.

(a) Stone Mountain, Georgia, is a core of unjointed granitic rock that rises boldly above jointed, chemically weathered granite in the Piedmont region of the Appalachians. (Warren Hamilton/U.S.G.S.)

(b) Chemical weathering in jointed granite in arid eastern California creates a jumble of exposed rock. In the distance is the steep wall of the Sierra Nevada, its crest splintered by frost weathering; Mt. Whitney is in the center. (T. M. Oberlander)

(c) Smooth rolling slopes in Devon, England, result from chemical weathering of limestone, accompanied by slow downward transfer, or "creep," of the resulting mantle of soil. (G. R. Roberts, Nelson, N. Z.)

(d) Precipitous slopes have been cut in basaltic lavas in the upland areas of the Hawaiian Islands. Strong vertical incision by torrential streams during heavy rains is accompanied by rapid sliding of the water-saturated soils on steep vegetated slopes (red scar at left). (Butch Higgins)

Mass Wasting

Gravity provides the energy for all slope-forming processes. Gravity itself can cause the material composing hillslopes to move downward without the assistance of any moving fluid in the process known as *mass wasting*. Such movements range from sudden catastrophic rockslides to the slow downslope creep of soil and rock fragments over hundreds of years.

Slopes composed of soil and rock fragments remain in place due to the friction between solid particles, while solid rock is held together by the attractive forces between neighboring atoms. Without such cohesive forces, hillsides would collapse under the pull of gravity. Weathering of the rock forming a slope may occasionally reduce the cohesive forces to the point where rocks break loose or a portion of the hillside avalanches downward.

For any loose material on a slope, whether soil or a layer of rock debris, there is a maximum angle, known as the *angle of repose*, that the material can maintain without slipping downward. At the angle of repose, gravitational stress is just balanced by the cohesion of the material on the slope. If new material, or even water, is added to a slope that is already at the angle of repose, a portion of the slope may detach and slide downward. Wet soils are most likely to slide because the absorbed moisture lessens the friction between soil particles at the same time that it increases the weight of the soil.

At the bases of rock cliffs there are normally cones of debris consisting of loose rock fragments that have fallen from the face of the cliff (Figure 12.8, pp. 310–311). This material, called *talus* (large chunks) or *scree* (small particles that shift underfoot), accumulates at an angle of repose of 34 to 39 degrees. Talus slopes resulting from rock falls are present wherever there are rock cliffs but are most common in alpine areas and dry regions where bare rock exposures are most plentiful.

In most areas there are no cliffs, and mass wasting acts slowly and invisibly. The soil of ev-

Figure 12.8
(a) Weathered rock that is saturated with water has little cohesive strength, and sometimes large bodies of rock debris break loose and flow rapidly downslope. The enormous earthflow in the photograph occurred in Idaho, near the headwaters of the Pahsimeroi River, visible in the foreground. It has a characteristic lobed appearance, and spread into the valley before coming to rest. Slope failures are frequently found in semiarid regions where sudden heavy rains saturate the soil and where the vegetation cover is too sparse to help bind the soil. (Dr. John S. Shelton)

(b) The hillside scar shown in the photograph was produced in 1955 by a rockslide that moved tons of soil, rock, and vegetation into Emerald Bay on Lake Tahoe at the border of California and Nevada. (Warren Hamilton/U.S.G.S.)

(c) Frost weathering of well-jointed granite rock at elevations above 3,500 m (11,000 ft) in California's Sierra Nevada loosens rock masses and produces rock falls, particularly during the spring, when ice is melting. This creates cones of rock rubble, known as *talus*, at the base of frost-shattered cliffs. (T. M. Oberlander)

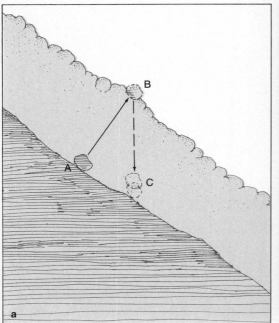

Figure 12.9
(a) Soil tends to creep downslope under the influence of gravity. This schematic diagram illustrates the mechanism responsible for soil creep. When soil becomes moist or freezes, it expands at right angles to the slope, but when the soil dries and contracts, the soil particles tend to move vertically downward, which results in a net downslope migration of the soil surface. (John Dawson)

    (b) The "needle ice" shown in this photograph, at much enlarged scale, is a mechanism that moves soil downslope. When water in damp ground freezes and expands outward, it carries soil particles with it at right angles to the surface. Soil particles tend to fall vertically downward or forward when needle ice melts, resulting in a net downslope displacement of the particles. (David Cavagnaro)

ery sloping pasture or forested hillside moves downhill a fraction of a centimeter per year (Figure 12.7c). This form of mass wasting is aptly known as *soil creep*. Soil creep on sloping ground often causes fence posts and tombstones to tilt conspicuously.

    The creep process is related to expansion and contraction of the soil by wetting and drying and freeze and thaw. Figure 12.9 shows how volume changes in the soil always result in slight downslope displacement of the soil mass. The rate of displacement is related to the slope angle but is generally less than a centimeter a year. The process of soil creep smooths hillslopes and causes their angles to become less steep through time.

A distinctive type of mass wasting is seen in subarctic and highland tundra regions where soils that are silty and moisture-retentive cover ground that remains frozen throughout the year. We have seen that in many tundra regions only the upper meter or so of the soil thaws during the spring and summer. The water released by thawing cannot drain downward through the still frozen subsoil. For a few days the water-saturated soil loses its cohesion. This allows the soil to sag slowly downslope. Different sections move at different rates, which makes the slope appear to be covered with overlapping scales or lobes of soil (Figure 12.10). This process is known as *solifluction* (soil flowage). Despite their active appearance, solifluction lobes seem to move only a few centimeters each year. Like

Figure 12.10
Solifluction lobes, shown here on a slope in Alaska, develop where silty, water-retentive soils cover permanently frozen ground. The lobes exhibit irregular margins because the rate of downslope movement of the thawed soil varies from point to point. (P. S. Smith/U.S.G.S.)

soil creep, solifluction fills in depressions and tends to smooth the landscape.

Slopes covered by silt or clay soils that have an unusual capacity to absorb rainwater often collapse suddenly in the form of *slumps*. Like solifluction lobes, slumps result from loss of cohesion and increase of weight in water-saturated soils. A slump is a rapid rotational movement in which a mass of soil slips downward and tilts backward, leaving a conspicuous scar on the slope. A hillside slump usually terminates in an *earthflow* consisting of a bulge of the collapsed soil that is pushed out and down the slope (Figure 12.11, p. 314). Slumps and earthflows are often destructive and can be triggered by human activity, such as lawn watering or slope steepening during construction of roads or buildings. Slumping also occurs along stream banks, especially during the late stages of floods when the banks are saturated with water.

The most awesome of all gravitational transfers of material are the immense *rockslides* that occur in mountainous country (Figure 12.8). In these slides as much as several cubic kilometers of rock may fall and "splash" outward surprising distances with incredible speed and destructive force. Nearly all rockslides are triggered by unusual wetting of weathered rock or by the sudden shock of an earthquake. Mountain regions are particularly prone to earthquakes, which are an expression of the geological movements that have raised the mountains. Mountains are also periodically drenched with moisture due to heavy orographic rainfall and spring melting of deep snow. If an earthquake occurs when weak rocks are saturated with moisture, rockslides are to be expected. Large rockslides have occurred in historic time in all the earth's high mountain regions, and slide scars and deposits are a part of all high mountain landscapes.

Water Erosion on Slopes

The effect of gravity as an agent of landscape change is greatly multiplied when it sets water

Figure 12.11
(above) A destructive slump and earthflow in Oakland, California, destroyed 14 homes. The structures visible have moved downslope from street level.

(below) A smaller, more typical slump and earthflow about 50 years old in California's Coast Ranges. Slump and earthflow phenomena are widespread in central California due to the presence of mechanically weak rock and heavy clay soils with high water-absorbing capacity. (T. M. Oberlander)

in motion. Slope erosion by running water probably removes 50 to 100 times as much material as mass wasting processes.

In many areas slope erosion commences with *rainsplash*. When large raindrops strike bare soil, their impact can splash soil particles several millimeters laterally (Figure 12.12). On slopes, more than half the particles splashed into the air land downslope from their previous location, thereby causing a net downslope shift of soil particles. Because vegetation and plant litter intercept rainfall, rainsplash erosion is most effective in dry regions where bare soil is widely exposed.

Where a cover of vegetation and soil exists, rainfall usually produces no surface runoff of water. Runoff normally occurs where rain falls on areas with poorly developed or nonexistent vegetation and soils and, occasionally, where the downpours on vegetated surfaces are torrential. In such instances, water builds up on the surface and begins to slide down any available slope, initiating *overland flow* of water. The initial water flow is in the form of a slowly moving sheet that has no erosive effect. However, as it gains speed, the sheet breaks into turbulent threads, or *rills*. Turbulent flow in rills is required to lift soil or rock particles into the flow, initiating erosion by the process of *slopewash*. Slopewash is the main process of slope erosion in regions lacking a vegetative cover.

On rare occasions during extremely heavy rains, flows of water receive so much soil and rock debris from slopes that a viscous stream known as a *mudflow* or *debris flow* results. These flows are intermediate between water flows and earthflows, being much higher in density than the former but far more fluid and faster moving than the latter. Mudflows are composed of fine particles, whereas debris flows contain more coarse material, including large boulders. Since both flows result from exceptionally heavy rains on unprotected soils, they occur where the vegetation is poorly developed, as in arid or alpine regions. They are also a hazard where fire has recently destroyed the vege-

314

CHAPTER 12

Figure 12.12
These close-up, high-speed photographs of a raindrop splashing into bare soil show graphically the work that can be done by water.

(left) The raindrop is 3 mm in diameter and is traveling at a speed of 11 meters per second.

(right) When the drop strikes the ground, particles of soil are thrown several centimeters by the impact. The impact of raindrops on a bare slope tends to produce a net movement of soil particles down the slope. (U.S. Navy Office of Information)

tation. Although these flows subside quickly, they may travel distances of several kilometers and may destroy roads, residences, and even whole towns.

Slopes that are vegetated, and even some bare slopes, show little evidence of rill erosion. Such slopes are able to absorb most rainfalls, which percolate into the soil. This soil moisture then moves downslope between soil particles as subsurface *throughflow*. Throughflow is capable of removing dissolved substances as well as fine particles that can filter between the coarser particles composing the soil. In this way well-vegetated slopes in humid areas can gradually be eroded, although runoff rarely is seen on the slopes themselves.

MODELS OF LANDFORM DEVELOPMENT

The interplay of diverse gradational processes, earth materials, and geological structures creates an enormous variety of individual landforms. These combine to produce a great number of general landscape types. To unify the study of landforms, geographers have sought to find some common principle in their development. Two general models of landform development stand out as being widely applicable. They focus on the effects of time and the equilibrium tendency in the development of landforms.

The Cycle of Erosion Concept

By the late 1800s, the theory of organic evolution presented by Charles Darwin in 1859 was finally gaining acceptance. At this appropriate moment, the American geographer William Morris Davis wrote a series of essays in which he proposed that landforms, like organisms, could be described and analyzed in terms of their evolutionary development through time. Thus Davis classified landforms according to their stage in a theoretical cycle of development, using the terms "youth," "maturity," and "old age." He then inferred the processes by which the landforms of each stage would slowly evolve into those of the next stage.

Davis focused first on the evolution of landscapes dominated by the effects of stream erosion. To explain his *cycle of erosion* in the simplest way, he visualized a sudden uplift of the land surface from a lower to a higher elevation, after which erosion begins to affect the uplifted mass. Rainfall and runoff on the raised area initiate erosion as streams flow down the newly created slope toward the sea. At first these streams would erode their beds downward rapidly, cutting narrow valleys with V-shaped cross-sections. As long as areas of the original uplifted land surface remained visible between the new valleys, the landscape would be in the *youthful* stage of landform development (Figure 12.13). In this stage, settlement and most human activity are located on the plateau-like surfaces between the narrow, steep-sided valleys.

As streams deepen their valleys, they decrease their altitude above the sea. This decreases their potential energy and, therefore, the kinetic energy available to produce erosional work. Finally, in "late youth" the streams become "graded," with no excess energy to convert

to the work of downward erosion. At this point valley deepening ceases. This permits slope gradational processes and lateral stream erosion to widen the valleys and create flat floodplains. In the stage Davis called *maturity*, the original uplifted land surface is converted entirely into hillslopes leading down to flat-floored valleys. In the valleys, streams meander back and forth over continuous floodplains veneered with stream-deposited alluvium. Since the only flat land is now in the valleys, this is where human activity is concentrated.

Further evolution from maturity to *old age* was assumed to proceed very slowly, requiring tens of millions of years. In the old age stage visualized by Davis, the uplifted surface has been worn down close to sea level, producing a lowland with faint relief, called a *peneplain* ("almost a plain"). Isolated areas of higher ground that are remote from the larger valleys are called *monadnocks*, after solitary Mount Monadnock in New Hampshire. Davis stressed that his simple model could include variations, such as further uplift during any of the intermediate stages in the cycle of erosion.

Figure 12.13 (opposite)
This sequence of diagrams illustrates stages of the Davisian cycle of landscape evolution. This cycle applies to moist regions where erosion is accomplished primarily by flowing water. The diagrams assume that the underlying rock is uniform and exerts no controls over landform development.

(a) The initial surface is a landscape of low relief. After uplift of the region, streams have energy to begin cutting downward.

(b) In the stage of youth, the streams have cut narrow valleys downward, and much of the initial surface is preserved between the valleys.

(c) In the stage of maturity, the uplifted area has been eroded into a mass of hills. The streams have widened their valleys, and little trace of the initial plain remains.

(d) In old age, mass wasting and flowing water have eroded the region to a plain of low relief (*peneplain*) with isolated hills, or *monadnocks*. (T. M. Oberlander)

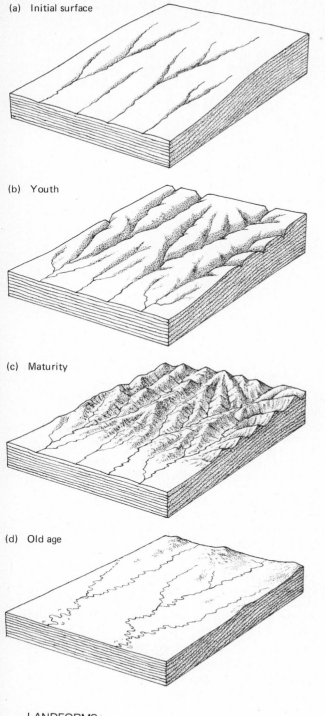

(a) Initial surface

(b) Youth

(c) Maturity

(d) Old age

Davis and his students devised other cycles of erosion for landscapes in which stream erosion was not the dominant process. There were cycles of evolution for deserts, coasts, glaciated areas, and regions of soluble limestone bedrock. In general, Davis viewed landforms as products of *geologic structure, geomorphic process,* and *stage of evolution.* Stage did not imply any fixed amount of time but referred only to the development of forms in a sequence. Form evolution is slow where rocks are resistant or processes are weak, and rapid where materials are weak and processes are vigorous.

The importance of Davis's work is that it provided an easily understood basis for the organization and classification of landforms and caused geographers to begin to focus on the relationships between different landforms. However, the cycle of erosion concept originated at a time when the processes by which landscapes are transformed had not been fully investigated. Today the cycle of erosion concept is regarded as being of value mainly as a way of introducing students to the idea of landscape change. The terms "youth," "maturity," and "old age," introduced into geomorphology by Davis, have been retained, but only as descriptive terms for stream-dissected landscapes. While the cycle of erosion concept is a useful introduction to landforms, it does little to increase our understanding of the relationship between geomorphic processes and the forms they produce.

The Equilibrium Theory of Landform Development

Since about 1950 geomorphologists have increasingly emphasized the actual mechanics of landform development. This approach focuses directly on the detailed relationships between form and process in the landscape. It emphasizes the tendency for the form of the land surface to keep the energy of the erosional processes in balance with the resistance of the material they affect, so erosion continues in a uniform manner with the least expenditure of work.

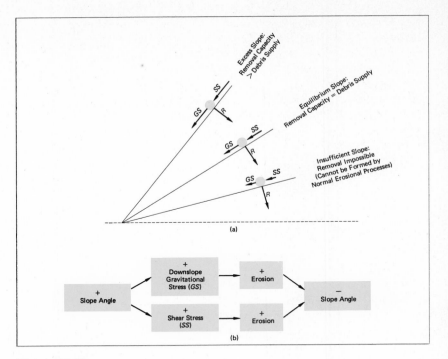

Figure 12.14
Principles of equilibrium slopes.

(a) Hypothetical slope angles indicate downslope gravitational stress *(GS)*, shear stress *(SS)* exerted by the gravitationally driven erosional agent (such as slopewash), and resistance to detachment of slope particles *(R)*. Resistance remains the same at varying slope angles, while *GS* and *SS* diminish with decreased slope. The steepest slope is unstable, as the sum of *GS* and *SS* far exceeds *R*, causing rapid erosional removal. The intermediate slope is stable, as the sum of *GS* and *SS* just balances *R*, so that removal can occur without changing the slope angle. The lowest slope cannot be produced by the stresses shown, as particle resistance *(R)* exceeds erosional stresses, so that erosional removal is impossible.

(b) Diagrammatic representation of a negative feedback relationship that causes slopes to be self-adjusting toward angles that equate stress to resistance. Any increase in slope increases the erosional stresses, which in turn decreases the slope angle until equilibrium is reestablished. (Vantage Art, Inc.)

Two forces cause material to be put in motion on a slope. One is *downslope gravitational stress*, which is proportionate to the slope angle. If you tilt a table, the objects on it slide off when the slope angle becomes steep enough. An object moves when the pull of gravity exceeds the friction between the object and the table top.

The other force that causes material to move is *shear stress*. This is the oblique downward and forward force exerted by one material rubbing against or flowing over another material. Shear stress is proportionate to the density and velocity of the moving material. Water, ice, or wind moving over any surface exerts shear stress on that surface and may cause loose particles on it to move. The shear stress exerted by water or ice is also related to slope angle, because water and ice are usually moving due to downslope gravitational stress. The magnitudes of both of these stresses, as well as the resistance of the material affected, can be expressed numerically and used in equations that explain or predict geomorphic processes and forms.

If erosion reduces the angle of a slope, it has the effect of reducing the downslope gravitational stress on the material forming the slope. It also reduces the shear stress exerted by the erosional agent. Then how far can erosion reduce a slope angle? Certainly not below the minimum that gives the erosional agent enough energy to continue to remove material from the slope.

On the other hand, any increase in slope angle increases the downslope gravitational force and the energy of the erosional agent. This produces more vigorous erosion, which tends to reduce the slope angle. This is an example of *negative feedback*, in which disturbance of a system that is in equilibrium triggers changes that tend to restore the original system (Figure 12.14). When a natural hillslope is made steeper

due to highway construction or some other artificial modification, a destructive slump may follow. This is the way the hillslope returns to its original equilibrium angle. Negative feedback is the principal means of maintaining equilibrium in physical systems and is a normal feature of many landform process-response systems.

Negative feedback causes many landform systems to be self-regulating, tending to maintain a steady state through time. In such a system, changes in form occur only when there is a change in the material or in the nature or intensity of the gradational processes.

Whenever the materials composing a slope vary in resistance, there are corresponding variations in the slope angle (Figure 12.15). Where rock material resists removal, slopes created by erosion will be steep to maximize downslope gravitational stress and sheer stress. Where ma-

Figure 12.15
Differential weathering and erosion due to variations in rock type are especially clear in arid regions. Here in Monument Valley, Utah and Arizona, massive sandstone forms vertical walls. The lower layers of thinly bedded sandstones and shales produce much gentler slopes, although even here the more resistant layers make small cliffs. (T. M. Oberlander)

terial is easily removed, slopes are gentler. Since steeper slopes tend to increase the erosional energy on resistant rocks, it is possible for the whole landscape, tough rocks and fragile ones, to wear away at the same rate over long spans of time.

The delicate adjustment of slope forms to the materials and gradational processes involved is important to understand. It is impossible to modify either landforms or geomorphic processes artificially without triggering a reaction in the natural system. This reaction tends to restore an equilibrium between process and form. We cannot make a change in a natural system and expect the system to remain passive; sooner or later it will respond with a change of its own to absorb the effect of the artificial change.

For example, straightening a river to combat flooding also shortens it and increases its slope in the shortened section. The river often responds to this disturbance by cutting downward in the straightened section. This restores its original slope in the disturbed section but leaves the upstream continuation of the channel "hanging," forcing adjustments there as well. The stream's downcutting produces sediment that may clog the channel downstream, increasing flooding there. Often these inevitable adjustments are unforeseen by those who make the initial change in the natural system.

Another aspect of equilibrium in landform systems is the balance between inputs and outputs of material in depositional landforms. Many depositional landforms persist only be-

Figure 12.16
One type of time-independent equilibrium landform. The maintenance of any beach, like this one at Cádiz, Spain, requires that sand arrivals balance sand losses. Seasonal fluctuations in the sand budget occur, but if there is no progressive change over the years, material income and outgo are in balance. (T. M. Oberlander)

cause periodic losses of material by erosion are balanced by arrivals of new material (Figure 12.16). Talus cones, beaches, river deltas, and volcanoes are a few examples. Such features enlarge or shrink as the balance between input and output changes. Any natural or artificial change in either the input or the output will produce a quick response in the feature, causing it to grow or shrink. But growth in one place means shrinkage somewhere else, for an output that is reduced always reduces the input to some neighboring system.

The value of the equilibrium concept of landform development is its focus on the exact relationship between geomorphic processes and surface forms. It is especially useful in assessing the impact of natural and artificial changes in landform systems. Finally, it can be described and analyzed mathematically. Consequently, the equilibrium model has largely replaced the cycle of erosion concept in landform analysis.

THE TEMPO OF GEOMORPHIC CHANGE

At this point it is appropriate to ask, how rapidly does the landscape change as a result of tectonic and gradational processes? In general, the maximum rates of crustal uplift are several times the most rapid rates of lowering of the land surface by erosion. Repeated precise surveys have shown that the rate of uplift in regions of mountain growth is about 5 to 10 m (15 to 30 ft) per 1,000 years. In mountain regions with steep slopes, the rate of sedimentation in reservoirs indicates that erosion is peeling away about 1 m (3 ft) each 1,000 years. Over very large areas of variable relief, the rate is much less. The rate of erosion over the entire area drained by the Mississippi River and its tributaries is only about 5 cm (2 in.) per 1,000 years. Even this is higher than the natural rate, since erosion in this region has been greatly augmented by human activities.

The landscapes we see around us are produced for the most part by small changes that occur frequently over vast spans of time. Mountain ranges are lifted a meter or less in a century or more. Though small in amount, and rare on the human time scale, mountain-building movements are frequent on the geological time scale in which we speak of years in terms of millions.

Some types of erosion and deposition do occur on a human time scale. Downward and lateral erosion by streams takes place mainly during floods that occur once every year or two. Measurements of the sediment loads carried by streams indicate that most sediment transport occurs during the five to ten days of greatest flood discharge each year. Only in the exceptional floods that occur at intervals of many years can streams completely clear out the sediment in their channels and erode downward into bedrock. Such events slowly change the landscape, but they are so uncommon that a greater quantity of work may be accomplished by smaller events that occur with greater frequency. It is suspected that this is a general principle in landform development.

SUMMARY

No known planet has landforms similar to those of the earth. This is largely because the other planets in the solar system lack liquid water, which is the main agent of the general process of gradation. Gradation is the constant transfer of rock material from high to low places on the earth's surface. The earth's internal energy keeps the planet from being totally smoothed by gradation. This energy is manifested in earthquakes, volcanic eruptions, and slow motions of the earth's crust.

The diversity of landforms on the earth's surface is a consequence of four general factors: geologic structure, tectonic activity, gradational processes, and the passage of time. Geologic structure includes both rock type and the geometry of rock masses. Tectonic activity creates geologic structures and causes the earth's crust to be elevated or depressed locally at varying rates. Gradation includes mechanical and chemical weathering of rock and the effects of agents of erosion that are set in motion by gravitational force. Water, in its various forms, is the chief agent of gradation. The type of gradational process is largely controlled by climate, and its intensity is often related to the vegetative cover. Where tectonic activity has ceased, long periods of time permit landscapes to be worn to very low relief. Response time and relaxation time refer to the disturbance of systems by changes in geomorphic processes and the subsequent restoration of equilibrium in the landform system.

Slopes of varying origins are the basic elements of landforms; therefore slope development is the basic concern of landform analysis. Mass wasting is the transfer of material down slopes by unassisted gravitational force. It includes soil creep, slumps and earthflows, solifluction, and rockslides. Slopes lacking vegetative protection are eroded by rainsplash and the overland flow of water, which initiates slopewash. Mudflows or debris flows are also initiated by water on bare slopes. Vegetated slopes are protected from surface erosion but lose material in the process of subsurface throughflow of water in the soil.

Two models of landform development have achieved popularity. The cycle of erosion concept originated by W. M. Davis suggested that landforms developed from stage to stage in an evolutionary cycle related to rock resistance and the passage of time. The terms "youth," "maturity," and "old age" were used to describe erosional landscapes of differing appearance. The more current equilibrium theory of landform development stresses the specific relationships between geomorphic processes and resulting forms. The emphasis is on the tendency of landforms to establish an equilibrium between erosional stress and material resistance. Many landform systems are self-adjusting through a negative feedback process, in which changes imposed on the system tend to be canceled by the response of the system. Artificial disturbances of equilibrium landform systems trigger responses that are predictable, but that often come as a costly surprise.

Strong crustal motion is localized, but its rate can greatly exceed regional rates of erosion. This accounts for the earth's varied relief. Most landscape changes are gradual and the result of events of moderate magnitude repeated over and over through vast spans of time.

APPLICATIONS

1. What are the landforms of your area on the macro, meso, and micro scales? Are there any unusual landforms or "textbook examples" of particular landforms in your vicinity?
2. Is there any landform evidence of climatic change in your region? If not, hypothesize as to the landform changes that would occur if the area became cooler, wetter, warmer, or drier than at present.
3. Look at the slope forms and slope angles in your local area. What do you think explains the differences from place to place?
4. What energy transformations occur (a) in the case of a rockslide that blocks a valley, damming the stream in the valley? (b) in the case of a wave that removes enough material from the base of a cliff to leave the higher part of the cliff without support?
5. It has been noted that for a beach to persist, particle input must be equivalent to erosional loss. What landforms in your region can be thought of in terms of a similar material budget?
6. W. M. Davis's cycle of erosion concept has been criticized on grounds that a full cycle from youth to old age could rarely be completed. What different factors could disturb the course of an erosion cycle?

7. Evaluate the relative roles of geologic structure, tectonic activity, gradational process, and time in the creation of the scenery of your region. Name other regions in which each of these factors is dominant.

FURTHER READING

Bloom, Arthur L. *The Surface of the Earth.* Englewood Cliffs, N.J.: Prentice-Hall (1969), 152 pp. This well-written paperback outlines the processes of landform development—brief but unusually good.

Bradshaw, Michael J., A. J. Abbott, and **A. P. Gelsthorpe.** *The Earth's Changing Surface.* New York: Wiley (1978), 336 pp. An abundantly illustrated account of landform development; unusual and interesting.

Brunsden, Denys, and **John Doornkamp, eds.** *The Unquiet Landscape.* Bloomington: Indiana University Press (1974), 171 pp. This is a magnificently illustrated collection of articles on various aspects of landform development, from a series appearing in the British periodical *The Geographical Magazine.*

Butzer, Karl W. *Geomorphology from the Earth.* New York: Harper & Row (1976), 463 pp. Different climatic regions are discussed in this textbook on landform development. It is not highly technical and uses a geographical approach.

Hunt, C. B. *Natural Regions of the United States and Canada.* San Francisco: W.H. Freeman (1974), 725 pp. This is a well-illustrated introduction to the regional landforms of North America. Fairly complete, but nontechnical.

Lobeck, A. K. *Geomorphology: An Introduction to the Study of Landscapes.* New York: McGraw-Hill (1939), 731 pp. Although dated, this text is included here because of its excellent photographs, maps, and diagrammatic illustrations of landforms of all types.

Shelton, John S. *Geology Illustrated.* San Francisco: W.H. Freeman (1966), 434 pp. This book is unsurpassed for crisp photographic illustrations of landforms. The text and organization are more imaginative than most—highly recommended; not advanced.

Twidale, Raoul C. *Analysis of Landforms.* New York: Wiley (1976), 572 pp. The best illustrated treatment of landforms of all types by a leading Australian geomorphologist.

Utgard, R. O., G. D. McKenzie, and **D. Foley.** *Geology in the Urban Environment.* Minneapolis: Burgess (1978), 355 pp. This paperback is a collection of articles concerning the significance of landforms and geomorphic processes in urban settings, demonstrating that landforms are not merely a "rural" topic.

CASE STUDY: Climatic Change and Landforms

We have noted that the sawtooth peaks of mountains like the Rockies in Colorado, Wyoming, and Montana are an obvious indication of climatic change. As we shall see in Chapter 15 there is no mistaking the meaning of such glacially produced landforms in areas that today have only tiny glaciers or none at all. Equally obvious indicators of climatic change are the extensive systems of dry canyons in the deserts of North Africa and Arabia. These so-called *fossil wadis* seldom carry water and are clogged with sand and boulders. However, they were obviously formed by stream erosion in the past. They too are a relict of the Ice Ages, which produced more rainfall and runoff in the subtropical regions. Less apparent effects of climatic changes have been imprinted upon virtually all land surfaces, though it may require an informed eye to detect them. Certain widespread types of scenery could only have been produced by climatic change, which is to say, by two dissimilar systems of geomorphic development acting at different times.

How did the major long-term climatic changes recorded by landforms come about? Several possibilities exist. We have seen that a change in solar output or the earth's relationship to the sun is thought by many to be the cause of the Ice Ages. The slow rise of mountain ranges gradually produces rainshadows extending far from the mountains, creating semiarid or desert areas. The changing elevations of the continents themselves may affect climates. Ocean currents are a strong climatic influence and must change as continents and ocean basins change shape and position due to sea floor spreading. Of course, sea floor spreading has caused the continents themselves to be moved across the planetary surface, from the poles to the tropics and from the tropics to the poles. Fossils of tropical plants are common in places like Spitzbergen, Greenland, and Antarctica, and ancient glacial deposits can be viewed in Brazil, South Africa, and the Sahara Desert.

French and German scholars were the first to comment, in the 1950s, on the red soils preserved under lavas in the central Sahara. Close examination of these soils in the midst of the world's largest desert revealed them to have been formed under summer-wet tropical savanna conditions. They indicate clearly that the Sahara has not been a permanent feature of North Africa. In fact, the erosional forms in granitic rocks in the central Sahara are very much like those seen in the summer-wet tropics, requiring intense chemical weathering in a moist climate some 20 to 50 million years in the past. Some French scientists have expressed the opinion that certain granitic domes and spires in the Sahara may be more than 50 million years old. Clearly these forms could not have originated in the present Saharan climate.

Large areas of the arid heart of Australia are dominated by flat-topped tablelands standing above a rough rocky relief typical of deserts. The level upper surface consists of a massive layer of iron- and aluminum-rich laterite produced by the soil-forming process that dominates in tropical savanna regions. Where the soil parent material lacked iron and aluminum, the capping layer is cemented by silica, being known as *silcrete*. It is impossible for either laterite or silcrete to form in a desert. Thus desert conditions must have become dominant after these soils had formed under a seasonally wet tropical climate. And long before the tropical phase, the same area, like the Sahara, was covered by an ice sheet.

Wherever deserts have been investigated, evidence has been found indicating a long earlier period of non-desert climate. In the Mojave Desert of southern California, as in the Sahara, a red-colored weathered mantle is preserved beneath lava flows having potassium/argon ages of 8 to 10 million years. This suggests a long predesert phase of deep bedrock weathering, with adequate vegetation to prevent immediate erosion of the weathered material. This is very significant in landform interpretation. Among the most common landforms in many deserts are smooth ramplike rock surfaces that slope outward from the foot of rocky hillslopes. Often there is no change in the rock type from the hillslope to the smooth ramp, which is known as a *pediment*. Pediments often extend 3 or more kilometers (2 or more miles) from associated hills and signify the trimming away of enormous volumes of rock, including whole mountain ranges in some areas. Because they are so obvious in deserts, pediments were long regarded as forms left by the wearing back of erosional hillslopes under arid conditions.

But pediments, though best seen in deserts, appear to have originated under non-desert conditions. In the California desert, lavas as much as 10 million years old rest on the pediments, indicating that the pediments were formed earlier. But there was no desert here 10 million years ago! The Mojave is largely a rainshadow desert, produced by mountains that did not rise strongly until about 6 million years ago. Thus the pediments are of predesert origin, as indicated by the fragments of red soils

preserved on them. This is reasonable, since the wearing back of hillslopes required to produce large pediments requires a far more efficient erosion system than that existing in deserts. Since all desert regions show evidence of a lengthy predesert history, all so-called "desert pediments" are most easily interpreted as relict landforms inherited from an earlier, more humid period.

The "normal" landforms of unglaciated, humid midlatitude areas may also be relict features. Characteristic hillslopes in the midlatitudes are smooth and ungullied, convex at the top and concave at the foot. Many are beginning to be gullied today, changing the slope form. Most European investigators interpret this form association as evidence of slope smoothing by periglacial solifluction during the Ice Ages, followed by renewed dissection by running water in the post-glacial period. The evidence for this is strong in Europe but less so in North America; however, in this hemisphere little research has been done to resolve the question. Further investigation may reveal that all of the earth's landscapes bear the imprint of changing climates and changing geomorphic processes.

Islandlike masses of granite rising above the bedrock erosional surfaces are known as *pediments*. This view is in California's Mojave Desert. Pediments are widely distributed in arid regions and were produced by erosion during a period of greater moisture availability millions of years ago. (T. M. Oberlander)

The Pass Through the Mountains by Fukaye Roshu. (The Cleveland Museum of Art, John L. Severance Fund)

Flowing water is one of the most important agents of landscape change, even in arid regions. The energy in moving water empowers it to pick up, carry, and redeposit rock waste. The landscape expresses the contest between gradation by flowing water and uplift caused by geologic processes.

CHAPTER 13

Fluvial and Aeolian Landforms

From 100 miles above the earth's surface, the organized patterns created by human activity are almost invisible. However, only from such an altitude can one begin to appreciate the organization of the earth's surface features. Two features in particular exhibit remarkable regularity. One is erosional, the other depositional. These are the systems of valleys carved out of the land surface by running water and the systems of sand dunes heaped up by wind in deserts that seldom experience rainfall or runoff.

Valley systems and dune systems are related. Both result from the friction of a fluid (something that "flows") passing over the material of the earth's surface. However, the flow of water is confined to channels, whereas the flow of air that concentrates sand and creates dunes has the whole landscape for its bed. The waves of sand raised by the wind as it passes over the land resemble the water waves raised by the wind as it blows across the sea. Despite the differences in the "channels" for running water and wind, the two fluids obey similar principles. They erode, transport, and deposit material in similar ways. For this reason we shall consider them together in this chapter. Since the domain of running water greatly exceeds that of wind erosion and deposition, and is much more the domain of human activity, our attention will center there.

Running water makes gradation possible by providing an effective transportation system connecting high and low places on the earth's surface. Processes related to channeled flows of water are known as *fluvial* processes, from the Latin *fluvius*, meaning river. From a high altitude the effects of fluvial processes are conspicuous over nearly all ice-free land areas. The only exceptions are the areas covered by sand dunes. From space one can see intricate valley systems carved by running water, large streams collecting the water and sediment delivered by smaller streams, and the mouths of the earth's great rivers, issuing plumes of sediment that discolor the sea. Clearly, rivers are the essential disposal system in the process of gradation.

In this chapter we shall first look at the way streams are organized in drainage networks; then we shall examine the mechanics of channeled flows of water. The vertical and horizontal patterns that channels make and the landforms that result from fluvial processes will be considered next. Finally, we shall look briefly at the effects of wind on landforms.

STREAM ORGANIZATION

Only a few streams, such as the Nile, flow far without being joined by other streams. Most streams are part of a *drainage network* that removes surface runoff from a *drainage basin*, which is the area drained by a stream and all its tributaries. All streams of every size have their own drainage basins, which can range in area from a fraction of a square kilometer to a sizeable portion of a continent (Figure 13.1).

The geometrical arrangement of streams within a drainage network is largely controlled by geological structure and is called the *drainage pattern*. The most common drainage patterns are illustrated in Figure 13.2 (see p. 330), which also demonstrates the remarkable uniformity of the drainage pattern in an area undergoing dissection by fluvial processes.

In all drainage networks small streams feed into successively larger streams. The characteristics of drainage networks can best be analyzed by using the idea of *stream order*. The smallest streams, which have no tributaries feeding into them, are *first-order* streams. Where two first-order streams join, the resulting larger single channel is a *second-order* stream. Similarly, *third-order* streams begin at the junction of two second-order streams, and so on, as shown in Figure 13.3 (see p. 331). In any drainage network, there are usually from 3 to 5 times as many streams in each order as in the next higher order.

There are consistent relationships between stream order and several measurable properties of drainage networks. Within any drainage network the streams of each order usually have a characteristic length, slope, and drainage basin area. These change in a regular manner as the stream order increases, as indicated in Figure 13.3. Studies of drainage networks suggest that the nature of their development is remarkably consistent from place to place. They seem to have evolved in a way that tends to minimize work, or the expenditure of energy, within the stream system.

Figure 13.1 (opposite)
The drainage basin of a large river such as the Mississippi contains a hierarchy of smaller nested drainage basins. In this figure, the basin of the Shoshone River is part of the basin of the Big Horn, which is part of the basin of the Yellowstone, which feeds into the Missouri, which is the major tributary of the Mississippi. The arrows indicate the downhill direction of water flow into the various basins. The Continental Divide along the Rocky Mountains separates basins draining eastward toward the Gulf of Mexico and the Atlantic Ocean from those draining westward toward the Pacific Ocean. (Doug Armstrong)

THE MECHANICS OF CHANNELED FLOW

To understand fluvial processes and the landforms they create, it is necessary to understand how water flows in channels. Take the example of water in a trough. It will not flow unless one end of the trough is higher than the other end. This permits the potential energy of the water at the high end of the trough to be converted to kinetic energy as the water flows to the low end of the trough. But not all the potential energy is converted to energy of motion. Some energy is lost in overcoming friction between the water and the trough walls and friction within the flow itself. Water flowing in a natural channel encounters friction with the stream bed and banks, which has a large retarding effect. This makes the speed of flow near the channel margins less rapid than in the center of the flow (see Figure 13.4, p.332). Boatmen heading upstream know that near the banks they will have less current to overcome than in midstream.

The way water flows in its channel determines the amount of energy it has for erosional and depositional work. Stream discharge and turbulence play important roles in providing this energy.

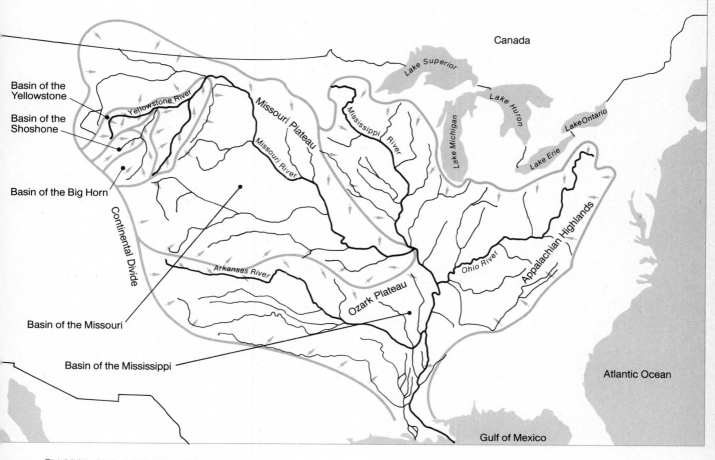

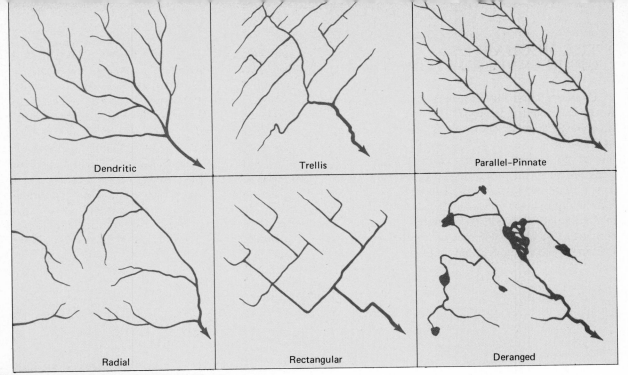

Dendritic

Trellis

Parallel-Pinnate

Radial

Rectangular

Deranged

Figure 13.2

(top) These are six of the most frequently encountered drainage patterns. *Dendritic* patterns are found in areas that lack strong contrasts in bedrock resistance, such as flat-lying sedimentary rock or massive crystalline rock that is deeply weathered. *Trellis* patterns develop where inclined layers of sedimentary rock of varying resistance to erosion are exposed at the surface. The parallel segments develop along the outcrops of the erodible layer of rock. *Parallel-pinnate* drainage reflects a topography of long parallel ridges. The short segments drain the flanks of the ridges, and the long segments drain the troughs between. *Radial* drainage indicates the presence of an isolated high mountain area and is common in individual volcanoes or dome-shaped mountainous uplifts. *Rectangular* patterns reflect strong jointing of resistant bedrock, with streams incising along the joint planes. *Deranged* drainage shows no geometrical pattern or constant direction and usually includes numbers of lakes; this pattern indicates destruction of prior drainage by the erosive effects of continental ice sheets. Such areas lack true valleys developed by fluvial erosion. (Vantage Art, Inc.)

(bottom) This side-looking radar image of a 32-km- (20-mile-) wide portion of eastern Kentucky portrays the nature of stream dissection of the land surface where there is little variation in rock type or geological structure. Note the large dendritic stream systems, the smaller tributaries, and the uniform density of channels. The northeast to southwest "grain" shows the influence of bedrock joints on stream development. Radar images like this penetrate clouds and haze, and often give better definition of surface form than do conventional photographs. (Raytheon Company and U.S. Army Engineering Topographic Laboratories)

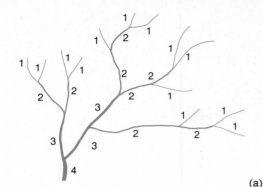

Figure 13.3
(a) This diagram illustrates the system of stream ordering devised by the engineer Robert Horton and simplified by the geomorphologist Arthur Strahler. A second-order stream arises at the junction of two first-order streams; a third-order stream is produced by the junction of two second-order streams, and so on.

(a)

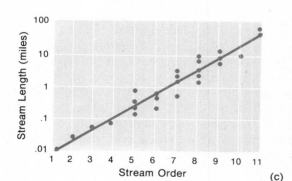

(b)

(b) (c) (d) These three graphs represent the Horton analysis of a drainage basin near Santa Fe, New Mexico. The graphs show the regularities that are typical features of drainage development, as revealed by a Horton analysis. (b) The numbers of streams of given order decrease regularly with increased stream order. (c) The average length of streams increases regularly with increased order. (d) The average area drained by a stream of a given order increases regularly as stream order increases. (Doug Armstrong after *Fluvial Processes in Geomorphology* by Luna B. Leopold, M. Gordon Wolman, and John P. Miller. W. H. Freeman and Company. Copyright © 1964)

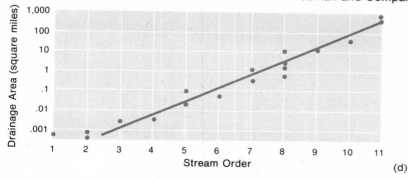

(c)

(d)

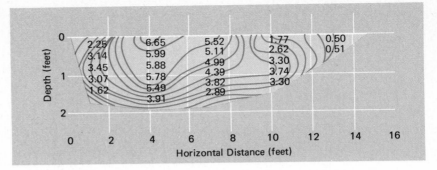

Figure 13.4
This diagram shows the average measured stream velocities, in feet per second, at various points in a cross section of Baldwin Creek, Wyoming. Note that the velocities tend to be lowest near the sides of the channel and highest near the center of the channel. The stream is therefore able to carry material suspended in its waters. Moderate velocities near the stream bed enable some material to be transported over the bed. (Doug Armstrong after *Fluvial Processes in Geomorphology* by Luna B. Leopold, M. Gordon Wolman, and John P. Miller. W. H. Freeman and Company. Copyright © 1964)

Discharge

The volume of water a stream carries past a given point during a specific time interval is called the stream *discharge*. Stream discharge is measured as cubic meters (or cubic feet) per second, and is equal to the cross-sectional area of the flow times the flow velocity (distance traveled per unit of time). The United States Geological Survey maintains more than 6,000 stream gauging stations to measure stream discharges in the United States. The data are used to make flood forecasts, to assess irrigation water supplies, to plan sewage disposal, and for engineering purposes, such as the design of dams and bridges. Usually stream discharge is inferred from the height of the stream's waters, using a previously determined relationship known as a *rating curve* (Figure 13.5).

Stream energy is closely related to stream discharge, because discharge is one of the controls of the flow velocity. The velocity of any segment of a stream reflects the balance between the downslope gravitational stress, which is controlled by stream slope, and the energy lost in overcoming friction at the boundaries of the flow. The larger the flow and the smoother the channel, the less the proportion of energy lost to friction and the greater the stream velocity. Streams normally increase in size in a downstream direction as tributaries and groundwater enter them. Other things being equal, this would cause streams to increase in velocity and energy in the same direction. However, as streams grow larger, their slope tends to decrease. This prevents a constant build-up of energy in the downstream direction and produces a more uniform distribution of stream energy. Even during floods, streams seldom have a flow velocity that exceeds 3 to 5 meters per second (7 to 11 miles per hour).

Turbulence

At very low velocities water can move as a smooth sheet, like the top card in a deck of cards that is pushed forward while the bottom card

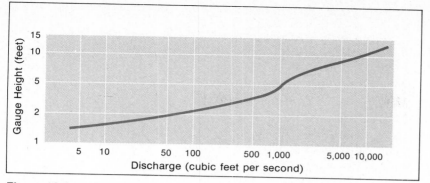

Figure 13.5
This *rating curve* for Seneca Creek, Maryland, allows the stream discharge to be inferred from a simple measurement of stream height at a gauging station. The average discharge of this stream is 100 cu ft per second, but discharges more than 50 times as great have been observed. (Doug Armstrong after *Fluvial Processes in Geomorphology* by Luna B. Leopold, M. Gordon Wolman, and John P. Miller. W. H. Freeman and Company. Copyright © 1964)

remains in place. Such flow is *laminar* (Figure 13.6, p. 334). Flows of this type lack the kinetic energy to put loose particles in motion. If the flow velocity increases, friction within the flow and at its boundaries soon causes it to break into separate currents that are no longer parallel. This irregular *turbulent flow* is normal in streams. In turbulent flow, the speed and direction of motion vary continuously, with currents in every direction, including upward. Nevertheless, the average motion is in the downslope direction. The constantly moving eddies that make quiet streams so fascinating to watch are evidence of turbulent flow, as are the boiling rapids of steeply descending streams.

In turbulent flow, most stream energy is consumed in overcoming the friction at the channel boundaries and between adjacent currents. Only a small portion of any stream's energy is available for the work of picking up rock debris and moving it through the channel. Even so, turbulent flows can be highly erosive. Rapid pressure variations near the stream bed and forceful upward-moving currents cause solid material to be lifted into turbulent flows and

carried away by them. This removal of material, in both solid and dissolved form, is an important aspect of stream flow, as we shall now see.

Scour and Fill

A rapidly moving stream may detach particles from its bed in the process of stream bed *scour*. Specific amounts of energy are required to put materials of different sizes and shapes in motion. The most easily scoured particles are those of sand size, having diameters between 0.05 and 1.0 mm. The flow velocity must be 15 to 30 centimeters per second (0.3 to 0.7 miles per hour) to put such particles in motion. Figure 13.7 (see p. 335) shows that smaller clay or silt particles resist detachment more than sand. This is due to the stronger molecular bonds between the smaller particles. Particles larger than sand size, including gravel, cobbles, and boulders, also resist motion more than sand because of their greater weight.

As Figure 13.7 reveals, it takes more energy to put a particle in motion than it does to keep it in motion, particularly in the smaller size

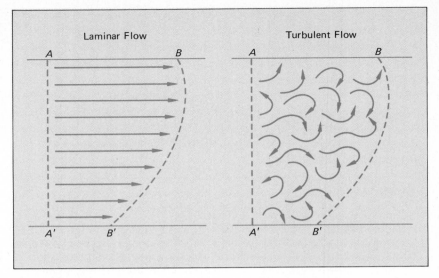

Figure 13.6
These cross sections compare laminar and turbulent flow, with overall motion forward from A−A' to B−B'. In laminar flow (left), the fluid moves like a deck of cards being deformed by internal shearing. The movement is slow and in a single direction. Under these conditions, sediment cannot be incorporated upward into the flow or supported within it. Thus the flow is nonerosive. Such flow is rare in streams but is common in slowly moving groundwater and glacial ice. Turbulent flow (right) is characterized by instantaneous velocities in all directions, which allows sediment to be picked up and supported within the flow. (Vantage Art, Inc.)

ranges. While a 0.1 mm particle requires a current velocity of about 20 cm/sec to be displaced, it will subsequently move at a velocity as low as 1 cm/sec. Nevertheless, a continued decrease in stream velocity causes progressively finer particles to settle out to become part of the deposited sediment in the process of stream bed *filling*, or fluvial *aggradation*.

Transport of Material by Streams

The effectiveness of streams as sediment carriers is one of the obvious facts of nature. China's Hwang Ho (Yellow River) received its name from the color of its sediment-laden waters, as did the Colorado (Red) River of the southwestern United States and the White River that drains the Badlands of South Dakota. The Mississippi River carries nearly 300 million metric tons of solid sediment to the sea each year, plus another 150 million tons of dissolved matter. To transport this much material by rail would require a daily train of 24,600 boxcars. Table 13.1 (p. 336) compares the sediment loads of a variety of large and small streams around the world.

Streams move their load of rock debris in three ways. Part of the transported material is dissolved in the water, forming the *dissolved load* (or *chemical load*) of the stream. Fine particles of clay and silt are carried in suspension within the flow of water as the *suspended load*. It is the suspended load that gives many streams their color and opaque appearance. Par-

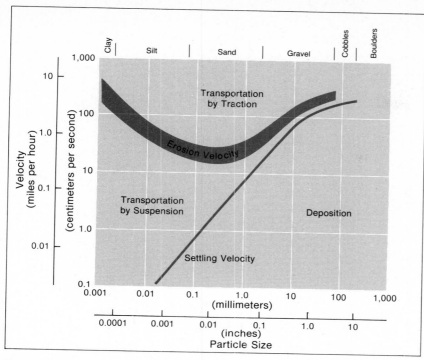

Figure 13.7
The ability of a stream to remove and transport material depends on the local velocity of flow, on the size of the particles, and to some extent on the shape of the particles. This diagram for uniform material relates particle size and flow velocity to mechanisms of erosion and deposition. Combinations of size and velocity located to the right of the settling velocity curve represent the regime of deposition; large particles settle out of a slow-moving stream and become deposited on the bed. Particles with velocities and sizes to the left of the settling velocity curve can be transported either by saltation (skipping or bouncing) along the bed or by suspension in the stream. The erosion velocity represents the minimum water velocity at which particles are carried away from the stream bed. Erosion velocity is shown as a band because it depends on particle shape, bed material, and other factors. The graph shows that once a particle becomes suspended it can be transported by water moving more slowly than the erosion velocity. However, particles larger than a few millimeters are transported largely by saltation along the stream bed. (After Morisawa, 1968, from Hjulstrom, 1935)

ticles too heavy to be carried in suspension are bounced and rolled along the channel bottom as the stream's *bed load*. The bed load and suspended load together constitute a stream's *solid load*.

The dissolved load of a stream is largely contributed by groundwater inflow into the stream. The amount of dissolved matter in groundwater varies according to the composition of rocks and soils, and the climate, weathering processes, and

FLUVIAL AND AEOLIAN LANDFORMS

Table 13.1
Characteristics of Selected Rivers

River and Location	Average Discharge at Mouth (thousands of cubic meters per second)	Length, Head to Mouth (kilometers)	Area of Drainage Basin (thousands of square kilometers)‡	Average Annual Suspended Load (millions of metric tons)§	Average Annual Suspended Load (metric tons per square kilometer of basin)
Amazon (Brazil)	180 (6,400)*	6,300 (3,900)†	5,800	360	63
Congo (Congo)	39 (1,400)	4,700 (2,900)	3,700	500	260
Yangtze (China)	22 (800)	5,800 (3,600)	1,900	296	91
Mississippi (U.S.)	18 (650)	6,000 (3,700)	3,300		
Yenisei (U.S.S.R.)	17 (600)	4,500 (2,800)	2,100		
Irrawaddy (Burma)	14 (500)	2,300 (1,400)	430	300	700
Bramaputra (Bangladesh)	12 (415)	2,900 (1,800)	670	730	1,100
Ganges (India)	12 (415)	2,500 (1,600)	960	1,450	1,520
Mekong (Thailand)	11 (390)	4,200 (2,600)	800	170	210
Nile (Egypt)	2.8 (100)	6,700 (4,200)	3,000	110	37
Missouri (U.S.)	2.0 (70)	4,100 (2,500)	1,370	220	160
Colorado (U.S.)	0.2 (6)	2,300 (1,400)	640	140	210
Ching (China)	0.06 (2)	320 (200)	57	410	7,200

*Numbers in parentheses indicate thousands of cubic feet per second.
†Numbers in parentheses indicate miles.
‡One square kilometer is equal to about 0.39 square mile.
§One metric ton is equal to 2,204.6 pounds.

Sources: Holeman, John N. 1968. "The Sediment Yield of Major Rivers of the World," *Water Resources Research,* 4 (August): 737–747. Fairbridge, Rhodes W. (ed.) 1968. *The Encyclopedia of Geomorphology.* Vol. III, Encyclopedia of Earth Sciences Series. New York: Reinhold. Espenshade, Edward B. (ed.) 1970. *Goode's World Atlas.* 13th ed. Chicago: Rand McNally. Curtis, W. F., Culbertson, J. K., and Chase, E. B. 1973. "Fluvial-Sediment Discharge to the Oceans from the Conterminous United States," *U.S. Geological Survey Circular 670.* Washington, D.C.

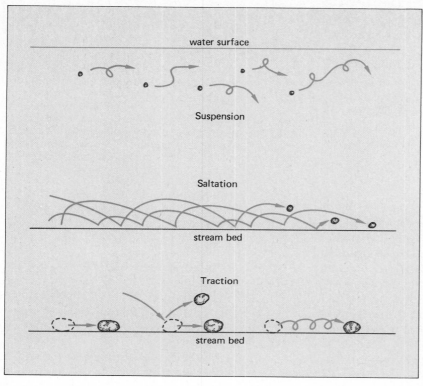

Figure 13.8
Solid material is transported in streams as suspended load and bed load. The finest particles are supported in suspension within the flow (top), while the bed load moves by both saltation and traction, in which particles are dragged or rolled by fluid shear stress, or are knocked forward by a saltating particle. (Vantage Art, Inc.)

vegetation cover in the local area. Human pollution of streams has added greatly to their chemical load in many areas, especially around industrial cities and where agriculture and mining are active. The average yearly rate of removal of dissolved matter over the entire United States is about 40 tons per square kilometer (100 tons per square mile). This is a little more than half the average annual rate of removal of solid matter (71 tons per square kilometer, or 185 tons per square mile).

To transport its solid load a stream must have forward and upward velocities greater than the constant downward pull of gravity on the particles carried in the flow. Material carried in suspension glides along irregular paths within the flow, while heavier material skips or rolls along the stream bed. Bouncing or skipping transport is known as *saltation*, while rolling or sliding transport is called *traction* (Figure 13.8).

As the particles of the bed move along, they collide with one another and with the solid rock of the stream bed. Impacts against the channel floor break loose new fragments, slowly lowering the stream bed. Because of the constant battering, the bed load particles become smaller and more rounded as they progress downstream—cobbles are reduced to pebbles, and pebbles are fragmented into sand and silt. Sand, silt, and the clay washed directly into the stream are the only solid particles carried far downstream by large rivers.

Any decrease in flow velocity reduces the forces supporting the suspended load and propelling bed load movement. This causes some of the solid load to drop out of the flow. First to settle out are the heaviest particles moved by saltation and traction. These come to rest in a new deposit that may be either temporary or permanent. Deposition of the finest particles carried in suspension occurs only when the water is almost still, as in a lake.

The transporting power of a large stream is enormous. In 1933 a flash flood in California's Tehachapi Mountains caught a train crossing a trestle. The locomotive and tender were carried a kilometer as part of the stream's bed load and were buried so completely in gravel that a metal detector had to be used to find them.

Channel Equilibrium

A stream is a sensitive dynamic system with the ability to adjust the form of its channel in a matter of hours in response to small changes in inputs of energy and material. By scouring and filling, a stream adjusts the slope of its bed and the shape of its channel so that stream energy remains in balance with the work of sediment transport. If a stream lacks the velocity to transport the sediment fed into it, sediment is deposited in the channel. This elevates the stream bed, which increases the potential energy at that point. Elevation of the stream bed increases the stream slope, which increases stream velocity. Thus by depositing sediment the stream increases its energy and ability to move future arrivals of sediment.

Conversely, where stream velocity exceeds that necessary for sediment transport, a stream will either scour its bed downward or lengthen its course by becoming increasingly sinuous. Either response will flatten the stream's slope and thereby reduce its velocity. In both cases negative feedback maintains equilibrium in the fluvial system.

During a flood, each section of a stream must transmit a discharge that is much larger than normal. As the discharge increases, the volume of water becomes too great for the normal channel size and shape. At the same time, the increase in flow volume decreases friction, which causes the stream velocity to increase. This enables the floodwaters to scour the bed and banks of the stream, thus enlarging the channel (Figure 13.9). However, scour of the stream bed re-

Figure 13.9 (opposite)
These channel cross sections for the Colorado River at Lees Ferry, Arizona, show the scouring and subsequent filling that occurred during a high-water period in 1956. Discharge is in cubic feet per second. The marked enlargement of the channel by scouring was accompanied by an increase in flow velocity to accomodate the greatly increased discharge. In August, when the discharge approached its initial February value, the cross section of the channel also returned to almost its original size. (Doug Armstrong after *Fluvial Processes in Geomorphology* by Luna B. Leopold, M. Gordon Wolman, and John P. Miller. W. H. Freeman and Company. Copyright © 1964)

duces potential energy and the downstream slope, restraining any further velocity increase. If the flood could persist, and if the stream bed were easily eroded, the stream slope and channel form would evolve to an equilibrium with the flood discharge.

But the flood crest eventually passes and the stream discharge falls. When this occurs, the stream is once more out of equilibrium. Its channel is now larger than the discharge requires, and it has too little slope to transmit the water and sediment of the weakening flow. In this way deposition of sediment begins. This elevates the stream bed, increases the stream slope, and reduces the size of the channel. These adjustments boost the velocity of the diminishing flow so that subsequent arrivals of sediment can be carried through, thereby reestablishing stream equilibrium.

The sediment underlying most stream beds is an important element in the equilibrium process. It permits channel enlargement and slope adjustment during floods and the restoration of normal channel characteristics after floodwaters pass. Complete equilibrium is not possible. Because flow conditions change constantly, most streams are in a state of perpetual readjustment, or "quasi-equilibrium." Whereas stream channels cut into erodible materials can adjust to changes quite rapidly, channels cut in hard

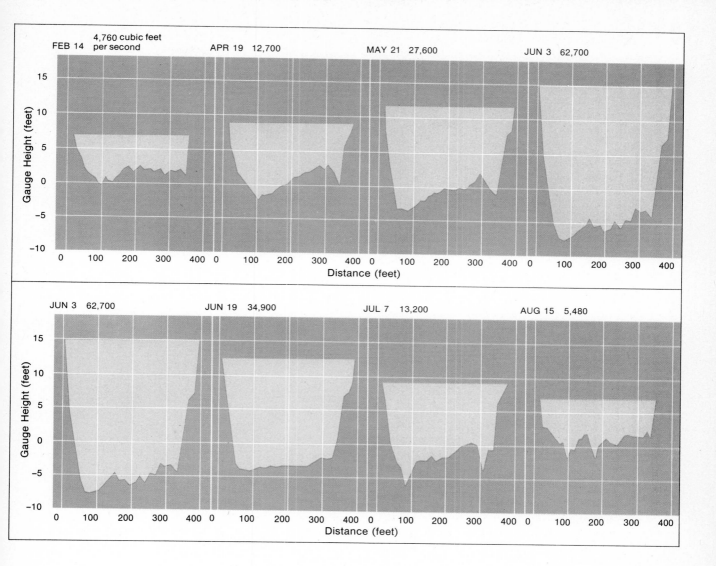

rock cannot be adjusted easily. Thus bedrock channels may never achieve an equilibrium condition.

A stream that is in equilibrium is said to be *graded*. Over the long term a graded stream neither scours its bed nor deposits sediment but maintains a rate of descent and a channel form that give it just enough energy to transmit its sediment load. Since absolute equilibrium is impossible, a graded condition really implies that no long-term change is occurring despite repeated short-term adjustments.

Stream Gradient, Longitudinal Profile, and Knickpoints

The slope of a stream channel, or the rate at which a stream channel descends from higher to lower elevations, is the *stream gradient*. This is measured as units of vertical fall per unit of

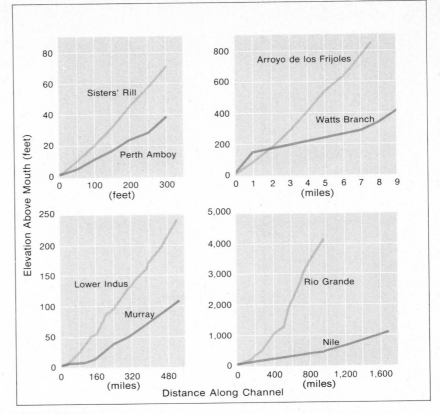

Figure 13.10
These graphs illustrate the longitudinal profiles of eight streams varying in length from a few hundred feet to several thousand miles. The vertical scales have been exaggerated for clarity. There are local irregularities in the profiles, perhaps because of local differences in rock structure and sediment load. Note that the downstream portions of the Rio Grande and the Nile River develop concave longitudinal profiles even though their discharges are effectively constant because of the lack of tributaries. (Doug Armstrong after *Fluvial Processes in Geomorphology* by Luna B. Leopold, M. Gordon Wolman, and John P. Miller. W. H. Freeman and Company. Copyright © 1964)

horizontal distance, expressed in like units (meters per meter) or as meters per kilometer or feet per mile. A graphic portrayal of a stream's gradient from its source to its mouth is called the stream's *longitudinal profile.*

Stream gradient is a major influence on stream velocity, and gradient changes are a means of balancing stream energy to the work of sediment delivery. Comparisons of the profiles of streams that are hundreds to thousands of kilometers long show that most are concave upward—steep near their sources and almost flat near their mouths (Figure 13.10). The initial steepness provides the energy to move coarse bed loads. The flatter downstream gradients indicate increased discharges and larger channels, as well as a reduction in bed load particle size. Point-to-point variations in stream profiles are a consequence of changes in discharge, sediment load, channel roughness (friction), and channel cross-section (size and shape). An outstanding example is seen on the Missouri, which doubles its gradient below the entry of the Platte River near Omaha. This large gradient increase is necessary to give the Missouri the energy required to transmit the enormous bed load of sand brought to it by the Platte.

Stream profiles are sometimes interrupted by sharp offsets where vertical waterfalls or steep cascades or rapids are present. Such profile interruptions are called *knickpoints*. Knickpoints in stream profiles have different origins. Some result from abrupt changes in bedrock resistance. Others may be related to crustal displacements, as when earthquake activity creates an escarpment that the stream must cross. However, knickpoints in stream profiles are most common in areas that have experienced severe glacial erosion. The many waterfalls and cascades seen in high mountain country usually indicate that glacial erosion has deepened the major valleys, leaving smaller tributary valleys "hanging" above them. Examples include California's Yosemite Falls and the many small waterfalls in New York's "Finger Lakes" region (see Chapter 15). Glacial erosion may create steps in the main valleys themselves, each the site of a waterfall or cascade, as in the case of Nevada and Vernal falls in California's Yosemite Valley.

Knickpoints can also form in coastal regions as a result of global lowering of sea level or local rise of the land over a short period of time. Newly exposed sea floor is always steeper in slope than the lower portions of large rivers. Therefore the exposure of new land by the retreat of the sea always produces a knickpoint near the old river mouth. This is the way the famous "Fall Line" of the Atlantic coastal region of the United States was formed. The Fall Line refers to rapids and cascades developed where streams pass out of the hard ancient rocks of the Piedmont region and into the weak sediments of the coastal plain. In colonial times the rapids on large streams at the Fall Line provided water power to turn grindstones in flour mills. Later the water power was used to generate electricity. The falls and rapids also necessitated the landing of ship cargoes and change to overland transportation. This stimulated the growth of such cities as Trenton (New Jersey), Philadelphia, Baltimore, Richmond (Virginia), Raleigh (North Carolina), Columbia (South Carolina), and Augusta and Macon (Georgia). Their growth was assisted by the fact that the rivers were narrower and easier to cross in the hard rocks above the Fall Line.

Increased stream energy in the steeper section gradually eliminates most knickpoints through channel scouring. However, some knickpoints wear back in an upstream direction a considerable distance before disappearing. This has been the case at Niagara Falls (Figure 13.11, p. 342), which has retreated 11 km (7 miles) since being formed by glacial modification of the landscape about 100,000 years ago.

Lateral Migration and Channel Patterns

It is well known that streams can shift position horizontally, as well as scouring and filling their beds. Near Needles, California, the Colorado River has moved laterally as much as 244 m (800 ft) in a single year, and at Peru, Nebraska, the Missouri River has shifted position from 15 to 150 m (50 to 500 ft) each year for more than 30 years. Where stream banks are erodible, channels may migrate distances of several kilometers over a period of years. This creates significant problems: on one side property is gnawed away, while on the other side it expands; political boundaries move with the stream when it shifts slowly, but stay in place when it changes course suddenly. Mississippi's boundaries with Arkansas and Louisiana show

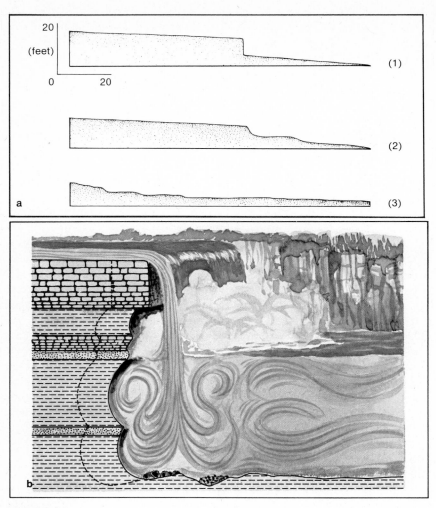

Figure 13.11
(a) Streams generally act in such a way as to reduce knickpoints and attain a smooth longitudinal profile. The diagram shows a knickpoint (1) that was formed on Cabin Creek, Montana, by an earthquake. The stream began to scour its channel above the knickpoint and to fill below the knickpoint. Because the material forming the stream bed was easy to erode, the knickpoint was significantly reduced in only a few months (2). Three years later (3) the profile showed no trace of the original knickpoint. (John Dawson after Marie Morisawa, *Streams: Their Dynamics and Morphology*, edited by P. Hurley, © 1968, McGraw-Hill Book Company)

(b) Since the Niagara River began flowing during the Pleistocene era, it has not been able to cut downward significantly into the cap rock of hard dolomite forming Niagara Falls. Instead, the river has cut a long gorge headward by undermining weak shale layers at the base of the falls, causing the collapse of the overlying layers. (From G. K. Gilbert, from O. D. Von Engel, *Geomorphology*, 1942, Macmillan Company)

these effects clearly, as portions of each state have been cut off from the remainder of the state by movements of the Mississippi River (see Figure 13.17).

When viewed from above, stream channels display three general patterns: meandering, braided, and straight, or nonmeandering (Figure 13.12). The normal pattern is a *meandering* one, in which the stream swings back and forth in either smooth or sharp curves. Meanders occur on streams of every size, from tiny rills on bare ground to kilometer-wide rivers. On streams of all sizes there is a relatively consistent relationship between the width of the stream and the *meander wavelength*, or distance between similar points on successive meanders (Figure 13.13). Meander wavelength is normally from 7 to 15 times the channel width.

Even streams that lack true meanders seem to show a succession of deeper pools and shallow-water "riffles" whose spacing is related to stream width (Figure 13.14, p. 348). In fact, laboratory experiments with artificial channels show that pool and riffle sequences in straight channels will in time evolve into sequences of

Figure 13.12
(top) The sinuous Animus River near Durango, Colorado, meanders over a broad flood plain produced by alluvial aggradation. (Dr. John S. Shelton)

(bottom) This braided tributary of the Yukon River in northwestern Canada carries outwash from the margin of a receding glacier. The St. Elias Mountains are in the distance. Braided channels develop where stream banks are gravelly and collapse easily due to lack of cohesion, and where streams receive excessive inputs of sand or gravel. (Larry W. Price)

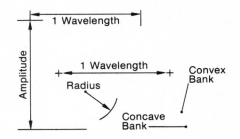

Figure 13.13
The diagram shows how the wavelength, amplitude, and radius of a meander are defined. The radius around a meander is not constant, however, because the form of a meander is usually more complex than a simple circular arc. The meander wavelength is usually 7 to 15 times the width of the channel. (Doug Armstrong after G. H. Dury from *Water, Earth, and Man*, edited by R. J. Chorley, © 1969, Methuen and Company)

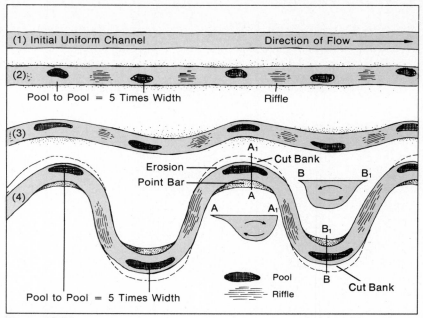

Figure 13.14

Streamflow in a straight channel of uniform cross section (1) seems to be unstable, and if the channel is readily erodible, the stream tends to develop a sequence of alternate deep pools and shallow gravel bars, or *riffles* (2). The straight channel becomes more sinuous, with pools forming toward the outer concave banks (3). In time, a meandering channel may be developed (4). A sinuous or meandering stream tends to scour its outer concave banks and to deposit sediment against its inner convex banks; the speed of flow is greatest near the concave banks and least near the convex banks. There may also be a lateral flow of water, as the cross-sectional diagrams A–A$_1$ and B–B$_1$ indicate. The lateral flow is thought to move sediment from the cut bank to the point bar. (Doug Armstrong after G. H. Dury from *Water, Earth, and Man,* edited by R. J. Chorley, © 1969, Methuen and Company)

meanders. Each pool becomes the site of an outward-pushing meander, with the riffles forming the points of current cross-over between successive loops. Thus the explanation of meander geometry lies in the nature of the flow that produces the initial pools and riffles.

The stream *thalweg*, which is the line that follows the deepest part of the stream channel, swings back and forth across the channel, moving toward the outer bank of each successive curve. The maximum flow velocity follows the thalweg, where friction is least. Scouring occurs at the concave banks, where the stream velocity is greatest, and deposition of sediment occurs at the inner convex banks, or "points," where flow velocity is least. The deposits of sand or gravel formed on the successive points are called *point bars*. The process of undercutting at the concave banks and deposition on the points directly across the channel causes the meander loops to migrate laterally in the process of *lateral planation*. Most scouring occurs during floods, when stream energy is at its peak. The greatest deposition of sediment occurs during the falling stages of floods, as stream energy declines.

Lateral planation increases the horizontal distance for each unit loss in stream elevation. Therefore, its effect is similar to that of vertical stream incision: both produce a flatter stream gradient, which reduces stream energy. An energetic stream that does not have the tools (bed load) to cut vertically into bedrock can achieve equilibrium by moving laterally instead. That is why meandering channels occur where the bulk of the stream load is fine material that is carried in suspension.

A stream load dominated by larger particles that move only by traction and saltation produces a different stream pattern: the *braided stream*, which consists of many intertwining shallow channels, as seen in Figure 13.12. Shallow channels are necessary so that the maximum flow velocity is close to the stream bed, where it can thrust against the bed load.

Braided streams form when sand or gravel is deposited in bars in broad shallow channels abundantly supplied with sand, gravel, and cobble-sized material. These bars divide the stream into many shallow channels and increase the downstream slope by elevating the stream bed. Braided streams must have exceptionally steep gradients to overcome the friction in shallow channels as well as to move their bed load of heavy particles.

Braided channels develop where stream discharge fluctuates widely from day to day or season to season, as in arid and semiarid regions. Wide fluctuations lead to bank collapse, channel widening, and sediment accumulation. The large sediment inputs required to initiate braiding occur where vegetation is sparse, where stream banks are composed of sand or gravel, and where glaciers feed great quantities of coarse waste to streams. Human activities that increase sediment movement into streams can cause meandering streams to become braided. Such activities include certain agricultural, mining, and logging practices, as well as urban and suburban construction projects. By creating shallower channels, these changes generally increase the frequency and magnitude of floods.

Straight stream channels are relatively uncommon and seldom extend far. Where present, they indicate that the channel is controlled by the underlying geological structure. Channels cutting through solid bedrock are occasionally confined by fractures in the rock. Some meander patterns are composed of straight segments that meet at sharp angles, indicating that joints control the channel. The channels of many river deltas are straight because cohesive clays have prevented lateral shifting of the stream beds (see Figure 13.20, p. 352).

FLUVIAL LANDFORMS

Valley Development

When a stream has more kinetic energy than it needs to carry its sediment load, the extra energy causes changes in the stream's environment. For example, streams flowing down a surface recently created by motions of the earth's crust, or by the construction of a volcano, commonly have excess energy. Streams in such settings will cut downward, flattening their gradients until their energy is in balance with their work of sediment transport. Streams already in equilibrium will begin scouring when environmental changes increase their discharge, giving them extra energy. This has been called stream "rejuvenation." The causes of stream rejuvenation are deformation of the land surface, drops in sea level, increases in stream discharge, and decreases in sediment load.

Uplift of the land surface, which increases the potential energy of streams, is the chief cause of stream rejuvenation and valley development. This rejuvenation causes valleys to deepen as stream profiles are regraded to restore equilibrium (Figure 13.15, p. 346). Continuous uplift and uplift that is more rapid than stream inci-

Figure 13.15
This image taken from an altitude of about 200 km (125 miles) shows the Colorado River's great canyon through the Colorado Plateau in northern Arizona (upper right). The river exits from its canyon at the escarpment of the Grand Wash Cliffs (see arrow), created by uplift along a major fault. Uplift of the Colorado Plateau relative to the area to the west has caused vertical incision of the river to a depth exceeding 1,500 m (5,000 ft). Healthy vegetation produces the red colors on computer-generated LANDSAT images such as this. Lake Mead appears in the left (west) part of the image. (Department of the Interior, U. S. Geological Survey)

sion both produce hilly or mountainous landscapes resulting from steady stream downcutting. Uplift in separate pulses results in stream terraces (discussed later in this chapter) and multi-story valleys, in which newer valleys are cut into the floors of successively older and higher valleys.

Lowering of sea level, or uplift of the land relative to the sea, causes valley deepening near coasts. The effects of world-wide sea level changes are discussed in Chapter 15.

Increases in stream discharges have been an important cause of stream rejuvenation at various times. Increased discharges can be produced by regional changes in climate and vegetation or by local stream piracy, in which an expanding drainage network "captures" the headwaters of a neighboring stream system. Flood discharges, during which valley deepening occurs, have been increased by such human activities as forest removal and urbanization, a

process that covers vast areas with impermeable concrete and asphalt. The fourth cause of stream rejuvenation, decreased sediment load, is also frequently a result of human activity, specifically dam construction, which traps both the bed load and suspended load of streams in artificial reservoirs. Immediately downstream from large dams, streams have excess energy since they have no sediment to transport. Such streams scour downward when large volumes of water are released from the reservoirs. This is occurring now along the Colorado River below the Glen Canyon Dam (completed in 1964) and along the Nile below the Aswan High Dam (completed in 1971).

Floodplains

When streams stop deepening their valleys, the valley floors gradually become wider. Valley

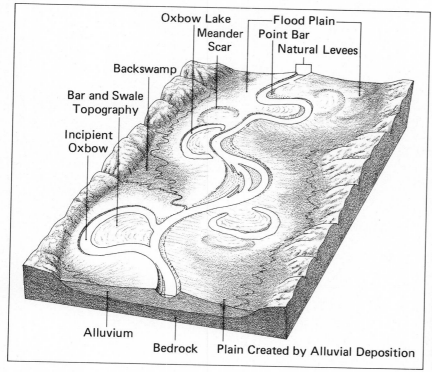

Oxbow Lake
Meander Scar
Flood Plain
Point Bar
Natural Levees
Backswamp
Bar and Swale Topography
Incipient Oxbow
Alluvium
Bedrock
Plain Created by Alluvial Deposition

Figure 13.16
Floodplains may result from either fluvial aggradation or lateral stream planation that wears back the valley walls. Oxbow lakes are the remnants of recently abandoned meanders. When the lake is eventually filled with sediment and vegetation, a *meander scar* remains. In times of flood, the overflowing river carries new sediment onto its floodplain and builds raised banks, or natural levees, by deposition of silt close to the river channel.

widening is caused by erosion of the slopes rising from the streams and also by lateral planation by the streams themselves. As meander planation shaves back valley walls and sediment is deposited on point bars, a continuous flat-floored trough is created. The stream swings back and forth across this open lowland (Figure 13.16). Since the stream channel is not large enough to hold the biggest flows that occur at intervals of a year or two, this lowland is occasionally flooded and is known as a *floodplain*.

Not all floodplains are produced by stream erosion. Many of the earth's largest rivers flow across vast lowlands created by crustal movements. Examples are the Amazon River of Brazil, the Ganges River of India, the Mississippi River, and the Sacramento and San Joaquin rivers of California. These streams are carrying sediment into the structural basins that they cross, which are not true valleys. Many other floodplains are products of stream aggradation that has filled deep V-shaped valleys with sediment. Such floodplains are especially common in coastal regions.

Floodplains are a hazardous environment, but they have great economic value. Each flood leaves a covering of fresh silt, often several centimeters thick. These overbank flood deposits cover the older sand and gravel point bar deposits and create nearly level agricultural land. This is usually very fertile since its nutrients are renewed periodically by fresh deposition.

River floodplains can contain a variety of fluvial landforms. When overbank flooding occurs, the largest amount of deposition is just outside the stream channel, where the flow velocity slows due to friction. The thickness of the overbank flood deposit decreases with distance from the channel. This wedge-like deposition produces paired *natural levees* that slope very gently away from the stream on both sides. Often there are low-lying *back swamps* between the natural levees and the valley walls that rise above the floodplain. Low ridges or cones of alluvium often extend from the levee into the back swamp. These are deposited where a jet of sedi-

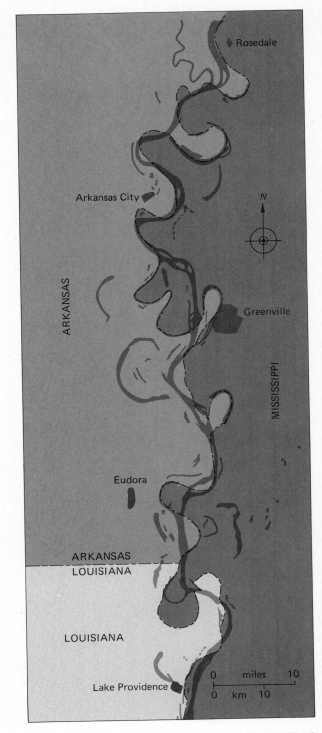

CHAPTER 13

ment-laden floodwater has broken across the levee crest. The break is called a *crevasse*, and the resulting alluvial deposit is a *crevasse deposit*. In Louisiana, the Mississippi River's natural levees rise 5 to 6 m (15 to 20 ft) above the back swamps. In the disastrous 1973 flood, the Mississippi River levees received as much as a meter (over 3 ft) of new sediment. Natural levees are prime locations for agricultural settlement because of their fertile soils and good drainage characteristics.

Where streams have considerable excess energy and the floodplain is composed of easily eroded sand rather than silt, meander shifting may be very active. In such places the meanders often expand into flaring "gooseneck" loops that press close against one another. Flood flows sometimes break through the narrow necks of land that separate adjacent loops, causing the loop between to be abandoned by the stream, forming an *oxbow lake* (Figure 13.16). Hundreds of present and past oxbow lakes are visible along the Mississippi River from Cairo, Illinois to Baton Rouge, Louisiana (Figure 13.17). Many of these were created at the same time during major floods. Oxbow lakes eventually fill with silt and clay but their traces remain clear long afterward.

Figure 13.17 (opposite)
This map of the Mississippi River from Rosedale, Mississippi, to Lake Providence, Louisiana, a distance of about 130 km (80 mi), shows how meander cutoffs during floods leave visible evidence in the form of detached political units. The state boundaries were originally fixed along the center of the Mississippi River. The boundaries shift with the slow migration of the channel, but stay in place when there are sudden channel shifts, or *avulsions*, during floods. The areas inside meander cutoffs, or *oxbows*, thus become isolated from the rest of their state. The three cutoffs closest to Greenville were made artificially by the Corps of Engineers in the early 1930s to shorten the river for navigation purposes.

Fluvial Terraces

Fluvial landforms often show that graded streams periodically adjust their profiles by either raising or lowering the stream bed. Where a stream cuts downward into a broad valley floor, the floodplain no longer acts as an overflow channel. Instead, it is left standing well above the level of the highest floodwaters to form a *fluvial terrace* (Figure 13.18, p. 350).

A temporary phase of aggradation can also produce fluvial terraces. In such a case decreased stream energy or increased sediment input causes deposition that raises the stream bed, forming an *alluvial fill*. Return to the original conditions causes the stream to cut back through the fill to its old level. This leaves a terrace composed of the alluvial fill. Fill terraces have been formed where streams were temporarily aggraded by sediment washed from glaciers and also where tectonic motions and climatic fluctuations have occurred.

River terraces are important indicators of environmental change. They reveal that tectonic, climatic, or hydrologic changes have forced streams to change their behavior to regain equilibrium. Some rivers are bordered by great flights of terraces, indicating repeated disturbances of equilibrium. The rivers of New Zealand are famed for their magnificent terraces, which reveal a history of strong tectonic activity and complex environmental changes.

Alluvial Fans

Where a steep-gradient mountain stream in a bedrock canyon flows out onto a flat plain, it loses much of its kinetic energy. This occurs because of the spreading of waters that were previously confined to a narrow channel. Where the flow spreads, the bed load is dropped, producing a fan-like deposit issuing from the canyon mouth and spreading over the plain. This deposit, shown in Figure 13.19 (p. 351), is an *alluvial fan*.

Alluvial fans are most common in arid re-

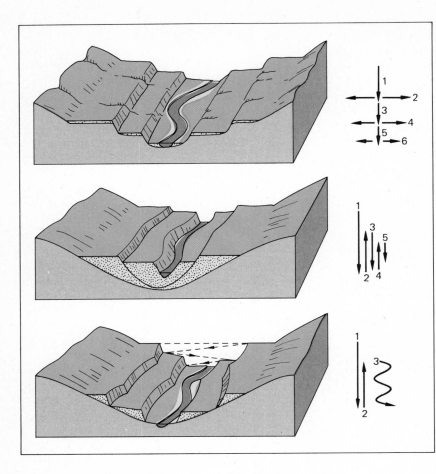

Figure 13.18
Stream terraces vary significantly in origin, as these diagrams illustrate. To the right of each diagram is an indication of the nature of the river movements necessary to create the associated terraces: downward arrows indicate stream incision; upward arrows, aggradation; horizontal arrows, lateral planation; sinuous arrows, simultaneous incision and lateral planation. **(top)** Terraces produced by periods of downward stream incision (1, 3, 5) separated by intervals of valley widening by lateral planation and slope erosion (2, 4, 6). These terraces are erosional and are said to be *paired* since those on opposite sides of the stream match in elevation. **(center)** Paired fill terraces produced by two separate phases of alluvial aggradation (2, 4) followed by renewed valley excavation (3, 5). **(bottom)** Valley deepening (1) is followed by aggradation (2) producing an alluvial fill. Subsequently the stream cuts downward at the same time that it is planing laterally in the fill (3), producing *unpaired* terraces. (Vantage Art, Inc.)

gions. The lack of vegetative protection permits the infrequent heavy rains to flush large quantities of rock debris from slopes. Stream floods, which may last only a few hours, are highly charged with sediment. Where streams flow out of their canyons, the coarsest part of their bed load tends to be dropped close to the canyon mouth, causing the apex of alluvial fans to be much higher than the fan margins. Many fans are composed largely of the deposits of mud

flows and debris flows. Since the density of these flows is many times that of water, they can transport large boulders, which can be seen scattered over fan surfaces. In mountainous deserts neighboring fans merge to form continuous ramps of sand and gravel, known as *alluvial aprons*, or *bajadas*. In California's Death Valley (Figure 13.19) these debris aprons rise as much as 600 m (2,000 ft) over a horizontal distance of 8 km (5 miles).

Figure 13.19
(top) This alluvial fan is located on the east side of Death Valley, California; its size can be judged from the road crossing it. Rains are infrequent in this region, but torrential storms occur in the mountains from time to time, carrying down large quantities of coarse sediment to be deposited over the fan. Alluvial fans **(left)** joining to form an alluvial apron sloping from bordering mountains toward the basin of Death Valley, California. The view shows the large fans on the west side of Death Valley. (Dr. John S. Shelton)

FLUVIAL AND AEOLIAN LANDFORMS

There is often abundant groundwater present at the base of alluvial fans in arid regions, which makes them favored locations for settlement and agricultural development. An outstanding example is Salt Lake City, which sprawls over the surface of a large fan adjacent to Utah's Wasatch Mountains. The Mormon settlers who founded Salt Lake City also established several other settlements in similar settings, including Las Vegas, Nevada.

Deltas

When rivers flow into lakes and the sea, their velocity is checked and they lose their load-transporting ability. Their bed load and much of their suspended load is dropped at the river mouth in the form of a *delta*. The term was first used some 2,500 years ago by the Greek historian Herodotus, who noted the similarity of the depositional form to the Greek letter Δ (delta). However, delta configuration is extremely variable. The Mississippi River delta, with its bird's-foot form, is quite unlike the classic examples of the Nile and Niger river deltas (Figure 13.20). Variations in delta form result from differences in the rate of sediment supply, the vigor of wave action and coastal currents, and the rate at which the deposit subsides as a result of com-

Figure 13.20
(left) The delta of the Nile River in Egypt has a different shape from the delta of the Mississippi because of more vigorous erosion by currents of the Mediterranean Sea and the fact that the delta is not subsiding. The construction of the Aswan Dam has changed the sedimentation pattern of the Nile, and its delta is in danger of accelerated erosion. (NASA)
(right) The delta of the Mississippi River in Louisiana resembles a bird's foot because the delta is subsiding, leaving only the crests of its natural levees above sea level. The subsidence is caused by compaction of sediments and also by general tectonic sinking of the floor of the Gulf of Mexico. (Doug Armstrong after S. M. Gagliano et al., "Hydrologic and Geologic Studies of Coastal Louisiana," Department of the Army, 1970)

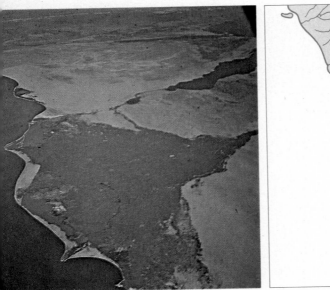

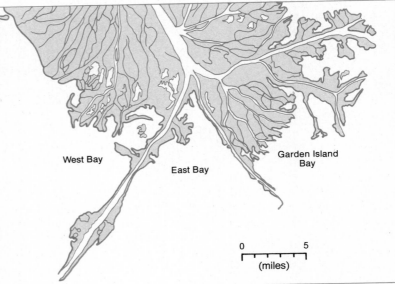

West Bay

East Bay

Garden Island Bay

0   5
(miles)

paction or the sinking of the sea floor under it.

When rivers enter deltas, their discharge usually becomes divided among several *distributary* channels. These are created by floodwaters that spill out across low natural levees to produce new channels. The unusual form of the Mississippi delta is primarily the result of slow sinking of the sea floor, which leaves only the crests of the stream's natural levees projecting above sea level. The present bird's-foot configuration developed after a change of the river's course that occurred at the site of New Orleans only about 400 years ago.

Like floodplains, deltas are fertile and often densely populated. The Nile delta is the home of some 26 million people—two-thirds of Egypt's population. Loss of life and property can be heavy in deltas during floods. Some of the worst disasters occur when hurricanes cause flooding along low-lying deltaic coasts. This has been extremely costly to the dense populations occupying the deltas of the Ganges and Brahmaputra rivers of India and Bangladesh. It is also a hazard to the inhabitants of the Gulf Coast from Florida to Texas.

AEOLIAN PROCESSES AND LANDFORMS

The greater part of this chapter has been devoted to fluvial processes and landforms because they are essential aspects of the environment in which most people live. Now we shall look briefly at the role of wind in erosion and deposition. While the effects of wind are distinctive and significant where they occur, their most spectacular manifestations are confined to certain sparsely inhabited regions of the world. But the wind can modify the land surface wherever the vegetation cover is weak or absent. Often human activities, by destroying the natural vegetation, can cause wind to become an important geomorphic agent in regions where its effects would be insignificant under natural conditions.

Wind as a Geomorphic Agent

Processes related to wind action are known as *aeolian* processes, after the Greek god of the winds Aiolis (Latin, Aeolus). Aeolian erosion and deposition are natural occurrences wherever loose, unprotected sediments are present, as along shorelines and in areas of recent glacial or fluvial deposition. The largest scale aeolian effects are seen in the world's desert regions, where lack of moisture results in vast expanses of bare dry soils.

The bare surface over which the wind sweeps may be thought of as a vast stream bed. Loose particles put into motion by the wind behave like those moved by running water. The smallest particles are carried in suspension as wind-blown dust. Heavier particles move by saltation and traction, as in the stream flow depicted in Figure 13.7.

The velocity of wind, which can greatly exceed that of running water, gives it the energy to move material. Dry sand having a diameter of 0.1 mm begins to move when the wind velocity near the ground reaches 16 km (10 miles) per hour. Coarse sand (2.0 mm) is put in motion by a 50 km (30 miles) per hour wind. Fine dust can be whirled up into the air by a slight breeze. But, while wind can achieve greater velocities than running water, its density is much less, giving it less buoyant force. Thus the wind cannot move rock debris larger than sand and can rarely lift sand more than a meter above the ground.

Aeolian processes mainly affect loose particles. It is uncommon for wind action to wear away solid rock, even in deserts. Wind erosion of rock essentially means sandblasting. This requires an exceptionally strong wind to blow hard quartz against soft rock. Wind erosion much more commonly consists of the removal, or *deflation*, of fine rock debris provided by other gradational processes. Aeolian deflation usually leaves no spectacular landforms. Since deflation slowly lowers large areas, there may be little by which to gauge its magnitude. Often the principal evidence of aeolian deflation is a tree or bush perched on a column of soil—all the surrounding soil having been blown away.

Loess

Deflation is chiefly visible in dust storms (Figure 13.21). These are usually produced by winds related to strong pressure gradients in arid and semiarid regions. The dust, composed of silt-sized particles, is lifted thousands of meters into the atmosphere. When it settles out far from its source, the deposit is known as *loess*.

Loess, which is centimeters to hundreds of meters thick, offers excellent agricultural opportunities. It has a high nutrient content, and soils developed on it are young and unleached. Loess is common in the American Midwest and Mis-

Figure 13.21
These two views illustrate erosion and deposition by the wind. On the left is a severe dust storm, photographed in Colorado in 1935. Here very strong winds associated with frontal disturbances are seen eroding bare agricultural land, stripping away the most fertile part of the soil—its A-horizon. This hazard exists wherever the natural vegetation has been removed from soil that is periodically dry. On the right is a 6-m (20-ft) bank of *loess* in Louisiana. Loess is composed of dust-size mineral particles that settle out of the atmosphere downwind from an area of aeolian deflation. This loess is typical in that its cohesiveness causes it to stand in near-vertical banks and to be gullied where vegetative protection is absent. The depth of loess in the Mississippi Valley region varies from centimeters to as much as 15 m (50 ft). Most Mississippi Valley loess was derived from aeolian deflation of barren late Pleistocene glacial deposits. (USDA Soil Conservation Service)

sissippi Valley and in a strip extending across Eurasia from northern France to northern China. The North American and European loess deposits originated by deflation of Pleistocene glacial deposits. Northern China has extremely thick loess deposits that result from deflation in the Gobi Desert of Mongolia and northwest China. This loess produces a spectacular landscape that is deeply trenched by gully erosion. Hundreds of generations of humans have lived in caves carved out of the soft material.

Dune Landscapes

The most impressive results of aeolian energy are the great sand seas of desert regions. It is a mistake to think of deserts in general as areas of sand dunes. Most desert terrain consists of gravelly plains, eroded rock surfaces, and stream-dissected relief. Nevertheless, deserts are distinctive in having large areas of dune topography as well (Figure 13.22).

In the deserts of North Africa and Arabia, individual sand seas, called *ergs*, with dunes resembling frozen waves, cover tens of thousands of square kilometers. Somewhat less spectacular dune landscapes also cover vast areas in India, western China, and Australia, and small ergs are present in every desert region. An ancient sand sea, comparable to those of North Africa but now grass covered, occupies much of western Nebraska.

Ergs form where sand-carrying winds lose velocity, causing their load of sediment to come to rest. Once an aeolian sand deposit is initiated, it is self-enhancing. It traps arriving sand, which does not saltate as well over sand as it does over rock or hard-packed soil. Ergs contain dunes of many types, reflecting variations in the sand supply and the nature of the sand-transporting winds. The most general division is between transverse and longitudinal dunes.

Transverse dunes form where sand-carrying winds blow mainly from one direction. The dune crests are at right angles to the sand-

Figure 13.22
Sand ripples (foreground) and dunes in Great Sand Dunes National Monument, Colorado. The sand is produced by weathering, carried a short distance by floods of water, and then moved along by strong winds. Sand accumulates in dunes when the transporting wind loses power because of loss of velocity as air currents either converge or diverge. (Grant Heilman Photography)

carrying winds (Figure 13.23 a, b, d, p. 356). The beds of sand-carrying streams also are often composed of small dunes of the transverse type, formed by flows of water rather than air. *Longitudinal dunes* form where the sand-carrying winds come from more than one direction. The dune crests may extend hundreds of kilometers and are believed to be the sum of the vectors of the different sand-carrying winds, as shown in Figure 13.23e.

The most recent direction of sand flow in a dune area can be found by looking at the smallest sand accumulations—the closely spaced *sand ripples* that crawl over the dunes. The migration of sand ripples brings sand up the dune *backslope* (Figure 13.23) to the dune crest. At

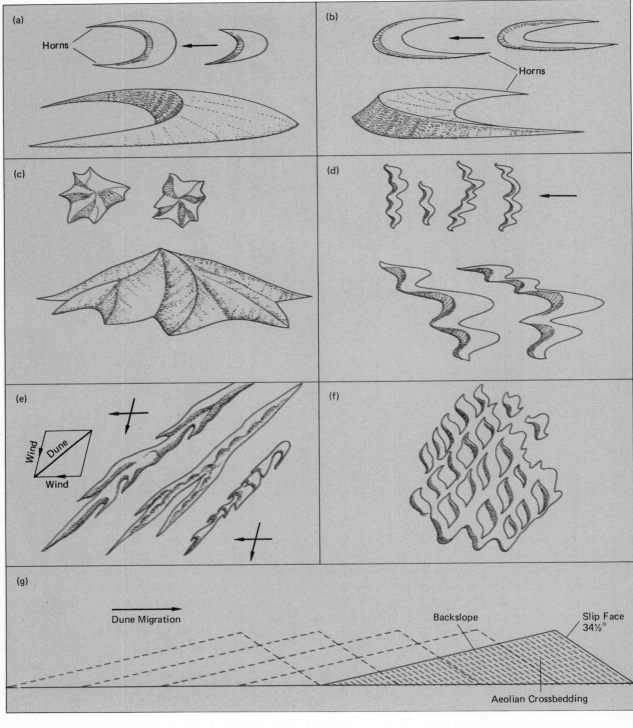

(a)

Horns

(b)

Horns

(c)

(d)

(e)

Wind

Dune

Wind

(f)

(g)

Dune Migration

Backslope

Slip Face
34½°

Aeolian Crossbedding

Figure 13.23 (opposite)
Morphology of sand dunes, with arrows showing the direction of the effective winds. In (a), (b), (c), and (d), typical map patterns of the dune types are shown at the top, with oblique views below; (e) and (f) show the map pattern only.

(a) Individual dunes with elongated tips, or *horns*, pointing downwind are known as *crescentic* or *barchan* dunes. This type of dune is found in dry, vegetation-free environments, and is usually seen migrating across a nonsandy surface of gravel or clay. Barchan dunes may move several tens of meters each year. Symmetrical barchans indicate a nearly constant wind direction.

(b) Single dunes whose horns point upwind are known as *parabolic* dunes. Often they assume a hairpinlike form. Such dunes form where sand blows into a moist area that has a vegetative cover. Vegetation or dampness in the lower part of the dune retards motion there, so that the dry crest pushes ahead of the base, causing the horns to "drag" behind. Parabolic dunes are found along coastlines and often push into forests, engulfing and killing the forest trees. Similar dunes also occur where older vegetation-covered sand accumulations become active again due to natural or artificial destruction of the vegetation. In such instances, the newly active parabolic dunes are called *blowouts*.

(c) Large erg areas sometimes include regularly spaced "sand mountains" as much as 200 m (650 ft) high; these are called *star dunes* or *rhourds*. Rhourds seem to be fixed in position, and their formation, which is not well understood, seems to require winds from several directions. They are seen in Arabia and the Sahara.

(d) Ergs commonly contain dune ridges transverse to the wind direction that appear to be composed of linked barchans. These transverse dune systems are known as *aklé* dunes. They frequently incorporate parabolic segments between the linked barchans. Aklé dune systems reflect effective winds from a single direction.

(e) Where sand supplies are especially abundant, *longitudinal* dunes are often encountered. Some of these are hundreds of kilometers long and as much as 200 m (650 ft) high. These dunes are thought to express the effect of multidirectional winds. Their morphology varies from smooth *whalebacks* to knife-edged *seif* dunes. Many have a fairly complex topography, including variously oriented subsidiary crests and slip faces.

(f) In some ergs, transverse and longitudinal dune systems appear to cross, producing *grid* dunes. Either seasonal or long-term changes in wind direction may be responsible for such patterns.

(g) This diagram shows the migration of a dune as sand is transferred from the backslope to the slip face, causing the dune to be displaced in the direction of the slip face. The internal structure of dunes consists of steeply inclined laminations of sand produced by accretion on the slip faces. The result is *aeolian crossbedding*, which can be seen in many ancient nonmarine sandstones. (Vantage Art, Inc.)

the crest the sand falls onto the steep *slip face* of the dune, which always has a steepness of about 34°. Sand ripples move in different directions from day to day, often approaching the dune crest obliquely. The ripples are asymmetrical, with their steep faces on the downwind side in the direction of ripple travel. Stream beds likewise commonly display ripples, showing that wind and water move particles in much the same way.

SUMMARY

Fluvial and aeolian processes are closely related. Both result from the shear stress of a fluid moving over a surface composed of movable particles. Running water forms streams that compose drainage networks. These evolve systematically in a way that seems to minimize energy expenditures. Stream energy results from the conversion of potential energy to kinetic energy. Stream discharge and turbulence play important roles in providing this energy. Although most stream energy is used to overcome friction, enough energy remains to permit the stream to scour its bed during floods and to move solid particles by suspension, saltation, and traction. Streams also carry a load of dissolved matter.

Streams tend toward an equilibrium condition through negative feedback processes. Insufficient stream energy causes deposition that raises the stream bed, which increases stream energy. Excessive energy causes channel incision or lateral planation that decreases stream slope and reduces stream energy. These adjustments can be seen in the course of a single stream flood. A graded stream has just sufficient energy to move its sediment load.

The longitudinal profile of a stream is a portrayal of its gradient, or relative steepness, from source to mouth. Knickpoints, or sharp offsets causing waterfalls or steep cascades, sometimes interrupt stream profiles but tend to be eliminated over time.

In general, stream flow tends to produce meandering channels. Meandering appears to be a way of equating stream energy to the work of sediment transport. Meanders seem to develop out of pool and riffle sequences that are maintained in the meandering channel. Where a coarse bed load is transported, shallow, braided channels evolve. Deposition of sediment in the channel gives braided streams steep gradients that maintain high flow velocities despite energy losses to friction. Straight channels are controlled by underlying geological structures.

The major fluvial landforms are valleys, floodplains, natural levees, terraces, alluvial fans, and deltas.

Landscape modification by the wind occurs only in areas of sparse vegetation and dry soils. Wind erosion is seen mainly in the form of deflation, which peels down surfaces composed of loose fine-grained material. Material removed by deflation sifts down elsewhere to form deposits of fertile loess. The vast ergs of certain desert regions form where sand-carrying winds lose velocity; they are composed of varying types of transverse and longitudinal sand dunes.

APPLICATIONS

1. Most urban areas are near a permanent stream. Look at the largest stream close to your campus. How has it been modified by human activity? What were its natural characteristics in the prehistoric period?

2. Looking at the same stream, where are current areas of bed or bank erosion? of deposition? Can you make any general statement about stream behavior: (a) upstream from your city? (b) in the urban area? (c) downstream from the urban area?

3. How much have lower order streams that enter the main stream (in question 1) changed in historic time? Can you see evidence of recent progressive change? What explains the tendencies observed?

4. To maintain constant stream velocity (and therefore transporting power), a decrease in stream depth must be offset by a change in one of the other factors affecting flow velocity. What sort of changes could compensate for a decrease in stream depth so flow velocity is constant?

5. Which requires a greater expenditure of energy, a gradient decrease produced by stream incision or a gradient decrease produced by increased stream sinuosity (increased lateral planation)?

6. The lower Mississippi River has followed several different courses to the Gulf of Mexico. Although the old channels have been silted up, the different locations of the ancient streams are clearly evident. What would be the principal evidence of the Mississippi River's former courses?

7. Why are the boulders found on some alluvial fans much larger than those of alluvial fill terraces?

8. Under what conditions would river deltas grow? shrink? be in a steady state?

FURTHER READING

Belt, C. B., Jr. "The 1973 Flood and Man's Constriction of the Mississippi River," *Science*, 189:4204 (August 1975): 681–684. This article analyzes the effects of human modification of the Mississippi River channel, which caused a flow of only moderately high magnitude to produce a flood of catastrophic proportions.

Chorley, Richard J., ed. *Water, Earth, and Man.* London: Methuen (1969), 588 pp. There are several chapters on channeled flows, river regimes, and floods in this collection of articles on water in all its forms, written by experts on the various topics.

Dury, George H., ed. *Rivers and River Terraces.* New York: Praeger (1970), 283 pp. This collection of influential writings concerning fluvial processes and resulting landforms shows the progress of thought on these topics over the past 100 years.

Leopold, Luna B. *Water: A Primer.* San Francisco: W.H. Freeman (1974), 172 pp. This is a simple but technically sound introduction to the general principles of hydrology and the action of water on slopes and in channels, by the former chief hydrologist of the U.S. Geological Survey.

————, **M. Gordon Wolman,** and **John P. Miller.** *Fluvial Processes in Geomorphology.* W.H. Freeman (1964), 522 pp. A landmark when first published, the book presents the findings of research over several decades on fluvial processes by U.S. Geological Survey field workers. It is somewhat technical but highly readable and most informative.

Morisawa, Marie. *Streams: Their Dynamics and Morphology.* New York: McGraw-Hill, Earth and Planetary Science Series (1968), 175 pp. This brief paperback presents most of the fundamentals of fluvial processes in nontechnical terms and in a concise format.

Russell, Richard J. *River Plains and Sea Coasts.* Berkeley and Los Angeles: University of California Press (1967), 173 pp. Many interesting facets of stream behavior are explored, with most examples being drawn from observation of the Mississippi River—especially good on flood plains and deltas.

U.S. Geological Survey. *The Channeled Scablands of Eastern Washington: The Geologic Story of the Spokane Flood.* Washington: U.S. Government Printing Office (1973), 25 pp. The greatest known flood discharge occurred in Pleistocene time when a glacially-dammed lake suddenly emptied. This account of that event is well illustrated and presented in simple language.

Ward, Roy. *Floods: A Geographical Perspective.* New York: Wiley-Halsted (1978), 244 pp. Flood causes, effects, forecasting, and human responses.

CASE STUDY: Water Management on the Mississippi

Historically, the best way to avoid the floods of the Mississippi River was to move to high ground. But as the valley became more settled, people were no longer content to adjust their affairs to the natural workings of the river. So the goal became to eliminate nearly all floods.

Until recently the principal defense against floods was the construction of artificial levees atop the natural levees formed by the river. However, confining the river and preventing its spread onto its normal floodplain during peak discharge raises the river height. A break in the levee under those conditions can produce a flood of disastrous proportions.

In the 1930s, the opening of the Bonnet Carré spillway just north of New Orleans, Louisiana, inaugurated a new approach to flood control on the lower Mississippi. The spillway is a controllable outlet 3 km wide and 7 km long leading from the Mississippi above New Orleans to the lower level of Lake Ponchartrain. During periods of high discharge, river water can be diverted to the lake, which eases the pressure downriver at New Orleans and in the lower delta. The Bonnet Carré spillway has been used several times since the 1930s, averting floods on each occasion.

By 1950 hydrologists and engineers reported that increasing volumes of flood water were flowing through the Old River channel into the Red and Atchafalaya rivers. The Atchafalaya River represents a much shorter and steeper route to the Gulf, and it threatened to divert much, if not all, of the Mississippi River into its own channel. So the Old River Diversion Control was built in the late 1950s, allowing only about one-third of the Mississippi to flow into the Atchafalaya at any time. However, during periods of very

high water, not all of the flow could be led past New Orleans safely, so the Morganza floodway was also built to divert more floodwater from the Mississippi into the Atchafalaya basin.

In 1973 residents along the river knew high waters were coming months in advance. The question was: How high would the water rise? The flood story of 1973 actually began in late 1972. Rainfall had been particularly high in the central Mississippi Valley during the summer and fall. Farther south, rainfall became heavy in the late fall and early winter. By winter, the lower Mississippi was unusually high and there was already talk of possible spillway openings. During the spring, excessive rain continued to fall over the entire basin. The water came dangerously close to the tops of levees. In February the Bonnet Carré spillway was opened, and in mid-April plans were made to open the Morganza floodway for the first time.

There is no perfect flood control system: relieving pressure in one area means adding pressure to another. Opening the Morganza floodway appeared to be the way to protect the greatest number of lives and property, but meant more water downstream in the Atchafalaya River. There was concern that sediment brought by silt-laden Mississippi River waters might fill in the shallow lakes and swamps of the Atchafalaya basin. Several hundred people as well as livestock were evacuated from the Morgan City area in the Atchafalaya basin. Oil and gas wells were flooded and their production was temporarily stopped. Farmland was under water, so crops of cotton and corn could not be planted.

Opening the Bonnet Carré spillway affected the ecology of Lake Ponchartrain, for a time at least, by adding sediment and fresh water that changed the salinity and damaged the oyster population. However, there was some speculation that over time the ecology of the lake would be improved by the flushing action of the water, because it washed out debris and water-clogging vegetation and brought in a fresh supply of nutrients.

The opening of floodways such as the Morganza and the Bonnet Carré prevented the main river from flooding. But water in the main stream was so high that the Mississippi's tributaries were unable to flow into it. Consequently, they overflowed locally, forming "inland seas" of backwater flooding, mainly in Mississippi and Louisiana. Farther north, the flooding was worse, the heaviest occurring around St. Louis, Missouri. In some places, levees broke or washed out, and the water poured through them; in other places, the levees held but the water cascaded over

them. Near St. Louis, people fled their homes, moved back to their sodden, muddy houses when the rivers dropped, and then had to leave again, and yet again, as new crests came down the river.

The rains and the flooding finally stopped in early summer. The flooding had lasted months, and four separate flood crests had flowed down the Mississippi. Parts of Kansas, Iowa, Illinois, Missouri, Kentucky, Tennessee, Arkansas, Mississippi, and Louisiana had been declared disaster areas.

Because the waters subsided so slowly, it was too late that season to plant a cotton crop in the fertile delta area of Mississippi. Soybean, corn, and sugarcane crops were also affected because planting was delayed. The heavy losses in food crops and livestock led to higher food prices throughout the country. Thousands of homes had to be rebuilt, roads and levees repaired, and massive clean-up projects undertaken.

But even with the great losses, many engineers believed that the flood control measures proved as effective as could be expected. Most of the flood warnings came early enough that people could be evacuated, and loss of life attributed to the flood was relatively small.

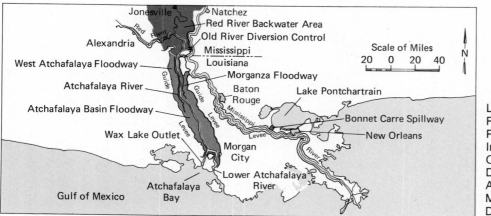

Lower Mississippi River Flood Control Plan. (Vantage Art, Inc. from U.S. Army Corps of Engineers, Department of the Army, Lower Mississippi Valley Division)

Kandahar by Victor Vasarely, 1951.

The land surface may be altered abruptly by a volcanic eruption or a sudden earthquake, or slowly by viselike compression. Such dramatic processes of change are powered by the earth's internal energy. Landforms created by movements of the earth's crust are subject to continued change by weathering and erosion driven by solar and gravitational energy.

CHAPTER **14**

Geologic Structure and Landforms

Differential Weathering and Erosion
Sedimentary Rocks
Igneous and Metamorphic Rocks
Volcanism
Magma Sources and Eruptive Styles
Basaltic Volcanism
Andesitic Volcanism
Volcanism and Humans
Crustal Deformation
Mountain Building
Crustal Extension by Faulting
Crustal Shortening by Faulting
Strike-Slip Faulting
Fold Structures and Landforms

From the window of an airplane, one can hardly fail to be impressed by the geometrical patterns made by landforms in many regions. Seen from a cloud-free flight across the United States, the linear mountains of Nevada, the sinuous lines of cliffs of central Utah, and the seemingly endless winding ridges of Pennsylvania stand out as highly distinctive forms. The scenic character of such areas results from their peculiar geologic structure. Resistant rocks project boldly above their surroundings, and weaker rocks have been worn to low relief. Jointing often controls drainage patterns and relief forms. Plutonic rocks produce particular types of landforms that are different from those developed in sedimentary rocks, and sedimentary rocks that are flat-lying display landforms unlike those seen where rock strata have been deformed by tectonic activity. The distinctive "personalities" of many local landforms, as well as entire regional landscapes, are largely a result of the type and arrangement of the rock masses present.

DIFFERENTIAL WEATHERING AND EROSION

Figure 14.1 reveals the complexity of the terrain in a small portion of the eastern United States. The variety of relief features seen in this region results from *differential weathering* and *erosion*. This refers to the varying responses of different rock types to the weathering processes that fragment rocks and the erosional processes that remove the fragmented material. In the region portrayed in Figure 14.1, the lower Hudson River valley, differential weathering and erosion are possible because various motions of the earth's crust have caused many different rock types to be exposed. West of this valley, in the Catskill Mountains, the rock strata are continuous and almost horizontal, so that the landscape has a more uniform character. Figure 12.15 (page 319) shows differential weathering and erosion on a smaller scale. In arid regions, even minor variations in rock resistance are revealed because the processes of weathering and erosion are even more selective where there is no cover of vegetation or soil.

Sedimentary Rocks

The layering of sedimentary rocks provides the best opportunity for differential weathering and erosion, as each rock stratum has its own

Figure 14.1
This *physiographic diagram* by the master cartographer Erwin Raisz reveals how geologic structure affects the scenery of the lower Hudson River region. North is at the top. In the upstream area, the cross section shows that gently dipping resistant sandstones form the Catskill Mountains, with metamorphic rocks underlying the Hudson River valley. In the cross section farther south, the highlands are shown to be produced by granitic rocks (darkest symbol), with basaltic lavas making ridges in the lowland west of the river. (From the Report of the International Geological Congress, Washington, D.C., 1933)

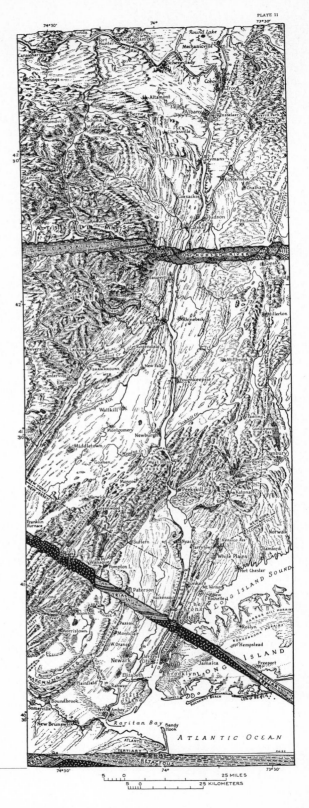

physical and chemical characteristics that may be quite different from those of adjacent strata. The various types of sedimentary rocks—shale, sandstone, conglomerate, and limestone—differ in their effect on landscapes.

Shale is mechanically weak, tending to fall apart in thin flakes. It produces clay soils that are slow to accept water. Therefore, it causes above-average runoff, resulting in a high number of stream channels per unit of area. Shale is, in fact, the most erodible of all rock types. It generally produces lowlands or dissected slopes fringing higher hills composed of other material. In dry regions the erosion of weakly cemented shales often creates a maze of closely spaced ridges and ravines known as *badland* topography. An excellent example has been set aside in South Dakota's Badlands National Monument, but similar topography occurs in many locations west of the Rocky Mountains.

Sandstone can be either weak or strong physically. Often it is porous, permitting water

Figure 14.2
Sandstone mesas and buttes dominate the landscape in Monument Valley in southeastern Utah. Erosion of shales under the sandstone caprock causes collapse of the caprock, maintaining the vertical cliffs. Where the shale is not exposed at the land surface, the sandstone is eroded into rounded forms and no vertical cliffs are present. (Dr. John S. Shelton)

to soak into it, or into the sandy soils that form on it. This reduces runoff and erosion. Since it erodes slowly, sandstone often forms ledges or cliffs (Figure 14.2) or hills rising above other rock types. Conglomerates, which are composed of cemented gravel or cobbles, are even more resistant than sandstone and almost always produce bold ledges.

Differential weathering and erosion in areas of sandstone and shale bedrock often produce a series of cliffs and slopes. These vary in form depending upon the tectonically produced tilt, or *dip*, of the rock, as shown in Figure 14.3 (see p. 366). Rock dip is measured as the angle between the plane of the rock strata and a horizontal plane (Figure 14.4, p. 366). Where the dip is gentle, a *cuesta* is formed. This is a plateau capped by resistant sedimentary rock that gradually dips below the land surface. Every cuesta has a steep erosional *scarp* and a gently dipping *backslope* capped by the resistant layer of rock. The cuesta scarp retreats due to erosion of the weak rock (usually shale) under the caprock, which causes the edge of the caprock to collapse from lack of support.

Flat-topped *mesas* may be detached from the cuesta as its scarp retreats in the direction of bedrock dip. Erosion eventually reduces the mesas to smaller *buttes* that no longer preserve extensive flat summits. Where the bedrock dips steeply, the result is a *hogback ridge*. The two slopes that meet at the sharp crest of a hogback are similar in angle, but one is an erosional feature that cuts across rock outcrops while the other is the top of the dipping bed that has created the ridge. The arid Colorado Plateau region of the western United States, including parts of Colorado, Utah, Arizona, and New Mexico, is famous for its cuestas, mesas, buttes, and hogbacks, but similar forms, softened by a forest cover, are also present in the eastern United States in Kentucky and Tennessee (cuestas, buttes, mesas) and in Virginia and Pennsylvania (hogbacks).

Limestone is peculiar in that its dominant mineral, calcite ($CaCO_3$), dissolves in soil mois-

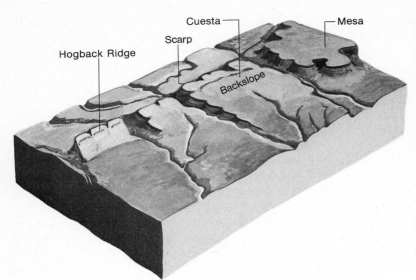

Figure 14.3
Mesas, cuestas, and *hogback ridges* are typical landforms that can be produced by differential erosion in regions where a resistant caprock overlies less resistant layers. A flat-topped mesa is formed where the rock strata are horizontal. A cuesta is formed where the strata dip at an angle of a few degrees, so that a cuesta possesses a steep face, or *scarp*, and a gently dipping *backslope*. A hogback ridge is formed where the strata dip steeply. One slope consists of the dipping caprock, and the other is the erosional scarp that cuts across the caprock and the underlying layers. (John Dawson)

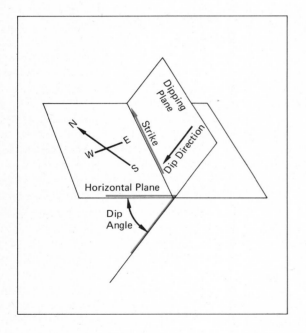

Figure 14.4
The attitude of any inclined plane, such as a fault or a tilted layer of rock, may be described precisely in terms of the *strike* and *dip* of the plane. The strike is the compass bearing of the line of intersection between the inclined plane and a horizontal plane. It is read in degrees east or west of a north-south direction. The dip is the angle the inclined plane makes with a horizontal plane; it is given in degrees together with a compass direction. In this diagram, the strike is about N 30° E and the dip is to the northwest at about 50°. (Vantage Art, Inc.)

ture, which is usually slightly acid because of its content of CO_2 and organic acids derived from decaying vegetation. Solution of limestone is concentrated where water enters, such as along joints and bedding planes. Where joints inter-sect, solution is especially rapid, producing underground voids of some size. These may eat through to the surface, or the surface rocks may collapse into them. In either case a depression known as a *sinkhole* results. Some areas of Indiana have 300 sinkholes per sq km (Figure 14.5). Portions of Kentucky and Florida are also pitted with sinkholes, which often divert small surface streams to underground routes.

The truly unique feature created by limestone solution is the *underground cavern*. Vast subsurface voids can form wherever thick masses of limestone are present in an area of high relief. High relief is required so that the

Figure 14.5
Limestone is susceptible to erosion by solution, particularly by acidic water. In humid regions where the bedrock consists of limestone, chemical erosion creates surface sinkholes and subsurface caverns. This landscape in southern Indiana is dotted with sinkholes, some of which contain small ponds. No surface streams are present in such areas; all drainage is underground. (Dr. John S. Shelton)

water table slopes toward streams, resulting in lateral flow of groundwater. This is necessary for large-scale solution to occur. Of the 48 coterminous states, all but Vermont and New Hampshire have limestone caverns, and some states have hundreds. Individual caverns may have more than 100 km (60 miles) of passages, including rooms tens of meters high and thousands of square meters in area. Many limestone caverns are beautifully decorated with hanging icicle-like *stalactites* and more massive upthrusting *stalagmites*, both of which are composed of calcium carbonate (Figure 14.6, p. 368). These and a variety of other types of cave deposits, known

collectively as *speleothems*, are formed where groundwater saturated with calcium carbonate seeps into the cavern and evaporates, leaving behind its dissolved load.

Variations in limestone solubility and the effects of different climates and geologic structures cause limestone landscapes to vary widely in appearance. Nevertheless, all landscapes whose form is dominated by solution effects in limestone are known as *karst* landscapes. The name is derived from the Karst region of Yugoslavia, where solution features were first described in detail in 1893.

Limestone can be either a resistant or a weak rock because it varies in strength and massiveness (degree of jointing and bedding). In humid areas it often dissolves to create lowlands surrounded by higher land on other rock types. However, hard limestones often form cliffs and even mountain peaks, such as those in the Alps and Himalayas. In arid regions there is little water available to dissolve limestone, and so it acts as a resistant rock, forming cuestas, mesas, and

Figure 14.6
This narrow passageway in New Mexico's Carlsbad Caverns shows the classic features produced by solution in thick beds of jointed limestone. The open passage is dissolved out along the plane of a vertical joint by groundwater draining laterally through the joint system at a time when the water table was much higher than at present. When the water table fell below the level of this passage, air entered and soil moisture percolating downward toward the lowered water table dripped from the cavern roof. This moisture carries calcium carbonate dissolved from the rocks above. Evaporation from water slowly dripping from the cavern ceiling caused precipitation of some calcite, which accumulated in downward-growing, icicle-like *stalactites*. Evaporation from water splashing on the cavern floor precipitated more calcite, producing upward-growing, columnar *stalagmites*. In time the stalactites and stalagmites shown may join to form columns that reach from the floor to the ceiling of the cavern. (T. M. Oberlander)

hogbacks. In the humid tropics aggressive solution along joints in thick limestone beds produces vertical-sided *karst towers*, which are the forms seen in Chinese and Japanese paintings.

Igneous and Metamorphic Rocks

Differential weathering and erosion are less conspicuous in igneous and metamorphic rock than in sedimentary rock. Nonetheless, there are significant variations in resistance in these rock types. Igneous and metamorphic rocks differ texturally and chemically, so that the entry of water and the nature of chemical reactions vary widely. In plutonic and gneissic rock, there are no bedding planes, so that jointing becomes the dominant factor in surface form. Areas of high joint density are weathered and eroded downward, while areas of few joints remain high-standing. Sometimes this leaves great spires or domes of massive unjointed granitic or gneissic rock jutting above surrounding decayed, jointed rock. Such a landscape is dis-

Figure 14.7
The spectacular rock cones, or "sugarloaves," of Rio de Janeiro, Brazil, result from weathering and erosion in gneissic rock with variable jointing. Unjointed masses resist subsurface chemical weathering and are left standing above surrounding decayed and eroded rock that is densely jointed. (© Louis Villota/The Image Bank)

played around Rio de Janeiro, Brazil (Figure 14.7).

In addition to rock type, the arrangement of rock masses plays an essential role in the appearance of landscapes. The varying configuration of rock bodies is the result of past tectonic activity, including volcanism and crustal compression and extension. We shall now turn to this aspect of structure as a control of landform development.

VOLCANISM

Volcanism refers to the intrusion of magma into the earth's crust and the extrusion of volcanic gases and molten material at the earth's surface. While volcanism's greatest effect is in the oceans, which are entirely floored by volcanic lava, its most spectacular displays are on the land.

Although volcanism is a widespread phenom-

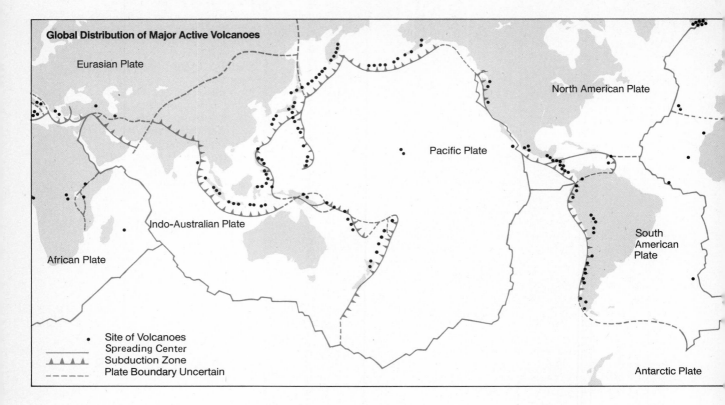

Global Distribution of Major Active Volcanoes

Eurasian Plate

North American Plate

Pacific Plate

Indo-Australian Plate

African Plate

South American Plate

Antarctic Plate

• Site of Volcanoes
— Spreading Center
▲▲▲▲ Subduction Zone
--- Plate Boundary Uncertain

enon, Figure 14.8 reveals that the global distribution of active volcanoes is quite systematic. Large-scale volcanism is related to crustal plate boundaries. Three types of plate boundaries exist: those where plates are pulling apart, those where plates are sliding past one another, and those where plates are pushing together, creating a subduction zone. The first two produce nonexplosive volcanism; the last is the location of violent explosive volcanic eruptions.

Magma Sources and Eruptive Styles

One can view eruptions of the earth's most active volcano, Hawaii's Kilauea, from close at hand in absolute safety. Other volcanoes devastate enormous areas with amazing suddenness and spare only those eye-witnesses who are far

away. This fact has been reaffirmed by the May 1980 eruption of Mt. St. Helens in the state of Washington, which in an instant destroyed all life in an area of 320 sq km (125 sq miles). The nature of a volcanic eruption and the form of an individual volcano are largely a consequence of the source of volcanic energy—the magma feeding the volcano.

There are two general sources of magma. One is the portion of the mantle that lies below the rigid crust of the earth. The permanent layer of fluid rock material here is chemically basic *sima*, having a silica content of 50 percent or less. The other source of magma is the melting of older rocks that have descended into the earth's furnacelike interior in the process of subduction. Magma produced by the melting of older rocks contains more than 65 percent silica. Silicic magma has a low density and is inflated by vol-

Figure 14.8
This map shows the locations of the known active volcanoes of the world; for clarity, some volcanoes have been omitted from regions where many are present. The map also shows the principal plate boundaries, oceanic spreading centers, and subduction zones associated with the earth's crustal plates. Regions of volcanic activity are frequently located near active plate boundaries, where molten rock from the earth's interior wells up through fissures. Note the large number of active volcanoes around the rim of the Pacific Ocean, which is ringed by oceanic trenches. The volcanoes in Iceland and some of the volcanoes in the Atlantic Ocean are associated with the spreading center along the Mid-Atlantic Ridge. Volcanoes in the Caribbean, in East Africa, and in the eastern Mediterranean are associated with plate boundaries as well. Conversely, the east coasts of North and South America, where there are no plate boundaries, are devoid of volcanic activity. Isolated volcanic regions far from plate boundaries, such as the Hawaiian Islands, are due either to "hot spots" in the earth's mantle, or fractures in the lithospheric plates, which allow molten rock from the earth's interior to reach the surface. (Andy Lucas after F. M. Bullard, *Volcanoes*, 1962, University of Texas Press)

canic gases. These gases explode outward when their pressure exceeds the weight of the rocks confining them. It is this silicic magma resulting from subduction that produces dangerous explosive volcanism.

When chemically basic magma reaches the surface, the magmatic gases are still dissolved in the melt, making it very fluid. It pours out freely, with little explosive activity, despite the presence of impressive steam clouds. The more explosive silicic magma is actually "cooler" than basic magma (about 900°C vs. 1,200°C, or 1,600°F vs. 2,200°F), and its gases have already separated out below the surface, where they are under enormous pressure. Thus silicic magma always erupts violently. The first eruption of gases and steam reduces the pressure within the magma, which allows more magma to flash into gas, causing more explosions, more pressure re-

lease, more gas, and so on in a chain reaction that can continue for hours or even days.

The force of some volcanic eruptions is incredible. When Krakatau, an island volcano between Java and Sumatra, exploded in 1883, the sound carried 3,000 km (1,900 miles) to Australia. Krakatau itself vanished. Ash falls from a similar explosion at Santorin in the Mediterranean Sea may have destroyed the ancient Minoan civilization of Crete about 3,500 years ago. The explosive eruption of Mt. St. Helens in 1980 leveled 400 sq km (150 sq miles) of forest and removed 2 cu km (0.75 cu mile) of the volcano itself.

Basaltic Volcanism

Where plates of the earth's crust are being pulled away from one another, basic magma rises to fill the opening between them. This erupts at the surface as basaltic lava. The rising magma creates the oceanic ridges, which occasionally push above the sea surface, as in Iceland. Basic magma is also sometimes erupted where portions of oceanic plates are sliding past one another. The ocean floor is studded with volcanoes that are now inactive. Many of these have been carried away from the ocean ridge systems by sea floor spreading.

Not all active volcanoes are associated with plate boundaries. Some lie at the ends of linear chains of submarine mountains. These volcanoes mark the position of so-called "hot spots" in the earth's mantle, where plumes of volcanic activity remain fixed in the mantle while lithospheric plates move over them. The extinct volcanoes that form the majority of such mountain chains have moved past the hot spot, so that the trend of the submarine mountain chain records the direction of plate motion. The Hawaiian chain is the best known example, its volcanic rocks increasing in age with distance from the hot spot, presently situated under the island of Hawaii itself (Figure 14.9, p. 372).

Volcanic eruptions on the sea floors build *shield volcanoes*, which have gentle slopes and

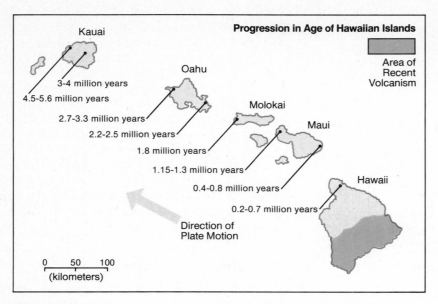

Figure 14.9
The ages of the Hawaiian Islands, as determined by the potassium/argon radiometric dating method, show a general progression from old to young moving southeastward along the chain. These observations were not explained until the development of the theory of plate tectonics, which attributes the formation of the Hawaiian Islands to a plume of volcanic activity fixed in the earth's mantle. As the Pacific plate drifts over the location of the plume, periods of volcanic activity cause the formation of islands of volcanic rock. (Andy Lucas after data by Ian McDougall, *Nature Physical Science*, Vol. 231, 1971)

broad rounded summits (Figure 14.10). Hawaii's Mauna Kea and Mauna Loa are the world's largest mountains, being built some 9,000 m (30,000 ft) above the sea floor. Their slopes rarely exceed an angle of 10 degrees. All shield volcanoes on the land and in the sea are composed of basaltic lava. The Hawaiian shield volcanoes clearly display two contrasting types of basalt (Figure 14.11, p. 374). The hottest flows issue in the form of very fluid *pahoehoe* (pah-ho-ay-ho-ay) lava, which solidifies with a satin-smooth skin that is often conspicuously wrinkled by the continued flow of the lava underneath. Somewhat cooler basalt issues as slaglike rough-surfaced *aa* (ah-ah) lava.

In addition to producing enormous volumes of free-flowing lava, nonviolent oceanic eruptions of the type seen in Hawaii and Iceland are still forceful enough to hurl solid particles hundreds of meters into the air. These rain down to form deposits of *pyroclastic* ("fire-broken") debris. This includes solidified clots of lava and fragments of rock torn loose during the eruption. The larger particles, or *scoria*, fall close to the vent and build up a *scoria cone* or *cinder cone*, such as Hawaii's Diamond Head. Sometimes the magma is so gaseous that it rises to the surface as a froth, like the foam on beer. This solidifies as spongy *pumice*, which is light enough to float in water.

Basaltic eruptions similar to the oceanic type also occur on the continents. Several land areas

Figure 14.10
(top) Basaltic *shield volcanoes* such as Hawaii's Kilauea (foreground) and Mauna Loa (horizon) have very gentle slopes, making their true sizes hard to appreciate. Mauna Loa rises almost 3,000 m (9,800 ft) above the summit of Kilauea. In the foreground is the summit crater of Kilauea volcano. (T. M. Oberlander)

(bottom) The flood basalts shown here comprise the east wall of the erosional trench of Grand Coulee in the state of Washington. Such great piles of superimposed lava flows occur in areas of crustal tension and rifting. (Dr. John S. Shelton)

are famous for their *flood basalts*, built up to a thickness of a thousand meters or more by repeated outpourings of lava. Areas of flood basalts include the Columbia Plateau region of Washington and Oregon and the Snake River Plain of Idaho, as well as portions of India, Brazil, Patagonia, South Africa, and Antarctica. Some of the lava floods on the land may have occurred during the break-up of continents as new oceans were being born. The lava flows forming the Hudson River Palisades probably originated in this way. The individual flows in areas of flood basalts may cover thousands of square kilometers, be tens of meters deep, and have volumes of tens to hundreds of cubic kilometers. A lava flood of this type occurred in Ice-

Figure 14.11
(right) Due to the gases dissolved in it, *pahoehoe* lava, shown here at Kilauea crater in Hawaii, is much more fluid than *aa* lava even though the general chemical composition is similar. Note the thin layer of solidified crust on the molten rock. When the lava solidifies, the surface is left in the form of smooth billows or ropes. (Frank Sojka © Island Pictures, Honolulu, Hawaii)

(bottom) This advancing *aa* lava flow on Hawaii is characterized by chunky, angular blocks mixed with still-molten lava. When an *aa* flow solidifies, an almost impassable field of jagged lava is formed. (Charles A. Wood)

land in 1783, producing some 12 cu km (3 cu miles) of new basaltic rock.

Most areas of flood basalts on the land have been somewhat dissected by stream erosion. The resulting canyon walls clearly expose the separate flows, one atop the other, as can be seen along portions of the Columbia River in Washington and the Snake River in Idaho (Figure 14.10).

Andesitic Volcanism

Despite the enormous volumes of lava generated by basaltic volcanism in the ocean basins and on the land, basalt eruptions are not violent when compared to volcanic activity where crustal plates are converging. Where rocks are forced down to melt in subduction zones, volcanism is far more dangerous, generating much smaller amounts of lava, but ejecting great volumes of coarse and fine pyroclastic material, called *tephra*, in explosive eruptions. The tephra (volcanic ash) and lava are principally andesitic to rhyolitic in composition. Explosive volcanism is confined to the continental edges and to volcanic *island arcs*, such as the West Indies and the Japanese, Philippine, and Indonesian archipelagos (island chains)—all of which lie above active subduction zones.

The first products of volcanic eruptions on the land are blankets of tephra and scoria cones like Paricutín, born in a Mexican cornfield in 1943 and built to a height of 400 m (1,300 ft) in eight months (Figure 14.12, p. 376). Continued eruptions may veneer a scoria cone with lava, making it a *composite cone*. Andesitic lava flows usually consist of a stiff pasty interior covered by a rubble of hardened blocks, unlike either aa or pahoehoe lava. As lava coats the cone, it is also injected into cracks in the cone, where it solidifies as lava *dikes*. These reinforce the cone's structure. As the cone grows higher, lava may issue from its side rather than from the summit crater. Over hundreds of thousands of years, giant steep-sided *strato-volcanoes*, such as Japan's Fujiyama, Washington's Mt. Rainier, and California's Mt. Shasta, are built—often rising 3,000 m (10,000 ft) above the surrounding countryside (Figure 14.12). All the earth's large strato-volcanoes have been built within the last one million years, and every one is potentially dangerous. Inactive older volcanoes are quickly removed by erosion.

A peculiar type of volcano is the so-called *plug dome*. These are seldom of large size, but their eruptions are extremely hazardous. Plug domes are formed when a mass of unusually stiff and highly silicic magma pushes to the surface. The pasty lava (dacite or rhyolite) congeals as it rises, jamming the surface vent so that enormous pressure builds beneath the volcano. This is released in catastrophic explosions that may hurl large blocks several kilometers. Plug domes can be recognized by the stubby masses of hardened lava that often ring their summits.

Associated with the plug dome is the *nuée ardente* (glowing cloud) type of eruption. A nuée ardente is an avalanche of glowing volcanic ash that moves down the flank of a volcano at speeds of more than 160 km (100 miles) per hour. Everything in the path of a nuée ardente is baked to a cinder or set afire instantly. In 1902 a nuée ardente issuing from Mt. Pelée on the Caribbean island of Martinique destroyed the town of St. Pierre, snuffing out almost 30,000 lives. Only two persons in St. Pierre survived—one in an underground dungeon.

Many giant strato-volcanoes have satellite plug domes, or have at times behaved like plug domes. They have erupted so violently that their summits have collapsed, producing vast craters known as *calderas*. This occurred in the 1980 eruption of Mt. St. Helens. These calderas often hold circular lakes. Crater Lake, Oregon, is an outstanding example (Figure 14.12). It partly fills a caldera 10 km (6 miles) wide and initially 1,200 m (4,000 ft) deep. Mt. Vesuvius, near Naples, was partially destroyed in its great eruption of 79 A.D., which buried Pompeii in volcanic ash. In the past 1,900 years Vesuvius has rebuilt itself out of the caldera created by that eruption.

It is difficult to say whether a volcano is active, dormant (potentially active), or extinct. Prior to its eruption in 79 A.D., Vesuvius had been inactive for centuries. Crater Lake was created by a violent eruption about 6,600 years ago, but subsequently much of the crater was filled by further volcanic activity. Wizard Island, at one side of the lake, is the summit of a volcanic cone built within the caldera. Still, in the volcano-studded Cascade Range of Washington, Oregon, and northern California, only Mt. St. Helens in Washington and Lassen Peak in California have produced violent eruptions in the past 200 years.

The Cascade Range is part of the "Pacific Ring of Fire"—a nearly continuous chain of volcanoes circling from South America through Alaska, Japan, the Philippines, and New Zealand. This gives clear evidence of crustal subduction around nearly the entire rim of the Pacific Ocean. The only interruptions in the volcanic ring are north of the Cascade Range in Canada and south of the Cascade Range in California and northern Mexico. In these gaps, crustal plates seem to be sliding past one another rather than converging.

All volcanoes eventually become inactive. When they do, erosion wears them away. All

Figure 14.12
These photographs show a variety of volcanic landforms produced by explosive eruptions.

(left) The simplest form is the *scoria cone*, produced by the fall of pyroclastic materials that have been hurled into the air by the force of escaping magmatic gases and steam during a volcanic eruption. The example here is Paricutín volcano, which rose from a Mexican cornfield in 1943 and erupted sporadically until 1952. (R. Segerstom/U.S. Geological Survey)

(top) *Strato-volcanoes*, such as Mount Hood in Oregon, are among the most impressive of all constructional landforms. Cones of this type are composed mainly of andesitic lava and pyroclastic material resulting from explosive eruptions. (T. M. Oberlander)

(bottom) Volcanic calderas are large circular craters produced by the collapse of the summits of volcanoes during extremely violent eruptions. Crater Lake in Oregon, shown here, was produced by cataclysmic eruptions of a large volcano of the Shasta type some 6,600 years ago. The caldera is about 10 km (6 miles) wide and 600 m (1,900 ft) deep despite the growth of new cones within the caldera. (Dr. John S. Shelton)

Figure 14.13
(top) Flat-topped *table mountains*, such as this one in southern California, form when the land surrounding a lava flow is removed by erosion. This occurs when lava enters valleys cut in erodible rocks. The lava itself erodes very slowly since its great permeability allows rainwater to soak into it rather than running off over its surface. (T. M. Oberlander)

(left) Shiprock, in northwest New Mexico, is a striking example of a *volcanic neck*. Note the extensive volcanic dikes radiating from Shiprock. (Dr. John S. Shelton)

GEOLOGIC STRUCTURE AND LANDFORMS

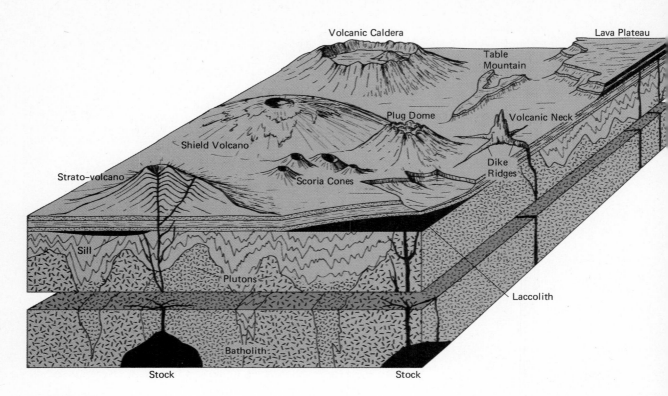

Figure 14.14
This diagram summarizes the variety of surface and subsurface forms produced by volcanism. The cross-section shows that a complex of intrusive igneous rock bodies underlies this area. The separate intrusions, called plutons, collectively form a *batholith* (such as the Idaho batholith or the Sierra Nevada batholith). Individual volcanoes are fed by small intrusions of magma that form subsurface *stocks* and *laccoliths*. A stock cuts up through the structure, whereas a laccolith is injected between older rock layers, doming up the surface above it. A *sill* is a sheet of magma injected between rock layers. A *dike* is a sheet of magma that cuts through older rock. Erosional lowering of the land surface may leave dikes standing up as walls of rock. Volcanoes often form where large dikes cross, concentrating the upward flow of magma.

Several types of volcanoes are shown. *Scoria cones* result from brief explosive eruptions that produce a ring of pyroclastic debris around the volcanic vent. *Plug domes* are produced where pasty silicic magma is forced out at the surface. They are steep-sided,

with a crest that may bristle with hardened masses of lava that have been pushed out as plugs. *Shield volcanoes* have gentle slopes and broad summits produced by huge floods of free-flowing basaltic lava. Often they have a large summit crater. *Strato-volcanoes* are high cones composed mainly of andesitic lava and pyroclastic material. Lavas and scoria are interbedded. *Volcanic calderas* are created by the collapse of strato-volcanoes during violent eruptions. Many have a summit lake and a new cone within the caldera, as shown. *Lava plateaus* result from lava floods fed by large eruptions from lengthy fissures rather than central vents. *Table mountains* are left when the land surrounding a lava flow is eroded away, producing a relief inversion in which an old lava-filled valley is transformed into a ridge. *Volcanic necks* are found where erosion has removed a volcano and lowered the land surface below the volcano's base. The volcanic neck is the lava-filled vent that fed the missing volcano. Volcano necks stand up as spires where the surrounding rocks are more erodible than the volcanic rock. (Vantage Art, Inc.)

that is left in some areas are landforms resulting from the differential erosion of the lavas and ash composing the volcano. The most impressive forms are *volcanic necks* that rise as pinnacles composed of the lava left in the "throat" of the eroded volcano; *dike ridges* that form freestanding rock walls, often radiating from a volcanic neck; and *table mountains* consisting of winding, flat-topped ridges of lava more resistant to erosion than the rocks over which it once poured (Figures 14.13 and 14.14).

Volcanism and Humans

Large volcanoes pose many kinds of hazards. Violent eruptions cause death by baking and suffocation. Volcanic ash can be toxic to humans, animals, and plants. Lava flows are de-

structive to property. Equally damaging are volcanic mud flows, known as *lahars*. These result from rainfall on fresh volcanic ash, lava eruptions under glaciers or snowfields, and slope failure on volcano summits weakened by eruptions, earthquakes, or acidic fumes vented from subsurface magma.

But there are also positive aspects to volcan-

Figure 14.15
(left) This schematic diagram shows the type of geologic structure in steam fields that can be readily exploited for power generation. A confining layer of rock covers an aquifer permeated with water. Water is heated by contact with hot subsurface rock, and steam is generated. The steam comes to the surface through fissures or is led to the surface through wells sunk into fissures or into the aquifer. The high-pressure steam may then be used to operate electric generators. Geothermal energy is a comparatively clean and inexpensive source of power, but its use is confined to regions having subterranean heat sources. (John Dawson)

(bottom) The hot water issuing from Castle Geyser in Yellowstone National Park, Wyoming, is driven upward by steam generated from groundwater in fissured hot rock deep below the surface. (David Miller)

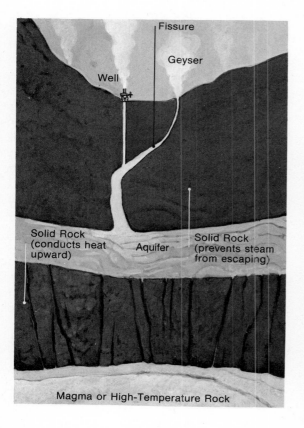

ism. Mineral-rich fresh volcanic ash and decomposed basaltic lava create excellent agricultural soils. This is especially important in the wet tropics, where leaching of nutrients makes older nonvolcanic soils infertile. The productivity of volcanic soils in Japan, the Philippines, Indonesia, and South America has, in fact, lured dense populations onto volcanoes, close to the hazards of explosive blasts, lava flows, ash falls, and lahars.

Another important aspect of volcanism is its role in the formation of metallic ores, such as those of iron, copper, lead, zinc, tin, tungsten, nickel, and chromium, as well as gold and silver. Metallic ores are deposited where hot, richly mineralized solutions move upward from magma bodies and enter older rock. Ore formation does not occur automatically, however; it depends on local rock and magma chemistry.

Finally, pockets of subsurface magma can be made to serve humankind. In a few areas of active, dormant, or extinct volcanism, we can see such natural phenomena as steam vents, geysers that periodically erupt boiling water, and sulphur springs with their smell of rotten eggs (hydrogen sulfide fumes). They all reveal a subsurface energy source that we can utilize as *geothermal energy* (Figure 14.15).

To bring geothermal energy directly from the zone of fluid magma that is present everywhere beneath the earth's crust would require wells many tens of kilometers deep. These are costly and hazardous to drill. But the hot springs and steam jets in volcanic regions bring heat energy to the surface naturally. Hot springs occur where groundwater rises to the surface after being in contact with hot rock or steam vented from magma. The steam itself may remain trapped in the subsurface within reach of wells, or it may vent to the surface through rock fissures. This steam is already used to heat water and generate electricity in Italy, Iceland, New Zealand, and some locations in the western United States. As the global energy crisis deepens, those nations having accessible geothermal energy supplies will be fortunate indeed.

Volcanism is a spectacular and obtrusive form of movement within the earth and creates impressive landforms on a variety of scales. But other movements, often unseen and occurring over a long time, have had much wider effects on the continental surfaces, having produced the structures that control the evolution of landscapes in many parts of the world. Thus we must turn our attention to the various motions of the earth's crust, the structures they produce, and the landforms that are associated with these structures.

CRUSTAL DEFORMATION

The earth's crust is subject to many different types of stresses as crustal plates move laterally—colliding, pulling apart, and scraping past one another. These stresses can always be resolved into tension and compression, both of which cause the deformation of rock masses (Figure 14.16).

Rocks are strained by tension where an area has been stretched or arched upward and also where two adjacent sections of the crust are moving parallel to one another in opposite directions. Tension causes rock masses to break, forming *faults*. Faults produced by crustal extension allow portions of the crust to tilt or sink vertically.

Where crustal plates are moving toward one another, rocks are strained by compression. Compression generally causes sedimentary rocks to be thrown into wrinkles called *folds*. Folds always indicate crustal shortening. Strong compression can also produce faults by causing rock masses to break into slabs that lap onto one another like roof shingles. We shall look more closely at the individual landforms associated with faulting and folding later in the chapter.

In addition to folding and faulting, there are very broad vertical motions of the earth's crust. Although these do not create striking geological

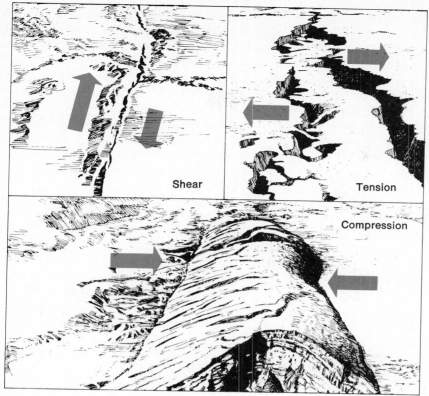

Figure 14.16
Shear, tensional, and compressional forces on the earth's crust. These three drawings show some of the possible landforms that can result from these forces or combinations of them.

(top left) Shear produces horizontal displacement, causing offset stream valleys, interruption of groundwater flow, and demarcations in vegetation.

(top right) Tension has produced a down-dropped block, forming a steep-sided *graben*.

(bottom) Compression has folded sedimentary rocks into a ridge. (John Dawson)

Shear

Tension

Compression

structures, they are important because they help control erosion and deposition. Rising areas are dissected by fluvial erosion, forming regions of hills, plateaus that preserve extensive flat areas between valleys or canyons, and mountains. Sinking areas are usually regions of low relief in which sediments are deposited. However, in a few areas hilly or mountainous country produced by earlier uplift has sunk, and erosional valleys have been invaded by arms of the sea, as in Greece, the Aegean islands, and western Turkey.

Broad uplift may result from localized heating deep in the crust or from the interaction of converging crustal plates. Smaller areas often rise because of a large subsurface intrusion of magma (Figure 14.14). As such an area rises, it is eroded. The result is usually a circular or elliptical area of plutonic rocks surrounded by an upturned rim of sedimentary rocks, which may form circular hogbacks. South Dakota's Black Hills are classic examples.

Some areas are rising today because of removal of load from the land surface. A load that has come and gone several times over the past two or three million years is glacial ice, which has been more than a kilometer thick over areas of millions of square kilometers (see Chapter 15). Both Scandinavia and eastern Canada are presently rising at rates up to 20 m (65 ft) per 1,000 years because continental ice sheets have withdrawn from them within the past 10,000 years.

Geological structures such as domes, folds,

and fault blocks are hardly ever seen intact; they are attacked by erosion as soon as they begin to form. But they are important to understand because they control the development of landforms as well as the location of mineral and fuel deposits. Subsurface structure can be deduced by studying the surface rock patterns and by measurements of local magnetism, gravity, and the behavior of seismic shock waves produced by explosions.

Having noted the major stresses affecting the earth's crust, we can turn our attention to the landforms resulting from crustal deformation, both on the large scale and in detail.

Mountain Building

Although volcanism has created some of our planet's most majestic individual mountains, all sizable mountain ranges—even those capped by volcanoes—have been formed by larger-scale motions of the earth's crust. Nearly all mountain systems began as deep troughs in the sea floors close to the margins of continents. These vast depressions collected sediment layers more than 10,000 m (over 33,000 ft) thick before being compressed and their sediments crumpled into the structures seen in mountain ranges. Until the 1960s it was not known how or why these great oceanic troughs, called *geosynclines*, were formed, or why their deposits were crushed together in the process that transformed them into mountains.

During the 1960s earth scientists came to understand the subduction process. It is the descent of crustal plates in subduction zones that first produces oceanic trenches that fill with sediment and then crushes the sediment mass collected in the trenches. The ocean trenches are, in fact, living geosynclines, serving as the birthplace of mountains. The sediment-filled geosynclines from which mountains are made also include the underwater continental shelves and the continental rises (see Chapter 11). Although every coast has a continental shelf and a continental rise, future mountain ranges will be built only where a trench system is also present.

The crushing of geosynclinal sediments does not itself produce mountain systems. Strong vertical uplift of the rock squeezed in the geosyncline seems to occur later, when the rate of compression slackens. A depth of 15 to 20 km (9 to 12 mi) of sedimentary rock has been removed by erosion during the slow rise of California's Sierra Nevada range, causing plutonic rocks to be widely exposed at the surface. Even greater uplift has occurred in portions of the Rocky Mountain region.

The most recent period of world-wide mountain building began about 20 million years ago with the first compression and uplift of the Alps, Himalayas, and coastal ranges of western North America. The structures of the Rocky Mountains and Sierra Nevada are somewhat older, but the present ranges were lifted by the latest mountain-building movements. The structures of the Appalachian Mountains of the eastern United States were formed earlier still, about 250 million years ago, when sea floor spreading drove the African continent against the edge of North America, eliminating an ancient Atlantic Ocean. The present Atlantic began re-opening about 200 million years ago, initiating subduction around the margins of the Pacific Ocean. These large-scale tectonic motions produce the individual geologic structures we can see expressed in landforms in regions with variable relief. We shall now turn our attention to these local structures.

Crustal Extension by Faulting

The Great Basin, centered in Nevada, offers one of the best displays of the effects of faulting to be seen on our planet. On a nineteenth-century map, the landscape of this vast region was said to resemble "an army of caterpillars crawling north." The topography consists of parallel mountain ranges separated by flat-floored basins (Figure 14.17). The basins are blocks of the earth's crust that have gradually sunk or tilted downward, leaving neighboring blocks standing

Figure 14.17
The scenery of Nevada is dominated by linear mountain ranges separated by flat-floored basins. These basins are strips that subsided along faults as the crust was stretched and fractured by large scale geologic movements beginning about 25 million years ago. The white patches in the basins are salt flats. (T. M. Oberlander)

as mountain ranges. This is clearly an area in which tension has stretched out the earth's crust and ruptured it on a massive scale.

The Great Basin is bordered on its east and west sides by two great tilted blocks, the Wasatch Range of Utah and California's Sierra Nevada. Each of these presents a bold wall, or *fault scarp*, facing the center of the Great Basin. Between the two are a host of *tilted blocks* with fault scarps on only one side, upthrust *horsts* with fault scarps on both sides, and down-dropped *grabens* bordered by horsts or tilted blocks.

Fault scarps are recognized by the abrupt rise of mountain slopes from a linear base line. The base line of a fault block often cuts cleanly across both resistant and erodible rock. Where the base line is irregular, with spurs projecting different distances, no active fault is present. Fault scarps are often dissected into *triangular facets* (Figure 14.18, p. 384) that sometimes reveal polished or grooved *slickensided* surfaces caused by the friction of one fault block rubbing against another. Since faults continue many kilometers down into the earth, they are often marked by hot springs and even volcanic cinder cones. Springs are common along all types of faults due to disruption of groundwater flow.

Faults that result from crustal extension, such as those of the Great Basin, are called *normal faults*. Such a fault forms a plane that passes into the earth at a steep angle known as the *fault dip*. Fault dip is measured as an angle

GEOLOGIC STRUCTURE AND LANDFORMS

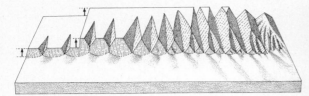

Figure 14.18
(left) This diagram shows the evolution of an escarpment produced by faulting. Two periods of uplift are shown at the left. The forms are successively older toward the right. Note how the growth of valleys cutting back from the fault scarp gradually converts the scarp into a series of spurs ending in blunt *triangular facets*. These become smaller and less apparent with time. Alluvial fans issue from the valleys, becoming larger as the valleys reach farther into the uplifted block. Most real fault scarps resemble the forms on the far right. (T. M. Oberlander)

(right) The east front of the Sierra Nevada is a fault scarp that has been dissected by erosion. Note the straight and regular appearance of the mountain range and its abrupt rise from the neighboring plain. Rising out of the lowland in front of the Sierran scarp is a second, much lower fault scarp. (T. M. Oberlander)

from the horizontal, like the dip of a rock stratum. In normal faulting the block resting on the inclined fault plane slips downward with respect to the adjacent block (Figure 14.19). Movement down the dip of the fault is termed *dip slip*. Normal dip-slip faulting has formed most of the mountain ranges in the Great Basin.

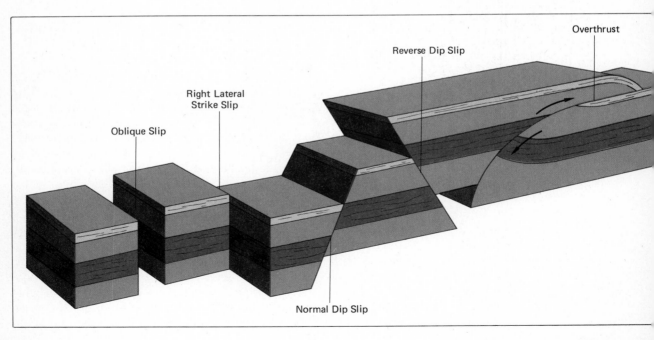

Crustal Shortening by Faulting

Where rocks fracture due to compression, fault planes vary from oblique to nearly horizontal. The block resting on the fault plane rides up and over the adjacent block. This shortens the crust. Such faults are known as *reverse faults* or *thrust faults*. Where the fault plane flattens to nearly horizontal, great slabs of rock may be forced tens of kilometers horizontally. Such far-traveled slabs are called *overthrusts*

Figure 14.19 (opposite)
Faults are the result of the rupture of rocks due to crustal tension, compression, and shearing. Fault motion is described by the relative displacement, or *slip*, of the rock units in contact along the fault. Motion may be parallel to either the dip or the strike of the fault plane, or oblique to both.

In the center of the diagram is a *normal fault*, in which the type of motion is *dip slip*, with the block resting on the fault plane slipping down relative to the adjacent block. This type of motion results from tension and produces crustal extension. The faults on the right of the diagram result from compression and produce shortening of the crust. In both *overthrusts* and *reverse* faults, the rock mass above the fault plane moves up and over the rock mass below the fault plane. In all three faults the motion is parallel to the dip of the fault plane, but in the latter two the direction is the reverse of normal dip-slip motion so that the type of motion is *reverse dip slip*.

Shearing stress involves side-by-side forces exerted in opposite directions. This results in horizontal motion parallel to the fault strike, or *strike slip*. The direction of the strike slip is designated in terms of the movement of the block on the far side of the fault. In this case, looking across the fault from either side, we see that the far block has moved to the right, so that the motion is *right lateral strike slip*.

The final fault on the left shows displacement oblique to both the strike and dip of the fault plane. Both strike slip and dip slip are involved in *oblique slip*. Whereas strike-slip faults are always vertical and dip-slip faults are always inclined, oblique slip can occur on both vertical and inclined faults. (Vantage Art, Inc.)

(Figure 14.19). Thrust faults and overthrusts are normal features of large mountain ranges formed by compression. They are almost always present in areas where crustal plates have collided. In the Alps three or four overthrusts, each having the area of a state like Massachusetts or Connecticut, are often piled atop one another. The incredible force required to produce such crustal deformation is vivid evidence of the enormous energy within the earth. The dynamic nature of the earth is also evident in the fact that the faults of Nevada recently produced by crustal extension cut through older thrust faults produced by strong crustal compression.

Thrust faults develop fault scarps that are hard to distinguish from those of normal faults. However, the scarps of thrust faults never retain the angle of the fault plane itself—to do so they would have to be overhanging. Overthrusts affect scenery mainly through the type of rock they carry into an area. Most have been folded after being displaced, so that their landforms are those developed in folded rocks. Being associated with high mountains, their topography has often been retouched by glacial erosion (Chapter 15).

Strike-Slip Faulting

Some of the earth's most important faults have been produced by motion in which adjacent segments of the crust move past one another horizontally rather than vertically. Faults of this type cannot produce mountains, but they are major sources of destructive earthquakes. Faults on which the motion is parallel to the strike rather than the dip of the fault plane are called *strike-slip* faults (Figure 14.19). One of the best known of these is the San Andreas fault, which extends 1,000 km (650 miles) northwestward through southern and central California. It was a sudden 3 to 6 m (10 to 20 ft) offset on this fault that produced the San Francisco earthquake of 1906.

A distinctive set of landforms known as *rift topography* is associated with strike-slip faults.

Rift topography is very well displayed along the San Andreas fault system. Three features are most evident: linear valleys eroded in crushed rock, elongated hills composed of huge "slivers" of rock dragged along in the fault zone, and closed depressions created by shifts in the ground. Small ponds form in the closed depressions. These *sag ponds* occur in lines that are the most unmistakable indication of the presence of a strike-slip fault.

One of the interesting features of strike-slip faulting is the horizontal offset of ridge lines and stream beds that cross the fault. Man-made features such as fences, orchard rows, streets, sidewalks, and buildings all may be similarly torn apart and offset. Some of this separation occurs in the slow process of *fault creep*, in which offset of one or two centimeters a year may occur without any earthquake activity.

Strike-slip faulting seems to be necessary to permit sea floor spreading to continue. The velocity of motion away from oceanic spreading centers cannot be uniform throughout a large crustal plate because of the spherical shape of the earth. For portions of crustal plates to move at varying rates, the plates must break into segments that can slip past one another. These strike-slip faults are called *transform faults* because they transform sea floor spreading into fault motion. The San Andreas fault seems to be a transform fault that has broken through the edge of the North American continent. Similar large transform faults are found in Turkey, Iran, and New Zealand, as well as throughout the oceans.

Fold Structures and Landforms

Everyone is aware that rock is brittle and breaks under sudden impact. But, under the temperature, pressure, and moisture conditions prevailing deep within an immense mass of geosynclinal sediments, rock under stress can undergo amazing deformation without breaking. Whatever ways a tablecloth can be rumpled when you push your hands over it, so rock can

Figure 14.20
The left side of this diagram illustrates the geometry of folded rock strata and the terms used to describe it. A terracelike flexure in bedded rocks is a *monocline*. A downfold whose sides, or *limbs*, are rotated upward on either side is a *syncline*, and an arch whose limbs are rotated downward is an *anticline*. Where folding is extreme, the limbs of the fold may be asymmetrically inclined, and the strata on one limb of the fold may be oversteepened or even overturned. If the rock strata are subject to great pressure, an overturned fold may rupture along a zone of weakness, and one limb of the fold may be thrust over the other limb, forming an *overthrust*. Fold structures are rarely seen intact as in this half of the drawing.

The right side of this diagram illustrates the variety of landforms that can be produced by erosion of folded rock strata. The thinner layers are assumed to be more resistant than the thicker layers. Erosion of the weaker rock overlying hard rock on an anticline forms an *anticlinal ridge*. If a portion of the top layer of resistant rock over an anticline has been worn away, *homoclinal ridges* are formed by the exposed edges of rock strata, with an *anticlinal valley* between them. A *synclinal ridge* is formed where a remnant of resistant rock in a syncline acts as caprock to retard the erosion of the underlying layers. (T. M. Oberlander)

also be deformed in the geologic process of folding.

Despite the complexity of form that is possible, some basic elements are recognizable in any system of folds in sedimentary rocks. Where rock strata have been forced into wrinkles, the ridges are called *anticlines*, and the troughs are *synclines* (Figure 14.20). The inclined layers between the anticlinal crests and the synclinal troughs are the fold *limbs*. These terms refer only to geologic structure, not to the visible surface landforms, which may reflect geologic structure in varying ways.

Before we discuss the surface effects of folding, it should be noted that areas of anticlines and synclines very often have great economic importance. The sediments that are compressed

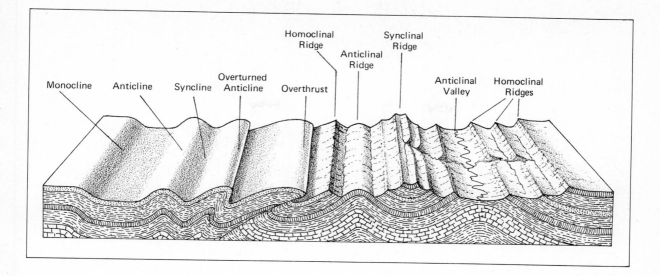

Monocline — Anticline — Syncline — Overturned Anticline — Overthrust — Homoclinal Ridge — Anticlinal Ridge — Synclinal Ridge — Anticlinal Valley — Homoclinal Ridges

into simple folds are usually deposits formed on continental shelves. These sediments often contain petroleum and natural gas produced by the decay of marine microorganisms. When the sediments are folded, the hydrocarbons, which are lighter than water, migrate toward the anticlinal crests and form "pools" there in porous rocks, especially sandstones. These rocks are the reservoirs from which we derive our major energy supplies. Plate tectonics help account for the fact that the Middle East is the world's greatest supplier of oil. In this case the Arabian crustal plate pushed into the Iranian plate, rumpling the continental shelf sediments between the two. This produced ideal reservoir structures for petroleum accumulation.

The terrain that develops in a region of folded sedimentary rock depends on the size and shape of the folds, the strata present (number, thickness, and relative resistances to erosion), the rate of tectonic uplift, and the climate. The landforms that develop at any point in such an area depend on the rock type, its thickness, and its dip at the location. Thus great diversity of form is possible.

In real landscapes, anticlinal structures rarely produce simple ridges, and synclines are seldom seen as simple troughs. Generally, the anticlines have been "unroofed" by erosion, revealing layer upon layer of rock—some resistant, some weak (Figure 14.21, p. 388). The scenery is dominated by edges of inclined rock strata that are resistant to erosion. These edges form *homoclinal ridges* (Figure 14.20), meaning that they are produced by rock layers that dip in one direction (rather than in two directions as along the axis, or center line, of an anticline or syncline). Often there are erosional basins in the centers of the anticlines, indicating that weak rock has been exposed there.

Synclines frequently form elongated plateaus or mountain peaks standing above anticlinal valleys. This is known as *relief inversion*. The structural highs form topographic lows, and the structural lows create topographic highs. Rock strata that are folded are stretched and weakened by jointing at anticlinal crests, but are compressed in synclines. This makes the rock at the crests of anticlines less resistant to weathering and erosion than the same rock in synclines, facilitating relief inversion.

Most anticlines are elongated domes that die out eventually. Homoclinal ridges wrap around the plunging "noses" of folds, as shown in Fig-

Figure 14.21
(left) Sheep Mountain, Wyoming, is an eroded anticline with concentric hogbacks, or homoclinal ridges, surrounding a core of resistant rocks.

(right) Another view of Sheep Mountain, showing the pattern of homoclinal ridges made by this fold, an adjacent syncline, and another anticline in the upper right of the photograph. (Dr. John S. Shelton)

Figure 14.22
Folded strata form a distinctive ridge and valley landscape in southeastern Pennsylvania. This aerial view near Harrisburg was made with side-looking airborne radar (see Appendix I); the image spans a distance of approximately 50 miles. The ridges are formed of resistant sandstone, and the valleys are cut into erodible shale or limestone. Note the nested hairpin ridges formed where anticlines and synclines rise and fall along the strike of the folding. The gaps in the ridges through which the Susquehanna River flows are clearly visible in the image. (SLAR Imagery, Grumman Ecosystems Corp.)

CHAPTER 14

ures 14.21 and 14.22. Where a series of folds press together, as in Figure 14.22, the homoclinal ridges loop back and forth, bending one way at the axis of an anticline and the other way at the axis of the neighboring syncline.

In most regions of fold structure there are two types of drainage. The smaller streams are controlled by geological structure and have a trellis-shaped pattern. Their longer segments lie in synclinal, anticlinal, and homoclinal valleys, with short connecting links across the grain of the topography. The main streams seem to ignore the geological structure. They cut through homoclines, anticlines, and synclines impar-

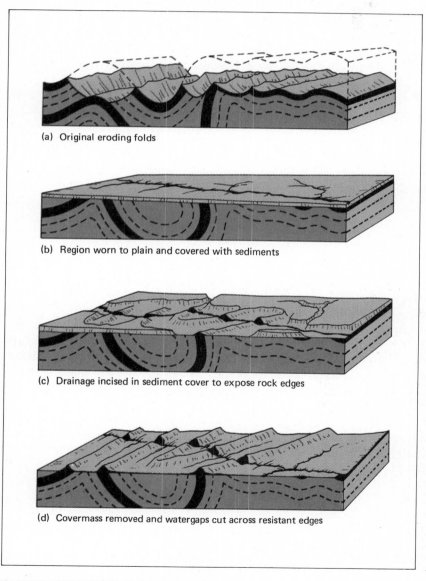

(a) Original eroding folds

(b) Region worn to plain and covered with sediments

(c) Drainage incised in sediment cover to expose rock edges

(d) Covermass removed and watergaps cut across resistant edges

Figure 14.23
These diagrams illustrate one hypothesis explaining how streams such as the Susquehanna River in the Appalachian Mountains of Pennsylvania have cut watergaps through bold ridges of resistant rock. A region underlain by folded rock strata (a) becomes worn to a plain of low relief (b). The plain becomes covered with sediments upon which streams develop independently of the rock structures beneath the newer sediments. Renewed uplift (c) causes these streams to incise through the sediment blanket, superimposing themselves across the edges of the tilted rock layers beneath. Erosion of the weaker rock (d) produces valleys between ridges of resistant rock. If the major streams are able to incise downward in pace with the removal of the weaker rock, they may maintain their original courses through notches they cut in the harder rock ridges. (Vantage Art, Inc.)

tially, often crossing resistant layers in deep gorges. Near Harrisburg, Pennsylvania, the Susquehanna River cuts through five successive massive ridges of resistant sandstone (Figure 14.22). The Indus and Ganges rivers of India cleave through the folded overthrusts of the Himalaya range in valleys more than 5 km (3 miles) deep.

Several hypotheses have been offered in explanation of streams that are transverse to geological structure. These include the suppositions that: (1) the streams were in place before the structures were formed and are *antecedent streams*; (2) the streams were let down on the structures from a covering mass of sediments and are *superimposed streams* (Figure 14.23); and (3) the streams extended headward through points of weakness by a process of *progressive stream piracy*. Seldom is there any proof of how transverse streams actually were established. They constitute the principal mystery associated with landforms controlled by geological structure.

SUMMARY

Much landscape diversity results from the nature of the bedrock and the large-scale arrangement of rock masses. These become expressed in the landscape through differential weathering and erosion, in which each body of rock responds in its own way to gradational processes. Differential weathering and erosion are clearest in sedimentary rock, because of frequent variations in rock type, but they are also a factor in igneous and metamorphic rocks as a result of variations in chemistry and jointing. Limestone is unique because of its solubility, which permits the development of karst phenomena, including surface sinkholes, disappearing streams, and large underground caverns.

Volcanic activity is a major source of new rock and an important influence on landscapes. Volcanism occurs at lithospheric plate boundaries, at hot spots, and along fractures within plates. Nonviolent basaltic volcanism occurs where plates are separating and basic magma rises to the surface. It creates the oceanic ridges and supplies the rock material of the sea floors. Such volcanism can also occur on the continents, especially where continents have broken up as new oceans were created. Eruptions of basaltic lava produce shield volcanoes and flood basalts. Explosive andesitic volcanism occurs above subduction zones, in which new silicic magma is formed by the melting of older rock. The violence of volcanic eruptions is a consequence of the explosive venting of gases separated from subsurface magma.

The first products of surface volcanic activity are scoria or cinder cones. In time these may develop into lava-veneered composite cones or towering strato-volcanoes. Plug domes are smaller than strato-volcanoes but extremely dangerous because of their violent behavior and release of nuées ardentes. Extremely explosive eruptions can cause the collapse of a volcano, producing a vast circular caldera. Basaltic plateaus, table mountains, dike ridges, and volcanic necks are erosional remnants of past volcanic activity.

Volcanic regions are hazardous, but soils derived from volcanic ash or basalt are usually very productive. Volcanism also is the source of metallic ores. Volcanic regions contain potentially valuable geothermal energy supplies. These have been developed for heating and electric power generation in a few areas and will be increasingly important in the future.

Horizontal crustal motions cause tension and compression of rock masses as adjacent sections of the crust thrust against each other, or pull apart, or move laterally past one another. Vertical movements of the crust result from local heating, the rise of plutonic rock, or changes in load, as when glacial ice builds up or melts. Mountain systems are produced by the compression and uplift of sediment-filled troughs, called geosynclines, which include the ocean trenches, continental shelves, and continental rises. The uplift of mountains occurs in pulses following the initial phase of compression.

Crustal deformation by compression and tension causes rock to fail by bending, producing fold structures, or by breaking, producing faults. Where the crust is stretched, it collapses in normal dip-slip faults, creating horsts, grabens, and tilted blocks. These are expressed in the landscape as elongated mountain ranges that rise very abruptly above flat-floored basins. Strike-slip faults have their own distinctive set of landforms, known as rift topography.

Strong compression of the crust produces thrust faults, overthrusts, and folds, all of which thicken the crust. Folds consist of anticlinal and synclinal structures. Erosion in fold structures produces linear valleys in strips of erodible rock, with resistant layers standing out in relief as homoclinal ridges. Trellis drainage patterns in fold regions feed into major streams that usually run transverse to the fold structure. In many gently folded areas, the anticlines are reservoirs for petroleum and natural gas.

APPLICATIONS

1. Draw north to south and east to west geological cross sections across your state, using information from state geological surveys, published articles, and maps.
2. How does geological structure affect the landscape in your area? If no specific structures can be seen nearby, where is the closest clear expression of a structural influence on the landscape?
3. In the library consult the periodical *Geotimes* and look at the section titled "Geologic Events" for a period of twelve consecutive months. Make a catalogue of all of the geological events noted over this period.
4. What is the earthquake history of your area? How often have earthquakes occurred there, and what magnitudes have been experienced? What geological feature has caused the earthquakes?
5. Volcanism seems to occur both where crustal plates are pulling apart and where they are pushing together. Explain this apparent contradiction.
6. How do the individual mountains in regions of folded strata differ from those produced by faulting?
7. Obtain a copy of Erwin Raisz's physiographic map *Landforms of the United States* and divide the nation into physiographic regions of fairly uniform characteristics. Write an essay on the problems of such an undertaking.

FURTHER READING

Green, Jack, and **Nicholas M. Short**, eds. *Volcanic Landforms and Surface Features.* New York: Springer-Verlag (1971), 519 pp. This is a large collection of spectacular high-quality photographs covering every aspect of volcanism.

Harris, Stephen L. *Fire and Ice.* Seattle: The Mountaineers and Pacific Search Press (1980). This fascinating paperback presents the eruptive history of each of the volcanoes of the Cascade Range. Written with authority and enthusiasm.

Iacopi, Robert. *Earthquake Country.* Menlo Park, Calif.: Lane Books (1964), 192 pp. Well-illustrated popular account of California's earthquake-prone fault landscapes.

Lobeck, A. K. *Geomorphology.* New York: McGraw-Hill (1939), 731 pp. Lobeck's textbook was constructed on the premise that a picture is worth a thousand words. This book includes about 200 pages on folding, faulting, and volcanism; these pages consist mainly of photographs and diagrams that have retained their value through the years. The material on folding is especially helpful.

Shelton, John S. *Geology Illustrated.* San Francisco: W. H. Freeman (1966), 434 pp. Nowhere is there published a better collection of photographs illustrating structurally controlled landforms. The text is very readable and original in approach.

Shimer, John A. *Field Guide to Landforms in the United States.* New York: Macmillan (1971), 272 pp. The first portion of this handbook outlines the nature of the various structurally determined physiographic regions of the United States, with portions of Erwin Raisz's detailed physiographic diagram of the United States used as index maps of each region. Following this is a catalogue of individual landforms, each illustrated by an excellent line drawing.

Sweeting, Marjorie M. *Karst Landforms.* New York: Columbia University Press (1972), 362 pp. This text by an English geomorphologist deals exclusively with the distinctive landforms that develop where limestone is the bedrock.

Twidale, C. R. *Analysis of Landforms.* New York: Wiley (1976), 572 pp. Twidale's massive geomorphology text contains many excellent illustrations of structurally controlled landforms.

Wilcoxson, Kent H. *Chains of Fire.* New York: Chilton (1966), 235 pp. This stimulating book on volcanism, written for the general public, includes many eye-witness accounts of volcanic eruptions of various types. Excellent reading.

Williams, Howell, and **Alexander R. McBirney.** *Volcanology.* San Francisco: W. H. Freeman (1979), 397 pp. This is the most recent of a large number of authoritative texts on volcanic processes and resulting landforms.

CASE STUDY: **The Cascades Awaken**

On March 27, 1980, the volcanic Cascades mountain range, extending from northern California to southern British Columbia, awakened from a 60-year sleep. The signal was a towering cloud of ash and steam rising from snow-covered Mount St. Helens in southern Washington—the first such event since the eruptions at Mt. Lassen in California between 1914 and 1921. The renewed activity in 1980 culminated on May 18 in a gigantic blast that thoroughly altered the form of Mount St. Helens, replacing its conical glacier-clad summit with a crater 750 m (2500 ft) deep.

In a single day, Mount St. Helens leveled a forest and killed everything in it, destroyed one lake and created fifteen others, destroyed one river and portions of several more, caused gigantic mud slides and floods, blanketed the Northwest with volcanic ash, paralyzed transportation facilities of every type, threatened the region's wheat crop, halted commerce on the Columbia River for weeks, and temporarily closed down the jobs of nearly half a million people.

Studies in the early 1970s had revealed that Mount St. Helens had a particularly violent past, having blasted away its summit 4,000 years ago and again only 500 years ago. Ash from explosive eruptions of the volcano reached as far east as Montana and Wyoming. These studies revealed that many man-made structures are far older than the mountain, much of which has been constructed in the past 500 years.

The great eruption came at 8:30 A.M. in the form of a terrific blast that removed 2.2 cu km (.55 cu mi) of the cone and flattened the forest over an area of almost 550 sq km (200 sq mi), stripping off the foliage and uprooting trees to a distance of almost 30 km (18 miles).

Within 8 km (5 miles) of the gaping crater produced by the blast, the ground was stripped as naked as a lunar landscape, even the soil being ripped away. The loss of animal life is difficult to imagine. People died in their tents, in their vehicles, and in the open, as much as 32 km (20 miles) from the blast—scorched, choked by ash, and buried by mud, rock, and uprooted trees. As could be expected, melted ice and snow from the devastated summit sent a series of immense scalding mud flows pouring down the valley of the Toutle River, blocking the valleys of smaller streams and converting them into lakes.

Terrible destruction at the scene of a violent volcanic eruption is expected. But most Americans had never considered the wider effects of such an event. The first inkling of the problems in store for distant areas was a blackening of the skies, producing total darkness in midmorning. Many believed an incredible thunderstorm was in store, until the ash began sifting down.

Unforeseen effects of falling ash were soon apparent. Automobile carburetors became clogged, stalling vehicles on highways and back roads throughout central and eastern Washington. Aircraft engines were affected similarly, causing emergency landings by commercial jets and light planes. Airports and highways were closed and buses and trains were halted, stranding thousands of people. Clouds of fine ash whirled up by moving vehicles choked pedestrians in cities, and attempts to hose away the ash on the streets turned it into slippery mud, causing vehicles to slither out of control. Electrical failures were widespread as the ash accumulated on power lines and transformers, shorting them out. Where power was available, machinery broke down as it was penetrated by abrasive ash particles that were harder than crushed glass. Sewer systems became clogged with ash and sewage treatment plants failed.

There was brief panic over the possible toxic effects of the ash to both humans and crops. Breathing was "like sticking your head in the fireplace and stirring up the ashes." Fortunately, the ash proved to be nontoxic.

In Pullman, Washington, 500 km (300 miles) from the eruption, the fallout of ash was as much as 8 tons/acre. The wheat crop in eastern Washington seemed to be threatened by a 10 to 15 cm (4 to 6 in) blanket of ash; the orchard crops of the Yakima and Wenatchee areas also seemed endangered; and growers of potatoes, peas, and lentils in Washington and Idaho expressed alarm. Most of these fears proved groundless, as vegetation recovered when the ash was shaken off or blown away.

A greater problem was damage to streams, large and small. The eruption buried the upper Toutle River under 150 m (500 ft) of mud and ash and sterilized the rest with steaming mud. The beds of other streams were paved by a concretelike mass of ash. The loss of the upper 400 m (1,300) feet of the mountain made it too low to support the glaciers whose meltwater kept the streams flowing vigorously throughout the summer.

The debris flood down the Toutle River reached the Cowlitz River and then spilled it into the Columbia River at Longview, Washington, shallowing the Columbia's channel from 12 m (40 ft) to less than 5 m (16 ft) over a distance of 6 km (3.7 miles). This stranded oceangoing ships upriver, loaded with grain for world markets, and blocked ships that had intended to move up the river to inland ports. The wheat crop of the Pacific Northwest was literally bottled up. Railroads and truckers diverted shipments of export cargoes intended for the blocked ports at Portland, Vancouver, and Kalama, increasing the costs of the goods and depriving the embargoed ports of jobs and income estimated at $5 million a week.

The gray wilderness of ash-covered tree trunks smashed down by the May 18 blast constituted some 800 million board feet of lumber, with a value estimated at between $250 million and $1 billion. Ironically, conservationists had spent years battling to save this very forest from the loggers' saws!

Meanwhile, the other young giants among the Cascade volcanoes—Baker, Rainier, Adams, Hood, Jefferson, the Three Sisters, and Shasta—tower over surrounding mountains, farmlands, hamlets, and cities— still asleep, as was shattered Mount St. Helens until March 27, 1980.

Mt. St. Helens following the great eruption of May 18, 1980. A gaping crater has replaced the previous symmetrical cone. Nearly 2 cu km of the pre-eruption cone slid to the north to form the vast debris flow seen in the foreground, and another 2 cu km of volcanic ash was blown into the sky to rain down over eastern Washington, Idaho, and Montana. The area's former lush forest was stripped off cleanly by the blast, with an estimated loss of life including 6,000 deer, 5,000 elk, 200 bears, some 50,000 small game animals, and 96 persons killed or presumed dead. (T. M. Oberlander)

Many of the earth's most spectacular landforms
were created by glaciers. Flowing ice is one of the
most forceful agents of geomorphic change, and
many regions owe their physical characteristics to
glacial erosion and deposition.

Maligne Lake, Jasper Park by Lawren S. Harris. (The National Gallery of Canada, Ottawa)

CHAPTER **15**

Glaciers and Glacial Landforms

If there had been no Ice Age, there would be no Matterhorn, no Denmark, no Great Lakes, and no incredibly productive agricultural heartland in Saskatchewan, Manitoba, the Dakotas, Iowa, Illinois, Indiana, and Ohio. The Norwegian fjords, Niagara Falls, Chesapeake Bay, and Long Island would also be missing from the map. All these were produced directly or indirectly by the slow advance of enormous masses of ice known as *glaciers*.

We live in an unusual near-glacial age today. Only 15,000 years ago the sites of Chicago, Detroit, Minneapolis, Seattle, Vancouver, Toronto, and Montreal were buried under a thousand or more meters of ice. The Mississippi River was then an enormous braided stream that carried milky glacial meltwater down to a sea whose level was a hundred meters lower than it is now. Some scientists predict that such conditions will occur again within the next 5,000 to 20,000 years. A moderate increase in winter precipitation along with a decrease of merely 5° to 7°C (9° to 13°F) in the earth's average temperature would be enough to cause ice sheets to invade the continents once more.

Conditions on earth today are so close to those of a glacial age that three-fourths of the earth's nonsaline water supplies are presently stored as ice, which covers 10 percent of the land surface. This is an exceptional condition that has occurred only occasionally throughout geologic time. The world's existing glaciers hold an amount of water equal to 5,000 years' flow of the earth's mightiest river (the Amazon) or 60 years of rain and snow over the whole planet. By comparison, the storage of water in lakes, rivers, swamps, and man-made reservoirs is trivial.

Most of this frozen water is locked in the immense Antarctic and Greenland ice sheets—85 percent in the Antarctic ice sheet alone. During much of the preceding 2 million years, two additional continent-sized ice sheets existed, covering all of northern Europe and Canada, as well as 2.6 million sq km (1 million sq miles) of the United States. Smaller ice fields blanketed portions of the world's great mountain ranges. Strangely enough, Siberia, widely regarded as almost glacial today, was too dry to sustain large glaciers during the Ice Age.

Glaciers of the past are of interest because they have significantly altered the environment we inhabit. Over vast areas in the higher latitudes, as well as at high altitudes, slowly moving currents of ice have scraped away all the soil and much of the weathered mantle that formerly covered the land surface. Elsewhere the melting of enormous masses of debris-laden ice has completely submerged the previous landscape under a blanket of boulders, gravel, sand, silt, and clay. This has created rolling or flat plains where once there were hills and valleys. In some areas the effect of glaciers was detrimental to later human use of the land; in others it was very beneficial. This chapter explores these effects and the processes responsible for them.

GLACIATION PRESENT AND PAST

Glaciers form where more snow falls each year than melts or evaporates. In Alaska, Greenland, Iceland, the Canadian Arctic Islands, the Canadian Rockies, the European Alps, the Himalayas of Asia, the Andes of South America, the New Zealand Alps, and Antarctica, winter snows do not melt completely during the succeeding warm season. The snow accumulates from year to year and is gradually transformed into ice. When a critical depth of ice accumulates—some 50 m (150 ft) or so—the weight of the ice and its low internal strength cause it to deform. A glacier is a mass of ice that is "flowing" because it becomes plastic and easily deformed under gravitational stress.

Glacier Types

Glaciers exist in several different forms (Figure 15.1). Dwarfing all other types are the great *continental ice sheets*. These are unconfined

Figure 15.1
(opposite top left) Glaciers form where winter snow persists throughout the year. If the annual accumulation of snow exceeds annual losses due to melting and evaporation, a glacier eventually is created. Glaciers may be unconfined, as *highland icecaps*, which submerge the older erosional topography in mountain regions, or confined, as *cirque* and *valley glaciers*. All glacial ice drains to lower elevations under gravitational stress. The combination cirque and valley glacier is known as an *alpine glacier*. The valley glacier descending from the unconfined icecap is an *outlet glacier*. In some regions, the ice from one or more valley glaciers spreads out over a plain and forms a *piedmont glacier*. (John Dawson)

(opposite top right) The surface of this valley glacier in the French Alps is broken by numerous crevasses. (Anitra Kolenkow)

(opposite bottom) The Malaspina Glacier in Alaska forms an extensive piedmont ice sheet. The ice flow is from right to left in this picture. (Austin Post/U.S.G.S.)

Highland Ice Cap

Cirque Glacier

Outlet Glacier

Valley Glacier

Piedmont Glacier

GLACIERS AND GLACIAL LANDFORMS

blankets of glacial ice that submerge the land surface over areas of millions of square kilometers. Only two remain today—the Antarctic ice sheet, which covers 12.5 million sq km (4.8 million sq miles), and the Greenland ice sheet, which has an area of about 1.7 million sq km (650,000 sq miles). Far smaller are the unconfined *highland icecaps* that blanket hundreds to thousands of square kilometers of mountainous terrain in the higher latitudes. Highland icecaps are conspicuous in Iceland, the Canadian Arctic Islands, and the Canadian Rockies. Often they leak outward through valley systems, producing confined *outlet glaciers* (Figures 15.1 and 15.18).

More common in high mountains are still smaller, confined *alpine glaciers*. This term includes both ice fields that occupy depressions below high mountain crest lines and streams of ice that are hemmed in by valley walls as they drain from mountain crest lines to lower elevations. Alpine glaciers originate in high-altitude rock-walled ice reservoirs called *cirques* (French, from Latin *circus*, or circle). Cirques are distinctive rock basins created by frost action and glacial erosion. In many mountain areas the only glaciers present are *cirque glaciers*—masses of ice that are restricted to cirques and do not enter valleys. A cirque glacier may occupy less than a square kilometer.

Where ice does spill from cirques into the valleys below, it forms channeled streams of ice called *valley glaciers*. These may have lengths of several tens of kilometers. Valley glaciers (as well as outlet glaciers) occasionally drain into open areas, where they spread out in pools of ice known as *piedmont glaciers* (Figure 15.1). Thus alpine glaciers can take three forms: cirque glaciers, valley glaciers, and piedmont glaciers. Alpine glaciers are best developed in Alaska, the Canadian Rockies, the European Alps, the Caucasus of the USSR, the Himalaya and Karakorum ranges of Asia, the Andes of South America, and the southern Alps of New Zealand. Confined and unconfined glaciers produce different effects on the landscape, as we shall see later in this chapter.

More than 1,000 glaciers exist in the coterminous United States, mostly in Washington's Cascade Range. Nearly all are small, having a combined area of barely 500 sq km (200 sq miles). Nevertheless, they are important water sources, providing about 2.1 billion cu m (1.7 million acre ft) of meltwater each year. Melting of glacial ice in the summer provides water for irrigation just when the demand is highest and streams fed by rainfall and groundwater inflow are lowest. North America's greatest glaciers are in Alaska, where they cover more than 50,000 sq km (20,000 sq miles). The Malaspina glacier system of Alaska's St. Elias Range is itself more than 4,000 sq km (1,500 sq miles) in area (Figure 15.1).

Pleistocene Glaciation

The present geologic epoch, known as the *Holocene*, or *Recent*, began about 10,000 years ago, when the great northern hemisphere continental ice sheets melted back rapidly, eventually to disappear about 6,000 years ago. The preceding *Pleistocene epoch* began almost 2 million years ago and included as many as seven or more major episodes of glaciation. Glacial periods, lasting tens of thousands of years, were separated by warmer *interglacial* periods when only Antarctica and Greenland maintained ice covers. There have been other Ice Ages in the earth's history, but those occurred hundreds of millions of years ago. We are interested in Pleistocene glaciation because its effects dominate many existing landscapes.

Figure 15.2 compares the present area of ice coverage in the northern hemisphere with the area covered about 15,000 years ago. At their maximum, Pleistocene glaciers sprawled over some 44 million sq km (17 million sq miles)— almost one-third of the earth's present land area. These glaciers left clear evidence of their advances and retreats in the form of ice-scoured bedrock landscapes and vast expanses blanketed by glacial deposits. The North American continental ice sheet, known as the Laurentide ice sheet, expanded from the vicinity of Hudson

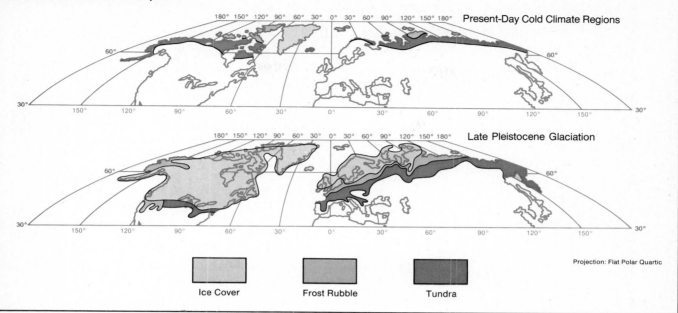

Glaciation in the Northern Hemisphere

Present-Day Cold Climate Regions

Late Pleistocene Glaciation

Projection: Flat Polar Quartic

Ice Cover   Frost Rubble   Tundra

Figure 15.2

(top) At present, active glaciation and cold climates are restricted to high latitudes and high altitudes in mountain ranges such as the Alps, the Himalaya and Karakorum ranges, the Rocky Mountains, and the Alaskan mountains. These mountain areas are not shown on the map due to their limited size.

(bottom) During the last major period of Pleistocene glaciation, approximately 15,000 years ago, large parts of northern North America and northern Europe were subjected to the direct or indirect effects of glaciation. Many sections of the northeastern and north-central United States exhibit landforms characteristic of glacier action, and periglacial regions not directly covered by ice sheets display the effects of cold climates in the form of smoothed slopes, solifluction deposits, patterned ground, rock streams, and dells, as well as accumulations of glacial outwash (alluvium deposited by streams of glacial meltwater). Frost rubble landscapes consist largely of bare rock with little vegetation. The tundra areas are covered by low-growing vegetation that is adapted to extreme cold. (Andy Lucas and Laurie Curran after J. L. Davies, *Landforms of Cold Climates, An Introduction to Systematic Geomorphology*, Vol. 3, 1969, A. N. U. Press, Canberra)

Bay and terminated close to the present line of the Missouri and Ohio rivers. Europe was invaded by the Fennoscandian ice sheet, which originated in northern Scandinavia and spread southward to the uplands of Central Europe.

All around the world Pleistocene glacial erosion completely transformed the scenery of the high mountains, destroying their previously rounded summits and giving them their picturesque "alpine" appearance.

The Pleistocene ice sheets and alpine glaciers left a magnificent legacy of landscapes especially suited for human recreation and enjoyment. The pinnacle of the Matterhorn in Switzerland, the fjords of Norway, the ski slopes of Tuckerman Ravine in New Hampshire, the

Figure 15.3

Cold climate landscape. Where temperatures remain below freezing the larger part of the year, landscapes take on a highly distinctive appearance. Below the top meter or so of the land surface, the water in all pore spaces in rock and soil remains frozen solid throughout the year. Permanently frozen rock and soil is known as *permafrost*. Permafrost reaches to depths exceeding 600 meters in northern Alaska and Siberia. Above the permafrost, the surficial blanket of soil thaws each summer, becoming saturated with water that cannot escape downward due to the frozen substrate. This water here and there lubricates the soil to the degree that the soil slips down as a mass, a few centimeters in a matter of a week or two. The water-saturated soil moves in lobes that are conspicuous on hillslopes, as the inset (bottom right) indicates. This type of movement is known as *solifluction*. Solifluction lobes are a few centimeters to a meter or so in height. Studies show that despite the appearance of great activity produced by these solifluction lobes, most are inactive, movement occurring only in a few places during each thaw. However, considered over a long period of time, the entire soil cover is draining downslope at a far more rapid pace than that produced in warmer regions by the creep process. The portion of the soil that freezes and thaws annually, expanding, contracting, and moving downslope, is known as the *active layer*. The long-term effect of the solifluction process is to smooth the landscape, filling preexisting depressions and peeling down projections. Small valleys are infilled with *colluvium* (slope deposits), forming flat-floored *dells* with poorly developed watercourses. Water leaks through the colluvial valley fills.

Bare rock within the active layer or projecting above it is subject to intense frost weathering during the long winter season. The expansion of water in passing from the liquid to the crystal state exerts enormous pressures on confining walls, causing rocks to split and, occasionally, to crumble. Thus projecting solid rock masses are rapidly reduced to rubble, which becomes incorporated into the solifluction lobes. High areas are therefore lowered effectively. Broad summits are covered with angular rock rubble produced by the intensity of the

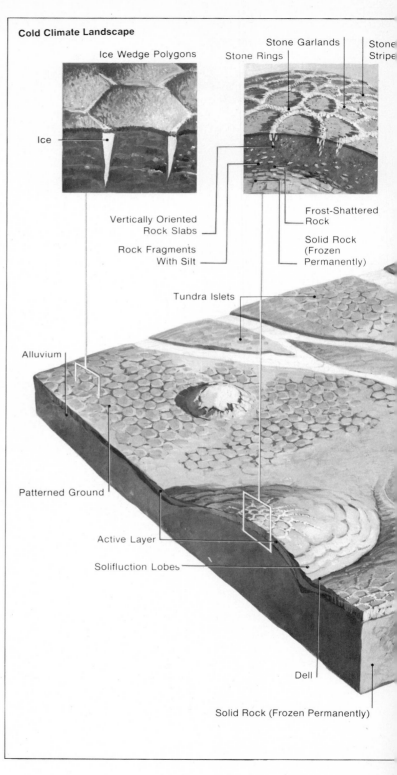

Cold Climate Landscape

Ice Wedge Polygons

Stone Garlands

Stone Rings

Stone Stripe

Ice

Vertically Oriented Rock Slabs

Rock Fragments With Silt

Frost-Shattered Rock

Solid Rock (Frozen Permanently)

Tundra Islets

Alluvium

Patterned Ground

Active Layer

Solifluction Lobes

Dell

Solid Rock (Frozen Permanently)

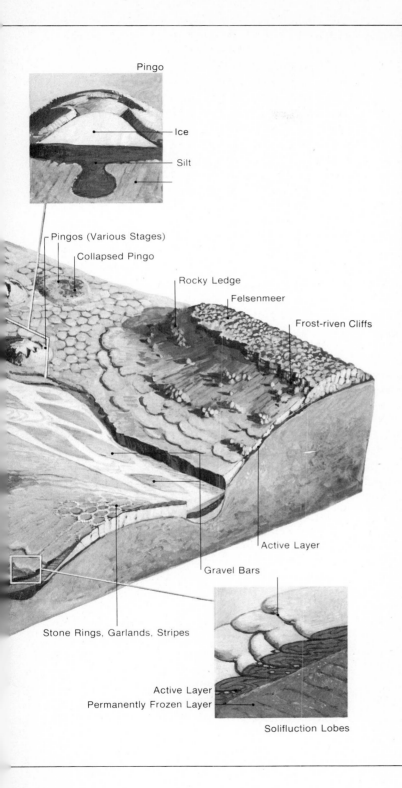

Pingo

Ice

Silt

Pingos (Various Stages)

Collapsed Pingo

Rocky Ledge

Felsenmeer

Frost-riven Cliffs

Active Layer

Gravel Bars

Stone Rings, Garlands, Stripes

Active Layer

Permanently Frozen Layer

Solifluction Lobes

freezing process in exposed sites. The result is a "sea of rocks," or *felsenmeer*. At the edges of rock exposures, large talus accumulations reflect the downslope movement of frost-riven blocks.

The annual freeze and thaw process produces a host of unusual minor landforms in addition to solifluction lobes. Repeated volume changes in the weathered mantle have the effect of sorting out coarse and fine material. Due to the efficiency of freezing in loosening slabs of rock, the weathered mantle on hills contains an abundance of coarse debris. This becomes shunted away from the fines, which accumulate in masses surrounded by rings of rocks. The result is the stone rings shown in the inset (top center). On slopes these rings are drawn out downhill into "garlands" and "stripes" by movements of the active layer. In lowlands composed of silty alluvial deposits, another type of *patterned ground is developed*. During the freezing process, water-soaked silt first expands, but at prolonged low temperatures, eventually begins to contract, cracking into a network of polygonal fissures. Ice forms in these and eventually produces wedges, often a meter across and 5 meters deep (see inset at top left). These are *ice-wedge polygons*. Once formed, they are permanent, continuing to grow very slowly.

Clear ice also forms in the ground in lenslike masses that heave up the overlying sod (see inset at top right). This phenomenon is known as a *pingo*. Pingos develop in old lake beds or marshes that are filling with sediment and vegetation. At a certain point, the encroachment of permafrost into the fill traps a lens of unfrozen water and fill. Freezing of the water in turn draws moisture from the fill to produce a growing mass of ice that eventually domes up the recently developed sod above it. If a pingo is destroyed by thawing, it leaves a depression ringed by an earth rampart. A large pingo may be 100 meters across and 30 meters high.

Other oddities of landscapes formed by *cryergic* (low-temperature) processes are overhanging (frozen) riverbanks and unusually smooth stream profiles due to the abundance of fresh rock waste, which provides streams with abrasive tools. (John Dawson after T. M. Oberlander)

deep gouge of California's Yosemite Valley, the sandy arm of Cape Cod in Massachusetts, and the ten thousand (and more) lakes of Minnesota—all were fashioned by Pleistocene glaciers.

Indirect Effects of Pleistocene Glaciation

The effects of Pleistocene glaciation are not limited to the area actually covered by ice. The glaciers themselves, their deposits, or the cold climate accompanying them had a variety of effects on landscapes. Wind erosion of loose, unvegetated glacial deposits coated the Midwest with loess, providing fertile parent material for soil development. Meltwater from the ice enlarged some valleys and filled others with sediment that was later carved into stream terraces extending far beyond the boundaries of the glaciers themselves. Drainage patterns were altered as glaciers blocked river systems and forced tributary streams to overflow into new paths. Portions of the Ohio and Missouri rivers were created as spillways that carried water from one blocked drainage to another along the front of the Laurentide ice sheet.

The cold climates of the Pleistocene intensified the break-up of rock by frost weathering. Over vast areas the ground became frozen solid, only the upper few inches thawing each summer. Consequently, slopes were smoothed by the process of solifluction, the movement of water-saturated soil in lobes. A periglacial (near-glacial) landscape evolved, consisting of many distinctive elements, as seen in Figure 15.3. Active examples of such landscapes exist today only in tundra regions and the cold margins of the coniferous forests, as in Arctic Canada, Alaska, Siberia, and certain highland areas.

The average thickness of the Pleistocene continental ice sheets is estimated to have been 1 to 2 km (0.6 to 1.2 miles), with a maximum thickness of about 4 km (2.5 miles). The weight of this ice—approximately 900,000 kg (1,000 tons) for each square meter of area—caused the earth's semiplastic crust to sink under the burden. The crust was depressed by as much as 1,200 m (4,000 ft) under the thickest parts of the ice sheets. Since the retreat of the ice, the crust has been slowly returning to its pre-glacial level. Parts of Scandinavia and North America are continuing to rise, or "rebound," as much as 2 cm per year, with the area at the center of the former Laurentide ice sheet having risen 170 m (560 ft) in the last 7,000 years.

For ice sheets to grow on the land, water must be removed from the seas. Consequently, a world-wide lowering of sea level accompanies every period of glacial advance. Several times during the Pleistocene, sea level was lowered 100 to 150 m (330 to 500 ft). This drop is evidenced by the fact that river valleys continue onto the continental shelves far below present sea level. The average depth of the submerged outer edges of the continental shelves coincides with estimates of low Pleistocene sea levels. Some oceanographers believe the continental shelves were beveled by wave erosion during Pleistocene sea level changes.

The lowering of the sea caused rivers to cut downward and expand their valleys near their previous outlets. The return of the sea to its present level drowned these enlarged valleys, producing broad, deep estuaries that now form some of the world's best harbors, such as those at New York, Norfolk, San Francisco, Seattle, Anchorage (Alaska), Amsterdam (Netherlands), Hamburg (West Germany), Liverpool (England), Leningrad (USSR), Lisbon (Portugal), Manila (Philippines), and Tokyo-Yokohama (Japan). If the Antarctic and Greenland ice sheets were to melt away completely, sea level would rise another 65 m (210 ft), which would submerge most of the earth's largest cities and send arms of the sea far up the heavily populated valleys of the Mississippi, Ganges, Indus, Nile, Yangtze, and Yellow rivers, as well as many other interior lowlands.

When sea level fell during the Pleistocene cold phases, the land area increased greatly, as shown in Figure 15.4. The Florida peninsula doubled in size, and land bridges joined Ireland,

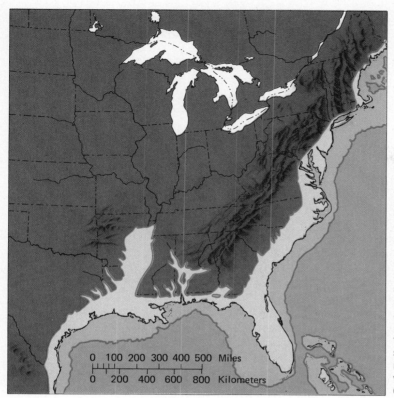

Figure 15.4
This map of the Atlantic and Gulf coasts of the United States shows the degree to which the storage of water as glacial ice affects the position of coastlines. The darkest blue tone represents ocean areas deeper than 130 m (430 ft), which were unaffected by Pleistocene ice volumes. The middle tone shows presently submerged areas that were subaerial plains during low stands of the sea that accompanied maximum advances of Pleistocene ice sheets. The lightest tone indicates land areas that would be submerged if the present continental ice sheets in Greenland and Antarctica were to melt, raising sea level by about 65 m (210 ft). (T. M. Oberlander)

England, and France, as well as most of the islands of Indonesia and Malaysia. The exposed continental shelves allowed humans, animals, and plants to migrate freely between areas now separated by water. They also provided low-lying refuge areas with moderate maritime temperature regimes, enabling plants to survive the rigors of Ice Age climates.

Origin of the Ice Ages

The mechanism that starts and stops periods of glaciation remains a puzzle. Ice sheets are in a delicate balance between growth and retreat, and relatively small changes in climate can cause the balance to shift either way. Some of the hypotheses advanced to account for the Ice Ages include changes in the positions of the con-

tinents and in ocean circulation patterns as a result of sea floor spreading; increased altitude of the land masses after a period of geologic upheaval; and variations in the amount of solar radiation the earth receives. The last could be caused by changes in the sun's energy output, the earth's relationship to the sun, and the atmosphere's content of carbon dioxide and volcanic dust.

A successful theory must account for the growth and subsequent decay of ice sheets several times in less than 2 million years. It must also explain glaciation that occurred hundreds of millions of years ago. Several mechanisms probably combine periodically to produce unusual climatic cooling. Scientists are still searching for a completely satisfactory explanation of the climatic changes causing the Ice Ages.

ANATOMY AND DYNAMICS OF GLACIERS

Glaciers are formed above the annual *snowline*, the elevation at which some winter snow is able to persist throughout the year. Presently, the snowline reaches sea level only in polar areas. Moving away from the poles, the snowline rises progressively higher and reaches a maximum of 6,000 m (20,000 ft) in the Peruvian Andes, close to the equator. In the Colorado Rockies and California's Sierra Nevada, the snowline on slopes exposed to direct solar radiation would be at about 4,500 m (15,000 ft), just above the highest peaks of both ranges. But on the shaded northeast-facing slopes, the snowline descends to between 3,600 and 4,000 m (12,000 to 13,000 ft). Thus glaciers are present only in shaded spots on the cool northeastern faces of the highest crests of both ranges. Further north, the descent of the snowline permits even the south- and west-facing slopes of Mt. Rainier in the Cascade Range of Washington to be glacier-clad, even though the summit of Mt. Rainier is lower than ice-free peaks in both the Rockies and Sierra Nevada.

The Making of Glacial Ice

Snowfall is what produces glaciers, but glaciers themselves are solid ice. How is this transformation accomplished? Snow usually arrives at the surface in the form of lacy hexagonal ice crystals. Eventually, compaction, vaporization, and local melting and refreezing convert the initial delicate snowflakes to rounded granules of ice. When the snow is wet and heavy, the transformation of snow crystals to ice granules requires only a few days. But in the extreme cold of Antarctica, the snow remains fluffy for years, and the transformation to ice is extremely slow.

Newly fallen snow contains much air space and has a specific gravity (weight relative to water) between 0.05 and 0.15. With time, and under the pressure of overlying snow, the ice granules formed from snowflakes gradually pack closer together. In midlatitude areas, the granular material begins to coalesce after one summer and attains a specific gravity of about 0.55. Such material is called *firn*. In tens or hundreds of years, depending on summer temperatures, the pore spaces in the firn gradually disappear until the firn has become solid ice with a specific gravity of about 0.85. Then, as air bubbles gradually disappear from the ice, the specific gravity increases to 0.9 or so, producing true glacier ice.

The Glacier Budget

Glaciers are dynamic systems—expanding or contracting, thickening in some sections and thinning in others, but always feeding ice forward to replace ice lost by melting or evaporation below the snowline. Whether a glacier advances or recedes depends on the glacier's *mass budget*—the balance between the input of snow above the snowline and the loss, or *ablation*, of glacial ice by melting or evaporation below the snowline.

All glaciers may be divided into two portions: an upper *accumulation zone*, which has a net annual gain of material, and a lower *ablation zone*, which has a net annual loss of material (Figure 15.5). The *equilibrium line* separates the two zones and is the line at which the input of snow exactly balances ablation losses. The equilibrium line more or less coincides with the *firn line*, which is the annual snowline on the glacier itself. During the winter, snow accumulation exceeds ablation on a glacier. In the summer, when snowfalls are less frequent or absent altogether, ablation is dominant (Figure 15.6, p. 408). Whether a glacier experiences net growth or decay depends on the balance between annual accumulation and annual ablation.

Ice is always moving forward to the end of an active glacier; this is true whether the glacier margin is advancing, retreating, or fixed in position. Where the ice margin (or "snout" in the case of a valley glacier) is stable in position from year to year, the amount of ice flowing to the margin is balanced by that removed by down-

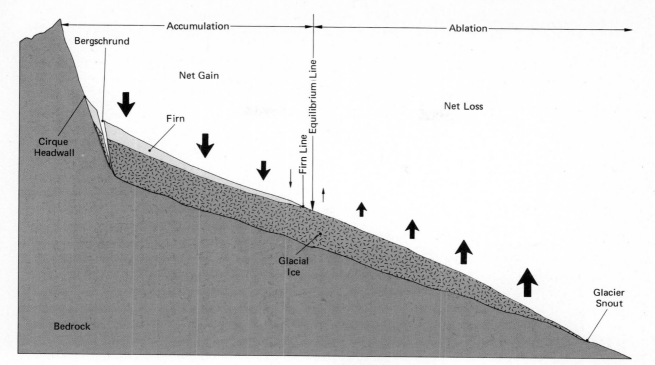

Figure 15.5
This schematic diagram illustrates the dynamics of a cirque glacier. Arrows show relative gains and losses due to accumulation and ablation. The *firn line* marks the boundary between the annual accumulation of firn and glacial ice; the glacier's surface is white upslope from the firn line and bluish on the downslope side. Alpine glaciers exhibit a deep crevasse, called a *bergschrund*, at their heads. It has been suggested that glacial erosion of the rock headwall occurs because water from melting snow and ice seeps into the crevasse and helps to pry rocks loose when it freezes. (Vantage Art, Inc.)

ward melting and evaporation. An increase in the rate of ice arrival or a decrease in the ablation rate will cause the glacier snout to advance. Such a glacier would have a *positive mass budget*. If the rate of ablation exceeds ice replacement, the ice front retreats and a *negative mass budget* exists. Glaciers in the Canadian Rockies, Alaska, and Scandinavia lose a depth of about 12 m (40 ft) of ice to ablation each year. To hold their position, they must replace this loss by inflow from above the equilibrium line. Most of these glaciers have been retreating since about 1900, indicating that negative mass budgets have prevailed throughout the present century.

Movement of Glaciers

Because of its much greater viscosity (resistance to deformation), ice flows in quite a different manner than water. When a mass of ice attains a thickness of about 50 m (150 ft), it begins to spread and flow outward as a consequence of its own weight. It can move in two ways: by internal deformation and by slipping on its base. Internal deformation is very gradual, amounting to a few centimeters per day (Figure 15.7, p. 409), and depends on the fact that ice melts under pressure. The flow is generally laminar, with little internal mixing, and is produced by repeated partial melting and recrystalliza-

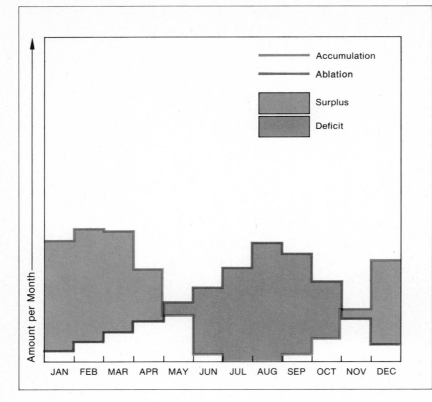

Figure 15.6
This diagram illustrates a typical glacier budget in the northern hemisphere. During the winter months, the accumulation of new snow on the glacier exceeds the loss of ice and snow by melting and evaporation. During the warm summer months, ablation exceeds accumulation, and the glacier experiences a deficit. Note that both accumulation and ablation occur during most months; the glacier may be losing ice near its snout by melting while gaining new snow in its upper, colder reaches. The mass of the glacier increases during periods of surplus and decreases during periods of deficit. If the net annual surplus exceeds the net annual deficit on the average, the glacier grows and advances. (Doug Armstrong)

tion of ice that is subjected to various stresses within the moving ice mass.

In addition to internal deformation, glaciers move by *basal slip*, in which the ice mass as a whole slides over its bed on a film of water. This water is produced by the melting of ice that is pressed against obstacles on the glacier bed (called pressure melting). Basal slip permits more rapid flow than does internal deformation alone. Still, the total rate of ice flow in a midlatitude alpine glacier is typically only a meter a day, or less.

The nature and rate of movement within a glacier are closely related to the temperature of the glacial ice. In *temperate glaciers*, meltwater is produced by pressure melting within or at the base of the ice or by summer melt on the glacier surface. When this water refreezes within the glacier, it liberates the latent heat of crystalliza-

tion, which keeps the glacier "warm"—near 0°C (32°F)—so that it can continue to melt when pressed against obstacles on its bed. Under such conditions water-lubricated basal slip is an important factor.

In polar regions, however, summer melting is minimal and temperatures remain far below the freezing point throughout the depth of the ice. The result is *cold*, or *polar*, glaciers, such as those in Antarctica. Cold glaciers are frozen to their beds, at least on their accessible margins. Because both basal slip and internal deformation are greatly reduced, glaciers of this type are extremely slow-moving. We shall shortly see that the distinction between cold and temperate glaciers is important in the development of landforms by glacial erosion.

Occasionally an alpine glacier moving by basal slip exhibits behavior called a *surge*, in which

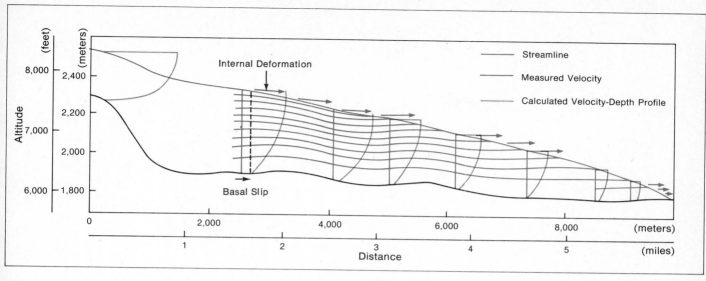

Figure 15.7
The movement of a temperate glacier is accomplished partly by basal slip over its bed and partly by internal deformation of the ice. As the diagram indicates, the process of basal slip causes the glacier to move as a whole, such that the distances traveled by the base and by the surface are the same. In the part of the motion due to internal deformation, however, the ice flows in such a way that the surface moves more rapidly than deeper portions of the glacier. Streamlines show paths of points within the glacier and indicate laminar flow of the ice.

This diagram, in which the vertical scale has been exaggerated, shows the ice velocity at various points in the Saskatchewan Glacier in the Canadian Rockies. Only velocities at the surface of a glacier can be measured directly; the internal velocities indicated in the diagram are inferred. The measured velocity of this glacier's surface in its thicker upper portion is about 400 m (1,300 ft) per year. Near the snout, the velocity is approximately 100 m (330 ft) per year. The streamlines, which show the trajectories of points within the ice, show that a point on the upstream portion of the glacier's surface tends to move slightly downward, whereas points on the downstream portion of the surface move slightly upward. Near the snout of the glacier, most of the movement is by water-lubricated basal slip. (Doug Armstrong after M. F. Meier, 1960)

the ice margin advances as much as 20 m (65 ft) a day. Glacial surges can have several explanations. Some occur when large pockets of water form under the ice or when lake water dammed by the ice seeps out under the glacier. Other surges occur following periods of unusually high snow accumulation, causing the glacier to thicken in the accumulation area. This produces a "bulge" or wave of ice that subsequently moves through the glacier at a speed that exceeds the normal ice velocity. When the bulge reaches the glacier snout, the ice front begins to push forward. This merely signifies that suddenly more ice is arriving than is melting away.

In confined valley glaciers, the fastest ice currents are at the center of the ice surface, with forward movement of the ice margins retarded by friction with the valley walls. The resulting shearing within the ice stream causes its surface to split open in a series of parallel fissures called *crevasses* (Figures 15.1, 15.8). Where steepening of the glacial bed causes a valley glacier to accelerate in speed, crevasses arc across the breadth of the ice surface. Sometimes the glacial bed steepens so abruptly that the ice surface becomes totally broken up in an *icefall* (Figure 15.8). Such areas are dangerous, since icefalls in-

Figure 15.8
These ice streams descending steeply from the summit of Mont Blanc in the French Alps are broken by crevasses. The glaciers appear to be passing over convexities in their beds, producing small icefalls. (T. M. Oberlander)

clude enormous blocks that frequently topple with crushing force.

GLACIAL MODIFICATION OF LANDSCAPES

Flowing ice is a far more powerful agent of landscape change than running water. A thousand-meter-thick alpine glacier can remove rock material from both the floor and walls of a valley, whereas a stream's work is confined to the valley floor. Furthermore, the special properties of glacial ice enable it to remove and transport debris by processes not available to running water. Like streams, glaciers have their greatest erosional potential when their depth and velocity are highest, but unlike streams, they do not drop their load of debris wherever their velocity decreases. In fact, large-scale glacial deposition

occurs only where the ice terminates or where great masses of ice stagnate and melt with no forward motion at all.

Glacial Erosion

Glacial ice by itself has little destructive ability. Ice cannot generate enough pressure to break away unweathered rock, since slowly moving ice yields by melting as it is pressed against any solid obstacle. However, a glacier that is frozen onto its bed will tear out rock and soil material as it is pushed forward by the ice behind it. Ice that freezes onto rocks and pulls them loose soon becomes armored with rock debris. It is similar to coarse sandpaper, with the ice acting as the glue that holds the abrasive particles in place. A rock-studded glacial bottom, or *sole*, is an effective file. The faster the glacier moves, the greater the number of abrasive tools scraped over a given surface and the greater the potential for erosion.

All glacial erosion can be attributed to the processes of tearing out, known as *plucking* or *quarrying*; filing down, known as *abrasion*; and *crushing* of rock projections in the glacier's path. Plucking is best accomplished by cold glaciers in which the ice is frozen to the glacier bed. Abrasion is favored under temperate glaciers where basal slip is caused by water at the base of the ice. Crushing occurs where there is basal slip and the subglacial surface is highly irregular. In general, plucking roughens the surface under the ice and provides the glacier with tools for abrasion. Glaciers probably remove far more material by plucking than by abrasion, particularly where the ice flow encounters unconsolidated (though frozen) soil and where well-developed jointing divides rock into blocks of movable size. Abrasion tends to smooth rock surfaces and to gouge out grooves parallel to the direction of ice flow (Figure 15.9).

The effects of glacial erosion can be seen on many scales, from an individual rock outcrop to an entire countryside. The first alteration a glacier produces when it invades a region is the removal of the soil and weathered mantle. This demonstrates that glacial erosion is more potent than the gradational processes that preceded it. The resulting exposure of the underlying bedrock enables the glacier to begin to remove unweathered rock, which would remain relatively untouched by normal processes of erosion.

The abrasion process causes bedrock to be scratched, chipped, gouged, and filed down. Crescent-shaped chips and gouges called *chatter marks* or *friction cracks*, produced by the pressure of rock upon rock, are characteristic small-scale effects of glacial abrasion on individual rock outcrops. Sometimes large expanses of rock are sculptured into smoothly contoured furrows

Figure 15.9
The ice of a glacier is armed with rock fragments of various sizes, which can abrade the bedrock under the flowing glacier. Glacial abrasion has polished and striated the rock shown in the photograph. The rock also exhibits a prominent break caused by the crushing and plucking action of the glacier. (G. K. Gilbert/U.S.G.S.)

and concavities. Smoothly sculptured forms may be created by either flowing meltwater under a glacier or a subglacial slush of sandy mud mixed with ice crystals. The difference between abrasion and plucking can be seen on single rock outcrops, which often display *stoss and lee* topography. As Figure 15.10 indicates, abrasion tends to smooth and streamline the upglacier (stoss) face of rock knobs, while plucking and crushing steepen and roughen the downglacier (lee) side.

Severe ice erosion is evident in landscapes with numerous rock basins that fill with water, as in Figure 15.11. Some rock troughs were deepened more than 600 m (2,000 ft) by glacial erosion. Examples are Norway's Sogne and Hardanger fjords and California's Yosemite Valley. Glacial erosion excavated some 400 m (1,350 ft) of bedrock to form the basin now occupied by Lake Superior. The ice-scoured areas of Canada, New England, and Minnesota contain countless

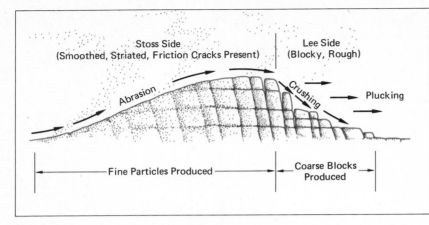

Figure 15.10
In this diagram illustrating the characteristic action of glaciers on outcrops of jointed rock, the direction of glacier flow is from left to right. The upglacier *(stoss)* side of the rock is smoothed and polished by abrasion, and the downglacier *(lee)* side is roughened as the glacier crushes projections and plucks large rocks from the outcrop. (Vantage Art, Inc., after T. M. Oberlander)

large and small lakes in basins produced by glacial quarrying.

The present rate of glacial erosion is not easy to measure because the area of active erosion is covered by ice. One method is to measure the rate at which sediment is carried away from the glacier by meltwater streams. Available data indicate that glacial erosion is 20 to 50 times more rapid than the rate of erosion of nearby unglaciated areas. Glacial abrasion of marks purposefully chiseled into bedrock, or of objects artificially placed in glaciers themselves, indicates removal rates of 1 to 40 mm (0.04 to 1.5 in.) per year, the variation being related to differences in rock resistance to abrasion.

Glacial Deposition

Glaciers have been called "dirt machines" because they resemble conveyor belts that steadily feed rock debris forward to the glacial margin.

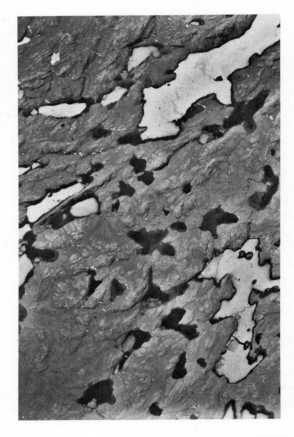

Figure 15.11
Ice scouring in crystalline rocks tends to produce multitudes of basins that later fill with water, as seen in this aerial photograph covering an area of about 26 sq km (10 sq miles) in northern Canada. White areas are frozen lake surfaces. (Air Photo Division, Energy, Mines & Resources © Canadian Government)

CHAPTER 15

Most of the debris carried by glaciers is concentrated in the lowest few meters of the ice, and along the ice margins in the case of alpine glaciers. Because flow in glaciers is laminar, there is little upward or lateral transport of debris into the body of a glacier except near its snout. Glacier snouts tend to develop thrust faults that push basal ice up over slower moving ice near the glacier margin, thus thickening the debris zone.

At the glacier margin the debris is melted out and is overridden by advancing ice or piled into a ridgelike glacial dump if the ice front is stationary. Even when the ice front is retreating, the "conveyor belt" remains in action, continuing to transport debris forward to the glacier margin. Forward transport of debris stops only when the ice thins to the point where flow cannot be maintained over obstacles in the glacial bed. When this occurs, the ice mass decays, or stagnates, by melting in place.

The general term for all types of material deposited directly or indirectly from glaciers is *glacial drift*. The term "drift" dates from about 150 years ago, when it was supposed that the rock debris covering vast areas in Europe and North America was left by the Biblical flood of the time of Noah. Glacial drift is quite variable in character, for it can be produced in many different ways. When ice advances, it eventually becomes so laden with debris that it begins to release some of its load, which is plastered over the subglacial land surface in the form of *glacial till*. Till is a distinctive mixture of coarse and fine material, including fragments of all sizes from boulders to clay; indeed, it was once called "boulder-clay." Till is most distinctive by the absence of any form of sorting or bedding (Figure 15.12 top). Usually some boulders within the till have smooth flat faces produced by glacial abrasion. Till can be a problem for engineers, since excavations in it are often delayed by unexpected encounters with large boulders or extremely tough clays.

When the ice front has pushed to its limit and finally begins to retreat, its marginal por-

Figure 15.12
(top) At this location in California's Sierra Nevada, *glacial till* lies atop sheeted granitic bedrock. The till contains fragments of all sizes, from boulders to clay, and is distinctive by its lack of bedding or sorting.
(bottom) In contrast to till is the deposit of outwash, in which discrete lenses of sand, gravel, and cobbles are visible. This deposit is in a recessional moraine in central New York State. (T. M. Oberlander)

GLACIERS AND GLACIAL LANDFORMS

tions sometimes stagnate, melting in place, with great volumes of debris collecting amid the irregular topography of the decaying ice. When the ice finally disappears, a rather disorganized mass of hills and depressions is left behind. This is known as *ice stagnation topography*. Much of the material has been deposited by water and is somewhat sorted by size. When the ice margin retreats, enormous volumes of debris are carried away from it by braided meltwater streams. The coarser particles are rounded and deposited in vast aprons of *glacial outwash* extending away from the margin of the ice (Figure 15.12 bottom). Till, ice stagnation deposits, and outwash all are regarded as glacial drift.

Occasional components of glacial drift are enormous surface boulders known as *glacial erratics*. These range up to the size of houses and are composed of rock types not otherwise found in the local area. Some have ridden hundreds of kilometers with the ice, and their known sources help establish the flow patterns of continental ice sheets.

Specific landforms are associated with the various types of glacial drift. The general term for a deposit of drift lodged by glacial ice is *moraine*. Moraine includes both till and ice stagnation deposits, but not outwash deposits. *End moraines* are ridges of drift that mark positions where the ice terminus remained stationary for a period of time, or where it pushed up a ridge of drift before retreating. End moraines may mark the terminus of an ice advance or a pause in the retreat of a shrinking ice mass. Thus end moraines are either *terminal* or *recessional moraines*. A morainic landscape that has no conspicuous linear features is simply called *ground moraine*, a *till plain*, or a *drift plain*. Where outwash collects around the margin of a continental ice sheet, the result is an *outwash plain*. Outwash plains that are confined by hillslopes in areas of both alpine and continental glaciation are termed *valley trains*.

The depositional topography produced by glaciers includes a host of small-scale landforms made by advancing, retreating, and stagnating ice. As we shall see in the following pages, those produced by the channeled flows of alpine glaciers differ noticeably from those created by continental ice sheets.

LANDFORMS RESULTING FROM ALPINE GLACIATION

Alpine glaciation has created some of our planet's most magnificent scenery—from the awesome form of Mount Everest and its satellite peaks to the ice-clad pinnacles and knife-edged summits of Switzerland's Alps. Alpine glaciation either partially or totally remodels the upland topography of mountain regions, as well as transforming the valleys leading away from the high summits.

Erosional Forms

The basic element in alpine glacial sculpture is the *cirque*, the amphitheater-like rock basin that is the first product of mountain glaciation. When climatic changes lower the snowline below mountain summits, hillside depressions and the heads of stream valleys become occupied by firn. As this firn is transformed into glacial ice, the depressions and valley heads are deepened and expanded into steep-walled bowls with ice-scoured floors. In most glaciated mountain ranges, there are areas in which glacial erosion never proceeded beyond this point (Figure 15.13). Such areas are pock-marked with cirques that are separated by flat-topped plateaus.

In areas of more severe glacial erosion, cirques have expanded toward one another from the opposite sides of ridges, reducing the ridges between them to jagged knife-edges of rock, known as *arêtes* (Figures 15.14, 15.15). Where three or four cirques intersect, the sharp-crested arêtes come together in a steep rock pinnacle, or *glacial horn*. The classic example is the Matterhorn in the Swiss Alps (Figure 15.15). It is the growth of cirques that creates the sawtooth peaks of alpine topography. Most of the rock

faces forming such peaks are remnants of cirque headwalls.

Where snowfall is abundant, nourishing large glaciers, ice streams cascade from the cirques into the valleys below. The results are nearly as impressive as glacial modification of uplands. Valley glaciers both deepen fluvial valleys and steepen their walls, converting them into *glacial troughs* (Figures 15.14, 15.16). The interlocking spurs created by the irregular paths of deeply incised mountain streams are trimmed back. Valleys are expanded and simplified into open grooves as they are deepened. The result is a semicircular or U-shaped valley cross-section.

Figure 15.13
The steep-walled basins in this photograph of the Wind River Mountains, Wyoming, are *cirques* formed by the quarrying action of glacial ice. The steep walls of a cirque are believed to be produced by collapse when frost action at the bottom of the bergschrund undercuts the cliffs. Small lakes, or *tarns*, are present in the cirques and appear in the glacially scoured area in the foreground. This part of the Wind River Range preserves large areas of the rolling preglacial topography between the separate cirque basins. (Austin Post/U.S.G.S.)

GLACIERS AND GLACIAL LANDFORMS

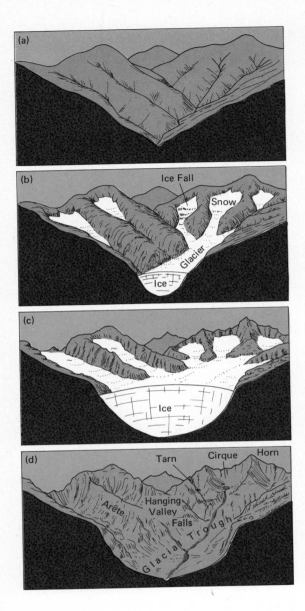

Figure 15.14
This series of diagrams shows the transformation of a fluvially dissected mountain region (a) into a glacially eroded landscape (d). At the highest levels, snowfields develop into *firn* basins that gradually produce glacial ice (b). Cirque growth by glacial scouring and frost action creates sharp crested *arêtes* that come together in horns (c). At lower elevations, ice streams convert sinuous fluvial valleys into open *glacial troughs*. Valleys are deepened and expanded, and their walls are made much steeper by glacial erosion (c). The greater modification of trunk valleys by large ice streams leaves smaller tributary valleys hanging along their margins, producing *hanging valley* waterfalls and cascades (d). Uneven glacial scouring of jointed bedrock excavates basins that later fill with *tarn lakes*. The result is alpine scenery (d). (Vantage Art, Inc. after P. E. James, *Geography of Man*, 3rd ed., 1966, Ginn and Company)

Because of the greater volume of ice channeled, glacial deepening of major valleys is more rapid than deepening of smaller tributary valleys. Retreat of the ice leaves the tributary valleys hanging along the margins of the major glacial troughs. Streams occupying these valleys descend to the floor of the trough by picturesque cascades, or *hanging valley waterfalls* (Figure 15.15).

Because glaciers do not excavate the underlying weathered or jointed rock evenly, valley glaciers commonly produce irregular longitudinal valley profiles. The result may be a *glacial stairway* with high steps and waterfalls in the center of the trough itself. Lakes known as *tarns* occupy basins quarried out by the ice. Strings of tarns, such as those visible in Figure 15.13, are quite common in alpine topography; they are called *paternoster lakes* because of their resemblance to the beads on a rosary.

Downward erosion by valley glaciers is best displayed in the *fjord* landscapes of Greenland, Norway, eastern and western Canada, Alaska, Chile, and New Zealand (Figure 15.16). Fjords are glacial troughs that have been cut far below present sea level, resulting in deeply penetrating arms of the sea hemmed in by rock walls 1,000 or more meters (3,300 ft) high. The glacial origin of fjords is demonstrated by their longitudinal profiles, which show deep basins that could not have been excavated by fluvial processes. Many fjords have depths of more than 1,000 meters

Figure 15.15
(top left) Severe frost weathering of the rock walls above cirque glaciers may erode the walls between adjacent cirques to the narrow, ragged rock ridges, or *arêtes*, seen here in the French Alps. (T. M. Oberlander)

(bottom left) The deepening of the principal stream valleys in mountain regions by glacial erosion is more rapid than deepening of tributary valleys that channel smaller ice streams. After the glaciers melt away, the tributary valleys may be left as *hanging valleys*, which join the deep glacial trough high above the valley floor. The waterfall in the photograph is a stream falling from a hanging valley in Yosemite National Park, California. (T. M. Oberlander)

(right) The Matterhorn in Switzerland is the classic example of a *horn* formed by erosion where the headwalls of three or more cirques intersect. (William Burkhardt)

below sea level. This is because a glacier with a density of 0.9 entering the sea, which has a density of about 1.0, continues to erode downward until it is nine-tenths submerged, after which it will float. Thus, an ice stream 1,000 m thick will still have the potential to erode its bed in 800 m (2,700 ft) of water.

Depositional Forms

The depositional landforms produced by alpine glaciation are no match in scenic grandeur for alpine erosional forms, but they are nevertheless quite conspicuous. The largest depositional features produced by valley glaciers are *lateral moraines* built along the sides of the ice streams. These are ridges of till banked against valley walls or issuing from mountain canyons. They frequently rise more than 300 m (1,000 ft)

Figure 15.16
(top) Valley glaciers follow preexisting fluvial valleys as they flow toward lower elevations. The glacier scours the valley floor and walls, removing irregularities and forming a smooth *glacial trough*. The glacial trough in the photograph is the Lötschental in Switzerland. (William Burkhardt)

(bottom) If a valley glacier reaches the sea, it can continue to scour out its channel below sea level until the ice is nine-tenths submerged, when it will finally float. When the glacier melts or retreats, it leaves a deep rock-walled ocean inlet, or *fjord*, such as Milford Sound, New Zealand. (Alvin Lynch)

above surrounding lowlands. The waxing and waning of glaciers during the Pleistocene created many smaller end moraines that commonly make a succession of arcs across valleys, impounding lakes between the morainic crests (Figure 15.17). Many campgrounds in the Sierra Nevada and Rocky Mountains are situated on these boulder-strewn end moraines.

Climatic cooling since the post-glacial temperature maximum about 6,000 years ago has produced glacial readvances in all the world's high mountains. These advances have created Neoglacial (post-Pleistocene) moraines, visible mainly in cirques. These moraines are usually ridges of coarse rock rubble. Present glaciers have retreated well behind their various Neoglacial moraines as a result of the warming trend of this century.

Valley and outlet glaciers also produce glacial outwash, which aggrades stream valleys beyond the glacial margin (Figure 15.18). The deposits consist of sand and gravel, changing to silt at greater distances from the sediment source. In periods between glaciations, such as the present, the sediment supply is cut off. The decrease in stream loads increases available stream energy, causing the streams to incise the outwash deposits, converting them into terraces.

LANDFORMS RESULTING FROM CONTINENTAL GLACIATION

The dual effects of continental glaciation are clearly illustrated in the very dissimilar landscapes of the Canadian Shield, an area dominated by the erosive action of an ice sheet, and the Canadian Prairie Provinces and American

Figure 15.17
Convict Lake, on the east
face of the Sierra Nevada,
California, lies in a glaciated
valley dammed by deposits of
glacially transported rock
debris, known as *moraines*.
(Dr. John S. Shelton)

Figure 15.18
Outlet glaciers from a highland icecap on
Baffin Island in the Canadian Arctic have
built the outwash fan in the foreground. Note
the discoloration of the sea produced by the
arrival of silt-laden glacial meltwater. (Air
Photo Division, Energy, Mines & Resources
© Canadian Government)

Midwest, north of the Ohio and Missouri rivers, an area of deposition by the same ice sheet.

The Canadian Shield is a vast expanse of ancient igneous and metamorphic rocks surrounding Hudson Bay and extending outward to the arc of large lakes—Great Bear, Great Slave, Athabaska, Winnipeg, Superior, Huron—and the St. Lawrence River. The Laurentide ice sheet stripped this area of its weathered mantle and subjected it to hundreds of thousands of years of glacial quarrying and abrasion. The resulting landscape is one of ice-scoured rock knobs and countless rock-bound lakes in quarried-out depressions (Figures 15.11, 15.19). Stream patterns are rambling and unsystematic, the old patterns having been deranged by ice erosion. In general, drainage lines are somewhat rectangular, following joints in the exposed bedrock. Within this glacially stripped region there are deposits of till, but they are thin and discontinuous. Most glacial drift consists of ice stagnation deposits and *fluted moraine* that has been scoured by active ice (Figure 15.19).

Denuded of soil, and with a short growing season, the region has little agricultural potential. Most human activity here concerns mining and the logging of undemanding coniferous forest. But the forests, streams, and countless lakes of this wilderness give it enormous recreational value. Similar landscapes occur south of the St. Lawrence River in New England and northern New York. Here the relief is greater than over much of the Canadian Shield because of a more mountainous pre-glacial landscape. Scenery and water, both largely the result of glacial processes, have long caused both of these latter areas to be prime tourist targets.

A somewhat different erosional topography appears in central New York State, south of Lake Ontario, where the continental ice sheet moved into a stream-dissected plateau composed of sedimentary rocks. In this region drainage divides were breached, and valleys aligned with the ice currents were greatly deepened. This produced New York's picturesque Finger Lakes. Even though 300 km (180 miles) inland,

Figure 15.19
Vast areas of northern Canada have a surface of *fluted moraine* produced by active ice flow that leaves a thin layer of striated drift. Note the many lakes in this landscape. Some of these result from erosion; some from irregular deposition of glacial drift. (Air Photo Division, Energy, Mines & Resources © Canadian Government)

the floors of some of these glacial troughs were deepened well below sea level. Glacial retreat from this landscape flooded the valleys and glacial troughs with outwash, producing much usable flatland that was not a part of the pre-glacial landscape.

In contrast to the glacial effects in the Canadian Shield area, the American Midwest and Canadian Prairie Provinces have benefited from glacial deposition on an immense scale. Lobes of the continental ice sheets spread glacial drift over the North American Midwest from Alberta and Saskatchewan to eastern Ohio (Figure 15.20). Drift deposits in this area are deep enough to cover the preglacial topography completely. An example of the hilly pre-glacial land-scape is preserved in the so-called Driftless Area of Minnesota, Iowa, and Wisconsin—a triangle with Dubuque (Iowa), Winona (Minnesota), and Wassau (Wisconsin) at its apexes (Figure 15.21, p. 422). This area remained unglaciated throughout the majority of the Pleistocene because of its slightly higher elevation, which caused the Lake Superior and Lake Michigan ice lobes to diverge around it.

The depth of the glacial deposits in some areas of Illinois can be measured in tens of meters, and surpasses 100 m (330 ft) where pre-glacial stream valleys have been submerged in glacial drift. The old stream gravels in large buried valleys are important as aquifers, supplying well water for towns and cities in the region. From Alberta to Ohio, vast areas are gently rolling till plains, while others are nearly flat outwash plains. Soils developed on the recent glacial deposits are fertile due to the short time

leaching has affected them. The largest relief forms are end moraines, which mark positions where the ice front remained stationary for long periods as debris was being conveyed to it. Both terminal and recessional moraines are present. They are arranged in concentric arcs, concave to the south, indicating that the Pleistocene ice front was lobate in form (Figure 15.22).

End moraines produced by continental ice sheets are usually rolling rather than hilly and include many depressions that hold marshes and lakes. Most end moraines rise less than 100

Figure 15.20
This map shows the southern limit of Pleistocene continental glaciation in North America, as well as the major areas of alpine glaciation. The limit of the latest Wisconsinan glaciation, which terminated about 10,000 years ago, is differentiated from the limit of earlier Pleistocene glaciations, the most recent of which ended about 120,000 years ago.
(Vantage Art, Inc. after R. F. Flint, *Glacial and Quaternary Geology*, 1971, John Wiley & Sons)

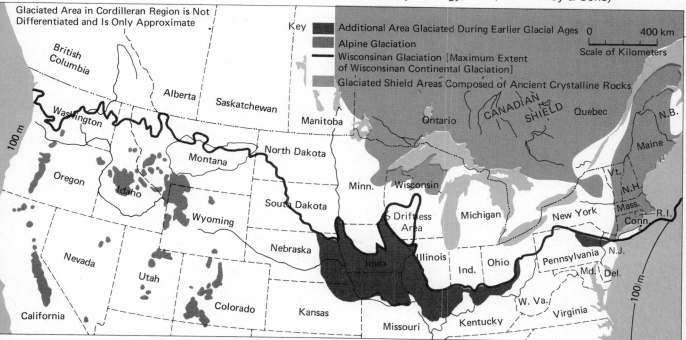

Figure 15.21
Castle Rock, in the Driftless Area of southwestern Wisconsin, would have been destroyed if the ice sheets of the late Pleistocene had invaded this region. The Driftless Area reveals the nature of the pre-glacial landscapes that were eroded by ice and buried by glacial drift during the Pleistocene. (Robert J. Kolenkow)

m (300 ft) and are 10 or more km (over 6 miles) wide. Thus end moraines produced by continental glaciation include huge volumes of drift but are much less conspicuous than the sharply defined lateral moraines built by alpine glaciers.

The rise in sea level at the end of the Pleistocene has fully or partly submerged portions of major end moraines, converting them into peninsulas or islands. Both Denmark and New York's Long Island are partly submerged terminal moraines and outwash plains built by the last major glacial advance, and Nantucket Island, Martha's Vineyard, and much of the Cape Cod peninsula are recessional moraines.

Drift plains include a variety of landforms smaller than the great end moraines. Particularly distinctive forms produced by advancing ice are *drumlins* (Figure 15.23, p. 424), which are elongated streamlined hills composed of every type of drift. Drumlins occur in great swarms, with the long axis of each one parallel to the direction of ice flow. Clearly they have been molded by actively flowing ice, but the relative roles of erosion and deposition in drumlin formation remains a question. The greatest drumlin fields in the United States are located in central New York and eastern Wisconsin, but the best known individual drumlins are Bunker Hill and Breed's Hill in Boston, Massachusetts, where British and Colonial troops clashed in 1775 in the first real battle of the American Revolutionary War.

Advancing ice also creates "rippled" till surfaces in many areas, known in North America as *washboard moraines*. Good examples are seen in North Dakota and many parts of Canada. The

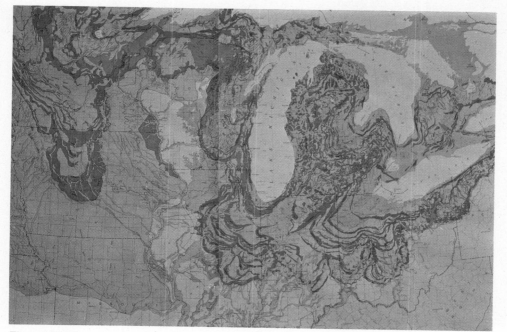

Figure 15.22
End moraines in the midcontinent region of
North America. Wisconsinan age deposits are
shown in green, Illinoian in pink, Kansan in buff,
Nebraskan in brown. Blue areas were
submerged by lakes during ice retreat. Yellow
indicates glacial outwash. Red indicates ice
stagnation deposits such as kames and eskers.
Each moraine, shown in the dark tone, marks a
position at which the front of the continental ice
sheet remained stationary long enough to build
up an exceptional accumulation of glacial drift.
Some of these end moraines are *terminal*,
marking the farthest forward advance of the ice
at a certain period; many are *recessional*,
indicating long pauses in the retreat of the ice
front. The lobate form of the ice front is
conspicuous in the pattern of end moraines.
(Geological Society of America)

ridges, which are transverse to the ice flow, are
only a few meters high and tens to hundreds of
meters apart. They have been interpreted in
various ways and may be complex in origin.

Areas of ice stagnation frequently covered
hundreds of square kilometers. In such areas the
irregular hills of sand and gravel are called
kames, and marshy or lake-filled depressions are
known as *kettles* (Figure 15.24). Kettles mark
places where ice masses have melted after being
wholly or partially buried by drift or outwash.
Many end moraines are *kame moraines* composed of *kame and kettle topography*. To convert kame and kettle topography to cropland requires large expenditures for artificial drainage
of the land. Where ice masses have been floated
out onto outwash plains and partially buried,
they eventually melt, producing *pitted outwash
plains*. Kame and kettle topography and pitted
outwash plains are widespread in the American
Midwest, especially in Michigan, Illinois, and
Indiana, and in the lowland portions of New
England.

Perhaps the most unusual forms related to
ice stagnation are *eskers*, irregular ridges of
well-sorted sand and gravel that resemble meandering railroad embankments (Figure 15.23).

Figure 15.23
(top) This hill, shaped like the reverse side of a spoon, is a *drumlin*, photographed in western Ireland. The blunt end of the drumlin faced toward the ice currents that modeled it. Drumlins are composed of drift that has been shaped by advancing glacial ice. (T. M. Oberlander)

(bottom) The long sinuous ridge in the center of the photograph is an *esker* in northern Canada near Great Bear Lake. This deposit marks the course of a subglacial stream channel and is composed principally of sand and gravel. (Canada Department of Mines and Resources)

Figure 15.24 (below)
The left half of this view, in the state of Washington, shows the irregular topography produced by the stagnation of the ice border in this region. The low hills, or *kames*, are composed of material collected between masses of "dead ice," which subsequently melted, forming the depressions, known as *kettles*, between the hills. (Dr. John S. Shelton)

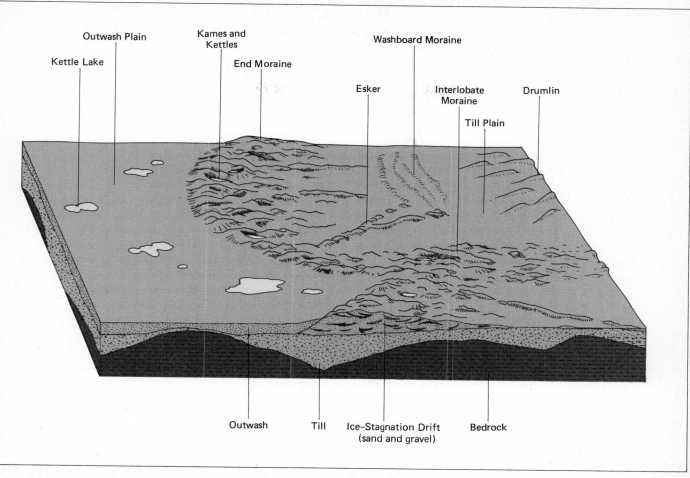

Kettle Lake   Outwash Plain   Kames and Kettles   End Moraine   Washboard Moraine   Esker   Interlobate Moraine   Till Plain   Drumlin

Outwash   Till   Ice-Stagnation Drift (sand and gravel)   Bedrock

Figure 15.25

This diagram represents the typical landforms encountered on drift plains in the midwestern United States and Canada. The length of this block would be on the order of 100 km (60 miles). The sizes of features are exaggerated for the sake of legibility (drumlins are only 1 to 3 km long). The drift covers an irregular bedrock topography. The largest features are the *end moraines* and *interlobate moraines* constructed of drift delivered to the ice margin during a long period in which the ice front is stationary. The topography of the end moraines is often of the *kame and kettle* type. In front of the end moraine is commonly encountered a flat *outwash plain* with a scattering of *kettle lakes*, indicating positions of melted ice masses that were partly or fully engulfed by sand and gravel outwash carried away from the ice front by meltwater streams. Behind the end moraine may be an area of ice stagnation topography, including *eskers* built of sand and gravel deposited by streams flowing under the melting glacier. The otherwise regular surfaces of *till plains* may be diversified by *washboard moraines*, consisting of inconspicuous low ridges transverse to the ice flow, and streamlined *drumlins* aligned with the ice flow and composed of glacial drift that has been scoured by advancing ice. In front of this block (to the left), the entire sequence might be repeated, indicating an older position of the ice margin. (Vantage Art after T. M. Oberlander)

Eskers are stream deposits formed in tunnels under or within stagnant glacial ice. Eskers many tens of kilometers in length and 3 to 30 m (10 to 100 ft) high are common in New England and eastern Canada, and many also can be seen in Michigan, Minnesota, and Wisconsin. They once provided horse and wagon routes through marshy glaciated regions. Today they are extremely valuable as sources of sand and gravel for construction use.

Only glacial deposits formed within the last 20,000 years preserve the original depositional topography (Figure 15.25), including multitudes of small depressions that hold lakes or marshes. Because the latest deposits—having ages up to 70,000 years but mostly less than 20,000 years—are well developed in Wisconsin, they are termed *Wisconsinan* in age. Older drift is known in order of increasing age as *Illinoian* (formed by approximately 140,000 years ago), *Kansan* (approximately 600,000 years old), and *Nebraskan* (perhaps 1.5 million years old). All pre-Wisconsinan drift forms undulating, well-drained, stream-dissected plains. Recognition of the ages of the various pre-Wisconsinan deposits is based on their degree of weathering and soil development. Agricultural use of soils on the older glacial deposits does not involve the expense of artificial drainage, but leaching and eluviation make the older soils less fertile than soils on deposits of Wisconsinan age.

SUMMARY

Our present global climate is only slightly warmer than the climates that brought on the Ice Ages, as can be seen by the fact that the majority of the earth's fresh water supplies are stored as glacial ice. At present glaciers are found only in high mountains or in the higher latitudes, but during cold intervals within the Pleistocene epoch, glacier ice covered vast areas of North America and Europe and also expanded in high mountains on all the continents.

While glacial erosion has stripped vast areas of their soil, the subsequent deposition of glacial till and outwash has created extremely favorable conditions for agriculture in warmer areas.

The major glacier types are continental ice sheets, highland icecaps, outlet glaciers, cirque glaciers, and valley glaciers. Glaciers form at high altitudes and high latitudes where cool summers permit some winter snow to persist through the following warm season. As snow accumulates, it is transformed into firn, which, in turn, is gradually metamorphosed into glacier ice. Gravitational stress causes glaciers to move both by internal deformation and by slipping over their beds. The glacier mass budget, including both accumulation and ablation, determines whether a glacier margin advances, retreats, or is stationary. Glacier retreat is by downward melting of the ice surface at a faster rate than the ice loss is replaced by forward ice flow.

Pleistocene glaciers left a distinctive imprint on the land, being responsible for such features as North America's Great Lakes, the fjords of Norway, and the drift plains of the North American heartland. The indirect effects of Pleistocene glaciation include loess deposits, landforms produced by glacial meltwater, intensified frost action and mass wasting, local depression of the earth's crust, and sea level changes.

Glaciers erode the land surface by the processes of abrasion, crushing, and plucking or quarrying. The debris carried by glaciers is deposited as glacial drift, which includes till, ice stagnation deposits, and outwash. Deposits of till and ice stagnation deposits are called moraine. The landforms produced by alpine glaciers and continental ice sheets vary considerably. Alpine glaciers create cirques, arêtes, horn peaks, glacial troughs, tarns, fjords, and lateral moraines. Continental ice sheets produce ice-scoured plains, drift plains, fluted moraines, end moraines, washboard moraines, kames, kettles, drumlins, eskers, and outwash plains. Only the deposits of Wisconsinan age preserve such landforms.

APPLICATIONS

1. It is usually possible to determine the limit of the major late Pleistocene glacial advance by merely looking at a map of moderate scale. Why?
2. If both drumlins and eskers are seen in the same area, which must have formed first?
3. How might the stones found in glacial till differ in appearance from those deposited by stream or wave action? If your campus is in a glaciated region, collect or photograph the most distinctively formed till stone you can find (often in or washed out of a road cut through a moraine). Compare your find with those of your classmates.
4. Obtain a topographic map of the closest area of unglaciated mountains. On such a map relief is shown by topographic contours. Redraw the bold index contours to show the area after transformation by alpine glaciation. Now redraw the contours again to represent the area after continental glaciation.
5. If your campus is in or close to an area of alpine or continental glaciation, make a catalogue of all the glacial landforms you can identify, indicating the specific location of each.
6. Obtain the *Atlas of American Agriculture* from your campus library. The atlas graphically displays agricultural statistics for the nation, using counties as data units. On how many of these economic maps can you identify the effects of continental glaciation? Wherever the glaciated areas stand out, attempt to explain the specific effect of glaciation on the particular economic phenomenon.

FURTHER READING

Andrews, John T. *Glacial Systems: An Approach to Glaciers and Their Environments*. North Scituate, Mass.: Duxbury Press (1975), 191 pp. This treatment of glaciers, glacial landforms, and glacio-eustatic and glacio-isostatic effects is concise and advanced; it is for students with an understanding of mathematics.

Embleton, Clifford, ed. *Glaciers and Glacial Erosion*. New York: Crane, Russak (1972), 287 pp. This collection of previously published articles contains many highly readable "classics" in the literature of glacial geomorphology.

——and **C. A. M. King.** *Glacial Geomorphology*. New York: Wiley (1975), 573 pp. This is a very complete treatment of all aspects of glacier structure and dynamics, processes of glacial erosion and deposition, and the landforms resulting from each.

Flint, R. F. *Glacial and Quaternary Geology*. New York: Wiley (1971), 892 pp. Flint's book is the standard American source of glaciation and the Pleistocene. The treatment is broader but less detailed than that of Embleton and King.

Post, Austin, and **E. R. LaChapelle.** *Glacier Ice*. Seattle: The Mountaineers (1971), 110 pp. The highlight of Post and LaChapelle's book is its excellent large-format photographs of glaciers. The text, dealing with the various types and structures of glaciers, is also excellent.

Schultz, Gwen M. *Ice Age Lost*. Garden City, N. Y.: Anchor Press (1974), 342 pp. Written for the general public by a geographer, this book is an account of all aspects of the Ice Ages, including glaciation, periglacial activity, sea level change, and effects on the biotic world, including man.

Sugden, David E., and **Brian S. John.** *Glaciers and Landscape: A Geomorphological Approach*. London: E. Arnold (1976), 376 pp. Although the subjects covered in this book are generally similar to those treated by both Flint and Embleton and King, Sugden and John's more selective approach is original and thought-provoking. Available in paperback.

CASE STUDY: Midwestern Glacial Deposits

It is hard to associate a windswept desolation of glacial ice with the wonderfully productive farmlands of Iowa, Indiana, central Illinois, and western Ohio. Yet these lands, among the most ideally suited to agriculture of any on earth, are the great gift of North American continental glaciation. Nearby areas beyond the glacial limit tell us what was here before the ice sheets came—stream-dissected terrain, with soils that were thin and stoney or leached and underlain by a hardpan. Without glacial modification, this region would have been useful for pasture and the small-scale production of some commercial crops, but it could not have become the agricultural cornucopia that sends meat to the rest of the continent and grain to much of the world.

Here, in the incredibly productive "corn belt" from Iowa to central Ohio, everything came together. In this area of warm summers and adequate moisture, a rough landscape was fertilized, smoothed, and made ideal for the use of agricultural machinery by the deposition of a blanket of nutrient-rich glacial drift and glacio-aeolian loess.

But even in this bountiful area, physical conditions and agricultural opportunities are not uniform. Depressions in the drift plains contain heavy clays and are typically waterlogged; to be useful, they must be drained. Beds of former ice-dammed lakes are flat, silty, and productive, except for gravelly beach ridges. Kames and outwash plains composed of sand and gravel are quite porous and cannot support water-demanding crops. But they, along with eskers, are valuable in their own way—as sources of aggregate used in construction.

The texture and chemical qualities of glacial drift are related. Soils developed on sandy till or outwash are likely to be acidic, requiring repeated treatment with lime to be kept productive. Such are the better drained soils of much of Michigan, Wisconsin, and Minnesota, north of the corn belt. Here the till was produced by glacial erosion of the crystalline rocks of the nearby Canadian Shield. Acid soils in cool climates are best suited to the growth of pasture and fodder grasses; thus these are the great dairy states. Close to urban areas, soils developed on sandy till and outwash are heavily fertilized to support truck farms feeding city populations.

The soils with the greatest natural fertility developed on calcareous loess or till derived from glacial erosion of limestone or dolomite. Ironically, much of the parent material for the corn belt's calcareous soils was transported from limestone and dolomite outcrops to the north that are now blanketed by nutrient-poor sandy till. Here, as elsewhere, the ice sheets shunted surface materials along, carrying away one type and replacing it with another, often thoroughly changing the raw material for soil development as well as the topography.

The age of the drift exposed in areas of glacial deposition also affects agricultural productivity. In the areas of most recent deposition, poor drainage is a problem; many lakes and marshes are present and stream systems to evacuate surface water are poorly integrated. However, this is where we find the least leached and eluviated soils. As a consequence, in Ohio, Indiana, Illinois, Iowa, Michigan, and Minnesota, some 50 million acres of land have been artificially drained. Where the surface is composed of older drift—as in eastern Nebraska, northeastern Kansas, and the southern portions of Iowa and Illinois—mild fluvial dissection has produced gently rolling, well-drained surfaces, but the older soils covering them have already been leached of some nutrients and have begun to develop unfavorable hardpan structures.

Surface topography seldom gives a good indication of what may be encountered at depth in a mass of glacial drift. Where the drift cover is thick, several different layers of till, stratified drift, loess, and outwash can be expected. Unpredictable variations in materials at depth create difficulties in the construction of roads, wells, canals, bridges, dams, reservoirs, and large buildings. Enormous erratic boulders or tough clays often lurk unsuspected in the way of intended excavations, slowing progress and raising costs. In the 1950s, the cost overrun due to improper assessment of construction difficulties in glacial deposits on the route of the St. Lawrence Seaway amounted to more than $27 million.

Loose materials of different textures and water contents behave differently under loads; thus glacial deposits must be analyzed by deep boreholes before structures can be placed on them. Sand and gravel lenses within drift produce elevated, or "perched," water tables and aquifers that can be drawn upon agriculturally or even for urban water supplies, but which create difficulties for construction projects. Waste disposal is a particular problem on drift plains. Deep excavations risk intersecting aquifers and perched water tables, causing flooding of disposal sites. Liquid waste disposal is complicated by the danger that toxic fluids may find pathways into aquifers that feed wells.

Although subsurface deposits of sand or gravel may be valuable enough to warrant

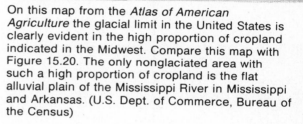

On this map from the *Atlas of American Agriculture* the glacial limit in the United States is clearly evident in the high proportion of cropland indicated in the Midwest. Compare this map with Figure 15.20. The only nonglaciated area with such a high proportion of cropland is the flat alluvial plain of the Mississippi River in Mississippi and Arkansas. (U.S. Dept. of Commerce, Bureau of the Census)

considerable excavation to expose them, the major significance of the subsurface sands and gravels continues to be their role as aquifers supplying well water for human use. The aquifers consist of glacial outwash deposited in preglacial bedrock valleys that are deeply buried in drift and also in interglacial valleys cut into old drift and subsequently filled with younger drift. The depth of glacial drift over the old valley floors is commonly 100 m (300 ft) and sometimes as much as 200 m (600 ft). Where the drift cover is deep, aquifers may be present at several levels. Cities drawing their water supplies from drift-filled valleys include Champaign-Urbana, Illinois; Canton and Dayton, Ohio; and Peru, Indiana. Many smaller communities depend upon this water supply, which at present remains underutilized. Thus a thick drift cover not only increases agricultural opportunities, it also provides the reservoirs of water that large populations require.

Atmospheric energy generates water waves that transport energy over vast distances, expending that energy against coastlines, driving them back in some places and building them out elsewhere, molding them to accord with prevailing wave climates. The shoreline is a place of constant activity and change—one of the best locations for observing the dynamic relationship between landforms and the processes that create them.

Early Morning After a Storm at Sea by Winslow Homer. (The Cleveland Museum of Art; gift from J. H. Wade)

Marine Processes and Coastal Landforms

One of the most challenging tasks for classical scholars is to reconstruct the geography of the ancient world. Since the time of the Trojan Wars described by Homer in the *Iliad*, the coastlines of the Aegean Sea have changed so much that some portions can hardly be recognized today. The land has advanced miles in many locations and has been chewed back in others. The sites of towns have been carried away by the wave action, and ancient harbors are now agricultural plains.

We do not have to go back 2,500 years to be aware of changes along coastlines. They often occur overnight. The coastal zone is a particularly dynamic interface on our planet, a zone of continual energy conversion where wind-initiated water waves and currents ceaselessly transform the edges of the land, molding shorelines to accord with motions of the sea surface. Two-thirds of the world's population resides near the edges of the land, close to the sea, constantly affected by its behavior and its many influences. In this chapter we shall examine the nature of the sea and its interaction with the land.

MOTIONS
OF THE SEA SURFACE

Much of the sea's fascination as well as its power to shape coastlines (Figure 16.1) proceeds from the fact that it is never still. Since it is a liquid with little resistance to deformation, its surface is easily set in motion. The friction created by wind that blows over water surfaces produces *waves* that roughen the surface and *currents* that transport water in slow-moving streams (Chapter 4). The rotation of the earth-moon system and the gravitational attractions of the sun and moon produce the regular rise and fall of the sea surface known as *tides*. Waves, currents, and tides are daily phenomena along the earth's coasts. However, there have been much larger and longer lasting changes in the level of the sea, significantly altering the sizes and shapes of the oceans themselves.

Long-Term Changes in Sea Level

Sea level appears to have risen to its present position approximately 6,000 years ago, following the melting of the North American and Scandinavian continental ice sheets (Figure 16.2). We have seen that melting of the existing Antarctic and Greenland ice sheets would raise the level of the oceans another 65 m (210 ft) or so, drowning the coastal plains and river valleys regarded as home by the majority of the earth's population.

How stable is the level of the sea today? Since fluctuations in climate have caused changes in the size of alpine glaciers even over the past 500 years, slight changes of sea level are to be expected. The current trend is hard to establish firmly because many coastal regions are slowly rising, sinking, or tilting, as suggested in Figure 16.3. As a result, historic changes in sea level relative to the land vary considerably from place to place. Thus, two phenomena can affect the level of the sea relative to the land: *eustatic* changes, which are world-wide changes in sea level caused by changes in the volume of sea

Figure 16.1
Wave erosion of weak chalk formations along the coast of England has caused sea cliffs to retreat into a rolling, fluvially dissected landscape. In such settings, the rate of cliff retreat may be as much as 2 m (6 ft) a year. (Eric Kay/Östman Agency)

water or the capacity of the ocean basins; and *tectonic* changes, meaning local uplift or depression of the land itself.

The Tides

More significant to the everyday lives of coastal dwellers are the smaller daily changes in sea level known as the *tides*. Tides most commonly cause two periods of rising water, or *flood tides*, and two periods of falling water, or *ebb tides*, in every 24 hours and 50 minutes. In the

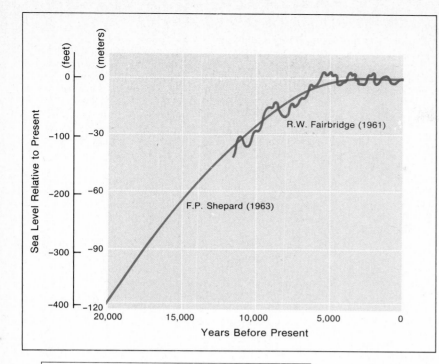

Figure 16.2
As this graph indicates, the level of the sea has risen by at least 120 m (400 ft) since the late Pleistocene glacial maximum. Melting of the ice sheets returned large quantities of water to the oceans. The estimates of two different scientists are shown, one (Shepard) assuming a fairly steady rise of sea level as the ice sheets melted, and the other (Fairbridge) suggesting that sea level fluctuated during its rising trend. Numerous coastal landforms that were exposed 20,000 years ago are now deep under water; many of the world's major harbors are located in what were once wide river valleys that were submerged by the rising seas. (Doug Armstrong after E. C. F. Bird, *Coasts, An Introduction to Systematic Geomorphology*, vol. 4, 1969, by permission of the M.I.T. Press)

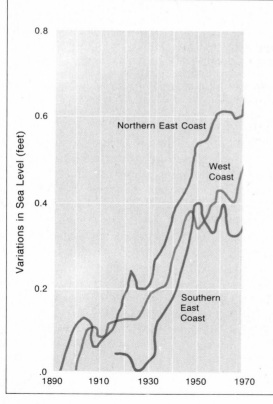

Figure 16.3
The level of the sea relative to the coastlines of the United States has been rising during this century. Part of the apparent increase in sea level is attributed to local lowering of the land. (Doug Armstrong after National Oceanic and Atmospheric Administration)

433

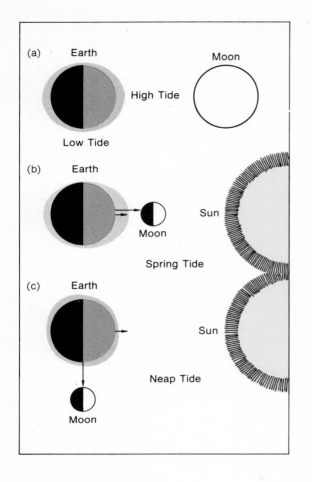

Figure 16.4

(a) Tidal bulges of water, here greatly exaggerated in height, are produced on the earth because of the gravitational effect of the moon and sun and the centrifugal force produced by rotation of the earth-moon system. As the earth rotates, two tidal bulges sweep across it, producing twice-daily high tides.

(b) The highest high tides and the lowest low tides, or *spring* tides, occur during new moon and full moon, when the earth, moon, and sun are aligned.

(c) The lowest high tides, or *neap* tides, occur during the moon's first and last quarters, when the gravitational pulls of the moon and sun are at right angles, as shown. (Tom O'Mary)

(opposite) This diagram illustrates how the gravitational attraction of the moon creates one tidal bulge on the side of the earth facing the moon, while the centrifugal force produced by rotation of the earth-moon system creates a tidal bulge on the side of the earth farthest from the moon. Three successive positions of the earth and moon are shown, numbered 1, 2, 3, with arrows indicating the path followed by the center of each. Note that the earth and moon rotate together about the center of gravity of the earth-moon system at *C*, which lies just below the earth's surface on the side closest to the moon. It is the centrifugal force *(CF)* created by the earth's rotation about this off-center axis that forces water toward the point on the earth that is farthest from the moon. (Vantage Art, Inc.)

ideal case, the sea rises slowly for a little more than 6 hours, then falls for another 6 hours, then repeats the cycle. The amount of tidal rise and fall, or *tidal range*, varies greatly from place to place and also from month to month in any one location. The timing of flood and ebb tides is consistent and predictable at each locality, but varies from one locality to another.

The tidal effect is actually created by two forces. One is the gravitational attraction of the moon and, to a lesser extent, the sun. The other is the centrifugal force generated as the earth and moon rotate about a common point at the center of gravity of the earth-moon system (Figure 16.4).

The point on the earth that is closest to the moon is most strongly affected by the moon's gravitational force. Water on that side of the earth is pulled toward the moon, creating a *tidal bulge*. The moon's gravitational pull diminishes toward the earth's center as distance from the moon increases. At the earth's center, lunar attraction becomes balanced by an opposing force that pulls in the opposite direction. This opposing force is the centrifugal effect of rotation of the earth-moon system. The earth and the moon revolve together about a common center of gravity—like a twirler's baton with a large and a small end. This common center of gravity falls just inside the mass of the earth (the large

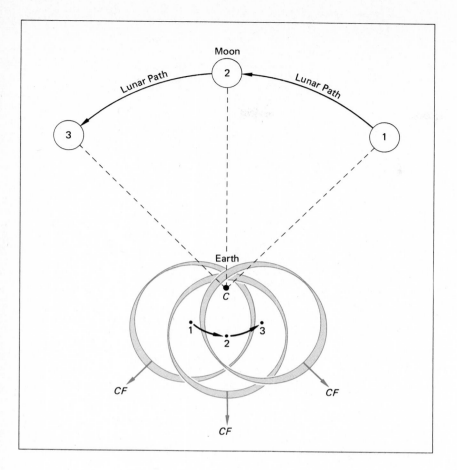

end of the baton) on the side facing the moon. The momentum of the mass of the earth on the far side of the uncentered axis of rotation produces centrifugal force directed away from the moon. This forces ocean water outward in the second tidal bulge, opposite to the tidal bulge created by simple lunar gravity. Since the two tidal bulges draw water from the areas between them, low tides occur in regions at 90° from the moon (or when the moon is near the horizon). The earth rotates through both watery bulges every 24 hours and 50 minutes, or once every 12 hours and 25 minutes.

The sun also has an effect on the tides. But the sun's great distance from the earth reduces its gravitational force to about half that of the much smaller but closer moon. The greatest tidal range in any region, known as a *spring tide*, occurs when the sun aligns with the earth and moon to increase either the gravitational pull in the direction of the moon (time of new moon— moon between the sun and earth) or the centrifugal effect (full moon—earth between the sun and moon). Conversely, the tidal range is at a minimum (*neap tide*) when the sun and moon make a 90° angle with the earth, so that the sun's gravity draws off some of the crests of both tidal bulges.

The oceanic response to these tide-raising forces is affected by many factors, including

MARINE PROCESSES AND COASTAL LANDFORMS

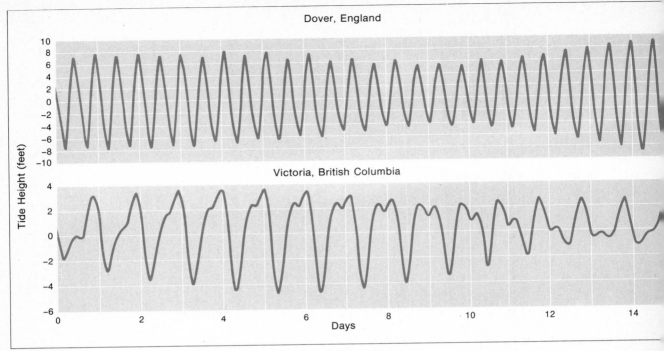

Dover, England

Victoria, British Columbia

Figure 16.5
The graphs show the tide heights relative to average sea level at Dover, England, and Victoria, British Columbia, for a period of 15 days. The tides at Dover exhibit the usual pattern of twice-daily high tides. Note that the heights of the daily tides vary through the month because the earth, moon, and sun change their relative positions. The tides at Victoria show effectively only one high and one low tide per day. Once-daily tides can occur where the twice-daily tides are unusually small because of special land configurations. If tides arrive at a location by way of two channels, with a relative time delay of 6 hours between the separate flows, the twice-daily tides may be canceled. The tides at many locations show a mixture of twice-daily and once-daily tides. (Doug Armstrong after U.S. Naval Oceanographic Office)

ocean currents, variations in water density and atmospheric pressure, and the shapes of land masses. As a result, there are many local peculiarities in tidal characteristics, as can be seen in Figure 16.5. Tidal ranges are greatest where the flood tide is channeled into narrowing estuaries; there the mass of water is crowded together so that it surges strongly upward. Where the tidal bulge pushes into the English Channel, its range increases from less than one-quarter of a meter in the open sea to 7 m (23 ft) at the narrowest part of the channel. The Bay of Fundy in Nova Scotia, Canada, is famed for the world's highest tidal range, the average spring tide range being 15.4 m (50.5 ft). This is a consequence of rocky walls that confine the water, a channel that abruptly splits into two narrow arms, and rapidly shallowing (or "shoaling") bottoms.

When very strong onshore winds or storms occur at the time of high tide, a *wind tide* or *storm surge* may occur, raising the water as much as 3 m (10 ft) above normal high tide. Storm surges can be tremendously destructive to life and property. America's greatest natural disaster occurred in 1900 when a hurricane-generated storm surge destroyed Galveston, Texas, and much of its population of 7,000.

Tides are extremely important in marine navigation. Some harbors can only be entered at high or flood tides. In some places powerful tidal currents constitute a hazard to ships. It was weather and tide conditions that determined the exact timing of the invasion of France by the Allies during World War II.

Waves in the Open Sea

The waves that make the sea surface so fascinating to watch are a clear example of an energy transformation. The significance of all waves, whether they are sound waves, light waves, or water waves, is that they carry energy from one place to another. The energy that initiates wave motion is transmitted by the waves to some other point where it is ultimately transformed into heat and work of some kind.

The vast majority of water waves are initiated by the kinetic energy of wind moving over a water surface. As the water surface deforms under wind stress, waves are initiated. The wave disturbance is passed from particle to particle in such a way that the passage of a water wave merely involves a rise in the water surface, pro-

ducing a *wave crest*, followed by a sinking of the surface to produce a *wave trough* (Figure 16.6). There is little forward displacement of the water itself, for what is transmitted is energy, not mass. This is obvious when a wave passes under a floating object, which merely rises and falls as the wave moves by it. The vertical distance between the crest and the trough is called the *wave height*. The distance between wave crests

Figure 16.6
The nearly circular motion executed by individual water particles is what transmits forward wave motion. When a wave crest passes, the surface water particles at that point are at the top of their orbits. When a wave trough passes, the same surface water particles are at the bottom of their orbits. Orbital motion of water particles is negligibly small at depths greater than half a wave length. Waves in water deeper than half a wave length, which are called *deep-water waves*, are not influenced by the presence of the ocean bottom. As the wave enters shallow water near a shore, the ocean bottom begins to interact with the wave motion. The wave becomes steeper and the wave length becomes shorter. The water particles at the top of the wave eventually reach a speed greater than the speed of the wave, and the wave breaks. The wave dissipates its remaining energy as it washes up on the beach. (Doug Armstrong)

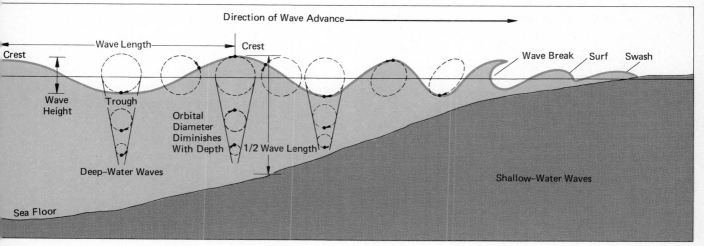

is known as *wave length*, and the time required for successive crests to pass the same point is the *wave period*. The speed of a wave in deep water is a function of the wave length: the greater the wave length, the more rapid the speed of wave transmission.

Water waves are of two types. The waves raised directly by wind stress on the water surface are sharp-crested *forced waves*. The round-crested linear waves traveling outward from the point of energy transfer are known as *swell*. The size of the forced waves produced directly by the wind increases with wind force, wind duration, and the extent of water—known as the *fetch*—over which the wind can build waves (Figure 16.7). In storms in the open sea, the height of forced waves is commonly about 6 m (20 ft); however, in 1933 a U.S. naval vessel caught in an unusually vast storm in the mid-Pacific measured a wave height of 34 m (112 ft).

The vast majority of the waves that persistently roll in against our shores are swell generated in some far-off place. When watching these waves come in, bear in mind that they very likely were born in storms raging in the seas off Japan, Alaska, Iceland, or even Antarctica. Typical swell with wave lengths of 100 m (330 ft) moves at about 15 meters per second (30 mph). Swell of moderate size is capable of traveling halfway around the world before all of its energy is absorbed. Nearly all moderate swell lives long enough to reach a coastline.

Seismic Sea Waves

Not all waves are initiated by wind stress on the water. The most hazardous of all waves are known popularly as "tidal waves," although they are quite unrelated to tidal action. They are also called *tsunamis*, the Japanese word for exceptionally large waves that strike the land. These greatest of all waves are actually shock waves in water. Thus they are *seismic sea waves*. They are triggered by earthquakes, landslides, and volcanic eruptions on the ocean floor.

Oceanic earthquakes are frequent, the largest originating in the oceanic trenches where subduction is occurring. Since the Pacific Ocean is ringed with these earthquake-producing oceanic trenches, the shores of the Pacific experience frequent seismic sea waves. Around the margins of the Atlantic, ocean trenches appear only in the area of the West Indies and east of Patago-

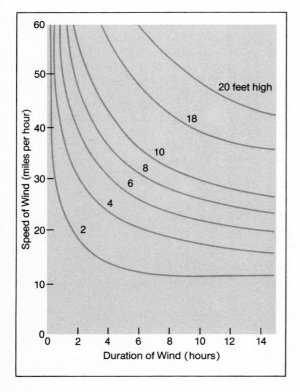

Figure 16.7
The height of waves on open water depends on the fetch, the speed of the wind, and the length of time the wind has been blowing. These calculations are based on a distance from the windward shore that is great enough to preclude the influence of fetch on wave height. (Doug Armstrong after Sverdrup and Munk, "Wind, Sea, and Swell: Theory of Relations for Forecastings," Technical Report of the U.S. Hydrographic Office, no. 1, Publication 601, 1947)

nia, so seismic sea waves in the Atlantic basin are infrequent. Most Atlantic coast tsunamis have been triggered by earthquakes or volcanic disturbances on the oceanic ridge system and associated transform faults.

Seismic sea waves are awesome phenomena. In the open sea they are barely detectable, having wave lengths exceeding 100 km (60 miles) and heights of less than a meter. But since the speed of water waves is always proportionate to the wave length, seismic sea waves have incredible velocities—as much as 800 km (500 miles) per hour. All water waves slow and bunch together approaching the land, when they begin to be retarded by friction with the sea floor. The principle of energy conservation causes the wave height to increase as wave speed slows. In the case of seismic sea waves, this causes a series of rapidly moving mountains of water to lift suddenly out of an ordinary sea, often rising to heights of 10 to 30 m (30 to 100 ft) as they strike the land. In some cases, a great wave trough appears before the first crest, causing the sea surface to sink and recede far from shore. This happened at Hilo, Hawaii, in 1923. Many people ran out to pick up fish stranded on the newly exposed land, and were engulfed by the gigantic wave crest that suddenly rose from the sea.

Historically, seismic sea waves have caused as much destruction and loss of life as have major earthquakes and volcanic eruptions on the land. The 30,000 people killed as a consequence of the eruption of Krakatau in 1883 died in the seismic sea waves the volcano generated rather than as a result of the eruption itself.

Hawaii lies in the path of tsunamis emanating from three different sources: the Peru-Chile trench, the Aleutian trench, and the oceanic trenches of the western Pacific. Since tsunamis are triggered by earthquakes, warnings can be issued and loss of life averted, although waterfront property damage can be reduced only by restricting building in the hazard area. Everywhere, of course, there are a few people who go down to the beach to see the predicted "tidal wave," and are never seen again.

WAVE ACTION AGAINST THE LAND

Analysis of coastal landforms in Greece reveals that in some areas wave erosion has driven back the shoreline some 800 m (0.5 mile) in the past 6,000 years. The softer rocks of the Yorkshire coast of England have been eroded back some 3 km (2 miles) since Roman times. How can waves, which roll harmlessly past boats, swimmers, and children in inflated innertubes, be so destructive to the land itself? Waves are, in fact, not destructive until they "break."

Wave Break

In the open sea, waves glide onward unaffected by the ocean floor, their velocities dependent on their wave length. But when waves move into shallow water and wave motion begins to be affected by friction against the sea floor, wave characteristics are modified. Waves then change from deep-water types with velocities dependent on wave length to shallow-water types with velocities dependent on water depth.

When incoming waves are slowed by friction with the sea floor, the waves begin to bunch together and to increase in height. Actually, as shown in Figure 16.6, friction causes the orbit of the surface water particles to be deformed from a circle to a forward-leaning ellipse. This happens when the water depth decreases to about half the wave length. Two things can happen to this ellipse. It can merely collapse in a mass of foam, producing a *collapsing*, or *spilling, breaker*, or it can topple over forward, with its crest arching down to produce a *plunging breaker* (Figure 16.8, p. 440). Surfers attempt to ride the smooth moving hill of water just ahead of the toppling crest, or "curl," of a plunging breaker or the foam of a spilling breaker. A fourth type, the *surging breaker*, does not really break at all; it merely surges onto the beach and then drains back quietly to the sea.

The way a wave breaks depends on a combination of factors, including the wave form

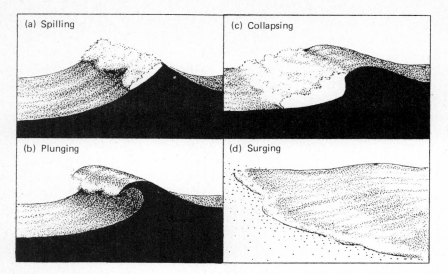

Figure 16.8
The manner in which waves break affects beach erosion as well as the stability of man-made structures such as breakwaters and seawalls. Breaker type is a response to wave height and steepness and beach slope.

Spilling breakers (a) result from the downward slumping of the crests of steep waves. In plunging breakers (b) steep wave crests curl over the front face of the wave and fall vertically. In collapsing breakers (c) the lower part of the wave front becomes vertical and the wave collapses in a mass of foam. When wave height is small, waves slide up the beach without collapse of the wave form, producing surging breakers (d). (T. M. Oberlander)

(length vs. height), the obliqueness of wave approach to the shore, and the steepness of the sea bottom. In general, an oblique approach and a gentle slope favor spilling breakers; a direct approach and steep slope produce plunging breakers. Collapsing breakers are produced by waves that are less steep than those that spill or plunge.

Wave Refraction

As waves enter shallow water, the line of the waves is seldom exactly parallel to the shore; one portion of each wave starts to break before the rest of the wave. But even before this portion of the wave breaks, it is affected by friction with the sea floor. Its forward velocity decreases

while the part of the wave that has not yet "felt bottom" moves on ahead. This change in velocity causes the wave to pivot slightly at the point where it is first affected by friction, bending the line of the wave, as illustrated in Figure 16.9. This phenomenon is known as *wave refraction*.

Wave refraction is very important because it either focuses or spreads the energy conveyed by incoming waves. If a coast is irregular, waves are first slowed by friction as they approach projecting headlands. As Figure 16.9 shows, the resulting wave refraction causes wave energy to become focused on the headland. In bays between coastal promontories, refraction causes the lines of wave approach to separate, spreading each unit of wave energy over a larger area. Given sufficient time and a constant sea level,

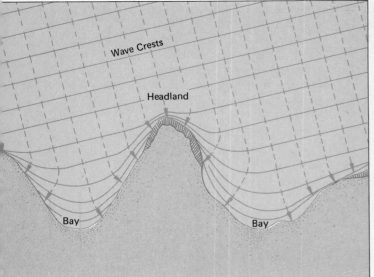

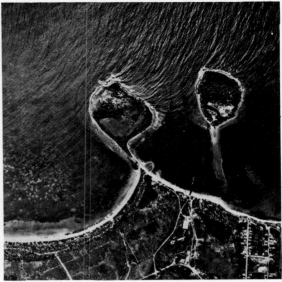

Figure 16.9
(left) Wave motion is altered in direction, or *refracted*, when it interacts with shorelines. The diagram shows waves refracted at a headland, or coastal promontory. Solid lines represent wave crests; dashed lines show direction of wave energy transmission. Compartments bounded by wave crests and energy vectors contain equal amounts of wave energy. Refraction causes waves to concentrate their energy on headlands, accelerating erosion and causing the formation of such features as sea stacks and sea caves excavated into the sides of the headland. In bays, wave energy is dissipated by divergence of wave trajectories. Therefore bays become areas of sediment deposition. (John Dawson)
   (right) This photograph shows wave refraction along a portion of an irregular coast. (Department of the Navy)

the above-average wave energy at headlands and below-average wave energy in bays would smooth the coast by wearing back the headlands and allowing sediment to accumulate in the bays.

Coastal Erosion

In many places waves do not break on broad sand or gravel beaches, but impact against steep slopes, which are progressively worn back by wave attack. This erosion is accomplished in several ways. Some removal is by chemical corrosion of rock that is constantly wetted, and some is a result of the crystallization and thermal expansion of salts in rock pores repeatedly dampened by salt spray. There is also abrasion by rock debris hurled against cliffs by wave impact. Lighthouse windows 30 m (100 ft) above sea level commonly have wire screens to protect them from wave-tossed rock shrapnel. Part of the break-up of rock along coasts results from the enormous impact of heavy masses of water crashing into the jointed rock during storms. One of the most effective processes for loosening

and detaching rock is the compression of air within fissures in the rock when a wall of water smashes against it—followed by the explosive expansion of the air as the water falls away.

A coast's vulnerability to wave erosion depends on the local geological materials, the fetch that controls the size of forced waves or storm surges, the pattern and magnitude of swell from distant storms, and the possibilities for wave refraction at the coast. In England the rate of coastal retreat under wave attack may be as much as 2 m (6.5 ft) a year where the rocks are composed of poorly cemented sand or gravels, or less than a meter each thousand years where resistant rocks are present. In unusually intense storms, sea cliffs composed of glacial drift have been cut back by more than 10 m (33 ft) in a few hours.

Coastal erosion is concentrated in the zone of wave impact, ranging from low tide to somewhat above high tide. The effect of wave erosion is to undercut sea cliffs by notching them at their base, as illustrated in Figure 16.10. The unsupported upper portion of the cliffs then retreats by collapse. The retreat of sea cliffs leaves behind a submarine rock platform called a *wave-cut platform* or *abrasion platform*, whose seaward sloping surface may be littered with debris removed from the land. Debris deposits may form an embankment, or *wave-built terrace*, extending beyond the abrasion platform. As the width of a wave-cut platform increases,

greater amounts of wave energy are lost by friction with the surface of the platform, until there is no further effective wave erosion at the shoreline. It is now believed that the broad abrasion platforms seen in some areas must have been produced during periods of slowly rising sea level, with the zone of effective wave action and frictional energy loss being always narrow and close to the shoreline.

Figure 16.10
This diagram illustrates the principal features that may be seen along cliffed coasts. The major elements are *sea cliffs* that are kept steep by being undercut at the base along a *wave-cut notch*, producing an erosional *wave cut platform* that extends seaward from the base of the sea cliff. The cliff and platform meet at the *shoreline angle*, which is usually at the high-tide level. Features occasionally seen are sea caves and erosional remnants such as *sea arches* and *sea stacks*. Depositional *wave-built terraces* may also be present. The existence and character of these subsidiary forms depend on the nature of the rock present. (Vantage Art, Inc. modified from Karl W. Butzer, *Geomorphology from the Earth*, 1976, Harper & Row)

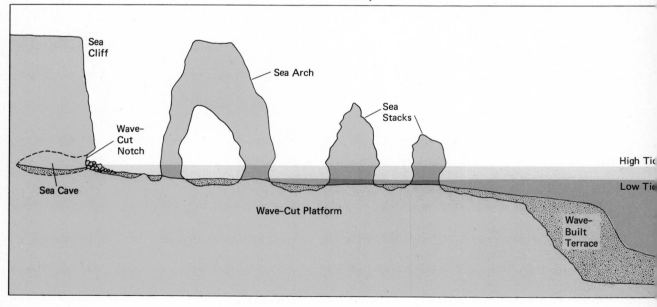

Figure 16.11
(top) The chalk cliffs at Étretat on the coast of Normandy, France, were cut by waves to form sea arches and sea stacks. Note the vertical face of the cliffs; wave action at their base undermines the rock and causes the collapse of the upper portions. The cliffs thus maintain a vertical face as they retreat. (Eric Kay/Östman Agency)

(bottom) This photograph of the Oregon coast shows the exposed surface of a wave-cut platform at low tide, and above it a marine terrace. This terrace was an older wave-cut platform produced at sea level and subsequently uplifted tectonically, initiating platform development at a lower level. As the low headland shown here is worn back by wave erosion, sediment is transferred into the bay in the background, where it accumulates in a broad beach. Thus this view illustrates the general process of coastal straightening by wave action. (T. M. Oberlander)

Where coasts are irregular in plan, with projections and bays, wave refraction at the headlands concentrates wave attack on the projections from two sides. Selective wave erosion along lines of weakness in these projections may separate them into clusters of rock stubs known as *sea stacks* (Figure 16.10). Undercutting along the flanks of rocky headlands often develops *sea caves*, which sometimes link together from the two sides to convert the promontory into a *sea arch* (Figure 16.11). Collapse of the arch leaves an isolated rock pinnacle, producing a single sea stack. Sea caves and stacks are common along the Pacific coast in Washington, Oregon, and California, with sea arches occasionally present. These features are also well developed on both sides of the English Channel.

It is not unusual to find unmistakable wave-cut platforms with former sea stacks well above sea level, forming a terrace along the coast. Such *marine terraces* could indicate either a drop in sea level or uplift of the land relative to the sea. Since marine terraces are local phenomena and do not maintain constant elevations, uplift of the land is the more likely cause. Marine terraces are especially common along tectonically active coasts, such as the Pacific coast of the United States, where many areas have clearly been lifted out of the sea (Figure 16.11). Some localities along the California coast exhibit as many as six clearly distinguishable terraces, arrayed like a giant staircase. The highest terraces that are definitely of marine origin in California lie some 600 m (2,000 ft) above sea level, and marine shells on the lowest terraces are more than 100,000 years old.

Beach Drifting and Longshore Drifting

Wave action not only beats back the edges of the land, but also causes a flow of sediment along the coast, nourishing beaches and associated marine depositional landforms. Where waves break close to beaches, the mass of water foams forward onto the beach as a turbulent sheet. This *swash*, or *uprush*, picks up fine sediment and pushes it up the beach slope. As the speed of the swash diminishes due to friction and gravity, a point comes at which forward momentum and gravitational pull down the beach slope are in balance. The uprush stops, much of the water sinks into the beach, and the remainder turns in its path and slides back down the beach slope as *backwash*, dragging fine sediment with it. When the waves strike the beach obliquely, each particle of water and every grain of sediment that is driven onshore by uprush and dragged down by backwash is also shunted laterally along the shoreline in zig-zag fashion— a process called *beach drifting* (Figure 16.12). Beach drifting is very important because it helps to provide the flow of sand necessary to maintain beaches and associated depositional features.

In the offshore zone of wave break, the water is kept in an agitated condition and is often clouded with fine suspended sediment that is swirled this way and that. The sediment may be suspended only momentarily after each wave impact, but while it is in suspension, it moves with the water. Because waves generally meet coastlines at a slight angle, they push water against the shore obliquely. This continual input of water cannot heap up vertically, so much of it drains off laterally, parallel to the coast in the general direction of wave advance. The result is a *longshore current*.

The oblique onshore piling of water by the arrival of one breaker after another causes the sediment-laden water in the zone of breaking waves to migrate slowly along the shore in the direction of the longshore current, producing *longshore drifting* of sediment. This may reverse direction from time to time, but most coasts have a preferred direction of longshore drifting during the season of maximum storminess and most vigorous wave action. This does

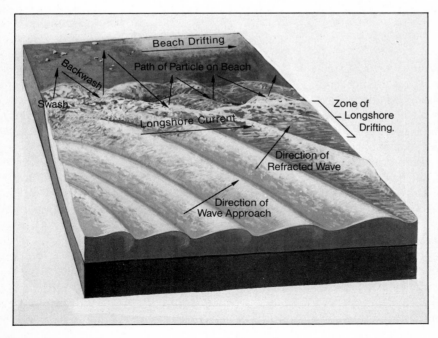

Figure 16.12
Waves that strike a beach obliquely cause transport of material parallel to the beach. Some of the material is shunted along by *swash* and *backwash*, as wave break causes water to wash up the beach obliquely and then return downslope. This produces the movement of particles known as *beach drifting*. The longshore current established by the oblique waves also transports material parallel to the beach in the process of *longshore drifting*. Most of this material moved in the surf zone is transported in suspension. (Tom O'Mary)

not always coincide with the major ocean currents described in Chapter 4.

Sediment transport by longshore drifting is particularly active when large waves are arriving at the shore. The combined processes of beach and longshore drifting produce a constant flow of sediment along smooth coastlines—a flow that waxes and wanes and even changes direction from day to day, depending on the state of the sea, but which creates and sustains all depositional landforms that are seen in coastal regions.

Coastal Deposition

Sediment transport under marine conditions follows the same principles as sediment transport by running water and wind. Transport occurs by suspension, saltation, and surface creep. It is easy to see (and hear) sand and gravel moving in all these ways in the swash zone. Deposition occurs where wave energy decreases, permitting sediment to come to rest.

Every marine depositional form has a characteristic sediment budget in which gains must balance losses if the form is to persist. Many marine depositional forms periodically grow and shrink as a result of shifts in the balance between sediment input and outgo. Although waves strike coasts from varying directions from day to day, every coast has an average "wave climate" and an average direction of sediment flow. Except in severe storms, sediment normally arrives at one end of a depositional form and leaves at the other end. Any interference with the flow of sediment at either end results in a change in the form of the deposit.

Beaches

Beaches are exposed areas of sand or rounded gravel (known as "shingle") deposited by wave action between the level of low tide and the highest levels reached by storm waves. On the landward side of the beach is material of a different type or a surface of different character from the beach, such as sand dunes, permanent vegetation, or a sea cliff. Beaches form along both irregular and straight coastlines. On irregular coasts, wave refraction causes sand or shingle to collect in the embayments between projecting headlands, creating *bayhead*, or *pocket*, *beaches*. Beaches also develop in continuous fringes along straight coasts where unresistant geological materials are present. Figure 16.13 (see p. 446) indicates the main features of beach morphology.

Beaches are among the most sensitive indicators of changes in geomorphic systems. We usually enjoy beaches in the summer, when the absence of destructive storm waves allows them to be well filled with sand (Figure 16.14, p. 446). But even in the summer, beaches must be steadily supplied with sediment. The sediment may be generated by wave erosion of headlands or the sea bed itself, or by rivers carrying sediment from the land. Along a coast that has a prevailing longshore drift, sand eroded from one beach is often passed along to the next beach. On the southern coast of Long Island, New York, sand produced by wave erosion of glacial drift moves from east to west along the beaches. Loss seems to be outrunning accumulation, and some beaches are retreating 1 or 2 m (3 to 6 ft) per year. Certain beaches elsewhere have been shrinking in recent years because the construction of upstream dams has seriously reduced the amount of sand supplied to the coast, or because residential construction on coastal dunes has cut off part of the sand supply for the beach system.

One solution to beach erosion is to interrupt the longshore drifting with a wood, concrete, or rubble dam, or *groin*, that extends into the water at right angles to the beach. The longshore current loses speed at the groin and drops part of its sediment—just as a stream drops part of its sediment when its velocity decreases. Inevitably, the deposition produced in one place by the groin robs beaches farther down the coast of incoming sediment. Those beaches erode all the more rapidly, to the anguish of resort owners. Miami Beach must import sand to balance ero-

Figure 16.13

A *beach* is the zone of transition between the land and sea. It extends from the low-tide level to the upper limit reached by the highest storm waves, which is the area subject to alternate erosion and deposition of sand. The actual profile of a beach is constantly changing. The diagram shows the principal geomorphic divisions of a beach. The *berm* is the nearly flat portion at the top of a beach; it is covered with material deposited by waves and constitutes the *backshore*. The *foreshore* extends from the edge of the berm to the low-tide line. Within this zone is the *beach face*, which is the area subject to swash and backwash. The *offshore*, which is permanently under water, contains *bars* and *troughs*. This is the zone of wave break and surf action. (Vantage Art, Inc. after Francis P. Shepard, *The Earth Beneath the Sea*, 2nd ed., 1967, Johns Hopkins Press)

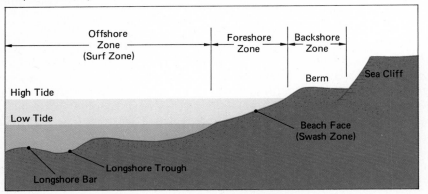

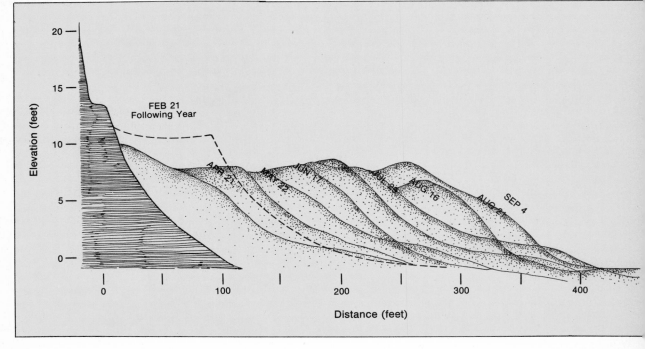

sional losses, despite efforts of individual hotels to maintain their beaches by constructing groins.

Changes in the size of breakers, or a change from long wave length waves of low height to steeper short wave length waves of greater height, greatly accelerates the removal of the particles composing the beach. Large steep waves can peel a beach downward two meters or more in a day. The beach material is deposited as submerged *offshore bars* in the zone of wave break (Figure 16.14). In areas affected by seasonal storminess, beaches tend to be stripped downward and steepened during the storm season—usually the winter in the middle latitudes. Some beaches periodically lose all their sand, retaining only large cobbles that are too heavy to be entrained even by the largest storm breakers. During the season of calms, the sand stored offshore slowly works back toward the beaches, refilling them to their former level.

Spits and Barrier Islands

Beach drifting and longshore drifting frequently produce deposits of sediment that extend outward from initial shorelines (Figure 16.15, p. 448). Such features fall into two broad categories: those that are attached to the land, and those that are detached, forming islands. Linear sediment accumulations that are attached to the land at one or both ends are termed *spits*. Sediment embankments completely separated from the land generally take the form of *barrier islands*. The term *bar* is reserved for submerged depositional forms.

Figure 16.14 (opposite)
The growth of the berm on the beach at Carmel, California, is shown in this series of measured profiles. By February of the following year, most of the berm had been cut away again. The vertical scale is exaggerated 10 times. (John Dawson after "Beaches" by Willard Bascom, *Scientific American*. Copyright © 1960 by Scientific American, Inc. All rights reserved)

Some spits form where sediment is moving along a coast that changes direction sharply, causing wave refraction and creating a low-energy environment in which sediment can accumulate. Where a coastline turns abruptly inland, as in a bay, the wave fronts pivot and diverge, decreasing wave energy and sediment transporting ability. Sediment moved by beach and longshore drifting comes to rest where the incoming waves begin to pivot around the corner. The resulting deposit becomes the new shoreline, which causes the point of wave pivot to be shifted laterally. This, in turn, produces more deposition. In time a linear spit forms, usually with a curved end, as in Figure 16.16 (see p. 449). At Cape Cod, Massachusetts, the spit has been extended westward from the end of a wave-eroded glacial moraine.

The construction of spits from one or both sides of small bays often closes off the bays, transforming them into *lagoons*. The lagoons eventually fill with fine sediment and are colonized by vegetation. This is one of the most important processes in the straightening of irregular coasts by marine action. Sand spits are common at river mouths, but river currents and floods normally maintain openings through such spits, preventing lagoon formation.

Sediment deposition also occurs in the low-energy wave environments in the lee of coastal islets or large sea stacks. Here wave refraction behind the obstacle sweeps sediment together from two sides, producing a type of spit known as a *tombolo*, which ties the island to the land (Figure 16.15).

Spits are especially subject to damage during hurricanes, which produce the largest waves experienced by most coasts in the low and middle latitudes. The topography of many spits reveals that they are compound features, having been partially eroded by storm waves and rebuilt many times during the course of their growth.

Many of the world's coastlines are paralleled by linear sandy islands known as *barrier islands*, some of them hundreds of kilometers long (Figure 16.15). These are often more regu-

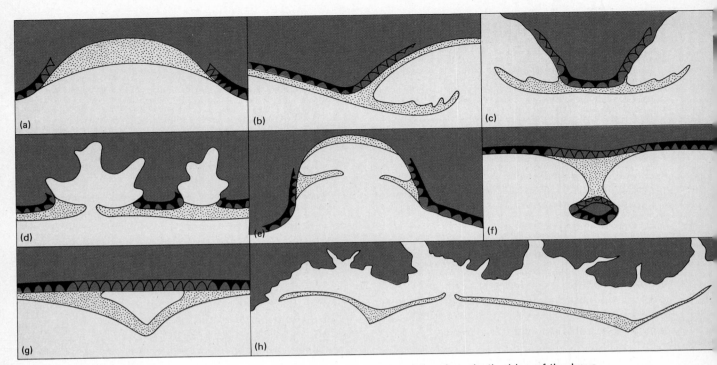

Figure 16.15
These diagrams illustrate the most common depositional forms produced by wave action. The dark sawtooth symbol indicates an active sea cliff; the open sawtooth symbol shows an inactive sea cliff. Land areas are dark. Depositional forms are yellow.

(a) *Bay head beach* with sediment deposition resulting from a low-energy wave environment.

(b) *Recurved spit* formed by sediment transport to the right, prolonging the previous line of the coast.

(c) *Winged headland* with sediment moved both ways from the eroding cliff as a consequence of changing directions of wave approach.

(d) *Bay mouth* or *barrier spits* straightening an initially irregular coast of submergence. Changes in wave approach produce spits extending from both sides of the bays, eventually closing them off and converting them into lagoons.

(e) *Mid-bay spits* with sediments accumulating before reaching the head of the bay.

(f) *Tombolo* formed by the deposition of sediment to the lee of an island and eventually linking the island to the mainland.

(g) *Cuspate spit* developed by reversals in the direction of longshore drifting along a straight coast.

(h) *Barrier island* developed along a coastline that is low-lying but irregular, suggesting recent submergence. The points on such barrier islands are called *cuspate forelands.* Cape Hatteras, North Carolina, is an outstanding example of such a form. (Vantage Art, Inc.)

lar than the coastlines behind them, as is the case around the Gulf of Mexico from Texas to Florida and along the Atlantic coast from Florida to New York. There is increasing evidence that barrier islands have a complex history, possibly originating as ancient dunes that formed behind beaches during the low sea levels of the Pleistocene epoch. Apparently, these sand banks were moved landward by wave action during the post-glacial rise in sea level and have

Figure 16.16
Deposition by longshore currents has caused the gradual extension of this curved spit near Puerto Peñasco in Mexico. The bay behind the spit may eventually be closed off, forming a *lagoon*. (Dr. John S. Shelton)

greatly increased in size since current sea level was attained. Occasionally barrier islands that are oriented differently, reflecting different wave sources or refraction patterns, link to form a prominent point, or *cusp*, as at Cape Hatteras and Cape Lookout, North Carolina (Figure 16.17).

Offshore barrier islands have great value as summer recreation areas, and they also create a sheltered waterway between themselves and the mainland. In the United States an intracoastal waterway system protected by barrier islands is navigable all the way from Massachusetts to the Mexican border, with only a short interruption along the west coast of the Florida peninsula, where barrier islands have not formed.

CORAL REEF COASTS

In some portions of the world, the coastline has been created by the activity of living organisms—corals and the marine animals associated with them. Corals are soft-bodied, tubelike marine animals that build a stony exterior skel-

Figure 16.17
This view from the Apollo spacecraft shows the barrier island system off the irregular submerged coast of North Carolina. Cape Hatteras is near the center and Cape Lookout is at the bottom of the picture. Outgoing tides are visibly washing fine sediment seaward through tidal inlets, which are a normal feature of barrier island systems. However, the effect of storm waves is to erode the seaward side of the islands and to move sediment across to the landward side, causing the islands to migrate toward the mainland (see the Case Study following this chapter). (NASA, Pilot Rock, Inc., © 1976)

Figure 16.18
This photograph shows the face of a typical coral reef developed in tropical seas. The plantlike forms are actually the exterior skeletons of lime-secreting marine animals. (John C. Hutchins/The Image Bank)

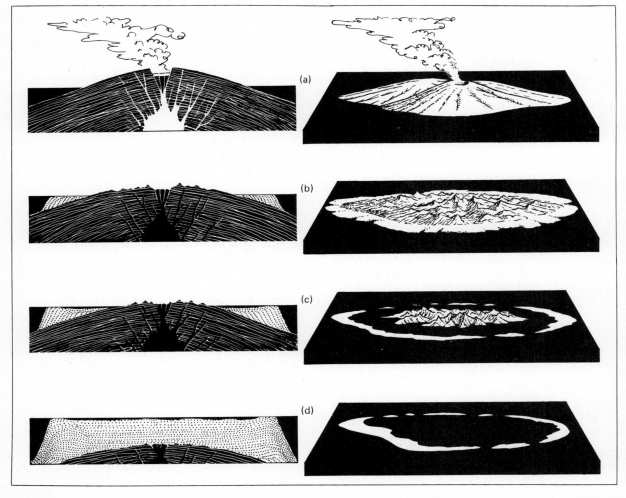

(a)

(b)

(c)

(d)

etal structure composed of calcium carbonate. Corals live in colonies comprising many species, ranging from massive mushroomlike masses to delicate twiglike structures (Figure 16.18). Vital to coral growth are photosynthetic algae that live in symbiotic relationship with the coral polyps. The algae extract CO_2 from sea water, providing the carbon necessary to form the corals' calcium carbonate skeletons. As layers of coral die, new layers form on top of them, building up solid masses of rocklike material known as *coral reefs.*

Coral reef coasts, along with fjord coasts (Chapter 15), are the only coastal types that have a distribution controlled by climate. None of the reef-building corals can survive in water temperatures lower than 18°C (65°F), and corals truly flourish only where water temperatures are between 25° and 30°C (77° to 86°F). Thus coral reefs are found only in tropical regions, seldom occurring poleward of the 30th parallel in the northern hemisphere. Because the associated algae require light, reef-building corals rarely grow at a depth exceeding 45 m (150 ft).

Figure 16.19 (opposite)
Deduced stages in the formation of a coral atoll based on observation of Pacific Ocean volcanic islands.

(a) An island is formed by volcanic eruptions that create a cone rising from the sea floor. Several volcanoes may be present. Molten lava is indicated within the volcanoes.

(b) The end of the volcanic phase is followed by erosion of the volcanoes, producing mountainous topography. A *fringing reef* grows outward from the shoreline.

(c) The beginning of subsidence of the island causes drowning of the erosional forms and upbuilding of coral to form a *barrier reef.* Storm waves toss up coral rubble to produce a rim higher than the lagoon behind it.

(d) Continued subsidence causes the volcanic island to become fully submerged and covered with coral. The result is a *coral atoll.* The ring surrounding the central lagoon is maintained by the action of storm waves. (T. M. Oberlander)

An additional requirement is clear water that is free of suspended sediment. The fact that corals cannot survive near the mouths of large sediment-carrying streams accounts for their preference for island rather than mainland locations. Coral reefs are present in the West Indies, the Florida Keys, and as far north as Bermuda, which lies in the warm Gulf Stream just north of the 32nd parallel. The greatest development of coral reefs is in the tropical Pacific Ocean, from the Hawaiian Islands to the Great Barrier Reef of Australia. Coral reefs are also widespread in the Indian Ocean.

All coral reefs are one of three types: *fringing reefs,* which are built out laterally from the shore; *barrier reefs,* which are separated from an island or land mass by a lagoon; and *atolls,* which enclose a lagoon, with no other land present (Figure 16.19). Atolls are often thought of as circular, but they are normally irregular in plan. In all coral reefs, the outer (seaward) slope is very steep, and even overhanging, whereas the inner slope is quite gentle, forming a horizontal platform in the case of fringing reefs. Barrier reefs and atolls are always pierced by several openings through which boats may pass from the interior lagoons to the open sea.

The formation of fringing reefs may be observed in countless locations where coral growth can be seen in reef platforms extending out from the shore. However, barrier reefs and atolls require somewhat more explanation. Although there have been several theories to account for them, modern investigations seem to substantiate Charles Darwin's belief that barrier reefs and atolls indicate subsidence of islands initially bordered by fringing reefs. If subsidence of the central island is gradual, the outer portion of fringing reefs can grow upward at a pace equal to island subsidence, so that the living coral remains in the sunlit layer of the sea. This produces barrier reefs. In the case of coral atolls, subsidence has lowered the island completely below sea level, but the coral continued to grow upward, sometimes hundreds of meters, as the island sank (Figure 16.20). Some of the con-

Figure 16.20
Coral builds outward from tropical coasts and upward when coasts begin to subside below the sunlit layer of the sea.

(top) A portion of the Great Barrier Reef off the northeast coast of Australia. This reef is growing upward as well as outward around the small island to keep pace with subsidence of the east coast of Australia. The majority of the picture area consists of reef coral. (John Lewis Stage/The Image Bank)

(bottom) These atolls photographed from the Apollo 7 spacecraft are in the Tuamotu Archipelago in the South Pacific. The view covers a distance of about 500 km (300 miles). The rings of coral have been able to grow upward fast enough to remain in the sunlit zone even though the volcanic islands forming their base have sunk far below sea level. Note that the atolls are irregular rather than circular in plan. (NASA)

structional activity in both barrier reefs and atolls seems to be due to hurricanes and tsunamis, in which large waves tear loose subsurface coral and toss it onto the reef or into the lagoon behind.

CLASSIFICATION OF COASTS

Because of the many variations in the nature of coastal scenery, a number of schemes have evolved for the classification of coastal landforms. These classifications are a good indication of the complexity of coastal landforms and the processes that create them.

The most general contrast in coastal form is between low-lying coasts that are fronted with continuous straight beaches and steep irregular coasts with rocky promontories and bold sea cliffs. The Atlantic and Gulf coasts of the United States are typical of the first type (except for the northern New England region), while the Pacific coast of North America clearly exemplifies the second type.

Coasts have been classified in terms of their relationship to the motions of lithospheric plates. *Collision*, or *subduction, coasts* are the Pacific type, in which plate convergence and tectonic activity create strong relief and a steep coast. *Trailing-edge coasts* are low-lying because there is no strong tectonic activity where the continents and sea floors are moving together as a single unit, as our Atlantic shores. *Marginal sea coasts* are coastlines separated from the deep ocean basins by volcanic island arcs, such as the Caribbean Antilles or the Japanese or Indonesian archipelagos (island chains). Generally, marginal sea coasts, such as those of the Gulf of Mexico, have broad continental shelves and low relief similar to trailing-edge coasts.

The most straightforward coastal classifications are purely descriptive, involving such categories as high-cliffed, low-cliffed, lagoonal, deltaic, estuarine, fjorded, barrier island, fringing reef, barrier reef, and so on. This type of classification is possible on any scale. The descriptive terms are general enough to permit differentiation of the world's coastlines on a single map.

Other classifications differentiate coasts that have *prograded* (advanced seaward) by deposition of deltas, sand spits, and lagoonal fills; *retrograded* (retreated) by erosion; or developed their characteristics as a consequence of *submergence* or *emergence* of the land relative to the sea. Coastal emergence, or rise of the land relative to the sea, clearly is required for the creation of marine terraces, which are common on tectonically active coasts. Where tectonic activity is weak, emerged areas of sea floor should be low-lying, producing a flat coastal plain. Along the Atlantic and Gulf coasts of the United States, there has been a general trend toward emergence, with Pleistocene sea level changes superimposed on this trend.

Two types of landforms have resulted from progressive coastal submergence during the 4,000 years following the end of the Pleistocene Ice Age. Clearly visible on a map of the eastern United States are deeply penetrating estuaries, or *rias*, where the sea has invaded Pleistocene river valleys. Chesapeake Bay is the classic example on a large scale. Between rias we find cliffed headlands and sometimes sea arches and stacks. As the headlands are battered back, their debris helps produce spits that close off bays, transforming them into lagoons.

Figure 16.21
This landscape was eroded by a continental ice sheet and then submerged by the postglacial rise in sea level. The long, parallel projections of land and elongated islands that jut into the ocean off the coast of Maine were continuous ridges before being submerged. Submergence of the coast filled the valleys with water and exposed the ridges to wave erosion. (Dr. John S. Shelton)

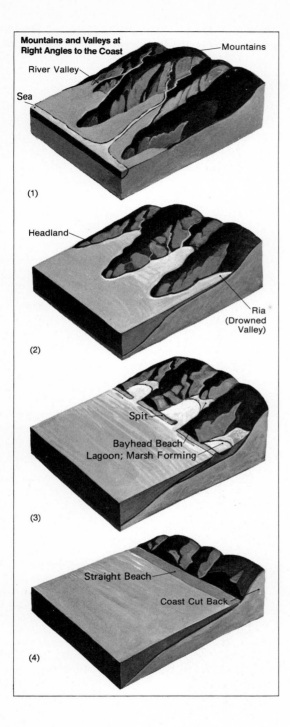

Mountains and Valleys at Right Angles to the Coast

Mountains

River Valley

Sea

(1)

Headland

Ria (Drowned Valley)

(2)

Spit

Bayhead Beach
Lagoon; Marsh Forming

(3)

Straight Beach

Coast Cut Back

(4)

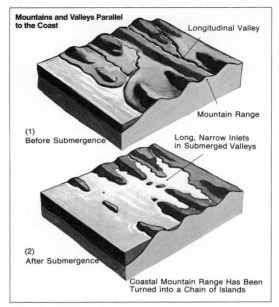

Mountains and Valleys Parallel to the Coast

Longitudinal Valley

Mountain Range

(1) Before Submergence

Long, Narrow Inlets in Submerged Valleys

(2) After Submergence

Coastal Mountain Range Has Been Turned into a Chain of Islands

Figure 16.22

(left) The sequence of diagrams shows the typical evolution of a coastline formed where valleys and ridges transverse to the coast are submerged by a rise of sea level relative to the land.

(1) The initial coastline consists of valleys and ridges behind a narrow coastal plain.

(2) The relative rise of sea level forms drowned river valleys, or *rias*, and headlands jutting into the ocean.

(3) Wave erosion cuts back the headlands and forms vertical sea cliffs. Deposition of sediment by currents builds spits and beach areas; spit growth eventually closes off the bays, forming lagoons.

(4) Continued erosion wears the coast back to a straight line bordered by sea cliffs and a narrow beach. Stream erosion will reduce the elevation of the ridges and highlands of the land area.

(above) (1) The diagram shows a coastal region where ridges and valleys are parallel to the coast.

(2) Submergence of the coastline by a relative elevation of sea level with respect to the land leaves numerous islands oriented parallel to the coast, such as those off the Dalmatian Coast of Yugoslavia. (Vantage Art, Inc. after R. B. Bunnett, *Physical Geography in Diagrams*, © 1968, Longmans Group, London)

Hudson River, the Columbia River, and the Zaire (Congo) River, but many others are not in the vicinity of any existing large stream. Submarine canyons are puzzling because many extend to depths of nearly 5,000 m (16,500 ft)—far too deep to be drowned river valleys. Some portions are excavated in soft sediments, but others are cut through hard rock in steep-walled gorges. All these canyons appear to channel great quantities of sediment from the continental shelves to the continental rises, ocean trenches, and deep ocean basins.

The most acceptable explanation for the origin of submarine canyons is that they are erosional forms created by streams of sand and silt that flow down continental slopes to the deep sea floor. These so-called *turbidity currents* have been seen and photographed. The fact that turbidity currents occasionally rupture telegraph cables on the sea floor indicates a destructive potential that over time could account for rock-cut submarine canyons. Thick masses of alternating sandstone and shale beds that are widely encountered on the continents are thought to be the deposits of turbidity currents in ancient oceans and are known as "turbidites." This suggests that submarine canyons have been present throughout most of the earth's history.

Guyots

Another puzzling feature of the sea floor is the *guyot*, a sizeable flat-topped submarine mountain. The form was discovered during World War II and has been named after a famous Swiss-American geographer of the last century. Hundreds of submerged volcanoes, or *seamounts*, are present in the world's oceans, particularly in the Pacific (Figure 11.5). A large proportion of these are flat-topped guyots. Although guyots have the appearance of having been planed off by erosion at the surface of the ocean, their summits are often 1,200 m (4,000 ft) or more below the sea surface—far too deep to be affected by wave action, even during times of lowered sea level.

It has recently been proposed that volcanic islands formed at oceanic ridges were planed off by wave attack, and then were carried laterally down the flanks of the ridges into the ocean depths as part of the sea floor spreading process. This hypothesis of subsidence by lateral plate motion seems to be verified by age determinations on the volcanic rocks of the guyots. They show increasing age with increasing depth and distance from the oceanic ridge systems. It is probable that many coral atolls have been built up from the summits of sinking guyots.

At present we have a more detailed knowledge of the surface of Mars than of the landforms of the earth's sea floors—our planet's last frontier. Future years are certain to turn up an ever-increasing list of mysteries of the deep and more problems to engage the attention of physical geographers and their colleagues in the earth sciences.

SUMMARY

The action of waves, currents, and tides keeps the sea surface in constant motion. The sea also undergoes long-term changes in level because of tectonic activity and changes in the storage of water in the form of glacier ice. Most waves and currents are produced by wind stress on the water surface, whereas tides are created by the gravitational attractions of the moon and, to a lesser degree, the sun, and by the rotation of the earth-moon system. Tides allow wave action to affect a considerable elevation range along coastlines.

Waves are of two types: forced waves and swell. Most of the waves breaking against the shores of the land are swell traveling outward from distant storms. Waves vary in size as a function of wind strength, wind duration, and fetch. The largest waves are produced by intense storms, such as hurricanes, and by submarine earthquakes and volcanic eruptions that produce seismic sea waves known as tsunamis.

Waves transmit energy from areas of oceanic storms to far-off coastlines, where the energy is expended in the work of erosion and sediment transport. Wave motion is transmitted by orbital motions of water particles. Where waves approach the land, they are slowed and refracted by friction with the sea floor. Along irregular coasts, wave refraction concentrates erosional energy on headlands and causes sediment to be shunted into bays. In this way the coast tends to become more regular. Coastal erosion by wave action creates sea cliffs, arches, stacks, and abrasion platforms. Where such features are found above sea level, uplift of the coastal region has occurred.

Beach drifting and longshore currents move sediment laterally along coastlines. Beaches, spits, and barrier islands constantly change in form according to shifts in the balance of sediment arrivals and sediment losses. Most of the sediment sustaining beaches is delivered by rivers that empty into the sea, rather than by coastal erosion.

In warm tropical seas, particularly in island locations, coastlines may be composed of coral reefs. Fringing and barrier reefs are associated with islands. Atolls are coral rings that have developed above sunken islands.

High-relief coasts are usually associated with converging oceanic and continental crustal plates, as around the margins of the Pacific Ocean. Low-relief coasts occur where the continent is welded to the adjacent sea floor, as is the case on both sides of the Atlantic Ocean. The world-wide rise in sea level at the end of the Pleistocene epoch has caused most coasts to be submerged, but in some areas this is a recent development superimposed on a long-term emergent trend.

Two perplexing features of the sea floors are submarine canyons and guyots. Submarine canyons are erosional features on the continental shelves and slopes and are thought to be produced by turbidity currents. Guyots are probably former wave-beveled volcanic islands formed along the oceanic ridge system and moved laterally into deeper portions of the oceans by sea floor spreading.

APPLICATIONS

1. What large cities around the world would be submerged if the Antarctic and Greenland ice sheets melted completely? What inland cities would be transformed into ports if this happened?
2. How would it affect the tides if the center of rotation of the earth-moon system were closer to the center of the earth? if it were between the earth and the moon?
3. In what way is a beach similar to a glacier?
4. If there is a beach near your campus, note the relationship between the slope of the beach face and the size of the particles forming the beach. Both the beach slope and the associated particle size differ from point to point as well as from day to day. What explains the relationship? Is it a constant one? Can you see any relationship between breaker type and beach slope? between beach plan (map view) and beach slope or breaker type?
5. If your department or campus has a collection of topographic maps of the United States (or other areas), locate map examples of each of the coastal depositional forms illustrated in Figure 16.15. What seems to be the source of the sediment in each case: a river, or coastal erosion? Can you find a location in which the growth of a beach or spit has stopped wave erosion of a former sea cliff?
6. What accounts for the striking difference in the form of the Atlantic coast of North America north and south of Cape Cod, Massachusetts? Why is the coast of California so much less regular than the coasts of Oregon and Washington?

FURTHER READING

Bird, E. C. F. *Coasts*. Cambridge, Mass.: M.I.T. Press (1969), 246 pp. Bird's small book offers a concise and well-illustrated introduction to coastal processes and landforms, with many examples being drawn from Australia.

Goreau, Thomas F., Nora I. Goreau, and **Thomas J. Goreau.** "Corals and Coral Reefs," *Scientific American*, 241:2 (1979), pp. 124–126. Excellent, well-illustrated article on all aspects of coral reefs, stressing the symbiosis between coral and photosynthetic algae that makes reef formation possible.

Heezen, Bruce C., and **Charles D. Hollister.** *The Face of the Deep*. New York: Oxford University Press (1971), 659 pp. This is a massive volume that deals in a thoughtful way with all aspects of ocean floors.

Inman, Doublas L., and **Birchard M. Brush.** "The Coastal Challenge," *Science*, 181 (July 6, 1973): 20–31. This article details modern findings concerning the physical processes in coastal systems, and shows how human activities have changed coastal environments, often diminishing both their utility and aesthetic qualities.

King, Cuchlaine A. M. *Beaches and Coasts*. 2nd ed. New York: St. Martin's (1972), 570 pp. King combines theory, model experiments, and field observation in detailed analyses of coastal processes and landforms.

Russell, Richard J. *River Plains and Sea Coasts*. Berkeley and Los Angeles: University of California Press (1967), 173 pp. About half of this engagingly written book is devoted to Russell's personal observations regarding shoreline processes. The author concentrates upon low-latitude coasts.

Shepard, Francis P. *The Earth Beneath the Sea*. 2nd ed. Baltimore: Johns Hopkins Press (1967), 242 pp. This very readable nontechnical treatment of the features of the sea floor is by a leading investigator of submarine canyons. As well as presenting facts and interpretations, Shepard indicates how data are collected beneath the sea.

————and **Harold R. Wanless.** *Our Changing Coastlines*. New York: McGraw-Hill (1971), 571 pp. This large book is a complete inventory of the coastal morphology of the United States, including Alaska and Hawaii. The text is superbly illustrated with aerial photographs.

Strahler, Arthur H. *A Geologist's View of Cape Cod*. Garden City, N. Y.: Natural History Press (1966), 115 pp. This small, nicely written book details how glacial deposition and wind and wave action have fashioned the landforms of a popular tourist area. Very well illustrated and nontechnical.

Problems at the Edge of the Land

Seacoasts are the summer playgrounds of the world. In any coastal region accessible to large numbers of people with leisure time and money to spend, each warm weekend triggers a mass migration to the shore. Local communities and national governments have catered to the desires of pleasure-seekers by developing beach areas, making them accessible by road, and creating protected marinas for boating enthusiasts. Is there anything wrong with this?

Plans for coastline development run into a hard fact—today's shoreline is not a permanent feature. Our planet has just emerged from the trauma of the Pleistocene, with its climatic changes and sea level fluctuations. Our coasts are not yet adjusted to the prevailing sea level, which was established but a moment ago on the geologic scale of time.

As noted in the preceding chapter, there is a marked contrast between the low-lying Atlantic and Gulf coasts of the United States and the steeply rising Pacific coast. Both environments are a lure to people in search of beauty and entertainment, and both coasts have been developed for recreation and residential construction. On both sides of the continent, the natural system is colliding with human desires.

The Atlantic and Gulf coasts are guarded by a chain of coastal spits and barrier islands, reaching from Cape Cod in Massachusetts to Matagorda and Padre islands in Texas and their prolongation in Mexico. All these were once strips of sand with beaches hundreds of meters wide facing the sea. The beaches were backed by low grassy dunes, and behind the dunes was a zone of grass and shrubland, succeeded by salt marsh on the side of the island facing the mainland. The barrier islands have protected the mainland, bearing the brunt of wave attack and storm surges (water piled against the shore by hurricane winds). The barriers were frequently washed over by high seas and sometimes broken through during hurricanes, when great quantities of sediment were moved across them and deposited in the salt marshes on their landward side. Clearly, their tendency is to be moved landward, by erosion on one side and deposition on the other. In some locations, their measured shift over the past century has averaged close to 10 m (33 ft) per year.

But how can roads be maintained along one of these shifting barrier beaches—and parking lots, boat landings, motels, restaurants, and beach cottages? The only way is to stop the shifting. This can be done, at least for a while, by preventing overwash during storms. The technique is to heighten the dunes artificially, using sand-trapping fencing or dredged sand, and to plant them with stabilizing shrubs and dense grasses, creating a sort of sea wall behind which roads and other facilities can be located.

The effect of this modification should have been foreseen. In the man-made landscape, the energy of storm waves is concentrated on the beach in front of the artificially elevated dunes instead of being exhausted in overwash across the full width of the barrier island. Thus beaches have become exposed to wave energies they have never before experienced. First, they lose their finer particles; the beach profile becomes steeper; and then intensified backwash drags away coarser particles. The process feeds upon itself, and the beaches become progressively narrower and steeper. Today the once broad beaches of many artificially "defended" barrier islands are strips only a few tens of meters wide in front of wave-steepened dunes. Migration of the islands has been halted for the moment, but their principal attraction, the beaches, are disappearing. If present processes continue, the dunes themselves will be undercut and washed away, and the islands will retreat at a faster rate than ever before.

Many solutions to this problem have been proposed, including elevated highways and other structures that would permit overwash, preserving the beaches, but this will not stop the inexorable migration of the island shorelines. The only real answer appears to be a more passive use of the barrier islands, maintaining them in their natural state as undeveloped beaches accessible by ferry from the mainland. This is the plan adopted for the Cape Lookout National Seashore Area, south of Cape Hatteras, North Carolina.

Along the Pacific coast, the problem is not the retreat of barrier islands, for such features do not exist along steep coasts, but the retreat of the mainland itself. Beach preservation is important here, for the surest way to minimize wave erosion of the land is to keep waves breaking on broad beaches rather than impacting directly against a sea cliff. Three types of human activities are causing increased erosion along the Pacific coast: artificial destruction of coastal dunes that are reservoirs of sand that is recycled to the sea during storm surges; interference with the longshore flow of sand required to maintain beaches in a quasi-equilibrium condition; and interference with the delivery of sand from rivers flowing down to the coast.

Destruction of dunes has occurred in connection with sand mining and residential development. Here, as on the east coast, dunes have sometimes been leveled merely to provide cottage owners with unobstructed views of the sea. The classic examples of interference with longshore sand flow are the Santa Barbara and Santa Monica (California) breakwaters, both built in the 1920s to create protected harbors. In both locations, the jetties reduced wave energy inside the harbors, causing sand to accumulate there rather than moving on. Interruption of the sand flow caused alarming shrinkage of beaches farther down the coast. This exposed the land behind the beaches to increased wave attack, producing accelerated shoreline erosion. The situation became so critical in both cases that permanent dredging operations had to be established in the harbors, with sand being pumped through pipelines and returned to the shoreline downcoast. Unfortunately, the rocks exposed along much of the Pacific coast are poorly consolidated and highly susceptible to wave erosion, with normal rates of cliff retreat being in the neighborhood of 15 to 30 cm (6 to 12 in.) per year.

The beaches of Coronado, near San Diego, have been noticeably affected by interference with a riverine sand supply. The sand nourishing the beaches formerly drifted northward from the Tijuana River. The mouth of the Tijuana is in California, but the major portion of the stream lies across the border in Mexico, where it has been dammed, trapping its abundant sediment in a reservoir whose life will be brief—though not brief enough for the citizens of Coronado. Plans to dam large rivers in northern California raise the specter of massive beach deterioration in the future.

And so the struggle between the human species, determined to bend nature to our will, and the seemingly irresistible natural forces, continues. But the real struggle is to understand nature's ways to prevent deterioration in those complex natural systems that have nurtured life on our unique planet.

Figure I.1
(top) This Babylonian map from 500 B.C. is one of the earliest attempts to portray the world. (The British Museum)

(bottom) Claudius Ptolemy's map of the world, reconstructed from his descriptions written in the second century, is a comparatively accurate representation that takes into account the spherical shape of the earth. (*Atlas of the Universe*, p. 13, © 1971, Mitchell-Beazley, Ltd.)

APPENDIX **I**

Tools of the Physical Geographer

Maps:
A Representation of the Earth's Surface

Modern maps exist in great variety, from simple street maps to complex navigational charts for jet aircraft. A map can convey a large amount of information in a way that is easily assimilated: a well-made map of climatic regions, for example, can make important climatic relationships much clearer than can lengthy written descriptions or tables of data.

A *map* can be defined formally as a two-dimensional graphic representation of the spatial distribution of selected phenomena. A map is *planimetric*; that is, it shows *horizontal* spatial relationships on the earth. In addition to specifying location, distribution, amount, distance, direction, sizes, and shapes, maps can also represent the form of the land surface or any statistical surface based on spatial data.

Scale and Distance A map's scale gives the relationship between length measured on the map sheet and the corresponding distance on the earth's surface. There are several ways to express the scale of a map. It may be given in a simple *verbal statement*, such as "1 in. equals 3 miles." The scale of a map also may be indicated by a *graphic scale* marked off in units of distance on the earth, as shown in Figure I.2a. One advantage of a graphic scale is that, unlike a verbal scale, it remains correct if the map is copied in a larger or smaller size.

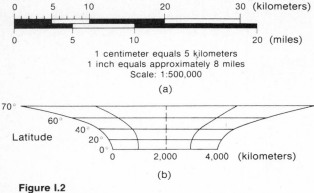

1 centimeter equals 5 kilometers
1 inch equals approximately 8 miles
Scale: 1:500,000

(a)

(b)

Figure I.2
(Andrea Lindberg)

Scale is often expressed as a fraction, called the *representative fraction*. A representative fraction of 1/5,000 (commonly written 1:5,000) means that a length of 1 unit on the map represents 5,000 units of distance on the earth. A representative fraction makes no reference to any particular system of units, because it represents simply the ratio between length on the map and a corresponding distance on the earth.

A verbal statement of scale can be restated as a representative fraction by converting both members of the statement to the same units; thus a scale of 1 in. to the mile is equivalent to 1:63,360 because there are 63,360 in. in 1 mile. Conversely, a representative fraction can be expressed as a verbal statement of scale by assigning units and applying conversion factors as required. A scale of 1:100,000 can be stated as "1 cm equals 1 km."

When a portion of a globe is represented on a flat map distortions inevitably occur. Consequently the scale of a map cannot be constant for every portion of the map. However, scale does not vary greatly on a flat map of a small region, and even the conterminous United States can be mapped in such a way that the scale does not vary by more than a few percent. Significant variations of scale occur on all global maps, however. The scale on a Mercator map of the world, for instance, is several times larger at higher latitudes than at the equator (Figure I.2b).

When the representative fraction is a small number, less than 1/1,000,000, a map is called a *small-scale map*. Small-scale maps are used when a large portion of the earth's surface, such as a continent or an ocean, must be represented on a map of limited size. If a map has a representative fraction larger than 1/250,000 (1:250,000), it usually is called a *large-scale* map.

Large-scale maps of small areas are capable of showing greater detail than small-scale maps of large areas.

Location The principal method for specifying location on the earth's surface is by the system of latitude and longitude. *Latitude*, the position of a place north or south of the equator, is expressed in angular measure relative to the center of the earth. The angle of latitude varies from 0° at the equator to 90° at the poles. *Longitude*, the position of a place east or west of a selected prime meridian, is expressed in angular measure that varies from 0° to 180° east or west of the prime meridian. The most commonly accepted prime meridian is the one on which the Greenwich Observatory in England is located, but other prime meridians have been used in the past. The framework of lines representing parallels of latitude and meridians of longitude on a map is called the *graticule* of the map. Depending on the method chosen to construct a map, the lines of the graticule may be straight or curved, and they may or may not intersect at right angles to one another, although they do intersect at right angles on a globe.

A more easily computed location reference system uses a rectangular grid composed of straight lines that do intersect at right angles.

The first step in constructing such a grid system is to choose a standard form of map that meets the needs of the user. (The advantages and drawbacks of different types of maps are discussed later in this appendix.) Once the map is chosen, a square grid is overlaid on the map and numerical coordinates are assigned to the reference lines of the grid. The coordinates are usually expressed in units of distance from a selected origin. The grid coordinates of any location can then be read from the map as illustrated in Figure I.3. By convention, the coordinate to the east, or the *easting*, is specified first. Then the coordinate to the north, or the *northing*, is specified. The rule is to read toward the *right* and *up*, following the same order used for giving the x and y coordinates of a point on a graph. The grid coordinates of a location are often given as one number consisting of an even number of digits; the first half of the number gives the easting, and the second half, the northing.

The United States National Ocean Survey (formerly the United States Coast and Geodetic Survey) has designed a rectangular grid system for each of the states, called the *State Plane Coordinate System*. The basic grid square of the state coordinates is 10,000 ft on a side; eastings and northings for the grid are listed in units of feet.

A modified grid system has been in use for many

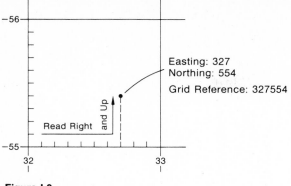

Easting: 327
Northing: 554

Grid Reference: 327554

Read Right and Up

Figure I.3
(Andrea Lindberg)

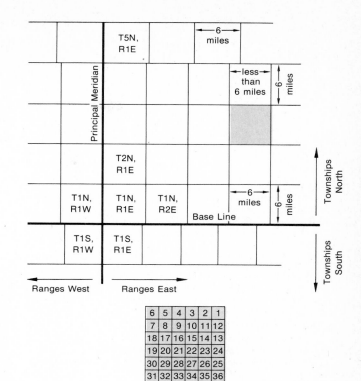

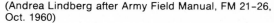

Figure I.4
(Andrea Lindberg after Army Field Manual, FM 21-26, Oct. 1960)

years in connection with the survey of public lands conducted by the Bureau of Land Management. The basic land unit of the survey, which was begun in the eighteenth century, is the *township*, a square plot 6 miles on a side. Townships are laid out with two sides along meridians and the other two sides along parallels of latitude. Because meridians converge toward the north, the north-south sides of the townships must jog eastward or westward every 24 miles to maintain the size of the 6-mile square.

Townships are laid out with respect to a north-south *principal meridian* and an east-west *base line*. Different land surveys established thirty-one sets of principal meridians and base lines for the conterminous United States and five sets for Alaska. The location of each township in a survey region is given with respect to the point at which the principal meridian and the base line intersect. The coordinates that specify a particular township are read as the number of townships north or south of the base line; the number of townships east or west of the principal meridian is called the *range*. The system for locating townships is an exception to the "right and up" rule of reading because northings are read before eastings. Townships are further subdivided into 36 squares, 1 mile on each side, which are called *sections*; sections are numbered 1 through 36 in a serpentine fashion, beginning in the upper right corner of the township (see Figure I.4).

Direction By definition, meridians of longitude are true north-south lines, and parallels of latitude are true east-west lines. Because of the distortions inherent in representing the surface of a sphere on a flat sheet of paper, meridians or parallels often vary in direction across a map. But for large-scale

maps that cover small areas the distortions are barely visible and the map sheet can be aligned with respect to a single standard direction to establish a sense of orientation.

Many large-scale maps indicate the direction of *true north* by means of a star-tipped arrow or the symbol *TN*. However, this direction is usually not the same as *magnetic north*, the direction in which a magnetic compass needle points. Large-scale maps usually indicate the direction of magnetic north by means of a half-headed arrow and the symbol *MN*. The earth's magnetic field is not uniform, and the magnetic poles do not coincide with the geographic poles, so the relation between magnetic north and true north must be specified separately for each region. The difference between magnetic north and true north is known as *magnetic declination* and is expressed in degrees east or west of the true meridian of a given location. Across the conterminous United States the magnetic declination varies from 0° to as much as 25° east or west, and in

TOOLS OF THE PHYSICAL GEOGRAPHER

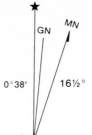

0°38'   16½°

Figure I.5
(Andrea Lindberg)

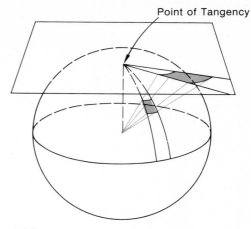

Point of Tangency

Figure I.6
Construction of the Gnomonic Projection (Andrea Lindberg)

polar regions a declination of 90° or more is possible. Furthermore, the direction of magnetic north at a given location varies with time, often by as much as 1° in 20 years. For precision map work, therefore, reference should be made to recent compilations of magnetic declinations, such as those prepared by the National Ocean Survey.

Direction on a map may also be specified as *grid north*, the northerly direction arbitrarily determined by a particular grid system, and symbolized on the map by *GN*. Grid north generally does not coincide with either magnetic north or true north. The grid north directions specified by two different grid coordinate systems usually differ from one another as well.

Directions other than north can be expressed in terms of *azimuth*, which is the angle of the desired direction measured clockwise from a chosen reference direction and expressed in degrees between 0° and 360°. According to the choice of true north, magnetic north, or grid north as a reference, the corresponding azimuths are termed *true azimuth*, *magnetic azimuth*, or *grid azimuth*.

Map Projections

A model globe is the only way to represent large portions of the earth's surface with accuracy, because only a globe correctly takes into account the spherical shape of the earth. A flat piece of paper cannot be fitted closely to a sphere without wrinkling or tearing, so small-scale maps that represent regions hundreds or thousands of miles in extent inevitably introduce noticeable distortions.

The fundamental problem of map making is to find a method of transferring a spherical surface onto a flat sheet in a way that minimizes undesirable distortions. Any method of relating position on a globe to position on a flat map is called a *projection*, or *transformation*. Numerous projections have been devised, each with its characteristic advantages and distortions. Because no projection is free of distortion, the choice of a projection should be made with regard to its proposed application.

The principles of projection are illustrated in Figure I.6, using the example of the gnomonic projection. This projection can be constructed by tracing the rays of light from a light source at the center of a transparent globe onto a plane that touches the globe at one point, called the *point of tangency*. Each point on the portion of the globe that is projected onto the plane constitutes a point on the map. However, only a few projections, such as the gnomonic, can be visualized geometrically. Many projections can be expressed only as a mathematical rule that relates points on a globe to points on a flat sheet.

Projections and Distortion Maps are commonly relied upon to show correct direction and distance from one location to another and the sizes and shapes of areas. A single flat map can depict one or another of these without distortion, but not all of them. The map user should realize where and to what extent inaccuracies are present in the projection being used. On the gnomonic projection illustrated in Figure I.6, for example, the scale is increasingly exaggerated with distance from the point of tangency, producing extreme distortion of shapes as well as sizes.

Scale distortion may lead to distortion of direction, shape, and size. On some projections, the scale in a small region is not the same in all directions, which necessarily leads to distortion of direction. Distortion of direction implies that shapes will not be geometrically accurate.

466

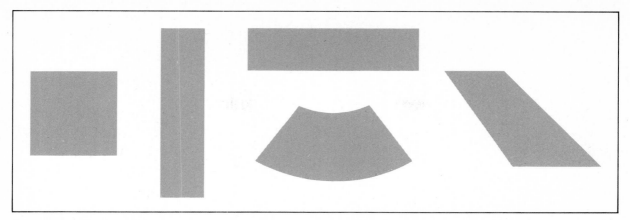

Figure I.7
(T. M. Oberlander)

A number of projections, called *conformal* projections, have been devised so that azimuths from point to point are correct. Shapes of small areas are portrayed accurately. However, on a conformal map the scale necessarily changes from one region to another. Regions that span a large portion of the globe exhibit distortion when mapped on a conformal projection. The well-known Mercator projection is conformal, but it represents areas in the higher latitudes several times larger than areas of the same size near the equator.

Another important class of projections includes the *equal area*, or *equivalent*, projections. On such projections, point-to-point azimuths are incorrect, but the scale is designed to vary over the map in such a manner that the relative sizes of all areas are correct. Figure I.7 illustrates how shapes can be distorted without altering their areas. Equal area projections should always be used when the geographic distributions of phenomena are displayed because regions are represented in their correct relative sizes. Many projections are neither conformal nor equivalent area but represent compromises to obtain adequate representation of shape without badly distorting size.

An impression of a projection's major properties can usually be obtained by observation of the lines of latitude and longitude. On a globe, parallels and meridians intersect at right angles. If a map shows them intersecting at right angles, the projection may or may not be conformal, but if they are shown intersecting at other angles, distortion of direction is present and the projection is not conformal. Distortions of shape or size can be seen by comparing the map compartments bounded by parallels and meridians to those on a globe.

The globe may be projected onto a plane, cylinder, or cone, which are the only surfaces that can be spread flat. Projections are conventionally classified into families: *azimuthal*, or *zenithal*, projections onto a plane, *cylindric* projections onto a cylinder, and *conic* projections onto a cone. A fourth family of *geometrical projections* is usually used to portray the entire globe. The projections in a given family tend to have similar properties and similar distortion characteristics.

Azimuthal Projections Azimuthal projections are projections of a globe onto a plane tangent to the globe at some point. The point of tangency is usually the north or south pole, but in principle any point on the globe may be used. Most azimuthal projections can depict only one hemisphere, or less, of the earth at one time.

The distortion of any azimuthal projection is least nearest the point of tangency and increases with increased distance away from the point of tangency. Figure I.8 exhibits some of the characteristics of standard azimuthal projections. Distortion patterns are depicted by degrees of yellow shading, with the deeper tints corresponding to greater distortion.

Cylindric Projections A cylinder closely fitted to a globe makes contact with the globe along a *great circle*, called the *circle of tangency*. Cylindric projections are usually designed so that the circle of tangency is the equator or a meridian.

Most cylindric projections with the equator as the circle of tangency show parallels of latitude and meridians of longitude as sets of straight parallel lines intersecting at right angles. The spacing of the

TOOLS OF THE PHYSICAL GEOGRAPHER

Figure I.8

Table of Azimuthal Projections

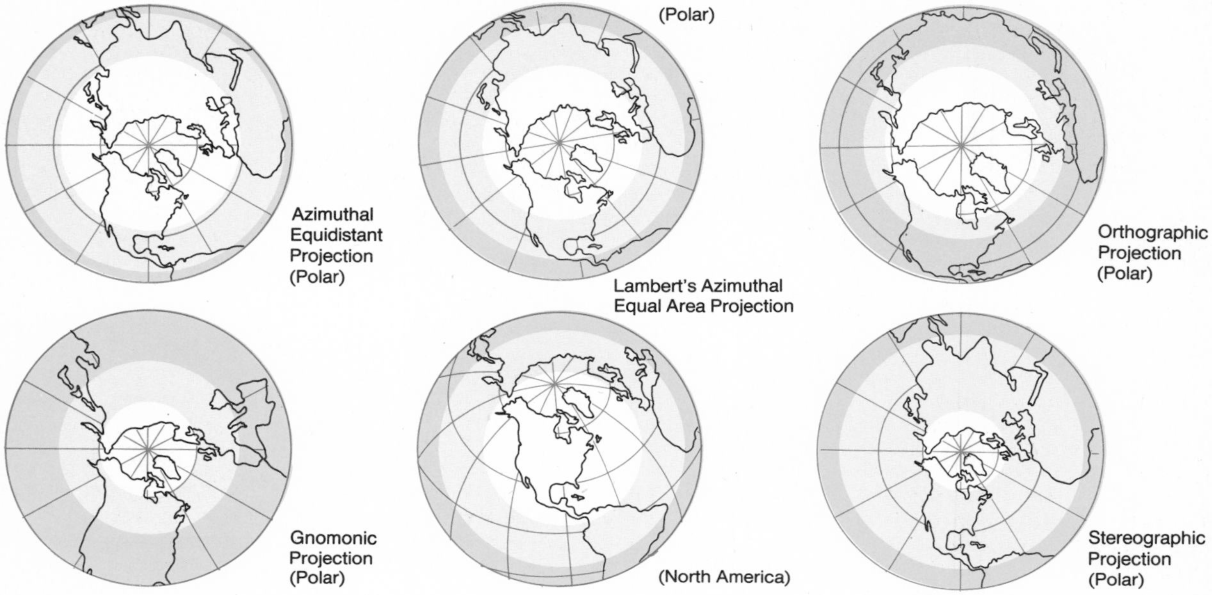

Azimuthal Equidistant Projection (Polar)

(Polar)

Orthographic Projection (Polar)

Lambert's Azimuthal Equal Area Projection

Gnomonic Projection (Polar)

(North America)

Stereographic Projection (Polar)

Projection	Appearance	Properties	Distortion Pattern	Best Uses
Azimuthal Equidistant (Polar)	Meridians are straight lines outward from the pole; parallels are equally spaced circles concentric about the pole.	Scale is constant and correct along meridians. Directions from central point are correct.	Distortion increases slowly with increased distance from the center. Shapes are represented comparatively well, but areas are distorted.	Directions and distance to the center are undistorted; useful for charts of radio propagation to a given location or for showing relative distance from a given location.
Gnomonic (Polar)	Meridians are straight lines outward from the pole; parallels are circles concentric about the pole with rapidly increasing spacing outward.	Great circles anywhere on the map are represented by straight lines.	Distortion increases rapidly with increased distance from the center. Shapes become badly distorted.	Polar navigation charts and great circle navigation. (The shortest route between two points on the earth's surface lies on a great circle.)
Lambert's Equal Area (Polar)	Meridians are straight lines outward from the pole; parallels are circles concentric about the pole with slowly decreasing spacing outward.	The only azimuthal equal area projection. Directions from central point are correct.	Distortion increases moderately with increased distance from the center. Shapes are represented well.	Polar maps and maps of one hemisphere, especially where distributions are to be represented.
Orthographic (Polar)	Meridians are straight lines outward from the pole; parallels are circles concentric about the pole with rapidly decreasing spacing outward.	Directions from central point are correct. Gives appearance of earth as seen from deep space.	Distortion increases moderately with increased distance from the center.	Used primarily for illustrations, shows how the earth looks from outer space.
Stereographic (Polar)	Meridians are straight lines outward from the pole; parallels are circles concentric about the pole with moderately increasing spacing outward.	The only azimuthal conformal projection. Directions from central point are correct.	Area distortion increases rapidly with increased distance from the center. Shapes are represented well.	Basis map for the UPS grid system, poleward of latitude 80°.

(Figures I.8–I.12 by Andrea Lindberg after Arthur Robinson and Randall Sale, *Elements of Cartography*, 3rd ed., © 1953, 1960, 1969 by John Wiley & Sons, reprinted by permission)

Figure I.9

Table of Cylindric Projections

Circle of Tangency

Equirectangular Projection

Standard Mercator's Projection

Lambert's Cylindrical
Equal Area Projection

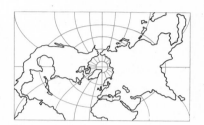

Transverse
Mercator's Projection

Projection	Appearance	Properties	Distortion Pattern	Best Uses
Equirectangular	Parallels and meridians are equally spaced and form a square grid. Often a parallel is chosen to be the circle of tangency.	No major properties.	Scale is correct along the standard parallel and along meridians, but shape and area distortion increase with increased distance from the standard parallel. Shapes and areas distant from the standard parallel are badly distorted.	Used only for large- or moderate-scale maps of limited areas.
Lambert's Cylindrical Equal Area	Parallels and meridians form a rectangular grid. Employs two parallels equidistant from the Equator as circles of tangency.	An equal area projection. Can depict the whole earth except the polar regions.	Shape distortion increases with increased distance from the standard parallels. Shapes at high latitudes are seriously compressed in the north-south direction.	Not widely used because of severe shape distortion at high latitudes, but it would be satisfactory for presenting distributions in the lower latitudes.
Standard Mercator's	Parallels and meridians form a rectangular grid, with the Equator as the circle of tangency.	A conformal projection. A straight line on the map is a line of constant azimuth on the earth. Can depict the whole earth except the polar regions.	Area distortion increases rapidly with increased distance from the Equator. Shapes of small regions are represented well, but there are gross distortions of area at high latitudes.	Navigation charts.
Transverse Mercator's	The circle of tangency is a meridian or portion of a meridian. Most meridians and parallels are curved.	A conformal projection. Most straight lines on the map are not lines of constant azimuth on the earth.	Area distortion increases with increased distance from the central meridian. Scale distortion is constant along lines parallel to the central meridian. Shapes of small regions are represented well.	Basis map for UTM grid system, for some State Plane Coordinate grids, and for the British Ordinance Survey maps.

TOOLS OF THE PHYSICAL GEOGRAPHER

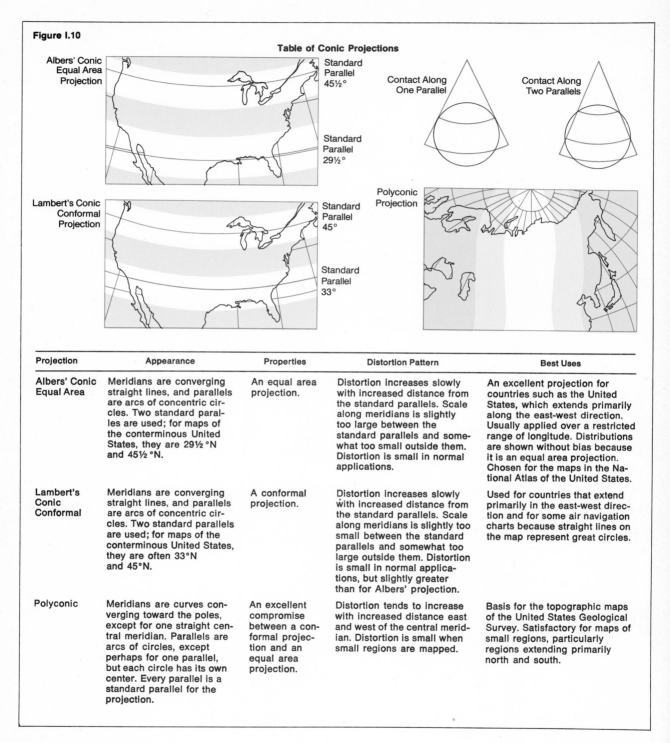

Figure I.10

Table of Conic Projections

Albers' Conic Equal Area Projection — Standard Parallel 45½° — Standard Parallel 29½°

Contact Along One Parallel — Contact Along Two Parallels

Lambert's Conic Conformal Projection — Standard Parallel 45° — Standard Parallel 33°

Polyconic Projection

Projection	Appearance	Properties	Distortion Pattern	Best Uses
Albers' Conic Equal Area	Meridians are converging straight lines, and parallels are arcs of concentric circles. Two standard parallels are used; for maps of the conterminous United States, they are 29½ °N and 45½ °N.	An equal area projection.	Distortion increases slowly with increased distance from the standard parallels. Scale along meridians is slightly too large between the standard parallels and somewhat too small outside them. Distortion is small in normal applications.	An excellent projection for countries such as the United States, which extends primarily along the east-west direction. Usually applied over a restricted range of longitude. Distributions are shown without bias because it is an equal area projection. Chosen for the maps in the National Atlas of the United States.
Lambert's Conic Conformal	Meridians are converging straight lines, and parallels are arcs of concentric circles. Two standard parallels are used; for maps of the conterminous United States, they are often 33°N and 45°N.	A conformal projection.	Distortion increases slowly with increased distance from the standard parallels. Scale along meridians is slightly too small between the standard parallels and somewhat too large outside them. Distortion is small in normal applications, but slightly greater than for Albers' projection.	Used for countries that extend primarily in the east-west direction and for some air navigation charts because straight lines on the map represent great circles.
Polyconic	Meridians are curves converging toward the poles, except for one straight central meridian. Parallels are arcs of circles, except perhaps for one parallel, but each circle has its own center. Every parallel is a standard parallel for the projection.	An excellent compromise between a conformal projection and an equal area projection.	Distortion tends to increase with increased distance east and west of the central meridian. Distortion is small when small regions are mapped.	Basis for the topographic maps of the United States Geological Survey. Satisfactory for maps of small regions, particularly regions extending primarily north and south.

Figure I.11

Table of Geometrical Global Projections

Flat Polar Quartic Equal Area Projection

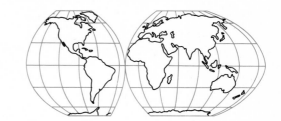

Sinusoidal Projection

Mollweide's Projection

Interrupted Flat Polar Quartic Projection

Projection	Appearance	Properties	Distortion Pattern	Best Uses
Flat Polar Quartic Equal Area	Parallels are straight parallel lines. Meridians are curves in general but the central meridian is straight. The poles are represented by straight lines one-third the length of the Equator. Meridians converge toward the poles and are equally spaced along each parallel. The spacing between parallels decreases slightly with increased latitude. The boundary of the map is a complex curve.	An equal area projection.	Distortion is least nearest the Equator and central meridian and greatest at high latitudes near the boundaries.	Generally useful as a world map and for depicting global distributions.
Mollweide's	Parallels are straight parallel lines. Meridians are elliptical in general, but the central meridian is a straight line half the length of the Equator. The meridians converge to a point at each pole and are equally spaced along each parallel. The spacing between parallels decreases slightly with increased latitude. The boundary of the map is an ellipse.	An equal area projection.	Distortion is least in midlatitude regions near the central meridian and greatest at high latitudes near the boundaries.	Generally useful as a world map and for depicting global distributions.
Sinusoidal (Sanson-Flamsteed)	Parallels are straight parallel, equally spaced lines. Meridians and the boundary are sinusoidal curves. The central meridian is straight and half the length of the Equator. Meridians converge to points at the poles. The length of each parallel is equal to its length on a globe of corresponding size.	An equal area projection.	Distortion is least nearest the Equator and the central meridian and greatest at high latitudes near the boundaries.	Somewhat inferior to other projections as a global map, but useful for maps of individual continents.

parallels can be manipulated mathematically to provide either conformality or equivalence. The best known such projection, the standard Mercator, is conformal; others are not. The standard Mercator projection is frequently used for navigation charts because a straight line on this projection represents a line of constant azimuth, which facilitates navigation by compass direction.

Distortions on a cylindric projection are least nearest the circle of tangency and increase with increased distance from the circle of tangency. On a standard Mercator projection areas in the higher latitudes are grossly exaggerated. Conversely, on a cylindric equal area projection based on the equator, shapes at higher latitudes are badly distorted. Figure I.9 shows the properties of several cylindric projections.

Conic Projections   A cone placed upon a globe contacts the globe along a circle of tangency. If the apex of the cone is above a pole, the circle of tangency coincides with a parallel of latitude, known as the *standard parallel* of the conic projection. Parallels of latitude are shown as curved arcs in such conic projections, and meridians of longitude as radiating straight lines. The distortions of a conic projection are least nearest the standard parallel and increase with distance from it. Conic projections are therefore useful in mapping mid-latitude regions, such as the United States, that have a considerable extent east-west and a limited range of latitude.

Conic projections of greater precision are produced by allowing the cone to intersect the globe along two standard parallels. If the standard parallels are not too far apart, the entire map area is displayed with good accuracy. Albers' equal area conic projection, which was chosen for the maps in *The National Atlas of the United States*, can show the conterminous United States with a linear scale distortion that does not exceed 2 percent. For the conterminous United States, the standard parallels of the Albers' conic projection are chosen to be 29 1/2°N and 45 1/2°N. For Alaska the standard parallels are 55°N and 65°N, and for Hawaii 8°N and 18°N. The properties of several conic projections are shown in Figure I.10.

Geometrical Global Projections   Equal area projections that show the entire earth are necessary to display global distributions of all types of phenomena. If the projection displays parallels of latitude as parallel straight lines, regions with the same latitude are aligned on the map, a useful property because many aspects of the physical environment as well as human activities are related to latitude.

On most of the commonly used global projec-

tions, the equator and the central meridian are shown as straight lines that intersect at right angles. The regions of greatest distortion lie near the outer margins. Distortion is least at the center of the projection. Figure I.11 shows the properties of several geometrical global projections.

The *flat polar quartic* projection, an equal area projection, is the basis for many of the global distribution maps in this text. In this projection, the poles are represented by lines one-third the length of the equator. To maintain equivalence, the stretching of the poles must be compensated by shrinking the lengths of meridians in the polar areas. Distortion on the flat polar quartic projection is greatest at high latitudes near the margins.

Interruption and Condensation of Projections On most maps showing global distributions, the landmasses are of greater interest than the oceans. In such a case, the projection can be *interrupted* in the oceans. The projection may then be reprojected to standard meridians through each major landmass so that no land area is far from a meridian. If ocean areas are of no interest in a particular application, portions of the Atlantic and Pacific oceans can be omitted entirely, producing a *condensed interrupted map*. Condensation permits the map to be at a larger scale without using more space.

Relief Portrayal

The geographer interested in landforms is especially concerned with the symbols that depict surface form and vertical relief on maps. *Contour lines*, special kinds of *shading*, and *color tints* are some of the methods for indicating form and relief on maps. Relief and surface form are represented most accurately by contour lines.

A contour line on a map represents a line of constant altitude above or below a chosen reference

Figure I.12

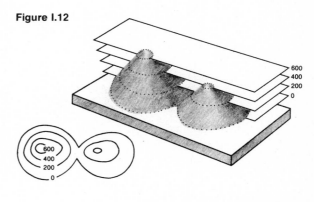

level, called a *datum plane*, such as the average level of the surface of the sea. Consider the hilly island in Figure I.12. The figure shows horizontal planes at a regular vertical spacing, or *contour interval*, of 200 ft. Each horizontal plane cuts the surface of the ground in a circle that is the contour line for that altitude; every point on a given contour line is the same altitude above the datum plane.

As shown in Figure I.13 contour lines can be used on a flat map to represent vertical relief. The contour interval should be chosen to be commensurate with the nature of the landscape being depicted; the contour interval will normally be larger for a mountainous region than for a gently sloping plain. The choice of contour interval also depends on the scale of the map. On a large-scale map the contour interval may be only a few feet, and the contour lines will show comparatively small changes in elevation. A small-scale map of the same region will employ a larger contour interval, and in addition the small kinks and bends of the contours will be smoothed and averaged, so that less detail will be represented. To make contour lines easier to read, index contours at regular intervals are thickened and labeled according to altitude. The elevations of prominent hilltops and depressions are indicated numerically as spot heights.

As Figure I.13 indicates, the horizontal spacing of contour lines can be interpreted in terms of the

Figure I.14
The relation between contour lines and topography is illustrated in this landscape model. The model was cut from a block of plastic by a cutter set to trace contour lines from a map; the depth of cut was adjusted according to the elevation of each contour line. Although individual layers are distinguishable on the model, the closely spaced contours give a good impression of the topographic relief. (Gerald Ratto Photography)

Figure I.13
(Andrea Lindberg adapted from Whitwell Quadrangle, Tennessee, U.S. Geological Survey)

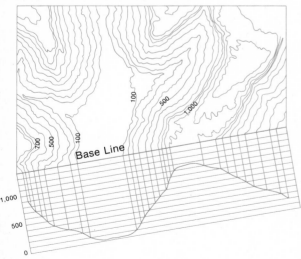

local slope angle. Contour lines that are relatively close together represent steeper slopes than do contour lines that are farther apart. The convexity or concavity of a slope and the form of ridge crests and valley floors can also be inferred by inspecting the horizontal spacing of the contour lines.

For quantitative purposes, a *topographic profile* of the landscape along a given direction can be prepared from a contour map according to the method illustrated in Figure I.13. The vertical scale of a topographic profile is usually exaggerated with

TOOLS OF THE PHYSICAL GEOGRAPHER

respect to the horizontal scale in order to portray minor relief features more clearly.

In addition to contour lines and spot heights, a variety of qualitative or partly quantitative artistic techniques can be used on a map to indicate relief. In *hachuring*, a method popular among cartographers in the nineteenth century, a slope is depicted by straight lines called *hachures*, drawn in the direction of slope descent. Hachures are at right angles to contours. They are drawn so that their darkness, due either to their width or spacing, increases with increased steepness (Figure I.15 *left*). Hachured maps present a pleasing appearance when well drawn, but they are seldom used now because of the difficulty of reading the specific slope angle and because of the labor involved in their construction.

A modern method of symbolizing relief on a map is to add shading that gives an impression of height and depth. Such *shaded relief* is commonly drawn as though the area were illuminated from the upper left, or "northwest," corner of the map (Figure I.15 *right*).

Altitude tinting, the use of color tints for successive ranges of altitudes, is often employed on small-scale maps of large areas. Green is usually used for altitudes near sea level, with colors for higher altitudes progressing through yellow, orange, red, and brown. The green tint used for low altitudes should not be taken as indicative of vegetation cover. Hachuring, shaded relief, and altitude tinting are sometimes used in combination on a contour map to depict the general character of a landscape while retaining the quantitative accuracy of contours.

Three-dimensional models with exaggerated vertical relief are mass-produced by molding thin sheets of plastic; on large-scale models, the quality of the molding is sufficiently good to represent the landscape accurately. Plastic "raised relief" maps are available for much of the United States at a scale of 1:250,000. Global relief maps are sometimes prepared from photographs of three-dimensional models, which are illuminated from the upper left to emphasize relief by light and shadows.

Figure I.15
Relief can be depicted on maps by hachuring **(left)** or by shaded relief **(right)**. (*Left*—Reprinted from Arthur Robinson and Randall Sale, *Elements of Cartography*, © 1953, 1960, 1969 by John Wiley & Sons, reprinted by permission; *right*—U.S. Geological Survey)

USGS Topographic Maps

A *topographic map* is a graphic representation of a portion of the earth's surface at a scale large enough to show human works such as individual buildings. Additionally, topographic maps show the configuration of the land surface. The United States has maintained a topographic map program since 1879 under the direction of the Geological Survey. The Geological Survey publishes the National Topographic Map Series, an invaluable source of information on the physical characteristics and human activities present in nearly all parts of the national territory. The maps are compiled primarily from aerial photographs, but field surveys are used to verify details of photo interpretation.

The Topographic Series is itself made up of several series of maps. Each of the maps is drawn so that its sides coincide with parallels of latitude or meridians of longitude, and a similar symbolism is used in all series. Topographic map series covering most of the United States are published at scales of 1:24,000; 1:62,500; and 1:63,360 (Alaska). The entire nation is also covered by map series at scales of 1:250,000 and 1:500,000. In addition, there are map series covering national parks, monuments, and historic sites; for certain metropolitan areas; for rivers and flood plains; and for Puerto Rico, other national territories, Antarctica, the moon, and Mars.

Figure I.16
(U.S. Department of the Interior)

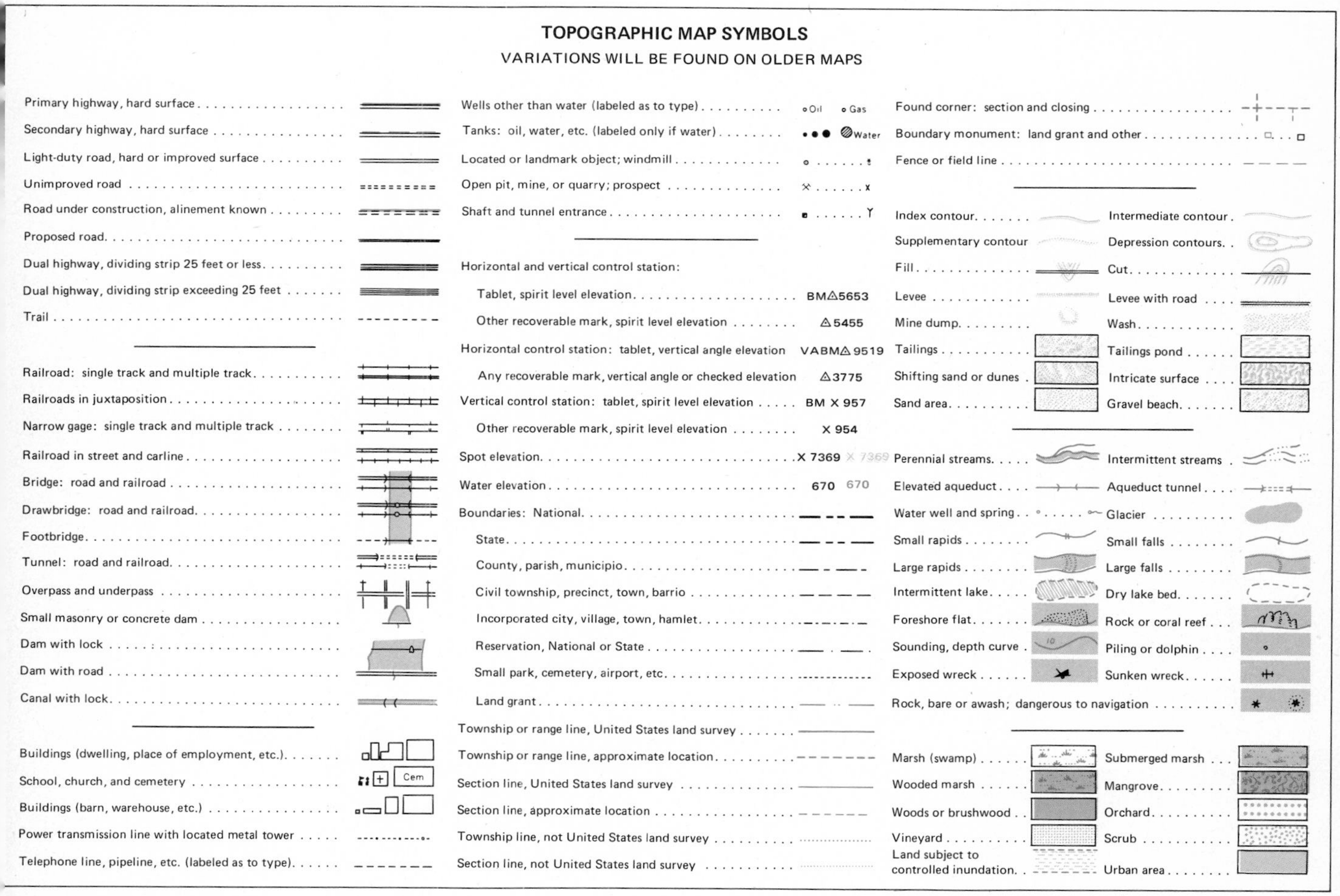

TOPOGRAPHIC MAP SYMBOLS
VARIATIONS WILL BE FOUND ON OLDER MAPS

Figure I.17
Remote sensing techniques utilize various portions of the electromagnetic spectrum. The visible and near-infrared regions are used for photography and television, and the infrared is used for scanning surface temperatures. Relief features are mapped by laser profiling and radar. (Calvin Woo)

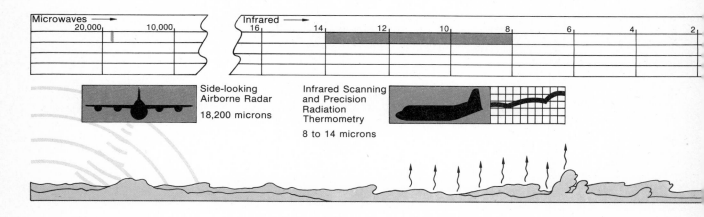

Microwaves ⟶
20,000   10,000

Infrared ⟶
16   14   12   10   8   6   4   2

Side-looking
Airborne Radar

18,200 microns

Infrared Scanning
and Precision
Radiation
Thermometry

8 to 14 microns

The 7 ½-minute quadrangle series (1:24,000) and 15-minute quadrangle series (1:62,500; 1:63,360) are large-scale maps suited to planning, engineering and landform studies on a local scale. Such maps contain a vast amount of information.

Topographic maps employ more than 100 different symbols to depict natural and man-made features. The table of USGS topographic map symbols reproduced in Figure I.16 is not printed on the map sheets but is available separately. The colors on a topographic map are an integral part of the symbolization, with each color restricted to a particular class of features.

Black is used for all man-made objects other than major roads and urban areas. All names and labels are in black. Blue is used for all water features, including streams, lakes, marshes, canals, the oceans, and water depths indicated by contours and numbers. Brown is reserved for land surface contour lines, indications of surface type, such as loose sand or mine tailings, and certain elevation figures. Green is used for vegetation symbols: woodlands, orchards, vineyards, brush, and wooded marshland. Red is used for major roads and land survey lines. Pink indicates densely built-up urban areas in which only landmark buildings are indicated individually. Purple is used on interim maps to symbolize changes that have occurred since the previous edition.

Remote Sensing

The space age has provided new techniques for obtaining information about the earth's surface. The new methods employ special sensing devices that can acquire data at a distance from the phenomena being studied, therefore the technique is known as *remote sensing*. The principal methods used are to sense and record electromagnetic radiation of sunlight, laser light, or microwave radar that is reflected from objects on the earth, or the infrared thermal radiation directly emitted by objects.

Remote sensing equipment is usually mounted in aircraft or satellites in orbit around the earth. The advantages of remote sensing include the ability to cover vast areas in a very short time, the ability to provide data from very large to very small scale, the possibility of repetitive measurements to follow the progress of selected events, and the ability to penetrate areas inaccessible to ground survey. The techniques of remote sensing have been greatly extended recently in an effort to obtain the information needed to manage the earth's resources more efficiently, and much of the information being gathered is of direct interest to physical geographers. The capabilities of remote sensing technology include the detection of vegetation types and their seasonal changes, the measurement of surface water distri-

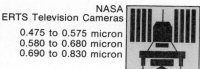

NASA
ERTS Television Cameras
0.475 to 0.575 micron
0.580 to 0.680 micron
0.690 to 0.830 micron

Near Infrared ——→   Wavelength (microns)

1.0   0.9   0.8   0.7   Visible ——→   0.6   0.5   0.4   0.3

Height

Color Infrared
Photography

0.51 to
0.89 micron

Aerial
Photography

0.4 to
0.7 micron

Laser Profiler
0.633 micron

bution and soil moisture, the depiction of landforms and surface geologic structures, and the monitoring of human activities that affect the environment.

Aerial photography, the first remote sensing method developed, is now used for most original mapping. Conventional aerial surveys employ the visible light reflected from the earth and detected with cameras using black-and-white film. Optical techniques allow the distortions inherent in a photograph taken at low or moderate altitudes to be rectified so that distant areas in the photograph have the same scale as areas directly below the camera. Two exposures can be used to record a stereoscopic (three-dimensional) image from which contour lines can be drawn. A single photograph may cover an area of a few square km to hundreds of square km, depending on the cameras and the altitude of the aircraft. Ground features as small as 1 cm in size can be detected on photographs made from aircraft flying at an altitude of a few kilometers, and satellite imagery can provide resolution of objects only a few meters in size from an altitude of several hundred kilometers. These advanced capabilities are gradually becoming available for civilian uses.

Aircraft fitted with detectors of longwave thermal infrared radiation can provide images of objects on the earth's surface with intensities proportional to their temperatures. Infrared imagery methods can be employed to detect cloud temperatures, ocean currents, or the mixing of warm water from industrial effluents with cooler river water.

Application of remote sensing methods to the study of the earth has been revolutionized by using remote sensors in conjunction with space satellites. Surveillance of the earth's cloud cover from weather satellites has greatly improved the accuracy of weather predictions and the ability to track potentially damaging hurricanes and typhoons. Weather satellites usually orbit at altitudes of several hundred kilometers.

Sunlight falling upon the earth consists of electromagnetic radiation in the visible region and in the near-visible infrared region of the electromagnetic spectrum. Different objects may reflect a different relative proportion of each wavelength, so that the light reflected from an object, when analyzed in terms of wavelength, forms a means of identification known as the *spectral signature* of the object. Vegetation, for example, usually reflects proportionately more short wavelength infrared radiation than does bare soil. Furthermore, different types of vegetation can usually be distinguished by their spectral signatures. Diseased plants or lack of soil moisture can also be detected.

Color infrared film is similar to ordinary color film, except that it depicts the previously invisible infrared radiation as red, the red colors as green, and the green colors as blue (see Figure I.18). How-

TOOLS OF THE PHYSICAL GEOGRAPHER

 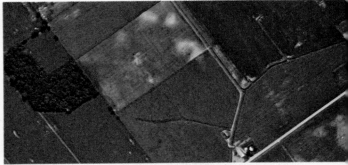

Figure I.18
(top, left and right) The aerial view on
the right was photographed with
ordinary color film. The same fields
photographed with color infrared film
are shown on the left; healthy
vegetation appears bright red.
(Courtesy of the Laboratory for
Applications of Remote Sensing—
LARS)

(opposite, top) Stereo pairs are
used to make topographic relief
apparent in aerial photographs. When
viewed with a stereoscope, relief
features in this view in eastern Utah
stand out as a three-dimensional
model. To see the stereo effect, the
left-hand photograph should be
viewed with the left eye, and the right-
hand photograph with the right eye. A
card held between the photographs
helps to separate the images. (U.S.
Geological Survey)

(right) Vegetation on irrigated land
is clearly differentiated from dry
desert in this color infrared
photograph of the Imperial Valley of
California taken from earth orbit. The
Salton Sea appears as a large dark
region at the upper left. The pattern
of agricultural fields is clearly
distinguishable; note in particular the
border between the United States and
Mexico that is evident toward the
bottom of the photograph. The well-
irrigated crops on the United States
side show as bright red, compared to
the bluish hues representing bare
ground or desert shrubs on the
Mexican side of the border. (NASA)

Another way to make use of spectral signatures is to employ several cameras, each fitted with a filter that allows only certain wavelengths of visible light to enter. Landsat, formerly Earth Resources Technology Satellite, ERTS-1, which views the earth from an altitude of nearly 1,000 km, employs four *multispectral scanners* and three television cameras, each sensing the same scene in a different range, or "band," of wavelengths. The resulting images can be assigned colors and combined to produce a "false-color" image (see Figure I.19). From its orbit between the poles, Landsat can scan a

ever, healthy vegetation appears red on a color infrared photograph. The intensity of the color is directly proportional to the amount of chlorophyll in the vegetation. Color infrared photographs of test fields and forests have proven that different trees and crops and their stages of growth can be distinguished by such means.

Figure I.19
(bottom left) Monterey Bay, California, appears at the left of this photograph taken by ERTS-1 from an altitude of 900 km (560 miles). The color print is made from a composite of three black-and-white photographs, each exposed to different spectral wavelengths; the image is similar to that produced by color infrared film. Vegetation areas stand out in red. A patch of fog hides part of the coast south of the Monterey peninsula. (NASA)

(bottom right) Relief features, including linear features in the San Andreas fault zone, show crisply in this side-looking radar image of part of the San Francisco peninsula. The thin linear feature at the lower right is the 2-mile-long linear accelerator used in atomic research at Stanford University. (J. R. P. Lyon, Stanford University)

given portion of the earth once every 18 days to allow the progress of crops to be followed through the growing season; to monitor streamflow, snow depth, and soil moisture; and to detect other short-term changes in the environment.

Many areas in the tropics are almost continuously covered by clouds, making normal photography and multi-spectral scanning unsuccessful. *Side-looking airborne radar*, which utilizes radio waves of short wavelength, can penetrate clouds and yield a high resolution picture of the surface. Radar waves emitted from the aircraft are partly reflected from the ground and detected by a receiving apparatus on the aircraft. Variations in the strength of the return signal are electronically translated into a picture of the terrain. Airborne radar has been used to map the Amazon Basin, Panama, and portions of Africa and New Guinea. Although radar methods do not differentiate between types of vegetation as clearly as multispectral methods, forest, range, and agricultural land can be separately identified on large scale radar images.

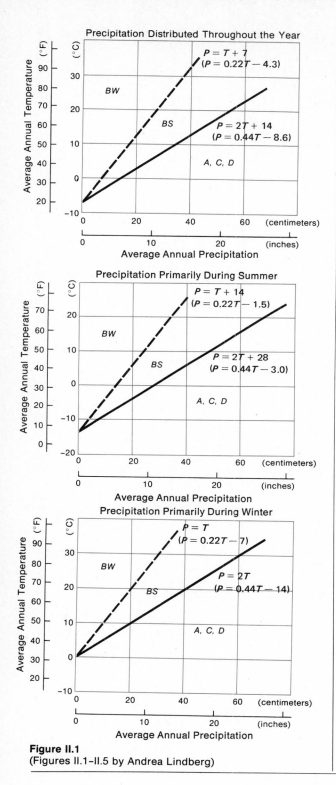

Figure II.1
(Figures II.1–II.5 by Andrea Lindberg)

APPENDIX II

Climate Classification Systems

Köppen System of Climate Classification

The Köppen system of global climate classification recognizes five major climatic types, symbolized by *A, B, C, D,* and *E.* The *B* climates, which are the dry realms, are further divided into *BS* (steppe) and *BW* (desert) types; and the *E* region is divided into *ET* (tundra) and *EF* (perpetual frost) types.

Figure II.1 shows Köppen's criteria for establishing the boundaries between the *A, C,* and *D* climates and the *BS* and *BW* climates. The solid line on each graph marks the boundary between the humid and dry climates. Each of the three sets of criteria is based on average annual temperature and precipitation. Which criterion is used depends on whether precipitation occurs evenly throughout the year or primarily during the summer or the winter. Köppen considered a region to have a dry winter if at least 70 percent of the precipitation occurs during the 6 summer months, and to have a dry summer if at least 70 percent of the precipitation occurs during the 6 winter months. Regions not fitting either category are considered to have an even distribution of precipitation. Summer is interpreted as the season when the midday sun is high in the hemisphere being considered, and winter is interpreted as the low-sun season.

The classification of a place into *A, C,* or *D* on the one hand, or *BS* or *BW* on the other, can be determined graphically by plotting the average annual temperature and precipitation on the appropriate diagram. The equation for each of the boundary lines is also given in the diagrams. In each case, the upper equation is in degrees Celsius and centimeters, and the lower equation in degrees Fahrenheit and inches.

481

The regions *A, C, D, ET,* and *EF* are distinguished from one another according to various criteria based on temperature, with the warmest being *A* and the coldest being *EF*. The criteria are as follows:

A: Average temperature of the coldest month exceeds 18°C (64.4°F).

C: Average temperature of the warmest month exceeds 10°C (50°F). Average temperature of the coldest month lies between 18°C (64.4°F) and −3°C (26.6°F).

D: Average temperature of the warmest month exceeds 10°C (50°F). Average temperature of the coldest month is below −3°C (26.6°F).

ET: Average temperature of the warmest month lies between 10°C (50°F) and 0°C (32°F).

EF: Average temperature of the warmest month is below 0°C (32°F).

H: Unclassified highland climates.

Principal Subdivisions of A Climates   The *A* climates are subdivided into *Af, Am,* and *Aw* types on the basis of the amount and seasonality of precipitation. If precipitation in the driest month exceeds 6 cm (2.4 in.), the climate is classified as *Af,* as indicated in Figure II.2. *Af* regions, such as equatorial lowland rainforests, receive abundant moisture for plant growth throughout the year. If there is a winter dry period during which precipitation for the driest month is less than 6 cm, the climate is classified as *Aw* or as *Am.* The distinction between *Aw* and *Am*

climates depends upon the relation between the amount of precipitation during the driest month and the annual precipitation, as Figure II.2 shows. Tropical wet and dry climates are classified as *Aw.* The *Am,* or monsoon, climate has enough annual precipitation that the moderate winter dry season does not exhaust soil moisture, and plant growth is not seriously affected.

A fourth subdivision, the *As,* would in principle be characterized by a summer dry season. However, *As* only occurs locally near the equator, and the *As* classification is not significant on the global scale.

Principal Subdivisions of BS and BW Climates
The *BS* and *BW* regions are divided on a thermal basis, as specified by the additional notation *h, k,* or *k′.*

h (hot):   Average annual temperature exceeds 18°C (64.4°F).

k (cold winter):   Average annual temperature is less than 18°C (64.4°F) and average temperature of the warmest month exceeds 18°C.

k′ (cold):   Average annual temperature is less than 18°C (64.4°F) and average temperature of the warmest month is less than 18°C.

The deserts of North Africa and central Australia are examples of *BWh* regions. *BWk* regions occur in the high plateaus of central Asia.

Principal Subdivisions of C and D Climates   The *C* and *D* climate regions are further differentiated according to the seasonality of precipitation. A second letter of notation, *s,* denotes regions with a dry summer (*Ds* seldom occurs); *w* denotes regions with dry winters; and *f* denotes regions with no marked dry season.

Cs, Ds: The driest month occurs during the warmest 6 months, and the amount of precipitation received during the driest month is less than one-third the amount received during the wettest month of the coldest 6 months. Also, precipitation during the driest month must be less than 4 cm (1.6 in.).

Cw, Dw: The driest month occurs during the coldest 6 months, and the amount of precipitation received during the driest month is less than one-tenth the amount received during the wettest month of the warmest 6 months.

Cf, Df: No marked dry season occurs.

Figure II.2

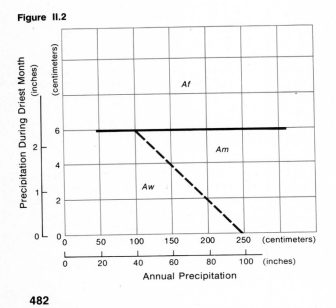

The subdivisions *Cs, Ds, Cw, Dw, Cf,* and *Df* may be further differentiated according to the seasonality of temperature by adding a third letter of notation—*a, b, c,* or *d.*

a: Average temperature of the warmest month exceeds 22°C (71.6°F).

b: Average temperature of the warmest month is less than 22°C (71.6°F), but at least 4 months have an average temperature greater than 10°C (50°F).

c: Average temperature of the warmest month is less than 22°C (71.6°F), fewer than 4 months have an average temperature greater than 10°C (50°F), and the average temperature of the coldest month is greater than −38°C (−36.4°F).

d: Same as *c*, except that the temperature of the coldest months is less than −38°C (−36.4F).

The Köppen system of climate classification possesses additional symbols to denote special features such as frequent fog or seasonal high humidity, but they are seldom used on the global scale. A global map of the Köppen climate classification is shown in Figure 8.6 on pp. 192–193.

Thornthwaite's Formula for Potential Evapotranspiration

The tables and graphs presented in this section make it possible to calculate potential evapotranspiration for a given month at a given location from Thornthwaite's formula, if the average monthly temperatures are known. Thornthwaite's formula is designed to be used with temperatures expressed in degrees Celsius, so temperature data on the Fahrenheit scale should be first converted to Celsius before beginning the calculation of potential evapotranspiration.

The first step in applying Thornthwaite's method is to calculate a monthly heat index, *i*, for each of the 12 months. The monthly heat index is defined according to the equation

$$i = \left(\frac{T}{5}\right)^{1.514}$$

where *T* is the long-term average temperature of the month in °C. Approximate values of the monthly heat index can be read from the graph in Figure II.3.

The sum of the twelve monthly heat indexes is the annual heat index, *I*. The annual heat index is representative of climatic factors at a given location because it is based on long-term averages. Thornthwaite found an empirical formula that gives poten-

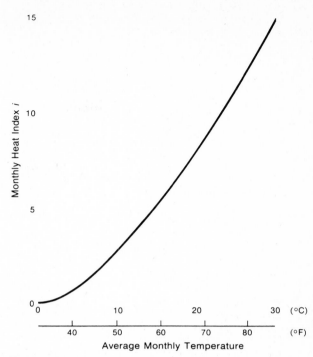

Figure II.3

tial evapotranspiration, *PE*, for a given month of a particular year in terms of *I*. His equation for *PE*, unadjusted for duration of sunlight, is

$$\text{Unadjusted } PE \text{ (centimeters)} = 1.6 \left(\frac{10\,T}{I}\right)^{m}$$

where *T* is the average temperature in °C for the specific month being considered, and *m* is a number that depends on *I*. To a good approximation, *m* is given by the equation

$$m = (6.75 \times 10^{-7})I^3 - (7.71 \times 10^{-5})I^2 + (1.79 \times 10^{-2})I + 0.492$$

With these equations, unadjusted potential evapotranspiration can be calculated using tables of logarithms or a computer. Alternatively, approximate values of unadjusted potential evapotranspiration can be read from the nomogram in Figure II.4 with enough accuracy for most purposes if the values for the average monthly temperature and the annual heat index are known.

If the average temperature of the month is below 0°C, potential evapotranspiration is taken to be 0. If

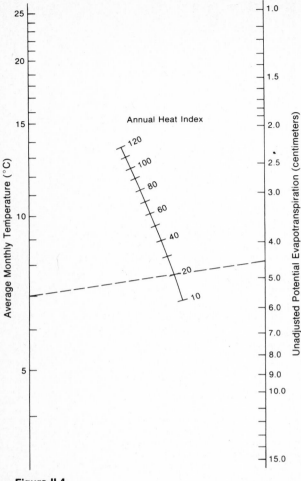

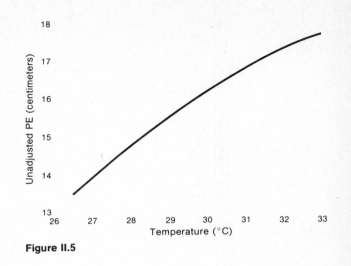

Figure II.5

the average monthly temperature exceeds 26.5°C, unadjusted *PE* is given directly in terms of temperature according to the graph in Figure II.5.

Values of unadjusted *PE* must be corrected for the duration of daylight in order to obtain the desired final values. Unadjusted *PE* for a given month at a given location should be multiplied by the correction factor listed in Table II.1. The correction factors for latitude 50°N are used for all latitudes farther to the north, and the factors for latitude 50°S are used for all latitudes farther to the south.

Figure II.4

Table II.1
Daylength Correction Factors for Potential Evapotranspiration

LATITUDE	JAN	FEB	MAR	APR	MAY	JUN	JUL	AUG	SEP	OCT	NOV	DEC
50°N	0.74	0.78	1.02	1.15	1.33	1.36	1.37	1.25	1.06	0.92	0.76	0.70
40°N	0.84	0.83	1.03	1.11	1.24	1.25	1.27	1.18	1.04	0.96	0.83	0.81
30°N	0.90	0.87	1.03	1.08	1.18	1.17	1.20	1.14	1.03	0.98	0.89	0.88
20°N	0.95	0.90	1.03	1.05	1.13	1.11	1.14	1.11	1.02	1.00	0.93	0.94
10°N	1.00	0.91	1.03	1.03	1.08	1.06	1.08	1.07	1.02	1.02	0.98	0.99
0°	1.04	0.94	1.04	1.01	1.04	1.01	1.04	1.04	1.01	1.04	1.01	1.04
10°S	1.08	0.97	1.05	0.99	1.01	0.96	1.00	1.01	1.00	1.06	1.05	1.10
20°S	1.14	1.00	1.05	0.97	0.96	0.91	0.95	0.99	1.00	1.08	1.09	1.15
30°S	1.20	1.03	1.06	0.95	0.92	0.85	0.90	0.96	1.00	1.12	1.14	1.21
40°S	1.27	1.06	1.07	0.93	0.86	0.78	0.84	0.92	1.00	1.15	1.20	1.29
50°S	1.37	1.12	1.08	0.89	0.77	0.67	0.74	0.88	0.99	1.19	1.29	1.41

Source: Thornthwaite, C. Warren, 1948. "An Approach Toward a Rational Classification of Climate." *Geographical Review.* 38: 55–94.

INDEX

Definitions of entries may be found on pages set **boldface**.

A horizon, 249, 250
aa lava, 372, 374
ablation, **406**, 407, 408
ablation zone, **406**, 407
abrasion, **411**, 411–412
abrasion platform, **442**, 443
absolute humidity, **104–105**
absorption, **50**
abyssal plains, **277**
accumulation zone, **406**, 407, 408
acidity, soil, **249**
active layer, 402
actual evapotranspiration, **165**, 168–170, 173
adiabatic cooling, **117**, 117–119
adiabatic heating, **119**
advection, **109**
advection fog, **109**
aeolian crossbedding, 356, 357
aeolian landforms and processes, **353**, 353–357
aggradation, **334**
agriculture:
  slash-and-burn, 238–239
  soil management and, 260–261, 264, 265
  terrace building, 268–269
air masses, **132**
  climate and, 184–190
  types of, 132–133
albedo, 54
Albers' conic equal area projection, 470, 472
alkaline soils, **249**
alluvial aprons (bajadas), 352
alluvial fans, **349**, 349–352
alluvial fill, **349**
alluvium, **255**
alpine glaciers, 398, 399, **400**, 414–420
altitude tinting, 474
altocumulus clouds, **111**, 113
altostratus clouds, **111**
aluminum, 280
amphiboles, 281
andesites, **284**
andesitic volcanism, 375–379
angle of repose, **309**
annuals, **210**
Antarctic Circle, 45
antecedent streams, 389, **390**
anticlinal ridge, **386**

anticlines, **386**, 386–388
anticyclones, **136**
aphelion, **44**
aquifers, **160**, 161–162
Arctic air mass, **132**
Arctic Circle, 44
arêtes, **416**
argon, 50
argon 40, 18–19
artesian springs, **162**
artesian wells, **162**
asthenosphere, **9**
atmosphere, **11**, 29–30
  circulation of, *see* atmospheric circulation
  cooling of, precipitation and, 117–123
  early, 11
  energy balance and, 52–60
  as energy transporter, 27–29
  gases in, 13–14, 16, 27–29, 50
  moisture in, 104–107
  solar radiation and, 48–52
  transfer of water to, 101–104
  water stored in, 155
atmospheric circulation, 77–93
  Coriolis force, **79**, 79–81, 83
  forces causing motion, 78–82
  friction, force of, **82**
  general, *see* general circulation
  ocean currents and, 90–93
  pressure gradient force, **79**
  secondary, *see* secondary circulations
autumnal equinox, 45
average lapse rate, **119**
axis (of the earth), 44–48
azimuth, 462
azimuthal equidistant projection, 468
azimuthal projections, 467, 468
azonal soils, **260**

B horizon, 249, 250
back swamps, 348
backslope, **355**, 356, 357, 365, 366
backwash, **444**
badland topography, **365**
Ballot Buys, 81
barchan dunes, 356, 357

barometer, 28
barrier islands, **447**, 447–449
barrier reef, **451**, 452
barrier spits, 448
bars, **446**
basal slip, **408**
basalt, 284, **284**, 285
basaltic volcanism, 371–375
base flow, **165**
batholith, 378
bauxite, **255**
bayhead beaches, **445**, 448
bay mouth, 448
beach drifting, 443–444, **444**
beaches, 30–31, 445–447, **446**
bed, **287**
bedding planes, **287**
bedload, **335**
Bergeron, Tor, 115
Bergeron ice crystal process, 114–117
bergschrund, 407
berm, **446**
big-bang hypothesis, 5–6
biomes, **213**
biosphere, **30**
biotite, 284
bituminous coal, 288
blowouts, 356, 357
bogs, soils of, 263
boulders, 288
braided stream, **345**
breccia, 288
budget (of a system), **33–36**
  see also energy budgets
bulk density, soil, **247**
Bureau of Land Management, 465
buttes, **365**

C horizon, 249, 250
calcic horizon, **257**
calcification, **257**, 257–259
calcite, 281
calcium, 280
calcium carbonate, 288
calcrete (calichein), 255, 263
calderas, **375**, 376, 378
California State Water Project, 176–177
calories, 26
canyons, submarine, 456